T0330039

Probabilistic Forecasts and Optimal Decisions

Probabilistic Forecasts and Optimal Decisions

Roman Krzysztofowicz
University of Virginia
Charlottesville, Virginia
USA

Registered Office(s)
John Wiley & Sons, Inc., 111 River Street, Hoboken, NJ 07030, USA
John Wiley & Sons Ltd, The Atrium, Southern Gate, Chichester, West Sussex, PO19 8SQ, UK

For details of our global editorial offices, customer services, and more information about Wiley products visit us at www.wiley.com.

Wiley also publishes its books in a variety of electronic formats and by print-on-demand. Some content that appears in standard print versions of this book may not be available in other formats.

Library of Congress Cataloging-in-Publication Data Applied for

Hardback ISBN: 9781394221868

Cover Design: Wiley
Cover Image: Courtesy of Roman Krzysztofowicz; © Kingline/Shutterstock

Set in 9.5/12.5pt STIXTwoText by Straive, Chennai, India

Printed and bound by CPI Group (UK) Ltd, Croydon, CR0 4YY

C9781394221868_141124

Those who fall in love with practice
without science
are like a sailor who enters a ship
without a helm or compass
and who never can be certain
whither he is going.

Leonardo da Vinci
(1452–1519)

Contents

Preface

This book has been written for upper-level undergraduate students and first-year graduate students of sciences (mathematical, statistical, decision, data, economic, management) and engineering (systems, industrial, operations research). The prerequisite knowledge includes the elements of calculus and probability — each at the level of an undergraduate course.

The book aims to be introductory yet specialized. Its first objective is to present the fundamentals of probabilistic forecasting and optimal decision making under uncertainty, which form the building blocks of *forecast–decision systems* for the real world. Its second objective is to teach mathematical modeling, probabilistic reasoning, statistical estimation, judgmental assessment, rational decision making, and numerical calculations — the skills which an aspiring modeler needs to develop and then to integrate into a systemic methodology for solving problems. Its third objective is to show the student with a quantitative bent the usefulness and the beauty of mathematics as a means of describing, understanding, and solving forecast–decision problems, ranging from personal to industrial.

Within the realm of established subjects, the book may be categorized as an introduction to probabilistic forecasting and statistical decision theory from a Bayesian perspective. But unlike many introductory texts, this one is unique and specialized in that (i) it integrates the two subjects into one, under the Bayesian principles of coherence, and (ii) it covers only selected models and procedures so that they can be treated in the depth necessary for fundamental understanding and rigorous application.

Following the definition of a forecast–decision system as the subject of study (Chapter 1), whose mathematical foundation is the postulates of rationality (Appendix A), Part I reviews the basic elements of probability theory (Chapter 2) and presents a methodology for modeling distribution functions of continuous variates (Chapter 3), together with parameter estimation methods (Appendix B) and a catalogue of 20 parametric families of special univariate distributions (Appendix C); this material is used throughout the book.

The main body of the book has two parts, which are independent yet parallel: Part II (Chapters 4–8) is devoted to discrete models; Part III (Chapters 9–15) is devoted to continuous models. Each part covers identical topics; thereby, it reinforces the terminology and the concepts learned in the other part, while it contrasts the mathematics of continuous models against the mathematics of discrete models. The topics covered are: judgmental forecasting (Chapters 4, 9), statistical forecasting (Chapters 5, 10), verification of forecasts (Chapters 6, 11), decision making (Chapters 7, 8, 12, 13, 14, 15).

The problems at the end of each chapter are grouped under two headings. Exercises are short and have a narrow scope. Mini-projects are long and have a broad scope intended to mimic, at least in a rudimentary way, real-world problems; many include real data, require substantial calculations, and give the teacher several options regarding the data or the models to be used. Some mini-projects continue through two or three chapters as they require forecasting, verification of forecasts, and decision making — thereby providing a vehicle for the student to develop the skill of integrative analysis and to acquire mental stamina, a trait essential for an aspiring analyst.

The overall scope of this book has been dictated by the feasibility of teaching from it at the senior–graduate level in one semester (while allowing a choice of advanced topics and of applications). Consequently, what is presented constitutes but a subset of a vastly larger body of knowledge, which I call the *Bayesian forecast–decision theory*. While its elements have been emerging since the 1940s through the works of many scholars, and applications of these elements have penetrated successfully into many fields, from signal detection theory of electrical engineers to flood forecast–response systems of hydrologists, its exposition as a comprehensive and coherent theory is still wanting.

Acknowledgments. Thanks are due to the teaching assistants who supported my pedagogy; to the students who took my course and then served as assistants — testing and grading the exercises and mini-projects; to Christopher M. Myers who made figures for Appendix C, and to those who made figures for the chapters; to Michael Gurlitz who coded the pilot version of the DFit software for Appendix B, and to Wray Mills who designed, coded, and maintained the current version.

I have been privileged to work with doctoral students who embraced the Bayesian theory and enthusiastically researched with me various aspects of forecast–decision systems: Dou Long, Karen S. Kelly, Ashley A. Sigrest, Coire J. Maranzano, Henry D. Herr, and Jie Liu. Our research was supported by the National Science Foundation and the National Weather Service.

Sherry Crane started typing the manuscript, then Margaret Heritage continued it through many versions; their expert and devoted work has been invaluable.

I dedicate the book to my wife Liana, and to our children Arman and Nayiri.

September 2023   *Roman Krzysztofowicz*
Charlottesville, Virginia

About the Companion Website

This book is accompanied by a companion website:

www.wiley.com/go/ProbabilisticForecastsandOptimalDecisions1e

This website includes:
Tables and Solutions Manuals

1

Forecast–Decision Theory

1.1 Decision Problem

1.1.1 Decision

Nothing is more difficult, and therefore more valuable and admirable, than to be able to decide.

(Napoléon Bonaparte)

What makes deciding difficult? Whereas a precise answer depends on the problem at hand, major sources of the decisional difficulty, and the accompanying decisional stress, may be grouped under four headings. (i) Complexity of the situation in the context of which one must decide. (ii) Multiplicity of objectives one wants to achieve. (iii) Multiplicity of alternative courses of action one can pursue. (iv) Uncertainty about the outcome, when it depends not only upon one's decision, but also upon inputs beyond one's control.

Since its origin in the eighteenth century, decision theory has developed a coherent logical-mathematical framework and effective analytical tools for dealing with all major sources of the decisional difficulty. This book focuses on mathematical models for solving basic decision problems in which uncertainty constitutes the major difficulty.

1.1.2 Uncertainty

Uncertainty is ubiquitous. Its existence is increasingly recognized. And the advantages of taking it into account in decision making, rather than ignoring it and relying on deterministic models (mental or mathematical), are progressively winning the argument among professionals of many disciplines. Here is a handful of examples.

Example 1.1 *(Accreditation Board for Engineering and Technology (ABET))*
In the 1990s, the ABET, which every 6 years reviews and accredits each undergraduate engineering degree program in the USA, began to require at least one probability and statistics course in each engineering curriculum, regardless of the discipline (e.g., computer, electrical, civil, mechanical). The premise was that every contemporary engineer must acquire at least a rudimentary appreciation of uncertainty and its quantification.

Example 1.2 *(American Meteorological Society (AMS))*
In 2002, the AMS issued a statement that endorsed making and disseminating probabilistic forecasts of weather elements (e.g., precipitation amount, temperature), in lieu of, or in addition to, traditional deterministic forecasts, which give only a single number (a so-called best estimate). The statement argued that forecasts in probabilistic terms "would allow the user to make decisions based on quantified uncertainty with resulting economic and social benefits".

Probabilistic Forecasts and Optimal Decisions, First Edition. Roman Krzysztofowicz.
© 2025 John Wiley & Sons Ltd. Published 2025 by John Wiley & Sons Ltd.
Companion website: www.wiley.com/go/ProbabilisticForecastsandOptimalDecisions1e

Example 1.3   *(Secretary of the US Department of the Treasury)*
In his 1999 commencement address at the University of Pennsylvania, Robert E. Rubin, former Secretary of the Treasury, recalled from his early career on Wall Street an incident in which he lost a lot of money on a stock. But another security trader, who had believed with "absolute certainty" that particular events would occur and had purchased a large volume of the same stock, "lost an amount way beyond reason — and his job", when his belief turned out to be wrong. Rubin's advice to the young graduates: "Reject absolute answers and recognize uncertainty … then all decisions become matters of judging the probability of different outcomes, and the costs and benefits of each."

1.2   Forecast–Decision System

1.2.1   Structure

To conquer the difficulty of making a rational decision under uncertainty, decision theory offers a way of structuring the problem as follows. The two major activities, quantifying uncertainty and making decisions, can be conceptualized as being performed by a *forecast–decision system* (F–D system) — a cascade coupling of two components (Figure 1.1): the forecast system (in short, the *forecaster*), and the decision system (in short, the *decider*). The coupling can be analyzed in each phase of the system's life: the design, the operation, the evaluation.

1.2.2   Design: Requirements

The design phase involves (i) specification of requirements for each system component, and (ii) formulation of models and procedures for the forecaster and the decider.

Requirements for decision system
The design begins with the decision system, which should meet the requirements of a client who wants to make rational decisions. To identify the requirements, four basic questions should be asked: (i) What is the purpose of the system? (ii) What is the decision to be made? (iii) What is the outcome of concern to the decider? (iv) What is the future input which is beyond the control of the decider and which, together with the decision, determines the outcome? When this input is uncertain at the decision time, it constitutes a random variable (in short, a *variate*), and it must be forecasted.

Requirements for forecast system
The design of the forecast system should meet the requirements of the decision system (Figure 1.1) with respect to (i) the *predictand* — the variate to be forecasted; (ii) the *lead time* of the forecast — the time that elapses from the moment the forecast is made to the moment the input occurs or can be observed; (iii) the *forecast frequency* — when the decision is to be made repeatedly (e.g., every day, week, month). Additional requirements may be specified, for instance, regarding information which should be used to prepare the forecast.

   Inasmuch as many books are devoted to the general problem of system design, this book does not address any further the first phase of the design process. Instead, system requirements are either specified, or implied through assumptions or problem descriptions, and the text concentrates on the second phase, which is modeling.

1.2.3   Design: Models

Given system requirements, models and procedures must be developed. This is accomplished in two steps. (i) A verbal description of the forecast problem or the decision problem is transformed into a *mathematical formulation*, which consists of symbols defining variables, sets, and functions (basically the notation which is given meaning

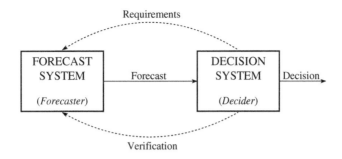

Taxonomy	Chapters	Chapters	Chapters
Discrete Models	4, 5	6	7, 8
Continuous Models	9, 10	11	12, 13, 14, 15

Figure 1.1 Forecast–decision system — a cascade coupling of the forecast system with the decision system in three phases of the system's life: (i) design (coupling through system requirements), (ii) operation (coupling through input–output), and (iii) evaluation (coupling through forecast verification). The topics covered in the book are: judgmental forecasting (Chapters 4, 9), statistical forecasting (Chapters 5, 10), verification of forecasts (Chapters 6, 11), decision making (Chapters 7, 8, 12, 13, 14, 15).

in the context of the problem at hand). A well-defined mathematical formulation has two advantages. First, it provides a structure which bestows the clarity and precision on the modeler's thought process. Second, it prescribes a decomposition which facilitates the next step. (ii) *Detailed modeling* of functions, estimation of parameters, and writing of procedures are undertaken, separately for each system component.

Forecaster

The major task is to model (to quantify) uncertainty about the predictand in terms of a distribution function. This *probabilistic forecaster* may be (i) a human expert preparing forecasts judgmentally based on the available information (quantitative and qualitative), or (ii) a statistical model calculating a forecast based on the realization of a quantitative *predictor* (a variate which is stochastically dependent on the predictand).

Decider

There are three major tasks. (i) To model (to assess) the preferences of a *rational decider* over possible outcomes in terms of a criterion function (a utility function, a profit function, an opportunity loss function). (ii) To model the outcome function that maps the decision and the input into an outcome. (iii) To formulate a decision procedure that integrates the probabilistic forecast, the criterion function, and the output function; and to find an *optimal decision* — one that optimizes (maximizes or minimizes) the expected value of the criterion function.

1.2.4 Operation

The forecaster prepares a probabilistic forecast of the predictand. Next, the decider makes the optimal decision under uncertainty as quantified by the forecast (Figure 1.1). The forecast and the realization of the predictand (once it becomes known) should be archived for the purpose of verification.

1.2.5 Evaluation

Periodically, a set of forecasts should be verified against actual realizations of the predictand. The purpose is to evaluate, and to track over time, the performance of the forecaster with respect to the needs of the decider (Figure 1.1).

The Bayesian *verification measures* quantify two attributes of probabilistic forecasts (*calibration* and *informativeness*) and provide (i) feedback to the forecaster, and (ii) statistics to the decider regarding the calibration, the informativeness, and the economic value of forecasts.

1.2.6 Coupling

The above overview of the design, operation, and evaluation phases of the F–D system reveals the intrinsic nature of the coupling between the forecast system and the decision system (Figure 1.1). We shall study this coupling, especially its economic implications, because it is illustrative of a more general interrelationship between information and decision, which is omnipresent.

1.3 Rational Deciding

1.3.1 The Procedure

"Jerky thinking is liable to give rise to jerky behavior." Coined by Good (1961), this is a witty antonym to the link between "methodical thinking" and "rational behavior".

Definition   Rational deciding under uncertainty is a *methodical procedure* for (i) analyzing all elements of a decision problem and then (ii) synthesizing them to reach a decision in a manner that adheres to the four *postulates of rationality* stated in Appendix A.

Depending on the complexity of the decision problem, this methodical procedure (i) may be as simple as thinking and doing arithmetic on a yellow pad, as Rubin (2023) recounts doing in his 50-year long career on Wall Street and in the Federal government, and as we shall learn to do in Chapters 7–8; or (ii) may require calculus, as we shall learn in Chapters 12–15; or (iii) may require complex Monte Carlo simulations and numerical calculations as, for example, a corporation must do to manage a global supply chain system — an extension of the topic of Chapter 13.

In the approach of this book: the analysis leads to a mathematical formulation of the decision model; the synthesis leads to the *rational decision procedure*. Its mathematically oriented synonym is the *optimal decision procedure* because it maximizes, or minimizes, the expected value of the criterion function in order to find the *optimal decision*. But this optimality is usually personal, not absolute, because, when faced with identical decision problems, different deciders may employ different criterion functions, and hence may reach different optimal decisions. This nature of optimality is explicit in the synonym the *most preferred decision*, which is apt for personal decisions, such as investment decisions — the topic of Chapter 14.

It follows that the term *rational decision* should be understood as a decision made via a rational decision procedure, not as any particular decision.

1.3.2 The Mind-Set

To become a wise decider, who copes well with uncertainty, one should adopt the mind-set of a Roman stoic (Seneca, 2004). For no matter how carefully the rational decision procedure is implemented, the decision is made *ex ante* — based on an imperfect forecast of the predictand. Ergo, one should be mindful of the logical implications.

1. The rational decision procedure does not guarantee a "good" outcome on any particular occasion. It only guarantees that, if it were applied consistently on a large number of occasions, then the resultant sequence of outcomes would be preferred.

2. On any particular occasion, the realized outcome may be "good" or "bad" because it is generated by a chance mechanism, which is beyond the decider's control. Although the rational decision procedure optimizes the trade-off between the probability of a good outcome and the probability of a bad outcome (as we shall learn in Chapters 7 and 14), there always remains the risk of a bad outcome.
3. One should not, therefore, judge the decision *ex post* based on the realized outcome. When the outcome is good, do not say the decision was good, and do not feel smart — you were lucky. When the outcome is bad, do not say the decision was bad, and do not feel depressed — you were unlucky. More philosophically: do not let the outcome touch your emotions. Just accept the outcome, like a Roman stoic would.
4. The record of past forecasts, decisions, and outcomes does provide feedback, which can be used constructively. Realizations of the predictand can be used to recalibrate the forecaster, if necessary, as we shall learn in Chapters 6 and 11. Realizations of the outcomes may trigger a change in the criterion function (e.g., by reassessing the utility function, and the implied risk attitude of the decider, as we shall learn in Chapter 14). The forecast model and the decision model may be refined. Ultimately, it is the quality of the models and procedures, which comprise the F–D system, that determines the degree to which the decisions are actually optimal (and, *ipso facto*, the degree to which the potential advantage of adherence to the rationality postulates of Appendix A is actually realized).

1.4 Mathematical Modeling

1.4.1 The Model

Toward our grand objective — to study models and procedures for the F–D system — let us picture the modeling task.

Pause for a moment, and imagine listening to beautiful, your favorite, music. A renowned Italian maestro, Riccardo Muti, once confided:

> "The most difficult thing in making music is to *express the complex* in the simplest possible way."

Now exchange "making music" for "developing a model". This states one of our challenges. Ergo, while learning the modeling we may cheer ourselves up with the thought that mathematics and music are twins.

Definition A model is an intellectual construct used to represent, in well-defined terms, a fragment of reality that contains the problem at hand (e.g., a forecast problem, a decision problem).

Four remarks are in order. (i) The means of modeling are not prescribed; they may include linguistic, logical, and mathematical expressions. (ii) It goes without saying that the reality is complex. (iii) The "well-defined terms" are what distinguishes an engineered model from a mental model. (iv) According to one school, the human mind does not analyze reality directly, only a model of reality created through perception. But this mental model is not explicitly stated and is often fuzzy; hence it is not communicable, not comparable, and not analyzable in any objective sense. To make it so is the intellectual challenge of mathematical modeling.

1.4.2 The Guideposts

From theories and experiences, systems engineers, decision scientists, and applied mathematicians deduced eclectic methodologies for modeling problems in various domains. An aspiring modeler should consult this vast literature. Here is a synopsis — seven modeling guideposts apt for the subject of this book.

1. The model should have a **purpose** and a **domain** (of its applicability), which should be negotiated with the client and then stated precisely.
2. Effective modeling requires a **trade-off** between the veracity (or the level of detail) of the model and the cost (or the duration) of the development.
3. Every modeling situation requires one or more **assumptions**. This is the nature of reductionism. Moreover, different modelers may choose different assumptions for different reasons. Hence, "modeling appears almost as a studio art" (Wymore, 1976, p. 53).
4. The model should satisfy its purpose in the simplest possible way. The challenge is "not creating complexity but retaining **informative simplicity**"(Howard, 1980, p. 8).
5. There may be more than one model suitable for the stated purpose. No perfect model may exist, but only different, more or less plausible, **approximations** to reality (Marschak, 1979, p. 172).
6. The theoretic principles plus experience … "combined with often laborious **trial and error** will yield suitable formulations" of models (Bellman, 1957, p. 82).
7. The value of a mathematical model derives not only (and not always) from the numbers it outputs, but also (and often foremost) from the **explanation** and the **understanding** (of the problem) that it affords us.

Being written for a first course on the subject, this book presents mathematical formulations of the basic, univariate, analytic models for selected types of ever-important forecast problems and decision problems.

1.5   Notes on Using the Book

These notes expand on the preface and explain the intentions behind some features of the book.

On organization
The organization of the book is diagrammed in Figure 1.2. Together with introductory remarks to the chapters, it should facilitate designing a course and tracking its progress. There is enough material to design several versions of the course: first, by deciding the format (lecture-based, project-based, seminar); second, by deciding the balance between breadth and depth of coverage; and third, by selecting the exercises and mini-projects. For example, in an undergraduate one-semester, lecture-based course, I always covered Chapters 3, 4–7, 9–12, and each year added one of Chapters 8, 13, 14, 15; Chapter 2 and Appendices B, C were assigned for self-study; Appendix A was assigned for reading and discussion. In a graduate course, I also covered advanced concepts, derivations, and proofs; these are relegated to the terminal sections of the chapters. (Most of the proofs are provided, either in the finished form within the text, or in the form of hints to the exercises.)

On "warm-up" tasks
Chapters 4, 9, 14, 15, and Appendix A include tasks to be performed by the student; they require judgments or intuitive decisions. I administer the task via a work sheet at the beginning of a lecture, and discuss the answers after the relevant material has been covered. The class answers may be juxtaposed with the normative answers, if they exist, and compared with answers by other groups, if the chapter reports them.

On applications
Only a few examples of applications appear within the text, but many are framed as exercises and mini-projects. These should be read, therefore, even if not performed, to get a sense of the possibilities.

On options
Many exercises and mini-projects include several options regarding the data or the models to be used. They multiply the number of assignments with distinct solutions to the same problem. The teacher needs to specify the option.

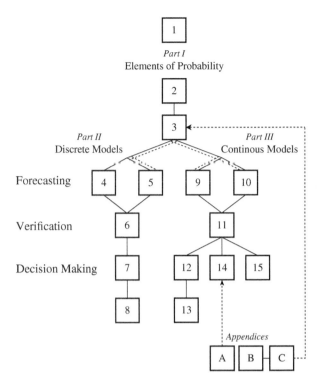

Figure 1.2 The main connections between chapters. Parts II and III are independent of each other. Appendix A supports mainly Chapter 14. Appendices B and C support mainly Chapter 3. Chapter 3 supports mainly Chapters 5, 9, 10.

On modeling

Exercises and mini-projects are stated mostly in words, purposely avoiding notation. The objective: to teach the art of mathematical modeling, which begins with mapping the words into the well-defined notation.

On numerical calculations and graphs

Almost all formulae and algorithms necessary for numerical calculations are provided, so that the student can implement them (by encoding in a spreadsheet or by writing computer code). The objective: to educate a versatile mathematical modeler, not just a software user. I make three exceptions. (i) Graphs of functions, which should look professional and be drawn to scale, can be created with any software. (ii) Numerical integrations, required in some mini-projects, can be done with any software. (iii) Modeling and estimation of parametric distributions, after the students have mastered the simple methods, can be performed with DFit software (see Appendix B).

On communication

Many exercises and all mini-projects end with directives such as: explain the graph, interpret the results, draw conclusions, make recommendations. The objectives: (i) to reinforce the learning of proper mathematical terms, (ii) to build intuitive understanding, and (iii) to practice the skill of technical writing, which is critical for success in a professional career and graduate study. A good answer to the above directives has the C^4 property: it is **correct**, **complete**, **clear**, **concise**.

Bibliographical Notes

The *forecast–decision system* is a paraphrase of the *information-and-decision system*, which was conceptualized by Marschak (1974). The original text attributed to Napoléon Bonaparte (1769–1821), the emperor of France

(1804–1815) and the exquisite decider on the battlefields of Europe, reads: "Rien n'est plus difficile, et donc plus précieux et merveilleux, que de pouvoir décider." The decisional stress, its symptoms, its sources, and the patterns that people follow while coping with it, have been researched by Janis and Mann (1977). A transcript of Robert E. Rubin's speech appeared under the heading "A Healthy Respect for Uncertainty" in *Decision Analysis Newsletter*, **18**(2), 9–10, Institute for Operations Research and the Management Sciences (INFORMS), August 1999.

The notion of rationality has been debated in philosophy, statistics, economics, and decision science. A collection of articles by Good (1983) conveys a flavor of many debates. I.J. Good was a statistician working on cryptanalysis in the British Foreign Office during World War II and later University Distinguished Professor of Statistics at Virginia Polytechnique Institute and State University.

The modeling guideposts are culled mainly from the writings of Wayne Wymore (1976), the founder of the first systems engineering department in the world, in 1960 at the University of Arizona; Ronald A. Howard (1980), the intellectual leader of the "modeling school" of decision analysis at Stanford University; and Richard Bellman (1957), professor of mathematics at the University of Southern California, the inventor of *dynamic programming*, a theory for modeling and solving sequential decision problems, and a prolific modeler in several disciplines.

The quotation from Riccardo Muti appeared in *The Wall Street Journal* (17 September 2010) when he became music director of the Chicago Symphony Orchestra, after he had led Milan's La Scala, Vienna Philharmonic, and the Philadelphia Orchestra.

Part I

Elements of Probability

2

Basic Elements

This chapter defines basic terminology and notation, basic concepts of mathematical analysis, and basic constructs of probability and statistics that are used freely throughout the remainder of the book. The exposition takes the form of a review, assuming that the reader is familiar with set theory and probability theory at the level of a good introductory text.

2.1 Sets and Functions

2.1.1 Sets

A *set* S is a collection of objects. Depending on the context, a set may be called a *class*, or *family*, and an object s may be called an *element*, *member*, or *point* of S, written $s \in S$. The symbolic definition of a set is

$$S = \{s : \ldots\},$$

which reads: "S is the set of elements such that ...". For instance, if $\Re$ is the set of real numbers, and an element $r \in \Re$ is selected, then $S = \{s : s \in \Re, s < r\}$ is the set of real numbers such that each is less than r; it is an *uncountable* set. A *countable* set is defined by listing all of its elements, when S is *finite*,

$$S = \{s_1, s_2, \ldots, s_N\},$$

or by listing a few initial elements, when S is *countably infinite* (or *denumerable*),

$$S = \{s_1, s_2, s_3, \ldots\}.$$

The *union* of two sets S and $\mathcal{T}$, denoted $S \cup \mathcal{T}$, is the set of elements that belong to either S or $\mathcal{T}$. The *intersection* of two sets S and $\mathcal{T}$, denoted $S \cap \mathcal{T}$, is the set of elements that belong to both S and $\mathcal{T}$. Set S is a *subset* of set $\mathcal{T}$, denoted $S \subset \mathcal{T}$ or $\mathcal{T} \supset S$, if every member of S is also a member of $\mathcal{T}$.

An *interval* $\mathcal{I}$ is a set which is a subset of the real numbers, $\mathcal{I} \subset \Re$, and such that (i) $\mathcal{I}$ contains at least two points, and (ii) if $x, y \in \mathcal{I}$ and $x < s < y$, then $s \in \mathcal{I}$. Let $a, b \in \Re$ and $a < b$. There exist four types of *bounded intervals*:

the *open interval* $(a, b) = \{x : x \in \Re, a < x < b\}$,
the *closed interval* $[a, b] = \{x : x \in \Re, a \le x \le b\}$,
the *semi-closed interval* $(a, b] = \{x : x \in \Re, a < x \le b\}$,
the *semi-closed interval* $[a, b) = \{x : x \in \Re, a \le x < b\}$;

the interval $(a, b]$ may be described as left-open, right-closed; the interval $[a, b)$ may be described as left-closed, right-open. There exist five types of *unbounded intervals*:

$$(a, \infty) = \{x : x \in \mathcal{R}, a < x\},$$
$$[a, \infty) = \{x : x \in \mathcal{R}, a \leq x\},$$
$$(-\infty, b) = \{x : x \in \mathcal{R}, x < b\},$$
$$(-\infty, b] = \{x : x \in \mathcal{R}, x \leq b\},$$
$$(-\infty, \infty) = \mathcal{R}.$$

The *Cartesian product* of $\mathcal{X}$ and $\mathcal{Y}$ is the set $\mathcal{X} \times \mathcal{Y}$, which consists of all ordered pairs (x, y) with x in $\mathcal{X}$ and y in $\mathcal{Y}$:

$$\mathcal{X} \times \mathcal{Y} = \{(x, y) : x \in \mathcal{X}, y \in \mathcal{Y}\}.$$

In the Cartesian coordinate system, (x, y) represents a point whose *abscissa* is x and *ordinate* is y.

2.1.2 Functions

A *function f* is a rule which associates with each element x in $\mathcal{X}$ (the domain of definition of f) an element $f(x)$ in $\mathcal{Y}$ (the range of f). A function may be also referred to as a *mapping, transform, transformation,* or *operator,* and is usually defined by writing

$$f : \mathcal{X} \to \mathcal{Y},$$

which reads: "function f from $\mathcal{X}$ into $\mathcal{Y}$" or "f maps $\mathcal{X}$ into $\mathcal{Y}$". The *graph* $\mathcal{G}$ of function f is a set

$$\mathcal{G} = \{(x, y) : y \in \mathcal{Y}, y = f(x), x \in \mathcal{X}\},$$

which is a subset of the product set $\mathcal{X} \times \mathcal{Y}$.

When $\mathcal{A} \subset \mathcal{X}$, the set of values which f takes on $\mathcal{A}$ is called the *image of $\mathcal{A}$ under f* and is denoted by

$$f(\mathcal{A}) = \{y : y \in \mathcal{Y}, y = f(x), x \in \mathcal{A}\}.$$

Of course, $f(\mathcal{A}) \subset \mathcal{Y}$. If the image of the domain $f(\mathcal{X})$ is all of the range $\mathcal{Y}$, so that $f(\mathcal{X}) = \mathcal{Y}$, then function f is said to map $\mathcal{X}$ *onto* $\mathcal{Y}$.

The function f is called *one-to-one* if for any $x_1, x_2 \in \mathcal{X}$ such that $x_1 \neq x_2, f(x_1) \neq f(x_2)$. If function f is one-to-one and onto, then there exists an *inverse function*

$$f^{-1} : \mathcal{Y} \to \mathcal{X}$$

as follows: for any $y \in \mathcal{Y}, f^{-1}(y) = x$ is the unique point $x \in \mathcal{X}$ such that $y = f(x)$.

The function f is said to be *monotone on $\mathcal{X}$* if it is either increasing ($f(x_1) \leq f(x_2)$ whenever $x_1 \leq x_2$) or decreasing ($f(x_1) \geq f(x_2)$ whenever $x_1 \leq x_2$). The function f is said to be *strictly monotone on $\mathcal{X}$* if it is either strictly increasing ($f(x_1) < f(x_2)$ whenever $x_1 < x_2$) or strictly decreasing ($f(x_1) > f(x_2)$ whenever $x_1 < x_2$). Only a strictly monotone function can be one-to-one.

Suppose $f : \mathcal{X} \to \mathcal{Y}$ and $g : \mathcal{Y} \to \mathcal{Z}$. The *composition* of functions g and f is the function

$$g \circ f : \mathcal{X} \to \mathcal{Z}$$

such that at each point $x \in \mathcal{X}$,

$$z = (g \circ f)(x) = g(f(x))$$

is a point in $\mathcal{Z}$.

2.2 Variates and Sample Spaces

2.2.1 Variates

An event whose occurrence is uncertain or a quantity whose value is uncertain will be modeled in terms of a suitably defined random variable. A synonym of a random variable is variate; both terms will be used.

A *variate* is denoted by an upper-case letter, say X. A *realization* of variate X is denoted by the lower-case letter x. In a specific context, a realization may mean an observation or a measurement.

"The uncertainty about variate X" is a shorthand for the uncertainty that exists at a specified time (e.g., the present time, the forecast time) about the realization of variate X that will be observed at some future time. To forecast a variate means to quantify the uncertainty about it. "A variate whose realization is used to forecast" another variate will be called a *predictor*, which is a customary statistical term. "A variate to be forecasted" will be called a *predictand*. This term has not entered the English lexicon yet, but it has been in use by meteorologists since at least 1972. It is succinct, and it parallels in form and function the term *integrand*, which has been in use since 1897 and means "a mathematical expression to be integrated".

2.2.2 Sample Spaces

The set of all possible realizations is called the *sample space* of variate X and is denoted by the upper-case script letter $\mathcal{X}$.

The sample space of a *discrete variate* X is *countable* and either *infinite*,

$$\mathcal{X} = \{x_1, x_2, x_3, \dots\},$$

or *finite*,

$$\mathcal{X} = \{x_1, \dots, x_I\}.$$

When $\mathcal{X}$ is finite, an element $x_i \in \mathcal{X}$ for some $i \in \{1, \dots, I\}$ is either a label attached to one event among the I mutually exclusive and collectively exhaustive events which are precisely defined (e.g., the type of precipitation: rain, snow, sleet), or a value from among I distinct values the quantity can take on (e.g., the number of cars stopped by a red light: $0, 1, \dots, 20$).

For any realization $x \in \mathcal{X}$ of a discrete variate, the *probability* of event $\{X = x\}$ occurring, or of variate X taking on value x, is denoted by $P(X = x)$.

The sample space of a *continuous variate* X is either the set of real numbers, the *unbounded interval*

$$\mathcal{X} = (-\infty, \infty) = \{x : -\infty < x < \infty\},$$

or an open interval *bounded below*,

$$\mathcal{X} = (\eta, \infty) = \{x : \eta < x < \infty\},$$

or an open interval *bounded above*,

$$\mathcal{X} = (-\infty, -\eta) = \{x : -\infty < x < -\eta\},$$

or a *bounded* open interval,

$$\mathcal{X} = (\eta_L, \eta_U) = \{x : \eta_L < x < \eta_U\},$$

where bounds are real numbers, with $-\infty < \eta < \infty$ and $-\infty < \eta_L < \eta_U < \infty$.

An *event* associated with realization $x \in \mathcal{X}$ of a continuous variate can be defined in many ways, for instance $\{X \leq x\}$, $\{X > x\}$, or $\{X \in \mathcal{X}_0\}$ where $\mathcal{X}_0$ is a subset of the sample space $\mathcal{X}$, symbolically $\mathcal{X}_0 \subset \mathcal{X}$. The probabilities of these events are denoted $P(X \leq x)$, $P(X > x)$, or $P(X \in \mathcal{X}_0)$, respectively.

2.2.3 Samples

A finite set of realizations drawn from the sample space $\mathcal{X}$ is called a *sample* of X and is denoted by

$$\{x(n) : n = 1, \dots, N\},$$

where n is the index of realization, N is the *sample size*, and $x(n)$ is a realization (an observation, a measurement, or an output of calculation or analysis), such that $x(n) \in \mathcal{X}$.

There are different kinds of samples; we shall characterize and name them as needed, in the context of applications.

2.3 Distributions

2.3.1 Discrete Distribution

Probability function

The uncertainty about a discrete variate X is characterized in terms of a *probability function h* on $\mathcal{X}$ defined at every $x \in \mathcal{X}$ by

$$h(x) = P(X = x).$$

That is, $h(x)$ is the probability of event $\{X = x\}$. Hence h is a nonnegative function and such that

$$\sum_{i=1}^{I} h(x_i) = 1.$$

A long name for h is the *probability mass function*.

Distribution function

If the order of realizations in the sample space has a meaning as, for instance, when X counts the number of items or occurrences so that $\{x_1, \dots, x_I\}$ is a subset of integers, then it is meaningful to define a *distribution function H* on $\mathcal{X}$ such that at every $x \in \mathcal{X}$,

$$H(x) = P(X \le x)$$
$$= \sum_{x_i \le x} h(x_i).$$

That is, $H(x)$ is the probability that the realization of variate X will not exceed the value x. It follows that H is a stepwise, nondecreasing function on $\mathcal{X}$, with $H(x_1) = h(x_1)$ and $H(x_I) = 1$. A long name for H is the *cumulative distribution function*.

2.3.2 Continuous Distribution

Density function

The uncertainty about a continuous variate X is characterized in terms of a *density function h* on $\mathcal{X}$. It is a continuous, nonnegative function such that for any subset $\mathcal{X}_0$ of the sample space $\mathcal{X}$, symbolically $\mathcal{X}_0 \subset \mathcal{X}$, the following holds:

$$P(X \in \mathcal{X}_0) = \int_{\mathcal{X}_0} h(x)\, dx.$$

Because it is certain that the realization of X is contained in the sample space $\mathcal{X}$ (i.e., $P(X \in \mathcal{X}) = 1$), function h is such that

$$\int_{\mathcal{X}} h(x)\, dx = 1.$$

When $\mathcal{X}$ is a bounded interval, it is tacitly assumed that at any point x outside this interval, $h(x) = 0$. A long name for h is the *probability density function*.

The value $h(x)$ does not have any practical interpretation. Just consider its unit: because probability is dimensionless and dx has the same unit as x has, the above integrals imply that $h(x)$ is dimensional — its unit equals the reciprocal of the unit of x. For example, when the unit of x is [ft], the unit of $h(x)$ is $[ft]^{-1}$. And what is a practical meaning of a number whose unit is $[ft]^{-1}$? (A not uncommon mistake is to interpret $h(x)$ as the probability $P(X = x)$. In fact, $P(X = x) = 0$ for any x whenever X is a continuous variate. This can be seen from the definition of $P(X \in \mathcal{X}_0)$ when the subset $\mathcal{X}_0$ degenerates to a point x, that is, $\mathcal{X}_0 = [x, x]$.)

Distribution function

The distribution function provides an equivalent, but directly interpretable, characterization of uncertainty about a continuous variate X. A *distribution function* H on $\mathcal{X} = (\eta_L, \eta_U)$ is defined at every point $x \in \mathcal{X}$ by

$$H(x) = P(X \leq x)$$
$$= P(X \in (\eta_L, x])$$
$$= \int_{\eta_L}^{x} h(t)\, dt.$$

That is, $H(x)$ is the probability that the realization of variate X will not exceed the value x. We call $H(x)$ the *nonexceedance probability* of x, or the *cumulative probability* up to x.

It follows that H is a continuous (from the right), nondecreasing function from $\mathcal{X}$ onto the open interval $(0, 1)$, with $H(\eta_L) = 0$ and $H(\eta_U) = 1$. It is tacitly assumed that $H(x) = 0$ at any point $x \leq \eta_L$, and $H(x) = 1$ at any point $x \geq \eta_U$. When $\mathcal{X} = (-\infty, \infty)$, the limits of H are $H(-\infty) = 0$ and $H(\infty) = 1$. A long name for H is the *cumulative distribution function*.

Graph of density function

When a distribution function H on $\mathcal{X}$ is given, the corresponding density function h on $\mathcal{X}$ is specified at every point $x \in \mathcal{X}$ by

$$h(x) = \frac{dH(x)}{dx}.$$

Accordingly, the value $h(x)$ has a technical interpretation as the derivative of the distribution function H at point $x \in \mathcal{X}$. Hence, a graph of the density function h depicts, at every point $x \in \mathcal{X}$, the rate $h(x)$ at which the nonexceedance probability $H(x)$ accumulates as x increases.

The point $x_M \in \mathcal{X}$ at which h attains the maximum on $\mathcal{X}$ is called the *mode* of X:

$$x_M = \arg\{\max_{x \in \mathcal{X}} h(x)\}.$$

It follows that x_M is the point of inflection of the distribution function H (the point at which H increases fastest and changes from a convex function to a concave function).

2.3.3 Quantile Function

Suppose that in addition to being continuous and onto, function $H : \mathcal{X} \to (0, 1)$ is strictly increasing (and hence one-to-one). There is then defined an inverse function $H^{-1} : (0, 1) \to \mathcal{X}$, called the *quantile function* of X, as follows: for any $p \in (0, 1)$,

$$x = H^{-1}(p)$$

is the unique point in $\mathcal{X}$ such that $p = H(x) = P(X \leq x)$.

The point $x_p \in \mathcal{X}$ having probability p of not being exceeded by a realization of X, namely $x_p = H^{-1}(p)$, is called the *p-probability quantile* of X. Certain quantiles have their own names:

$x_{0.1}$ is the first *decile* of X,   deciles: $x_{0.1}, x_{0.2}, \ldots, x_{0.9}$;

$x_{0.2}$ is the first *quintile* of X,   quintiles: $x_{0.2}, x_{0.4}, x_{0.6}, x_{0.8}$;

$x_{0.25}$ is the first *quartile* of X,   quartiles: $x_{0.25}, x_{0.5}, x_{0.75}$;

$x_{0.5}$ is the *median* of X.

For any $p \in (0, 0.5)$, the open interval (x_p, x_{1-p}) is called the $(1 - 2p)$-*probability central credible interval* of X. It is central because the numbers p and $1 - p$ are equidistant from 0.5, the center of the probability interval $(0, 1)$. It is labeled with the number $1 - 2p$ because this is the probability of X falling inside the credible interval (x_p, x_{1-p}). A central credible interval need not be symmetrical about the median $x_{0.5}$; of course, it is symmetrical when the density function h is symmetrical.

2.4 Moments

2.4.1 Mean

The *expectation* of a discrete variate X having the probability function h on the sample space $\mathcal{X}$ is defined by

$$E(X) = \sum_{x \in \mathcal{X}} x h(x).$$

The expectation of a continuous variate X having the density function h on the sample space $\mathcal{X}$ is defined by

$$E(X) = \int_{\mathcal{X}} x h(x)\, dx.$$

The synonyms of $E(X)$ are: the expectation of X, the *expected value* of X, the *mean* of X, and the *first moment* of X.

In general, for any positive integer n, the expectation $E(X^n)$ is called the nth *moment* of X.

Given a sample $\{x(n) : n = 1, \ldots, N\}$ of N realizations of X, the estimate of $E(X)$, called the *sample mean* of X, is given by

$$m = \frac{1}{N} \sum_{n=1}^{N} x(n).$$

2.4.2 Variance

The *variance* of variate X, which may be either discrete or continuous, is defined by

$$Var(X) = E[(X - E(X))^2] = E(X^2) - E^2(X),$$

where it is to be understood that $E^2(X) = [E(X)]^2$. The variance $Var(X)$ is also called the *second central moment* of X. In general, for any positive integer n, the expectation $E[(X - E(X))^n]$ is called the nth *central moment* of X.

Given a sample $\{x(n) : n = 1, \ldots, N\}$ of N realizations of X, several estimates of $Var(X)$ can be defined. The *sample variance* of X is given by

$$s^2 = \frac{1}{N} \sum_{n=1}^{N} (x(n) - m)^2 = \frac{1}{N} \sum_{n=1}^{N} x^2(n) - m^2.$$

The so-called *unbiased estimate* of the variance of X is

$$s_1^2 = \frac{1}{N-1} \sum_{n=1}^{N} (x(n) - m)^2 = \frac{1}{N-1} \sum_{n=1}^{N} x^2(n) - \frac{N}{N-1} m^2.$$

When variate X has a normal distribution with an unknown mean μ and an unknown variance σ^2 (see Section 2.6.2), the *maximum likelihood estimate* (MLE) of σ^2 is s^2. The estimate that has the smallest *mean squared error* (MSE) is

$$s_2^2 = \frac{1}{N+1} \sum_{n=1}^{N} (x(n) - m)^2 = \frac{1}{N+1} \sum_{n=1}^{N} x^2(n) - \frac{N}{N+1} m^2.$$

And s_1^2 has a larger MSE than s^2 has. Therefore, when X is normally distributed, the sample estimate s^2 is preferred over the unbiased estimate s_1^2, not only because it is the MLE, but also because it yields a smaller MSE.

The square root of variance $Var(X)$ is called the *standard deviation* of X. The sample standard deviation of X is s; however, the unbiased estimate of the standard deviation of X is not s_1. For this reason, the unbiased estimate of variance is unacceptable for models in which variance is subjected to some nonlinear operations. On the other hand, the MLE of variance, s^2, yields directly the MLE of standard deviation, s.

In summary, the analyst is faced with a problem of choosing the type of estimate for variance. The term "unbiased" has led over the years to a widespread preference for this type of estimate, but there is no general scientific basis for it. And oftentimes, the unbiased estimate is inappropriate or unacceptable. In this book, we shall use exclusively the sample variance s^2 because it ensures two properties: invariance and coherence (DeGroot, 1986, p. 348).

2.4.3   Invariance and Coherence

Suppose two parameters, θ and λ, are related to each other through a one-to-one function t, such that $\lambda = t(\theta)$, and an estimate $\hat{\theta}$ of θ is known. Can an estimate $\hat{\lambda}$ of λ just be calculated as $\hat{\lambda} = t(\hat{\theta})$? The answer depends on the nature of the estimate $\hat{\theta}$.

Theorem (*Invariance property*)   *If $\hat{\theta}$ is the MLE of θ, then $\hat{\lambda} = t(\hat{\theta})$ is the MLE of λ.*

Suppose an object (a function, a number) can be derived or calculated via a model or a procedure in more than one way. If each way yields an identical result, then *coherence* holds. We submit that coherence should be a *sine qua non* property of every model or procedure claimed to be "scientific". That is why the invariance of an estimate and the coherence of a model (or a procedure) go hand in hand, as the following example illustrates.

Example 2.1   (*Invariance ensures coherence*)
Consider a model t that outputs $Y = t(X) = X^2$. Given a sample of the input variate X, say $\{2, 6, 1\}$, estimate the expectation of the output variate, $E(Y)$. There are two procedures.

1. **Empirical procedure.** Transform the sample of X into the sample of Y, $\{4, 36, 1\}$, and find the sample mean: $m_Y = 41/3$.
2. **Model-based procedure.** Rearrange the expression for $Var(X)$ to obtain a theoretic relationship:

$$E(Y) = E[t(X)]$$
$$= E(X^2) = Var(X) + E^2(X).$$

Such a relationship is advantageous because it shows how the first two moments of the input X determine the expectation of the output Y. Now, for $E(X)$ take the sample mean $m_X = 3$, but for $Var(X)$ consider two alternative estimates: (i) The sample variance, which is $s_X^2 = 14/3$ and yields $m_Y = s_X^2 + m_X^2 = 14/3 + 9 = 41/3$. (ii) The unbiased estimate, which is $s_{1X}^2 = 14/2$ and yields $m_Y = s_{1X}^2 + m_X^2 = 14/2 + 9 = 48/3$.

The comparison of the three values of m_Y supports this fact: only the model-based procedure using the sample variance (equivalently, the MLE when X is normal) and the empirical procedure are coherent. The unbiased estimate of variance leads to incoherence.

Exercise 15 at the end of this chapter offers yet another example: when X is a discrete variate, the unbiased estimate of its variance is incoherent with the probability function of X estimated in terms of the relative frequencies.

2.5 The Uniform Distribution

A continuous variate X, with the sample space $\mathcal{X} = (\eta_L, \eta_U)$, has the *uniform distribution*, abbreviated $\text{UN}(\eta_L, \eta_U)$, if its density function is specified at every point $x \in \mathcal{X}$ by

$$h(x) = \frac{1}{\eta_U - \eta_L};$$

that is, $h(x)$ is constant and equal to the reciprocal of the length of interval $\mathcal{X}$. The corresponding distribution function is linear:

$$H(x) = \frac{x - \eta_L}{\eta_U - \eta_L}.$$

Table 2.1 summarizes the characteristics of this distribution, and below is one of its useful properties.

For any subinterval $\mathcal{X}_0 = (x_1, x_2) \subset \mathcal{X}$, the probability of X taking on a value inside $\mathcal{X}_0$ is

$$P(X \in \mathcal{X}_0) = H(x_2) - H(x_1)$$
$$= \frac{x_2 - x_1}{\eta_U - \eta_L};$$

in other words, the probability of event $(X \in \mathcal{X}_0)$ equals the proportion: the length of subinterval $\mathcal{X}_0$ to the length of the interval $\mathcal{X}$. Because this proportion is easy to visualize, the uniform variate offers a convenient *reference variate* for judgmental assessment of probabilities (Chapters 4 and 9).

Table 2.1 The uniform distribution.

Abbreviation: $\text{UN}(\eta_L, \eta_U)$

Parameters: bounds $-\infty < \eta_L < \eta_U < \infty$

Domains: $\eta_L < x < \eta_U, \quad 0 < p < 1$

Density function

$$h(x) = \frac{1}{\eta_U - \eta_L}$$

Distribution function $p = H(x)$

$$H(x) = \frac{x - \eta_L}{\eta_U - \eta_L}$$

Quantile function $x = H^{-1}(p)$

$$H^{-1}(p) = (\eta_U - \eta_L)p + \eta_L$$

Moments

$$E(X) = \frac{\eta_U + \eta_L}{2}, \quad Var(X) = \frac{(\eta_U - \eta_L)^2}{12}$$

2.6 The Gaussian Distributions

The Gaussian distribution, also called the normal distribution, originated with Abraham de Moivre (1667–1754), Pierre Simon de Laplace (1749–1827), and Carl Friedrich Gauss (1777–1855). It possesses mathematical properties that have elevated it to the preeminent position in probability theory. It plays the preeminent role in our forecasting theory as well. We shall describe first its standard form and then its four derived forms.

2.6.1 The Standard Normal Distribution

A continuous variate Z, with the sample space $\mathcal{Z} = (-\infty, \infty)$, has the *standard normal distribution*, abbreviated NM(0, 1), if its density function is specified at every point $z \in \mathcal{Z}$ by

$$q(z) = \frac{1}{\sqrt{2\pi}} \exp\left(-\frac{z^2}{2}\right), \qquad -\infty < z < \infty. \tag{2.1}$$

The corresponding distribution function, defined by

$$Q(z) = \int_{-\infty}^{z} q(t)\, dt, \tag{2.2}$$

does not have a closed form. The probability $Q(z) = P(Z \leq z)$ can be calculated for any z via either numerical integration according to (2.2), or some function that approximates Q and is easy to evaluate. Several approximating functions exist; one of them is a rational function (a quotient of polynomials) given in Table 2.2. (The more customary symbols ϕ and Φ, in lieu of q and Q, are not used in this book because we reserve them for other purposes.)

The p-probability quantile z_p of the standard normal variate Z is defined by

$$z_p = Q^{-1}(p), \qquad 0 < p < 1, \tag{2.3}$$

where Q^{-1} denotes the inverse of Q. The function Q^{-1} is called the *standard normal quantile function*. Because Q^{-1} does not have a closed form, z_p can be calculated for any p only approximately. Several approximations exist; one of them is given in Table 2.3.

The mean and the variance of the standard normal variate Z are

$$E(Z) = 0, \qquad Var(Z) = 1;$$

these values are the arguments in the abbreviation NM(0, 1). The mean of Z, the median of Z, and the mode of Z are all equal to zero.

Table 2.2 Rational function approximation $\hat{Q}(z)$ to the probability $Q(z) = P(Z \leq z)$ of the standard normal variate Z not exceeding value z.

If $0 \leq z < \infty$, then

$$\hat{Q}(z) = 1 - \frac{1}{2}(1 + a_1 z + a_2 z^2 + a_3 z^3 + a_4 z^4)^{-4}$$

$$a_1 = 0.196854 \qquad a_3 = 0.000344$$

$$a_2 = 0.115194 \qquad a_4 = 0.019527$$

Approximation error

$$|\hat{Q}(z) - Q(z)| < 2.5 \times 10^{-4}$$

If $-\infty < z \leq 0$, then $\hat{Q}(z) = 1 - \hat{Q}(-z)$

Source: Abramowitz and Stegun (1972).

Table 2.3 Rational function approximation $\hat{z}_p$ to the standard normal quantile $z_p = Q^{-1}(p)$ corresponding to probability p.

If $0.5 \leq p < 1$, then

$$\hat{z}_p = t - \frac{a_0 + a_1 t}{1 + b_1 t + b_2 t^2}$$

$$t = \sqrt{-2\ln(1-p)}$$

$$a_0 = 2.30753 \quad b_1 = 0.99229$$

$$a_1 = 0.27061 \quad b_2 = 0.04481$$

Approximation error

$$|\hat{z}_p - z_p| < 3 \times 10^{-3}$$

If $0 < p \leq 0.5$, then $\hat{z}_p = -\hat{z}_{1-p}$

Source: Abramowitz and Stegun (1972).

The density function q is symmetric about the median. Consequently, q is an even function,

$$q(z) = q(-z), \tag{2.4}$$

and $Q - \frac{1}{2}$ is an odd function, $Q(z) - \frac{1}{2} = -[Q(-z) - \frac{1}{2}]$; equivalently,

$$Q(z) = 1 - Q(-z). \tag{2.5}$$

This, in turn, implies that

$$z_p = -z_{1-p}. \tag{2.6}$$

2.6.2 The Normal Distribution

A continuous variate X, with the sample space $\mathcal{X} = (-\infty, \infty)$, has a *normal distribution*, abbreviated $\text{NM}(\mu, \sigma)$, with parameters μ and σ $(-\infty < \mu < \infty, \sigma > 0)$ if its density function is specified at every point $x \in \mathcal{X}$ by

$$h(x) = \frac{1}{\sigma\sqrt{2\pi}} \exp\left(-\frac{1}{2}\left(\frac{x-\mu}{\sigma}\right)^2\right), \qquad -\infty < x < \infty. \tag{2.7}$$

The corresponding distribution function is defined by

$$H(x) = \int_{-\infty}^{x} h(t)\, dt. \tag{2.8}$$

Like Q, function H does not have a closed form. The probability $H(x) = P(X \leq x)$ for any x is usually calculated in terms of the probability $Q(z)$, as shown in Section 2.6.3.

The mean and the variance of the normal variate X are

$$E(X) = \mu, \qquad Var(X) = \sigma^2. \tag{2.9}$$

The mean of X, the median of X, and the mode of X are all equal to μ. The density function h is symmetric about the median μ; consequently $h(x) = h(2\mu - x)$ and $H(x) = 1 - H(2\mu - x)$. Table 2.4 provides a summary. Figure 2.1 shows graphs.

In the density function (2.7), the mean μ plays the role of a *location parameter* — the parameter that locates the center of function h on the horizontal axis x, and the standard deviation σ plays the role of a *scale parameter* — the parameter that scales the function values $h(x)$ according to the unit in which x is expressed. Specifically, $h(x)$ is proportional to σ^{-1}, and because σ is in the unit of x, the probability density $h(x)$ has the unit equal to [unit of x]$^{-1}$.

Table 2.4 The normal distribution.

Abbreviation: $NM(\mu, \sigma)$

Parameters: location $-\infty < \mu < \infty$,　scale $\sigma > 0$

Domains: $-\infty < x < \infty$,　$0 < p < 1$

Density function

$$h(x) = \frac{1}{\sigma\sqrt{2\pi}} \exp\left(-\frac{1}{2}\left(\frac{x-\mu}{\sigma}\right)^2\right)$$

Distribution function $p = H(x)$

$$H(x) = Q\left(\frac{x-\mu}{\sigma}\right)$$

Quantile function $x = H^{-1}(p)$

$$H^{-1}(p) = \sigma Q^{-1}(p) + \mu$$

Linearized quantile function $v = bu + a$

$$v = x, \qquad u = Q^{-1}(p)$$
$$\sigma = b, \qquad \mu = a$$

Moments

$$E(X) = \mu$$
$$Var(X) = \sigma^2$$

Four other (special) distributions on the unbounded interval $\mathcal{X} = (-\infty, \infty)$ are catalogued in Section C.2. Each of them has a location parameter β and a scale parameter α. Thus, to compare the normal distribution with any of these special distributions, we let $\mu = \beta$ and $\sigma = \alpha$, so that the mnemonic of the normal distribution becomes $NM(\beta, \alpha)$.

2.6.3　The Standard Transform

Any normal variate X, with the given mean μ and variance σ^2, may be transformed into the standard normal variate Z. The procedure is called the *standardization* of X; the transform takes the form

$$Z = \frac{X - \mu}{\sigma}. \tag{2.10}$$

Inversely, the standard normal variate Z may be transformed into any normal variate X with the given mean μ and variance σ^2. The inverse transform takes the form

$$X = \sigma Z + \mu. \tag{2.11}$$

Each of these transforms is linear and thus amounts to changing the origin and the unit of measurement (of X or Z, respectively).

From the theory of derived distributions (Section 3.8), it follows that the density function of X can be expressed in terms of the density function of Z:

$$h(x) = \frac{1}{\sigma} q\left(\frac{x-\mu}{\sigma}\right); \tag{2.12}$$

and that the distribution function of X can be expressed in terms of the distribution function of Z:

$$H(x) = Q\left(\frac{x-\mu}{\sigma}\right). \tag{2.13}$$

The last expression provides a means of calculating the probability $H(x) = P(X \leq x)$. Given μ, σ, and x, calculate

$$z = \frac{x - \mu}{\sigma}; \tag{2.14}$$

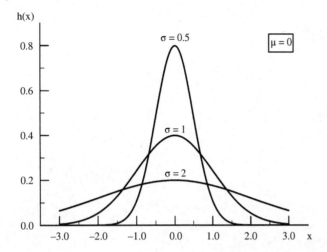

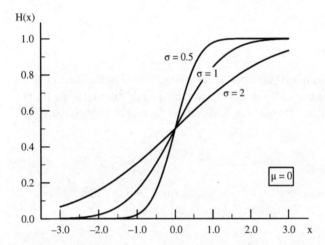

Figure 2.1 The normal distribution NM(μ, σ). With parameter values $\mu = 0$ and $\sigma = 1$, it is the standard normal distribution NM(0, 1).

next calculate $\hat{Q}(z)$ using the rational function approximation (Table 2.2); then set $H(x) = \hat{Q}(z)$.

Finally, for any number p ($0 < p < 1$), the p-probability quantile x_p of the normal variate X can be expressed in terms of the quantile z_p of the standard normal variate Z:

$$x_p = \sigma z_p + \mu. \tag{2.15}$$

Given p, calculate $\hat{z}_p$ using the rational function approximation (Table 2.3); next insert the values of μ, σ, and $\hat{z}_p$ into the above expression and calculate x_p.

2.6.4 The Log-Normal Distribution

Density and distribution functions

Suppose a continuous variate Y has the sample space $\mathcal{Y} = (\eta, \infty)$, which is a bounded-below open interval with a known value of the lower bound η. This is the case, for instance, when Y represents household income or body weight. Such a variate has a *log-normal distribution*, abbreviated LN(μ, σ, η), with the location parameter μ, the scale parameter σ, and the shift parameter η ($-\infty < \mu < \infty, \sigma > 0, -\infty < \eta < \infty$), if its density function is specified at every point $y \in \mathcal{Y}$ by

$$h(y) = \frac{1}{(y - \eta)\sigma\sqrt{2\pi}} \exp\left(-\frac{1}{2}\left(\frac{\ln(y - \eta) - \mu}{\sigma}\right)^2\right), \qquad \eta < y. \tag{2.16}$$

The corresponding distribution function is specified by

$$H(y) = Q\left(\frac{\ln(y - \eta) - \mu}{\sigma}\right), \qquad \eta < y. \tag{2.17}$$

Logarithmic transform

As the name suggests and a comparison of functions hints (compare (2.16) with (2.7), and (2.17) with (2.13)), the log-normal distribution is derived from the normal distribution through the following transform. (i) Take the natural logarithm of a shifted variate:

$$X = \ln(Y - \eta), \qquad \eta < Y. \tag{2.18}$$

(ii) Assume that the resultant variate X has a normal distribution NM(μ, σ). (iii) Apply the theory of derived distributions (Section 3.8). It turns out that the original variate, which is now given by the inverse of transform (2.18),

$$Y = e^X + \eta, \qquad -\infty < X < \infty, \tag{2.19}$$

has the log-normal distribution LN(μ, σ, η), in which η is the specified lower bound on Y, and

$$\mu = E(X) = E[\ln(Y - \eta)], \tag{2.20a}$$

$$\sigma^2 = Var(X) = Var[\ln(Y - \eta)]. \tag{2.20b}$$

The mean of Y and the variance of Y can be derived in terms of these parameters; the equations are given in Table 2.5. Graphs of h and H are shown in Figure 2.2.

Table 2.5 The log-normal distribution.

Abbreviation: LN(μ, σ, η)

Parameters: location $-\infty < \mu < \infty$,   scale $\sigma > 0$,   shift $-\infty < \eta < \infty$

Domains: $\eta < y < \infty$, $0 < p < 1$

Density function

$$h(y) = \frac{1}{(y - \eta)\sigma\sqrt{2\pi}} \exp\left(-\frac{1}{2}\left(\frac{\ln(y - \eta) - \mu}{\sigma}\right)^2\right)$$

Distribution function $p = H(y)$

$$H(y) = Q\left(\frac{\ln(y - \eta) - \mu}{\sigma}\right)$$

Quantile function $y = H^{-1}(p)$

$$H^{-1}(p) = \exp(\sigma Q^{-1}(p) + \mu) + \eta$$

Linearized quantile function $v = bu + a$

$$v = \ln(y - \eta), \qquad u = Q^{-1}(p)$$
$$\sigma = b, \qquad\qquad \mu = a$$

Moments

$$E(Y) = \exp(\sigma^2/2 + \mu) + \eta$$
$$Var(Y) = [\exp(\sigma^2) - 1]\exp(\sigma^2 + 2\mu)$$

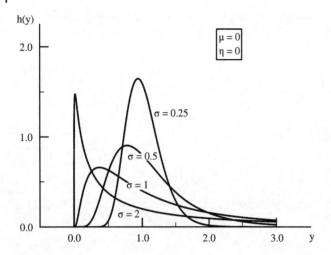

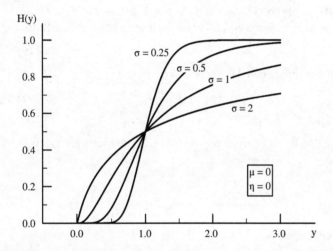

Figure 2.2 The log-normal distribution $\text{LN}(\mu, \sigma, \eta)$. The density value $h(y_M)$ at the mode y_M attains a minimum when $\sigma = 1$.

Measures of center

The common measures of the center of a distribution are the mean, the median, and the mode. Under the log-normal distribution, the mean of Y is

$$E(Y) = e^{\sigma^2/2 + \mu} + \eta.$$

The median of Y can be found by evaluating the quantile function (Table 2.5) at $p = 0.5$. To wit, $y_{0.5} = H^{-1}(0.5)$, and since $Q^{-1}(0.5) = 0$, the solution is

$$y_{0.5} = e^{\mu} + \eta.$$

The mode of Y can be found by differentiating the density function h (equation 2.16), setting $dh(y_M)/dy = 0$, and solving the resultant equation for y_M; the solution is

$$y_M = e^{\mu - \sigma^2} + \eta.$$

Comparison of the above three expressions leads to the conclusion that

$$y_M < y_{0.5} < E(Y).$$

That is, the three measures of the center of a distribution (the mode, the median, the mean) have a fixed order (which is the *reversal* of their *alphabetical order*), regardless of the values of the parameters (μ, σ, η).

Interestingly, this order of the mode, the median, and the mean is common under density functions which are skew to the right (i.e., have a long right tail). The reverse order is common under density functions which are skew to the left (i.e., have a long left tail).

2.6.5 The Log-Ratio Normal Distribution

Suppose a continuous variate Y has the sample space $\mathcal{Y} = (\eta_L, \eta_U)$, which is a bounded open interval with known values of the lower bound η_L and the upper bound η_U. This is the case, for instance, when Y represents the fraction of ore which is copper (in terms of weight), or the percentage of population diagnosed with seasonal flu. Such a variate has a *log-ratio normal distribution*, abbreviated LR1(η_L, η_U)-NM(μ, σ), with the location parameter μ, the scale parameter σ, and the shift parameters η_L and η_U $(-\infty < \mu < \infty, \sigma > 0, -\infty < \eta_L < \eta_U < \infty)$, if its density function is specified at every point $y \in \mathcal{Y}$ by

$$h(y) = \frac{\eta_U - \eta_L}{(y - \eta_L)(\eta_U - y)\sigma\sqrt{2\pi}} \exp\left(-\frac{1}{2\sigma^2}\left(\ln\frac{y - \eta_L}{\eta_U - y} - \mu\right)^2\right), \qquad \eta_L < y < \eta_U. \tag{2.21}$$

The corresponding distribution function is specified by

$$H(y) = Q\left(\frac{1}{\sigma}\left(\ln\frac{y - \eta_L}{\eta_U - y} - \mu\right)\right), \qquad \eta_L < y < \eta_U. \tag{2.22}$$

As the name suggests and a comparison of functions hints (compare (2.21) with (2.7), and (2.22) with (2.13)), the log-ratio normal distribution is derived from the normal distribution through the following transform. (i) Take the natural logarithm of the ratio of shifted variates

$$X = \ln\frac{Y - \eta_L}{\eta_U - Y}, \qquad \eta_L < Y < \eta_U. \tag{2.23}$$

(ii) Assume that the resultant variate X has a normal distribution NM(μ, σ). (iii) Apply the theory of derived distributions (Section 3.8). It turns out that the original variate, which is now given by the inverse of transform (2.23),

$$Y = \eta_U - \frac{\eta_U - \eta_L}{e^X + 1}, \tag{2.24}$$

has the log-ratio normal distribution LR1(η_L, η_U)-NM(μ, σ), in which (η_L, η_U) are the specified bounds on Y, and

$$\mu = E(X) = E\left(\ln\frac{Y - \eta_L}{\eta_U - Y}\right), \tag{2.25a}$$

$$\sigma^2 = Var(X) = Var\left(\ln\frac{Y - \eta_L}{\eta_U - Y}\right). \tag{2.25b}$$

The mean of Y and the variance of Y can be calculated numerically only according to the equations given in Table 2.6. Graphs of h and H are shown in Figures 2.3 and 2.4. (The numeral 1 after the abbreviation LR signifies the particular type of the ratio — ratio of type one — in transform (2.23).)

Table 2.6 The log-ratio normal distribution.

Abbreviation: $LR1(\eta_L, \eta_U)\text{-}NM(\mu, \sigma)$

Parameters: location $-\infty < \mu < \infty$, scale $\sigma > 0$, shift $-\infty < \eta_L < \eta_U < \infty$

Domains: $\eta_L < y < \eta_U$, $0 < p < 1$

Density function

$$h(y) = \frac{\eta_U - \eta_L}{(y - \eta_L)(\eta_U - y)\sigma\sqrt{2\pi}} \exp\left(-\frac{1}{2\sigma^2}\left(\ln\frac{y - \eta_L}{\eta_U - y} - \mu\right)^2\right)$$

Distribution function $p = H(y)$

$$H(y) = Q\left(\frac{1}{\sigma}\left(\ln\frac{y - \eta_L}{\eta_U - y} - \mu\right)\right)$$

Quantile function $y = H^{-1}(p)$

$$H^{-1}(p) = \eta_U - \frac{\eta_U - \eta_L}{\exp(\sigma Q^{-1}(p) + \mu) + 1}$$

Linearized quantile function $v = bu + a$

$$v = \ln\frac{y - \eta_L}{\eta_U - y}, \qquad u = Q^{-1}(p)$$

$$\sigma = b, \qquad \mu = a$$

Moments

$$E(Y) = \eta_U - (\eta_U - \eta_L)E\left(\frac{1}{e^X + 1}\right)$$

$$Var(Y) = (\eta_U - \eta_L)^2 Var\left(\frac{1}{e^X + 1}\right)$$

2.6.6 The Reflected Log-Normal Distribution

Suppose a continuous variate Y has the sample space $\mathcal{Y} = (-\infty, -\eta)$, which is a bounded-above open interval with a known value of the upper bound $-\eta$. This is the case, for instance, when Y represents temperature of water in either solid or liquid phase at a fixed pressure, so that $-\eta$ is the boiling temperature at a fixed pressure; for example, $-\eta = 100°C$ at 1.00 atm (the normal boiling point), or $-\eta = 373.99°C$ at 217.75 atm (the critical point: water cannot exist in the liquid phase above that temperature, no matter how much pressure is applied).

Variate Y, whose sample space is $\mathcal{Y} = (-\infty, -\eta)$, has a *reflected log-normal distribution*, abbreviated $LN(\mu, \sigma, \eta, -1)$, with the location parameter μ, the scale parameter σ, and the shift parameter η ($-\infty < \mu < \infty$, $\sigma > 0$, $-\infty < \eta < \infty$), if the *reflected* variate

$$X = -Y, \tag{2.26}$$

whose sample space is $\mathcal{X} = (\eta, \infty)$, has the log-normal distribution $LN(\mu, \sigma, \eta)$.

By applying the theory of derived distributions (Section 3.8) to the *reflection transformation* (2.26), one can derive the expressions for the density function h and the distribution function H of variate Y on the sample space $\mathcal{Y} = (-\infty, -\eta)$; these expressions are given in Table 2.7. Graphs of h and H are shown in Figure 2.5, which may be compared with Figure 2.2 to visualize the working of the reflection transformation. (For a more general and complete characterization of the reflection transformation, see Section C.5.)

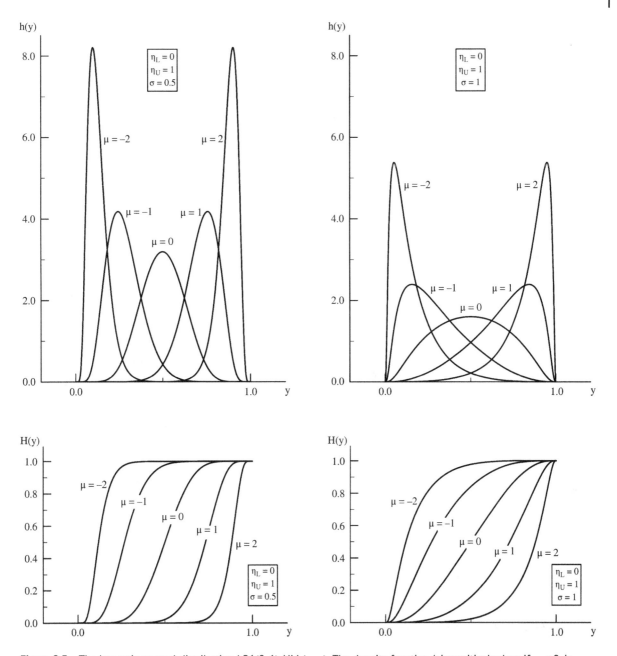

Figure 2.3 The log-ratio normal distribution LR1(0, 1)-NM (μ, σ). The density function h is positively skew if $\mu < 0$, is symmetric if $\mu = 0$, and is negatively skew if $\mu > 0$.

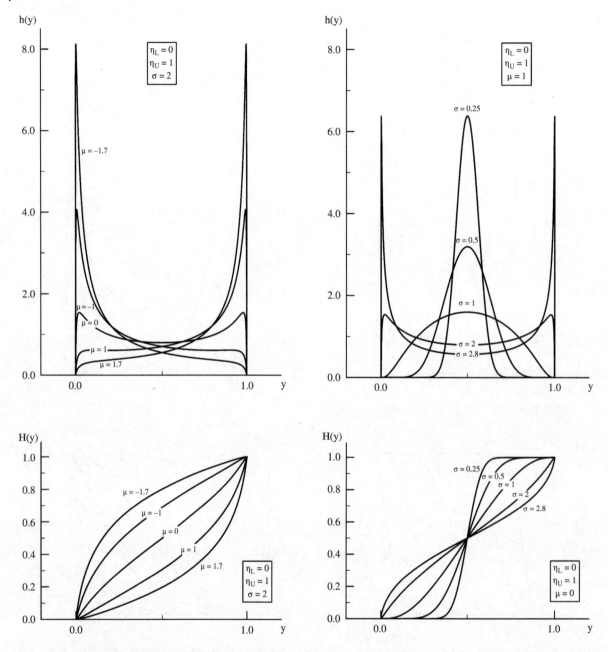

Figure 2.4 The log-ratio normal distribution LR1(0, 1)-NM(μ, σ). The density function *h* can be unimodal or bimodal.

Table 2.7 The reflected log-normal distribution.

Abbreviation: $LN(\mu, \sigma, \eta, -1)$

Parameters: location $-\infty < \mu < \infty$, scale $\sigma > 0$, shift $-\infty < \eta < \infty$

Domains: $-\infty < y < -\eta$, $0 < p < 1$

Density function

$$h(y) = \frac{1}{(-y-\eta)\sigma\sqrt{2\pi}} \exp\left(-\frac{1}{2}\left(\frac{\ln(-y-\eta)-\mu}{\sigma}\right)^2\right)$$

Distribution function $p - H(y)$

$$H(y) = 1 - Q\left(\frac{\ln(-y-\eta)-\mu}{\sigma}\right)$$

Quantile function $y = H^{-1}(p)$

$$H^{-1}(p) = -\exp(\sigma Q^{-1}(1-p) + \mu) - \eta$$

Linearized quantile function $v = bu + a$

$$v = \ln(-y-\eta), \qquad u = Q^{-1}(1-p)$$
$$\sigma = b, \qquad \mu = a$$

Moments

$$E(Y) = -\exp(\sigma^2/2 + \mu) - \eta$$
$$Var(Y) = [\exp(\sigma^2) - 1]\exp(\sigma^2 + 2\mu)$$

2.7 The Gamma Function

2.7.1 Definition and Properties

The *gamma function* Γ is defined at every point $x > 0$ by Euler's integral

$$\Gamma(x) = \int_0^\infty t^{x-1}e^{-t}\,dt, \qquad x > 0. \tag{2.27}$$

A graph of Γ is shown in Figure 2.6. The function has the *recurrence property*

$$\Gamma(x+1) = x\Gamma(x), \qquad x > 0. \tag{2.28}$$

At any positive integer x, it is identical to the *factorial function*

$$\Gamma(x) = (x-1)!, \qquad x = 1, 2, 3, \dots; \tag{2.29}$$

in particular, $\Gamma(1) = \Gamma(2) = 1$. At fractional values $x = \frac{1}{2}, \frac{3}{2}, \frac{5}{2}, \dots$ it has a closed form:

$$\Gamma\left(\frac{1}{2}\right) = \sqrt{\pi} \simeq 1.77245,$$
$$\Gamma\left(\frac{3}{2}\right) = \frac{1}{2}\sqrt{\pi} \simeq 0.88623,$$
$$\Gamma\left(\frac{5}{2}\right) = \frac{3}{2\cdot2}\sqrt{\pi} \simeq 1.32934, \tag{2.30}$$
$$\vdots$$

In general, for any positive integer $n = 1, 2, 3, \dots$

$$\Gamma\left(n + \frac{1}{2}\right) = \frac{1 \cdot 3 \cdot 5 \cdots (2n-1)}{2^n}\sqrt{\pi}. \tag{2.31}$$

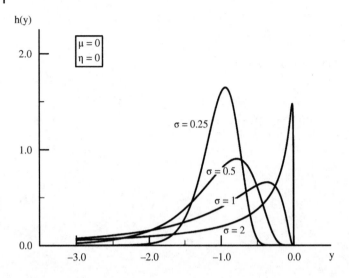

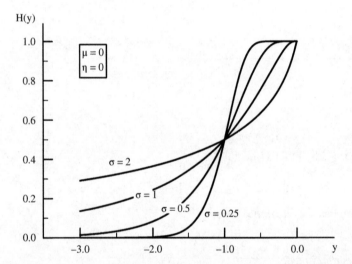

Figure 2.5 The reflected log-normal distribution $LN(\mu, \sigma, \eta, -1)$. The density value $h(y_M)$ at the mode y_M attains a minimum when $\sigma = 1$.

2.7.2 Polynomial Approximation

To calculate $\Gamma(x)$ at any point $x > 0$, one may either apply a numerical integration method to (2.27) or resort to an approximation, whereby function Γ is approximated by another function which is closed-form and easy to evaluate. One such a function is a polynomial. The algorithm for calculating a polynomial approximation to $\Gamma(x)$ is this.

Step 1. Given any $x > 0$, express $\Gamma(x)$ in terms of $\Gamma(z + 1)$, where z is the decimal part of x, so that $0 \leq z < 1$. This is accomplished by applying the recurrence formula (2.28) as follows. If $0 < x < 1$, then $z = x$, and

$$\Gamma(x) = \frac{1}{z}\Gamma(z + 1).$$

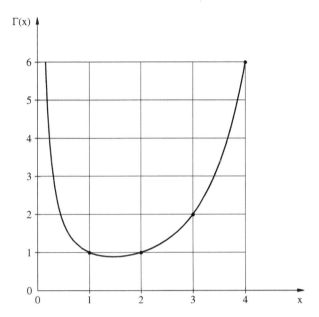

Figure 2.6 The gamma function Γ.

If $1 \leq x < 2$, then $z = x - 1$, and

$$\Gamma(x) = \Gamma(z + 1).$$

If $2 \leq x$, then $z = x - n$, where n is the greatest integer in x, and

$$\Gamma(x) = (z + n - 1)(z + n - 2) \cdots (z + 1)\Gamma(z + 1).$$

Step 2. Given z, calculate $\hat{\Gamma}(z + 1)$ according to the formula in Table 2.8. This formula specifies a fifth-degree polynomial which approximates $\Gamma(z + 1)$ at any $0 \leq z \leq 1$; the absolute error of this approximation does not exceed 5×10^{-5}.

Step 3. Return to the appropriate equation in Step 1; insert the value $\hat{\Gamma}(z + 1)$ in lieu of $\Gamma(z + 1)$ and calculate $\Gamma(x)$. Obviously, the calculated value is approximate.

Table 2.8 Polynomial approximation to the gamma function Γ.

$0 \leq z \leq 1$
$\hat{\Gamma}(z + 1) = 1 + c_1 z + c_2 z^2 + c_3 z^3 + c_4 z^4 + c_5 z^5$
$c_1 = -0.5748646 \quad c_2 = 0.9512363$
$c_3 = -0.6998588 \quad c_4 = 0.4245549$
$c_5 = -0.1010678$
Approximation error
$\|\hat{\Gamma}(z + 1) - \Gamma(z + 1)\| \leq 5 \times 10^{-5}$

Source: Abramowitz and Stegun (1972).

2.8 The Incomplete Gamma Function

2.8.1 Definition and Properties

The *incomplete gamma function* γ is defined at every point $x > 0$ and every upper limit $a > 0$ of the incomplete Euler's integral

$$\gamma(x, a) = \int_0^a t^{x-1} e^{-t} dt, \qquad x > 0, \qquad a > 0. \tag{2.32}$$

The function has the *recurrence property*:

$$\gamma(x + 1, a) = x\gamma(x, a) - a^x e^{-a}, \qquad x > 0, \qquad a > 0. \tag{2.33}$$

At any positive integer x, it has the closed form

$$\gamma(x, a) = (x - 1)! \left[1 - e^{-a} \sum_{j=0}^{x-1} \frac{a^j}{j!} \right], \qquad x = 1, 2, 3, \ldots, \qquad a > 0; \tag{2.34}$$

in particular, $\gamma(1, a) = 1 - e^{-a}$, which is the solution to integral (2.32).

The *complementary incomplete gamma function* γ^c is defined by the integral

$$\gamma^c(x, a) = \int_a^\infty t^{x-1} e^{-t} dt, \qquad x > 0, \qquad a > 0. \tag{2.35}$$

Now, the gamma function Γ, defined by integral (2.27), equals the sum $\Gamma = \gamma + \gamma^c$. Therefrom

$$\gamma(x, a) = \Gamma(x) - \gamma^c(x, a), \qquad x > 0, \qquad a > 0. \tag{2.36}$$

This equation is used below to calculate $\gamma(x, a)$ for any positive x.

2.8.2 Asymptotic Approximation

The function γ^c has a closed-form approximation $\hat{\gamma}^c$, given by its asymptotic expansion, truncated after J terms (Table 2.9). The approximation error is bounded above; the bound depends on x, a, J, and is proportional to the absolute value of the first neglected term, u_{J+1}. The optimal J, which minimizes $|u_{J+1}|$, may be found via a line search because the series $|u_1|, |u_2|, \ldots$ behaves thusly: (i) it is zero for $j \geq x$ when x is an integer; (ii) it is unimodal when x is not an integer — nonincreasing initially and increasing afterward.

The algorithm for calculating an approximation to $\gamma(x, a)$, at any point (x, a) in the positive quadrant of the plane, is this.

Table 2.9 Asymptotic approximation to the complementary incomplete gamma function γ^c.

$x > 0$, $\qquad a > 0$, $\qquad J \in \{1, 2, 3, \ldots\}$
$\hat{\gamma}^c(x, a) = a^{x-1} e^{-a} \left[1 + \sum_{j=1}^{J} u_j \right]$
$u_j = \dfrac{(x-1)(x-2) \cdots (x-j)}{a^j}, \qquad j = 1, 2, \ldots, J$
Approximation error
$\|\hat{\gamma}^c(x, a) - \gamma^c(x, a)\| \leq a^{x-1} e^{-a} \|u_{J+1}\|$

Source: Abramowitz and Stegun (1972).

Step 1. Given $x > 0$, calculate the approximate value of $\Gamma(x)$ by following the algorithm of Section 2.7.2.

Step 2. Given the above x and $a > 0$, find the optimal number J of the terms thusly: (i) If x is an integer, then $J = x - 1$. (ii) If x is not an integer, then calculate u_j for $j = 1, 2, \ldots$ until the first inequality $|u_j| < |u_{j+1}|$ is found; set $J = j - 1$.

Step 3. Given the above J, calculate the approximate value of $\gamma^c(x, a)$ according to the formula in Table 2.9.

Step 4. Insert the calculated values of $\Gamma(x)$ and $\gamma^c(x, a)$ into equation (2.36) to obtain an approximate value of $\gamma(x, a)$.

Step 5. Calculate the *relative error bound* for the approximate value $\hat{\gamma}^c(x, a)$; it is the upper bound of the approximation error, as defined in Table 2.9, expressed as the percentage of the approximate value:

$$REB = \frac{a^{x-1}e^{-a}|u_{J+1}|}{\hat{\gamma}^c(x, a)} 100\%. \tag{2.37}$$

Step 6. Decide if this asymptotic approximation to $\gamma^c(x, a)$ is acceptable. When x is a positive integer, $REB = 0\%$. Otherwise, REB may be large, especially when $x < 1$ or $a < 1$; a more specific guide is provided in the notes below. (For exercises in this book, $REB < 17\%$ may be considered acceptable. The polynomial approximation to $\Gamma(x)$ has a relatively small error bound, which is specified in Table 2.8 and is acceptable for exercises.)

Notes.

(i) When x is a positive integer, $|u_{J+1}| = 0$ and $\hat{\gamma}^c(x, a)$ is exact; that is, it equals $\gamma^c(x, a)$.

(ii) When $J = 0$, the first neglected term is u_1; consequently, the bracketed term in the equation for $\hat{\gamma}^c(x, a)$ equals 1.

(iii) When J is optimal (as set in Step 2) and x is fixed, $\hat{\gamma}^c(x, a)$ decreases with a, and the REB generally (but not always) decreases with a.

(iv) A guide to the magnitude of REB (Table 2.10) was constructed thusly. The REB was calculated at every point (x, a) of the grid having coordinates $x = 0.25, 0.5, \ldots, 4$ and $a = 0.25, 0.5, \ldots, 6$. Then for every x, two thresholds (a^*, REB^*) were selected: the smallest a for which $REB < 20\%$, and the smallest a for which $REB < 10\%$.

Table 2.10 A guide to the magnitude of REB: for a given x, there are two thresholds a^* such that if $a \geq a^*$, then $REB \leq REB^*$.

x	a^*	REB^*	x	a^*	REB^*
0.25	2.75	19%	2.25	0.75	17%
	3.5	9%		1	9%
0.5	2.25	19%	2.5	1	12%
	3	8%		1.25	7%
0.75	1.5	17%	2.75	0.75	14%
	2	9%		1	8%
1.25	1	15%	3.25	1.25	4%
	1.5	7%			
1.5	1	17%	3.5	1.5	4%
	1.5	9%			
1.75	0.75	17%	3.75	1.75	1%
	1.25	8%			

(v) When *REB* is unacceptable, it may be possible to reduce it by taking advantage of the recurrence property (2.33). For example, for point (0.25, 1.5), the algorithm yields

$$\hat{\Gamma}(0.25) = 3.6254, \qquad \hat{\gamma}^c(0.25, 1.5) = 0.1646, \qquad \hat{\gamma}(0.25, 1.5) = 3.4608,$$

and *REB* = 50%. For point (1.25, 1.5), the algorithm yields

$$\hat{\Gamma}(1.25) = 0.9064, \qquad \hat{\gamma}^c(1.25, 1.5) = 0.2881, \qquad \hat{\gamma}(1.25, 1.5) = 0.6183,$$

and *REB* = 7%. Via the inverse of the recurrence equation (2.33),

$$\hat{\gamma}(0.25, 1.5) = [0.6183 + 1.5^{0.25} e^{-1.5}]/0.25 = 3.4609.$$

The two values of $\hat{\gamma}(0.25, 1.5)$ are almost identical, but we are assured that the approximation error is acceptable, as *REB* = 7%.

Exercises

1 **Uniform distribution**. Suppose variate X has a uniform distribution, $\mathrm{UN}(\eta_L, \eta_U)$. Given the expression for the density function of X (Table 2.1), derive the expressions for the distribution function of X, the quantile function of X, the mean of X, and the variance of X.

2 **Normal distribution**. Show every step in the following derivations.
   2.1 Given expressions (2.1) and (2.12), derive expression (2.7) for the normal density function.
   2.2 Given expressions (2.2) and (2.8), derive expression (2.13) for the normal distribution function.

3 **Log-normal distribution**. Given is the logarithmic transform (2.18).
   3.1 Derive the transform between the standard normal variate Z and the log-normal variate Y.
   3.2 Use the above transform to derive the expression for the quantile function of Y.
   3.3 Write the expressions for the first quartile, the second quintile, and the seventh decile of Y; the expressions should be in terms of the distribution parameters: μ, σ, η.

4 **Log-ratio normal distribution**. Show every step in the following derivations.
   4.1 Given the log-ratio transform (2.23), derive its inverse (2.24).
   4.2 Given the inverse of the log-ratio transform (2.24), derive the expressions for $E(Y)$ and $Var(Y)$, which are given in Table 2.6.

5 **Reflected log-normal distribution**. Given the expressions in Table 2.5 for the log-normal distribution $\mathrm{LN}(\mu, \sigma, \eta)$, the reflection transformation (2.26), and the relationships in Section C.5.1, write the expressions for the *reflected* log-normal distribution $\mathrm{LN}(\mu, \sigma, \eta, -1)$ of variate Y: (i) the density function, (ii) the distribution function, (iii) the quantile function, (iv) the mean, and (v) the variance. Compare your expressions with those listed in Table 2.7.

6 **Linearized quantile function**. Study Section 2.6.3 and Table 2.4 to understand the derivation of the linearized quantile function of the normal variate X. Next perform an analogous task: take the quantile function of variate Y specified in the option, and show every step in the derivation of its linearized quantile function.

   Options: (LN)   The log-normal Y.
               (LR1)   The log-ratio normal Y.
              (R-LN) The reflected log-normal Y.

7 **Examples of Gaussian variates**. Consider each distribution type — LN, LR1-NM, reflected LN — as a hypothesized model of some variate X. For each of these models, (i) come up with an example of variate X such that the hypothesis is plausible, and (ii) list the attributes of X that support the plausibility of the hypothesis.

8 **Rational function approximations**. Study Section 2.6.1 and the rational function approximations to the standard normal distribution function Q (Table 2.2) and the standard normal quantile function Q^{-1} (Table 2.3). Next complete the following exercises.

    8.1 Write a computer code that calculates $\hat{Q}(z)$ for any $z \in (-\infty, \infty)$, and $\hat{z}_p$ for any $p \in (0, 1)$.

    8.2 Validate the code and the approximations by calculating each $\hat{Q}(z)$ and $\hat{z}_p$ at 10 judiciously selected values of the argument, and by comparing the calculated values with values either tabulated in a book of your choice or calculated using a library software of your choice. In particular, for each pair of values being compared, calculate the absolute difference. Is this difference within the approximation error listed in Table 2.2 or Table 2.3?

    Note. In the comparison, account for the truncation of tabulated or library values. Reference the source of the tabulated or library values.

9 **Recurrence formula**. Using the recurrence property of the gamma function Γ, prove that for any positive integer $n \geq 2$, and any $0 \leq z < 1$,

$$\Gamma(z + n) = (z + n - 1)(z + n - 2)\cdots(z + 1)\Gamma(z + 1).$$

10 **Polynomial approximation**. Write a computer code implementing the algorithm for the polynomial approximation to the gamma function Γ (Section 2.7.2). Validate the code by (i) calculating the approximate and the exact values of $\Gamma(x)$ at points $x = n + \frac{1}{2}$ and $x = n + 1$ for $n = 0, 1, 2, 3, 4$; (ii) calculating the approximation errors; and (iii) checking that each error does not exceed the upper bound.

11 **Closed-form formula**. Using the recurrence property (2.33), prove that for any positive integer $x \geq 1$ and any $a > 0$, the incomplete gamma function γ has the closed form (2.34).

12 **Error bound**. The algorithm of Section 2.8.2 applies equation (2.36) to obtain an approximate function $\hat{\gamma} = \hat{\Gamma} - \hat{\gamma}^c$. Given the upper bounds on the errors of functions $\hat{\Gamma}$ (Table 2.8) and $\hat{\gamma}^c$ (Table 2.9), prove that the upper bound on the error of function $\hat{\gamma}$ is

$$|\hat{\gamma}(x, a) - \gamma(x, a)| \leq 0.00005 + a^{x-1}e^{-a}|u_{J+1}|.$$

13 **Asymptotic approximation**. Write a computer code implementing the algorithm for the asymptotic approximation to the incomplete gamma function γ (Section 2.8.2).

    13.1 Test the code by calculating, at all points (x, a) specified in the option, the approximate values of $\Gamma(x)$, $\gamma^c(x, a)$, $\gamma(x, a)$, and the value of *REB*. Report these values and also the optimal number of terms J used for calculating $\hat{\gamma}(x, a)$.

    13.2 Are the calculated values of *REB* consistent with the guide values *REB** specified in Table 2.10? Explain your answer.

    13.3 For points (x, a) with integer x, calculate the exact values of $\gamma(x, a)$ from formula (2.34). Do these values match the values calculated by your computer code?

Options for the set of points $\{(x, a)\}$:

(A) $\{(0.25, 3), (1.25, 1.25), (2, 0.5), (2.25, 1.25), (3, 1.5), (3.25, 1.5), (4, 5.25)\}$;

(B) $\{(0.5, 2.5), (1.5, 1.25), (2, 1.5), (2.5, 1.5), (3, 1.75), (3.5, 1.75), (4, 5.5)\}$;

(C) $\{(0.75, 1.75), (1.75, 1), (2, 2.5), (2.75, 1.25), (3, 2), (3.75, 2), (4, 5.75)\}$.

14 Probability calculus. Perform this exercise for each of the special parametric distributions selected by the instructor (or by yourself) from the set:

LG, LP, GB, RG; EX, WB, IW, LW, LL; P1, P2.

The distributions are catalogued in Appendix C, where you may find the expression which is given as well as the expression which you are asked to derive.

14.1 Given the expression for the distribution function H, derive the expression for the density function h.

14.2 Given the expression for the distribution function H, derive the expression for the quantile function H^{-1}.

14.3 Given the expression for the quantile function H^{-1}, derive the expression for the linearized quantile function.

14.4 Given the expression for the density function h, show that the integral of the density function h over the sample space equals 1.

14.5 Given the expression for the density function h, derive the expression for the distribution function H.

14.6 Given the expression for the density function h, derive the expression for the mean and the variance.

15 Incoherence of unbiased variance. Let X denote the number of people walking into a bus stop during a 5-minute interval between 0600 and 0615 on a weekday. Counts from three days yielded the sample $\{2, 3, 1, 3, 2, 1, 1, 1, 2\}$.

15.1 Calculate the sample mean m, the sample variance s^2, and the unbiased estimate s_1^2 of variance.

15.2 Estimate the probability function h of X. To wit, identify the sample space $\mathcal{X}$ of X; and for every $x \in \mathcal{X}$, calculate $h(x)$ as the relative frequency of event $X = x$.

15.3 Use the probability function h to calculate the expectation $E(X)$ and the variance $Var(X)$ according to their definitions.

15.4 Compare the values of: (i) $E(X)$ with m; (ii) $Var(X)$ with s^2 and s_1^2. Draw conclusions regarding the coherence of the estimates.

3

Distribution Modeling

Univariate distribution functions play a key role in forecast models and in decision models. They are the elementary models of individual variates; they are the main carriers of predictive information; and they are the means of quantifying and communicating the uncertainty. Their modeling and estimation are partly a science and partly an art. And the success in these tasks is a necessary condition for developing well-performing models of complex problems. This chapter presents a modeling methodology—constructs and procedures needed to obtain a good parametric model for the distribution function of a continuous variate.

3.1 Distribution Modeling Methodology

Let H denote the distribution function of a continuous variate X whose sample space is $\mathcal{X}$. Given a sample of X, the objective is to obtain a parametric model for H. The general modeling methodology consists of six steps.

1. Construct the empirical distribution function $\check{H}$ of X (Section 3.2).
2. Specify the sample space $\mathcal{X}$ of X (Section 3.3).
3. Hypothesize parametric models for H (Section 3.4).
4. Estimate the parameters of each hypothesized model (Section 3.5).
5. Evaluate the goodness of fit of each model and choose the best one (Section 3.6).
6. Test statistically the chosen parametric model (Section 3.6).

3.2 Constructing Empirical Distribution

Let X denote a continuous variate having a fixed but unknown distribution function H. Let $\{x(n) : n = 1, \ldots, N\}$ denote a sample of N realizations of variate X. Suppose the experiment or process that generated the sample is understood enough to justify the assumption that each realization $x(n)$ was generated randomly, and independently of other realizations, from the same distribution function H. The objective is to construct an estimate of the unknown distribution function using solely the given sample, and without making any assumptions about the form or the properties of H. Such an estimate is called the *empirical distribution function* of X and is denoted by $\check{H}$. The construction proceeds as follows.

Step 0. Given is a sample of N realizations of a continuous variate X:

$$\{x(n) : n = 1, \ldots, N\}.$$

Probabilistic Forecasts and Optimal Decisions, First Edition. Roman Krzysztofowicz.
© 2025 John Wiley & Sons Ltd. Published 2025 by John Wiley & Sons Ltd.
Companion website: www.wiley.com/go/ProbabilisticForecastsandOptimalDecisions1e

Step 1. Arrange the realizations from the sample in ascending order of numerical values:

$$x_{(1)} \leq x_{(2)} \leq \cdots \leq x_{(N)}, \tag{3.1}$$

where $x_{(n)}$ denotes the nth realization and $(n) \in \{1, \ldots, N\}$. When two or more realizations have an identical value, their order is irrelevant.

Step 2. Define N nonexceedance probabilities, also called the *plotting positions*:

$$p_n = \check{H}(x_{(n)}) = P(X \leq x_{(n)}), \qquad n = 1, \ldots, N; \tag{3.2}$$

these probabilities form a strictly increasing sequence:

$$p_1 < p_2 < \cdots < p_N. \tag{3.3}$$

Step 3. Calculate the plotting positions p_n $(n = 1, \ldots, N)$.

- The *standard plotting positions* are

$$p_n = \frac{n}{N}. \tag{3.4}$$

They are suitable for theoretic analyses and some statistical tests. But they should not be used otherwise because they do not give a reasonable estimate of the distribution function H unless the sample size is very large. (See the example below.)

- The *Weibull plotting positions* are

$$p_n = \frac{n}{N+1}. \tag{3.5}$$

They are suitable, and identical to the plotting positions defined by equations (3.6)–(3.7), when $N > 20\,000$.

- The *meta-Gaussian plotting positions*, which we shall use, are

$$p_n = \left[\left(\frac{N-n+1}{n} \right)^{t_N} + 1 \right]^{-1}, \tag{3.6}$$

where t_N is a constant whose value is a function of the sample size N as follows:

$$
\begin{aligned}
t_N &= 3.0193 N^{-1.1018} + 1, & 2 \leq N \leq 3, \\
t_N &= 2.4035 N^{-0.9096} + 1, & 4 \leq N \leq 5, \\
t_N &= 2.1408 N^{-0.8423} + 1, & 6 \leq N \leq 10, \\
t_N &= 1.9574 N^{-0.8039} + 1, & 11 \leq N \leq 20\,000, \\
t_N &= 1, & 20\,000 < N.
\end{aligned}
\tag{3.7}
$$

Step 4. Pair each realization with its plotting position to form a set of N points:

$$\{(x_{(n)}, p_n) : n = 1, \ldots, N\}. \tag{3.8}$$

These points specify the empirical distribution function $\check{H}$ of X.

Example 3.1 Let $N = 5$. Thus the appropriate formula for plotting positions is (3.6) with $t_5 = 1.5560$ calculated from equation (3.7). Figure 3.1 shows a graph of $\check{H}$ constructed from a sample and the meta-Gaussian plotting positions. This $\check{H}$ is the basic construct for modeling H. Nonetheless, let us compare the plotting positions calculated according to each of the three formulae. The results reported in Table 3.1 illustrate three facts. First, for small N each formula yields drastically different probability values. Second, the standard formula violates Cromwell's rule (Section 4.2.4) — it assigns probability 1 to the event that X does not exceed the largest realization $x_{(5)}$. In effect it implies the impossibility of observing in the future a value of X greater than $x_{(5)}$. This is obviously false

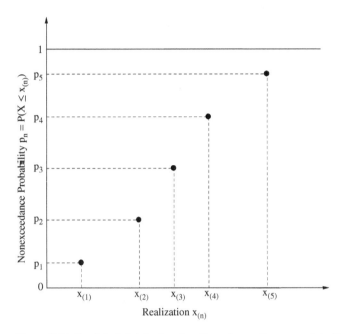

Figure 3.1 Empirical distribution function of a continuous variate X constructed from a sample of size $N = 5$ using the meta-Gaussian plotting positions.

Table 3.1 Plotting positions $\{p_n : n = 1, \ldots, 5\}$ calculated from three formulae for a sample of size $N = 5$.

n	1	2	3	4	5
Standard p_n	0.2	0.4	0.6	0.8	1.0
Weibull p_n	0.1667	0.3333	0.5	0.6667	0.8333
Meta-Gaussian p_n	0.0756	0.2538	0.5	0.7462	0.9244

unless the sample space of X is bounded above, and $x_{(5)}$ happens to be the bound. Third, the variance of X implied by the Weibull probabilities is larger than the variance of X implied by the meta-Gaussian probabilities. This is a shortcoming of the Weibull plotting positions: they are biased in that they overestimate the variance of X. The bias decreases as N increases and becomes insignificant when $N > 20\,000$.

3.3 Specifying the Sample Space

3.3.1 Types of Sample Space

The sample space $\mathcal{X}$ of a continuous variate X is assumed to be an interval—one of four types:

the unbounded interval $\mathcal{X} = (-\infty, \infty)$,
the bounded-below interval $\mathcal{X} = (\eta, \infty)$,
the bounded open interval $\mathcal{X} = (\eta_L, \eta_U)$,
the bounded-above interval $\mathcal{X} = (-\infty, -\eta)$,

where bounds are real numbers, with $-\infty < \eta < \infty$ and $-\infty < \eta_L < \eta_U < \infty$. Because the type of the sample space $\mathcal{X}$ determines the group of applicable parametric models (as is evident from Section 2.6 and Appendix C), $\mathcal{X}$ should be specified thoughtfully.

3.3.2   Assessing the Bounds

The task is to specify the sample space $\mathcal{X}$ of the variate X. It can be accomplished via a logical analysis guided by three criteria.

1. *The definition of the variate.* In many cases, the definition of the variate suggests a logical specification of the sample space. For instance, relative humidity expressed as a percentage is naturally bounded below by 0 and above by 100; thus $\mathcal{X} = (0, 100)$. River stage is naturally bounded below by a value η at which the flow ceases; thus $\mathcal{X} = (\eta, \infty)$. Travel time is naturally bounded below by 0; however, in a specific application, say the travel time by taxi from a city center to an airport, it may be feasible to determine a lower bound $\eta > 0$ such that $P(X > \eta) = 1$; thus $\mathcal{X} = (\eta, \infty)$. In the latter two examples, the absence of a logical upper bound leads by default to an interval unbounded above. In general, when no *natural bound* can be specified, one may leave the sample space unbounded (on that side) or specify an *artificial bound* based on the two criteria that follow.

2. *The extreme realizations of the variate.* Obviously, when a lower bound η_L is desirable, it should be specified below the smallest realization $x_{(1)}$ in the sample, and when an upper bound η_U is desirable, it should be specified above the largest realization $x_{(N)}$ in the sample. While specifying the bounds in this manner, it is prudent to take into consideration the possibility of future realizations falling outside the range of those in the sample currently available. Thus after $\mathcal{X} = (\eta_L, \eta_U)$ is specified, one should verify that one's judgmental probabilities are $P(X \leq \eta_L) = \varepsilon$ and $P(X \geq \eta_U) = \varepsilon$, where ε is a positive number near zero.

3. *The tails of the empirical distribution function.* When $\mathcal{X} = (\eta, \infty)$ or $\mathcal{X} = (-\infty, -\eta)$, the bound η becomes the *shift parameter* of a distribution function of X. When $\mathcal{X} = (\eta_L, \eta_U)$, the bounds η_L, η_U become the *shift parameters* of a distribution function of X. There are two implications: (i) the specification of a bound alters the group of applicable parametric models, and (ii) the value of a bound may affect the goodness of fit of the parametric model being applied. (Goodness of fit is explained in Section 3.6.) It follows that the analyst has a considerable latitude in affecting the choice of a parametric model and the goodness of fit. A heuristic guideline for the initial decision is this: if a tail of the empirical distribution function is short, then it may be advantageous to bound the sample space on that side; if a tail of the empirical distribution function is long, then it may be advantageous not to bound the sample space on that side.

   Finally, when $\mathcal{X}$ is naturally bounded, on one side or on both sides, the unbounded interval $\mathcal{X} = (-\infty, \infty)$ still may be specified — for an approximate model. Once such a model has its parameters estimated and the goodness of fit evaluated, one must ponder: Is this model satisfactory, especially in the tails where the natural bounds would be located?

3.4   Hypothesizing Parametric Models

3.4.1   All Models

There are many parametric models for distribution functions. This book contains 24 parametric models; they offer a rich selection of shapes, while being easy to estimate and to use. Each model has at most two parameters and admits a linearization of the distribution function and the quantile function (the inverse of the distribution function). The models are of two kinds.

Gaussian distributions

These are the four families of distributions derived from the standard normal distribution and presented in Section 2.6. Whereas only the density functions have closed-form expressions, the distribution and quantile functions have simple and accurate numerical approximations.

Special univariate distributions

A catalogue of 20 special univariate distributions of continuous variates is provided in Appendix C. These distributions are special because they are easy to estimate and to use: each of these distributions is parametric, has closed-form expressions for the density function, the distribution function, and the quantile function, and admits a linearization of the distribution function and the quantile function.

3.4.2 Narrowing the Hypotheses

While in concept every one of these parametric models can be hypothesized for the distribution function H of a given variate X, in practice one wishes to consider one, or two, or at most a few "suitable" hypotheses. Thus the question: How to judge whether or not a parametric model is suitable as a hypothesis? This task, admittedly subjective and seemingly difficult without experience, can be guided by four criteria.

1. *The sample space of the variate.* It may be one of four types, as defined in Section 3.3. Once the sample space $\mathcal{X}$ of the given variate X is specified, it suffices to consider only the parametric models applicable to the particular type of the sample space. In Appendix C, the parametric models are grouped according to the type of the sample space.
2. *The shape of the empirical distribution function.* Each parametric model can represent a family of shapes; this family is distinct for each model. The task is to match the shape of the empirical distribution function with a shape that can be represented by a parametric model. It is not an easy judgmental task, but with experience one can learn to recognize the shapes that can be represented by a particular parametric model (e.g., the exponential distribution) or that require a parametric model with some particular property (e.g., a symmetric distribution or a heavy tail distribution). To aid in the task of "matching the shapes", every model included in Appendix C is accompanied by graphs of its distribution functions for selected values of the parameters.
3. *The analytic convenience.* It is easy to use a parametric model that has closed-form expressions for the density function, the distribution function, and the quantile function (the inverse of the distribution function). Following the principle of Occam's razor, one should try first the easy-to-use models. Perhaps one of them will fit well to the empirical distribution function, thus obviating the need for considering models which require numerical approximations to evaluate either the density function or the distribution function.
4. *The empirical precedents.* The literature should be reviewed for studies that used the same or similar kind of variate. If a study reports good performance of a particular parametric model, then this model is a suitable hypothesis. If several studies confirm a particular parametric model as their best choice, then this model may be considered as the sole hypothesis, at least until it is not rejected by a goodness-of-fit test on the sample at hand. For instance, it is known that the precipitation amount accumulated anywhere in the eastern United States during a period of up to 72 h is often suitably modeled by a Weibull distribution; thus the Weibull model is a suitable first hypothesis.

> **Principle** (Occam's razor) for constructing explanations in philosophy and sciences. As stated by William of Occam (1285–1349), a Franciscan frier and philosopher, in Latin: *Entia non sunt multiplicanda praeter necessitatem.* In translation: "No more things should be presumed to exist than are absolutely necessary." As interpreted broadly, Occam's razor is the principle that the simplest of alternative models, which explains a phenomenon, should be preferred to the more complex models. Still more broadly, complexity for its own sake is not a virtue.

3.5 Estimating Parameters

3.5.1 Types of Parameters

All distributions in the catalogue (Appendix C) are parameterized in a consistent manner. There are four types of distribution parameters: *location, scale, shape,* and *shift* parameters. One shift parameter η or two shift parameters (η_L, η_U) are present when the sample space is bounded; these parameters are to be assessed judgmentally before other parameters can be estimated. With three exceptions, all distributions have two parameters to be estimated, a location parameter β and a scale parameter α, or a scale parameter α and a shape parameter β. The three exceptions are the exponential distribution, and the two power distributions (each of which has one parameter to be estimated).

Note on shift parameters
There exist statistical methods for estimating shift parameters from a sample, but they are generally cumbersome and are not considered herein. However, one can always consider several values of the shift parameter near the initial value assessed judgmentally, and find the value which, together with the optimal values of other parameters, gives the best fit.

3.5.2 Estimation Problem

Suppose the type of the sample space $\mathcal{X}$ of variate X has been specified, the shift parameters (if any) have already been assessed, and it is hypothesized that variate X has the distribution function $H(\cdot|\alpha, \beta)$ of a particular form, with two parameters α and β, whose values are fixed but unknown. When all feasible parameter values are considered, one may speak of the *family of parametric distribution functions* of a particular form:

$$\{H(\cdot|\alpha, \beta) : all\, \alpha, \beta\}.$$

The objective is to select a particular distribution function from the family by finding estimates $\hat{\alpha}, \hat{\beta}$ of the parameters α, β which are optimal with respect to some well-defined measure. Once the estimates $\hat{\alpha}, \hat{\beta}$ have been found, the conditioning of the distribution function on the parameters is usually not shown; we shall write simply H instead of $H(\cdot|\hat{\alpha}, \hat{\beta})$.

There are several methods for estimating parameters of distribution functions. We shall employ three methods:

LS — the method of *least squares*,
UD — the method of *uniform distance*,
MS — the method of *moments*.

The details of each method, the rationale for using it, and the reasons for not using other methods are presented in Appendix B.

3.6 Evaluating Goodness of Fit

Let H be a distribution function with fixed (estimated) parameter values, which is hypothesized as a model of a continuous variate X. After the parameters of every hypothesized model have been estimated, two steps remain.

1. Choosing the best parametric model.
2. Testing the chosen parametric model.

Each step requires answering the question: How well does the hypothesized distribution function H fit to an empirical distribution function $\check{H}$ constructed from the sample? Three procedures for evaluating the goodness of fit are detailed.

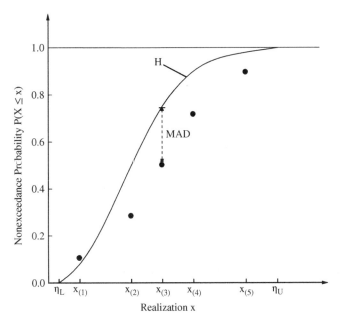

Figure 3.2 A hypothesized distribution function H of a continuous variate X having the sample space $\mathcal{X} = (\eta_L, \eta_U)$ superposed on the empirical distribution function constructed from a sample of size $N = 5$, as shown in Figure 3.1; the MAD is the maximum absolute difference between the two functions.

3.6.1 Graphical Comparison

First, one should prepare a figure with a graph of the hypothesized distribution function $\{(x, H(x)): x \in \mathcal{X}\}$ superposed on the graph of the empirical distribution function $\{(x_{(n)}, p_n) : n = 1, \dots, N\}$. Figure 3.2 shows an example. The figure helps to judge whether or not the fit of the hypothesized distribution function H to the empirical distribution function $\check{H}$ appears satisfactory. This is an apt way to develop one's intuition about the association between the visual perception of goodness of fit and a quantitative measure of goodness of fit, such as the MAD calculated according to expression (3.9).

This is also a good way to visualize the trade-offs that come with using the particular parametric distribution function, which has a particular form (e.g., exponential, Weibull, inverted Weibull, log-Weibull, log-logistic). An oft seen trade-off is this: two hypothesized distribution functions yield nearly equal MADs, but one H fits well the tails of $\check{H}$ but not the center, whereas the other H fits well the center of $\check{H}$ but not the tails. The *preferred trade-off* may depend on the application. For example, when the primary objective is to forecast an extreme event, it is the fit to one of the tails that matters most and thus should dictate the choice of H. (For an example of such an application, read "Challenger accident forecast", Exercise 10, Chapter 5.)

3.6.2 Uniform Distance

Distance measure

A measure of goodness of fit is the *uniform distance*. It is calculated as the *maximum absolute difference* (MAD) between the empirical distribution function $\{(x_{(n)}, p_n) : n = 1, \dots, N\}$ and the hypothesized distribution function H:

$$\text{MAD} = \max_{1 \leq n \leq N} |p_n - H(x_{(n)})|. \tag{3.9}$$

Figure 3.2 provides an illustration. Based on our experience, the association between the MAD and the visual perception of goodness of fit is roughly this:

$$0 \leq \text{MAD} < 0.05 \qquad \text{excellent to good fit,}$$
$$0.05 \leq \text{MAD} < 0.10 \qquad \text{good to adequate fit,}$$
$$0.10 \leq \text{MAD} \qquad \text{adequate to poor fit.}$$

Model choice

The graphical comparisons and the MADs for all hypothesized models provide the basis for choosing the "best" one. Sometimes there is no decisive winner as several models yield MADs close to the minimum MAD. Then choose the model based on (i) visual judgment of the goodness of fit (considering the trade-offs), (ii) model simplicity (Occam's razor, Section 3.4.2), (iii) your preference (for model's aesthetics or some other attribute).

Note on distance measure

The MAD comes from a branch of numerical analysis called *uniform approximations* (Szidarovszky and Yakowitz, 1978, Section 2.2). In particular, minimization of the MAD serves as a criterion for determining the "best approximating polynomial" with respect to a finite set of points in $\Re^2$. Besides possessing useful properties, uniform approximations of functions often appear to the eye as the best *vis-à-vis* approximations determined by other criteria. An MAD is also used in the following statistical test.

3.6.3 The Kolmogorov-Smirnov Test

When the choice of a model is reported to the scientific community (e.g., via a technical paper), it is customary to report also the results of a *statistical goodness-of-fit test*. Such a test accounts for the size N of the sample from which the empirical distribution function $\breve{H}$ is constructed; it recognizes that $\breve{H}$ is not the true distribution function of X, but only an estimate, subject to the sampling variability caused by the finite N.

Let H denote the hypothesized distribution function of a continuous variate X whose sample space is $\mathcal{X} = (\eta_L, \eta_U)$ with $-\infty \leq \eta_L < \eta_U \leq \infty$. The Kolmogorov-Smirnov (K-S) test considers two hypotheses.

Null hypothesis: variate X has the distribution function H.
Alternative hypothesis: variate X has a distribution function other than H.

The test employs an empirical distribution function $\breve{H}$ constructed using the standard plotting positions as follows:

$$\breve{H}(x) = \frac{n}{N} \quad \text{for} \quad x_{(n)} \leq x < x_{(n+1)}, \qquad n = 0, 1, \dots, N, \tag{3.10}$$

where $x_{(0)} = \eta_L$ and $x_{(N+1)} = \eta_U$. This is a *stepwise* increasing function which has ordinate 0 for $\eta_L \leq x < x_{(1)}$, jumps to ordinate $1/N$ at $x = x_{(1)}$ and remains constant for $x_{(1)} \leq x < x_{(2)}$, then jumps to ordinate $2/N$ at $x = x_{(2)}$ and remains constant for $x_{(2)} \leq x < x_{(3)}$, and so on. Finally, it jumps to ordinate 1 at $x = x_{(N)}$ and remains constant for $x_{(N)} \leq x < \eta_U$. Figure 3.3 shows an example.

Test statistic

The *Kolmogorov-Smirnov statistic D* is the supremum (the least upper bound) of the absolute difference between the empirical distribution function and the hypothesized distribution function over the sample space:

$$D = \sup_{x \in \mathcal{X}} \left| \breve{H}(x) - H(x) \right|. \tag{3.11}$$

Because H is a continuous and strictly increasing function, whereas $\breve{H}$ is a stepwise increasing function with jumps at a finite number of points, the supremum over the sample space $\mathcal{X}$ can be found as the maximum over a

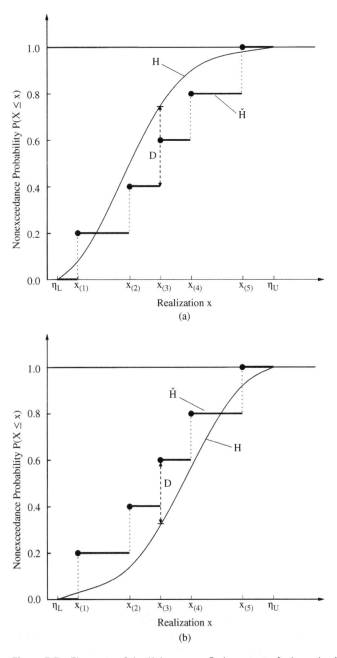

Figure 3.3 Elements of the Kolmogorov-Smirnov test of a hypothesized distribution function H of a continuous variate X, using a sample of size $N = 5$: (a) an example of H for which $D = D^L$; (b) an example of H for which $D = D^A$.

finite set of points. Figure 3.3 illustrates the idea. First, find the maximum of the absolute differences at all points just to the left of the jump points:

$$D^L = \max_{1 \leq n \leq N} \left| \frac{n-1}{N} - H(x_{(n)}) \right|, \qquad n = 1, \dots, N. \tag{3.12}$$

Second, find the maximum of the absolute differences at all jump points:

$$D^A = \max_{1 \leq n \leq N} \left| \frac{n}{N} - H(x_{(n)}) \right|, \qquad n = 1, \ldots, N. \tag{3.13}$$

Finally, find the overall maximum of the absolute differences:

$$D = \max \{ D^L, D^A \}. \tag{3.14}$$

Test procedure

The null hypothesis is rejected at a significance level α if

$$D > C, \tag{3.15}$$

where C is the *critical value* that depends upon the *sample size N* and the *significance level α*. The critical values are listed in Table 3.2.

The significance level α represents the size of the *Type I error*, which is the probability of rejecting the null hypothesis when the null hypothesis is true. For a fixed N, an increase of α results in a decrease of the critical value C. Hence two practical guidelines for selecting the significance level:

- If you want to accept model H, then choose α as large as possible and such that $D \leq C$; a larger α makes the acceptance more convincing.
- If you want to reject model H, then choose α as small as possible and such that $D > C$; a smaller α makes the rejection more convincing.

The smallest level of significance α at which model H is rejected (i.e., at which $D > C$) is called the *P-value*. Thus, when the P-value is reported, one may conclude that model H is rejected at every significance level α such that $\alpha \geq$ P-value.

Even with the above guidelines, the selection of the significance level remains a matter of subjective judgment and established tradition (which dictates the values of α being listed in tables and reported in publications). For this reason, the K-S test (as any other statistical test) should not be regarded as the ultimate arbiter who accepts or rejects a model, but rather as an aid that provides standardized information (N, D, α, C) which may help the scientist to evaluate a model and to compare it with other models.

Example 3.2 Let $N = 5$ and let the hypothesized distribution function H be as shown in Figure 3.3a. Table 3.3 lists the ordinates n/N and $H(x_{(n)})$ for $n = 1, \ldots, 5$, and shows the absolute differences calculated at all points according to equations (3.12) and (3.13). From this table, we find $D^L = 0.35$ and $D^A = 0.15$; thus $D = 0.35$. For $N = 5$ and $\alpha = 0.20$, Table 3.2 gives $C = 0.446$. Since $0.35 < 0.446$, we conclude that the hypothesized distribution function is not rejected at the 20% significance level.

Example 3.3 Let $N = 10$ and $D = 0.35$. We scan the row for sample size 10 in Table 3.2, and find the largest significance level α at which $0.35 \leq C$. This is $\alpha = 0.10$, at which $C = 0.368$. Hence we conclude that the hypothesized distribution function is not rejected at the 10% significance level (or lower). Of course, it is rejected at the 15% significance level (or higher). Thus the P-value is 0.15.

3.6.4 Insights into the Kolmogorov-Smirnov Test

Convergence of empirical distribution

Under the null hypothesis, the distribution function of the Kolmogorov-Smirnov statistic D is independent of the hypothesized distribution function H. For this reason, D is called a *distribution-free statistic*. This statistic has an important property, which is the cornerstone of the K-S test. To state it, let us consider an infinite

Table 3.2 Critical values *C* for the Kolmogorov-Smirnov goodness-of-fit test.

Sample size *N*	Significance level α				
	0.20	0.15	0.10	0.05	0.01
1	0.900	0.925	0.950	0.975	0.995
2	0.684	0.726	0.776	0.842	0.929
3	0.565	0.597	0.642	0.708	0.829
4	0.494	0.525	0.564	0.624	0.734
5	0.446	0.474	0.510	0.563	0.669
6	0.410	0.436	0.470	0.521	0.618
7	0.381	0.405	0.438	0.486	0.577
8	0.358	0.381	0.411	0.457	0.543
9	0.339	0.360	0.388	0.432	0.514
10	0.322	0.342	0.368	0.409	0.486
11	0.307	0.326	0.352	0.391	0.468
12	0.295	0.313	0.338	0.375	0.450
13	0.284	0.302	0.325	0.361	0.433
14	0.274	0.292	0.314	0.349	0.418
15	0.266	0.283	0.304	0.338	0.404
16	0.258	0.274	0.295	0.328	0.391
17	0.250	0.266	0.286	0.318	0.380
18	0.244	0.259	0.278	0.309	0.370
19	0.237	0.252	0.272	0.301	0.361
20	0.231	0.246	0.264	0.294	0.352
25	0.21	0.22	0.24	0.26	0.32
30	0.19	0.20	0.22	0.24	0.29
35	0.18	0.19	0.21	0.23	0.27
Asymptotic formula	$\dfrac{1.07}{\sqrt{N}}$	$\dfrac{1.14}{\sqrt{N}}$	$\dfrac{1.22}{\sqrt{N}}$	$\dfrac{1.36}{\sqrt{N}}$	$\dfrac{1.63}{\sqrt{N}}$

For $\alpha = 0.01$ and $\alpha = 0.05$, asymptotic formulae give values which are too high—by 1.5% for $N = 80$.
Source: Lindgren (1993, Table VIII, p. 594). © 1993 Chapman & Hall, Inc, with corrections for $N = 9$
based on the 3rd edition from 1976.
Reproduced by permission of Taylor & Francis Group.

sequence of samples with sizes $N = 1, 2, 3, \ldots$, the corresponding sequence of the empirical distribution functions $\{\breve{H}_N : N = 1, 2, 3, \ldots\}$, and the corresponding sequence of the K-S statistics $\{D_N : N = 1, 2, 3, \ldots\}$.

Theorem (*Glivenko-Cantelli Lemma*)   *Suppose the null hypothesis is true. As the sample size N increases indefinitely, the sequence of the Kolmogorov-Smirnov statistics $\{D_N\}$ converges to zero; that is,*

$$\lim_{N \to \infty} D_N = 0.$$

Ergo, the sequence of the empirical distribution functions $\{\breve{H}_N\}$ converges uniformly to the distribution function H of variate X.

Table 3.3 Calculation of the absolute differences for the Kolmogorov-Smirnov test when the sample size is $N = 5$; the hypothesized distribution function H is shown in Figure 3.3a; the values of the statistics D^L and D^A are in bold type.

n	1	2	3	4	5
$\dfrac{n}{N}$	0.2	0.4	0.6	0.8	1.0
$H(x_{(n)})$	0.08	0.49	0.75	0.90	0.98
$\left\|\dfrac{n-1}{N} - H(x_{(n)})\right\|$	0.08	0.29	**0.35**	0.30	0.18
$\left\|\dfrac{n}{N} - H(x_{(n)})\right\|$	0.12	0.09	**0.15**	0.10	0.02

In essence, the Glivenko-Cantelli Lemma implies that when the sample size N is large enough, it is near certain that the empirical distribution function $\breve{H}$ is arbitrarily close to the unknown distribution function H of variate X at all points of the sample space $\mathcal{X}$. This justifies using $\breve{H}$ as the benchmark for measuring the goodness of fit of a model.

Critical value

In all applications considered in this book, the hypothesized distribution function H which enters the null hypothesis is parametric, has a specified form, and has parameters estimated from a sample. The Kolmogorov-Smirnov test requires that the parameters of H be not estimated from the same sample from which the empirical distribution function is constructed. To meet this requirement, one would have to split the available sample of X into two subsamples, one to be used for parameter estimation and the other to be used for the test. In practice, this requirement is usually violated as one wishes to use the entire sample for estimation of the parameters. (To compensate for this violation, the critical values listed in Table 3.2 would have to be reduced; but this is usually not done.)

Alternative hypothesis

Under the alternative hypothesis, the distribution function of variate X is any distribution other than H. For example, using the abbreviations from Appendix C, if H is WB($\alpha = 0.7, \beta = 2.4, \eta = 30$), then the alternative hypothesis includes Weibull distribution functions with all feasible values of the parameters other than those listed above, as well as all feasible forms of the distribution functions other than Weibull, each form with all feasible values of the parameters. Any of these alternative distribution functions may enter the null hypothesis when the distribution function H is rejected.

Note on statistical tests

When a continuous distribution function H is hypothesized as a model of a continuous variate X, and a formal statistical test of the goodness of fit needs to be applied, it should be one that is appropriate for continuous distribution functions. For instance, Pearson's chi-square test is not appropriate because its conclusion depends on the number and widths of the class intervals into which the sample space of a continuous variate must be subjectively discretized; two different discretizations may lead to contradictory conclusions (Benjamin and Cornell, 1970, p. 466). The Kolmogorov-Smirnov test (Lindgren, 1993, p. 479) is appropriate and is also consistent with the uniform distance measure. The Anderson-Darling test (Johnson, 2005, p. 332) is also appropriate, but is inconsistent with the uniform distance measure.

Note on empirical distributions

The Kolmogorov-Smirnov test employs the *stepwise* empirical distribution function (3.10), which is specified at every point $x \in \mathcal{X}$ by assuming a constant ordinate $\breve{H}(x)$ between the realizations of X. The parameter estimation methods employ the *pointwise* empirical distribution function (3.8), simply because only a set of points is known, and fitting a function to the points requires no assumption about the ordinates $\breve{H}(x)$ between the realizations of X. The three formulae for plotting positions (Section 3.2) possess an important necessary property: as the sample size N increases indefinitely, the meta-Gaussian plotting positions (3.6) converge to the Weibull plotting positions (3.5), which in turn converge to the standard plotting positions (3.4). Ergo, the conclusion of the Glivenko-Cantelli Lemma applies to the sequence of the empirical distribution functions $\{\breve{H}_N\}$ constructed from any of the three formulae. There exist many other formulae for plotting positions. They may not possess the necessary convergence property, and are not necessarily well calibrated. Exercise 3 expounds the notion of calibration, and directs you to show that only the meta-Gaussian plotting positions are well calibrated against the standard normal variate.

3.7 Illustration of Modeling Methodology

The travel time by car from home to work during the same hour of every weekday varies randomly from day to day; hence, it is considered to be a continuous variate, X. The measurements of X taken in one week are 15, 11, 19, 10, 25 minutes. The shortest feasible travel time (at the speed limit, on empty streets, with all traffic lights set to green) is 9 minutes. We begin modeling by listing:

Sample size:	$N = 5$.
Sample:	$\{x(1), x(2), x(3), x(4), x(5)\} = \{15, 11, 19, 10, 25\}$.
Sample space:	$\mathcal{X} = (\eta, \infty) = (9, \infty)$.

Hypothesized parametric model for H : $EX(\alpha, \eta)$ with scale parameter α, to be estimated from the sample, and shift parameter $\eta = 9$.

LS estimation method

Table 3.4 summarizes the empirical distribution function $\breve{H}$ of X specified by the set of points $\{(x_{(n)}, p_n)\}$ (Section 3.2); the transformed set of points $\{(v_n, u_n)\}$ calculated from the equations for the linearized quantile function of the exponential variate (Section C.3.1),

$$v_n = \ln(x_{(n)} - \eta), \qquad u_n = \ln[-\ln(1 - p_n)];$$

the calculation of the estimate $\hat{\alpha} = 8.06$ of the scale parameter α via the least squares (LS) method (Section B.1),

$$\bar{v} = \frac{1}{N}\sum_{n=1}^{N} v_n, \qquad \bar{u} = \frac{1}{N}\sum_{n=1}^{N} u_n,$$

$$\hat{a} = \bar{v} - \bar{u}, \qquad \hat{\alpha} = \exp(\hat{a});$$

the ordinates of the hypothesized distribution function H, calculated from the expression (Section C.3.1)

$$H(x) = 1 - \exp\left(-\frac{x - 9}{8.06}\right), \qquad 9 < x;$$

the calculation of the MAD (Section 3.6.2); and the execution of the K-S test (Section 3.6.3). With MAD = 0.062 implying a good fit, and the K-S test not rejecting H at the 20% significance level (or lower), EX(8.06, 9) is a good model of X. Figure 3.4 shows the fit of the line $v = u + 2.09$ to the points $\{(v_n, u_n)\}$; Figure 3.5 shows the fit of H to $\breve{H}$.

Table 3.4 Total procedure for modeling distribution function H of variate X, travel time [min]; the hypothesized H is $EX(\alpha, \eta)$, with $\eta = 9$ min; the parameter estimation method is the least squares (LS).

	n	1	2	3	4	5
Ordered sample	$x_{(n)}$	10	11	15	19	25
Empirical DF	p_n	0.0756	0.2538	0.5	0.7462	0.9244
Transformed set of points	v_n	0.0000	0.6931	1.7918	2.3026	2.7726
	u_n	−2.5433	−1.2284	−0.3665	0.3157	0.9487
Hypothesized DF	$H(x_{(n)})$	0.1167	0.2197	0.5250	0.7108	0.8626
MAD	$\left| p_n - H(x_{(n)}) \right|$	0.0411	0.0341	0.0250	0.0354	**0.0618**
K-S Test:	$\dfrac{n}{N}$	0.2	0.4	0.6	0.8	1.0
D^L	$\left| \dfrac{n-1}{N} - H(x_{(n)}) \right|$	0.1167	0.0197	**0.1250**	0.1108	0.0626
D^A	$\left| \dfrac{n}{N} - H(x_{(n)}) \right|$	0.0833	**0.1803**	0.0750	0.0892	0.1374

$\bar{v} = 1.5120$	MAD $= 0.062$	Sample size $\qquad$ $N = 5$
$\bar{u} = -0.5748$		K-S statistic $\qquad$ $D = 0.180$
$\hat{a} = 2.0868$		Significance level $\qquad$ $\alpha = 0.20$
$\hat{\alpha} = 8.06$		Critical value $\qquad$ $C = 0.446$

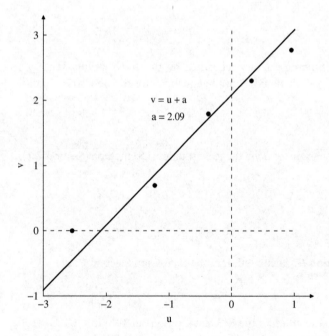

Figure 3.4 Linearized quantile function $v = u + a$ of the hypothesized distribution, which is $EX(\alpha, \eta)$, with $\eta = 9$ min, fitted via the method of least squares to the transformed set of points $\{(v_n, u_n) : n = 1, \ldots, 5\}$ of the empirical distribution function; see Table 3.4 for other details.

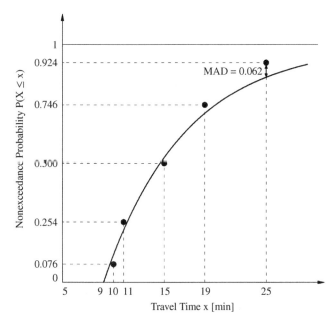

Figure 3.5 Empirical distribution function of travel time X constructed from a sample of size $N = 5$ using the meta-Gaussian plotting positions, and the hypothesized distribution function H, which is EX(8.06, 9) and for which MAD = 0.062; see Table 3.4 for other details.

Moments

Under the model EX(8.06, 9), the mean and the variance of the travel time X are (Section C.3.1):

$$E(X) = \alpha + \eta = 8.06 + 9 = 17.06 \text{ min},$$
$$Var(X) = \alpha^2 = 64.96 \text{ min}^2.$$

Because they are consistent with the model chosen for the distribution function H of X, these values of $E(X)$ and $Var(X)$ should be used henceforth, even though they are different from the sample estimates. (The sample mean is 16 min, the sample variance is 30.4 min^2.)

UD estimation method

The fit of H to $\check{H}$ could be improved by employing the uniform distance (UD) method (Section B.2). With the LS estimate $\hat{\alpha} = 8.06$ as an initial solution, the optimization algorithm DFit outputs $\alpha = 7.63$, MAD = 0.047, and $D = 0.169$.

3.8 Derived Distribution Theory

3.8.1 Transformation of Variate

Imagine a simple system defined by three elements (Figure 3.6): X is the input variate; Y is the output variate; and t is a *transformation* (a function) from the sample space $\mathcal{X}$ *onto* the sample space $\mathcal{Y}$. Thus every realization $x \in \mathcal{X}$ of the input variate is transformed into a realization $y \in \mathcal{Y}$ of the output variate,

$$y = t(x); \tag{3.16}$$

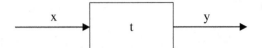

Figure 3.6 A simple system with input x, output y, and transformation t such that $y = t(x)$.

and the sample space $\mathcal{Y}$ includes only realizations y resulting from all $x \in \mathcal{X}$. Suppose the transformation t is differentiable (and therefore continuous) and one-to-one (and therefore has an inverse t^{-1}). The *inverse transformation* t^{-1} is such that $y = t(x)$ if and only if

$$x = t^{-1}(y).\tag{3.17}$$

Finally, suppose that the distribution function H_X and the density function h_X of the input variate X are known. The task is to derive the distribution function H_Y and the density function h_Y of the output variate Y. This task arises in two types of problems.

Modeling a system

This is a frequent problem in sciences and engineering. The behavior of a system being analyzed is known with certainty. Thus it can be described by a function t which transforms input x into output y. The function t is a *natural transformation* in that it is constructed by applying a known law (e.g., physical, chemical, economic, geometric) that governs the behavior of a system. When the input into the system is uncertain, it makes sense to define an input variate X and to determine its distribution function H_X and its density function h_X. It follows that the output is now uncertain as well. Thus, it is necessary to consider an output variate $Y = t(X)$, to derive its distribution function H_Y from H_X and t, and to derive its density function h_Y from h_X and t.

Example 3.4 *(Volume of a drop)*

A fluid medication is administered in drops. The diameter of the drop, x, determines the volume (equivalently, the dose) of the medication per drop, y, according to the transformation

$$y = \frac{\pi}{6}x^3.$$

In application, the diameter varies from drop to drop and so does the dose. Thus for any future application, the diameter is a random variable X, the volume is a random variable Y, and the transformation t between them is

$$Y = \frac{\pi}{6}X^3.$$

In a laboratory, drops were generated and the diameter of each drop was measured by a sensor. From the obtained sample, a distribution function H_X of the diameter X was estimated. The task now is to derive the distribution function H_Y of the volume Y.

Deriving a new distribution form

This is a frequent problem in statistics and probability. Take a variate X whose distribution function H_X has a specified and convenient form. Next imagine that the variate X is transformed into a variate Y by function t of a specified form. The function t is an *artificial transformation* in that it is invented. As a result, the variate $Y = t(X)$ has a distribution function H_Y whose form is different than the form of H_X. What is the form of this new distribution function? If H_Y allows for shapes different than H_X does, has distinct properties, and is analytically convenient, then it may be a useful new model.

Example 3.5 *(Weibull distribution)*

Let variate X on the sample space $\mathcal{X} = (0, \infty)$ have a distribution function H_X which is exponential (see Section C.3.1) with scale parameter equal to 1, and shift parameter equal to 0. Suppose each realization $x \in \mathcal{X}$ is transformed by a power function

$$y = \alpha x^{1/\beta} + \eta,$$

where $\alpha > 0$, $\beta > 0$, and $-\infty < \eta < \infty$. Thereby the sample space $\mathcal{X} = (0, \infty)$ is mapped onto the sample space $\mathcal{Y} = (\eta, \infty)$. The variate Y defined on the sample space $\mathcal{Y}$ is

$$Y = \alpha X^{1/\beta} + \eta.$$

What is the form of the distribution function H_Y of variate Y? It turns out that H_Y is Weibull (see Section C.3.2) with scale parameter α, shape parameter β, and shift parameter η. Section 3.8.5 shows how to prove this fact.

3.8.2 Derived Bounds

When the sample space $\mathcal{X}$ of the input variate X is an open interval bounded below, or above, or on both sides, the sample space $\mathcal{Y}$ of the output variate Y is bounded as well. Let $\mathcal{X} = (\eta_{LX}, \eta_{UX})$ be given. To derive $\mathcal{Y} = (\eta_{LY}, \eta_{UY})$, two cases, depicted in Figure 3.7, must be considered.

(i) When the transformation t is strictly increasing,

$$\eta_{LY} = t(\eta_{LX}), \qquad \eta_{UY} = t(\eta_{UX}). \tag{3.18}$$

(ii) When the transformation t is strictly decreasing,

$$\eta_{LY} = t(\eta_{UX}), \qquad \eta_{UY} = t(\eta_{LX}). \tag{3.19}$$

3.8.3 Derived Distribution Function

Again, two cases must be considered (see Figure 3.7).

(i) When the transformation t is strictly increasing,

$$P(Y \le y) = P(t(X) \le y)$$
$$= P(X \le t^{-1}(y));$$

therefore

$$H_Y(y) = H_X(t^{-1}(y)). \tag{3.20}$$

(ii) When the transformation t is strictly decreasing,

$$P(Y \le y) = P(t(X) \le y)$$
$$= P(X > t^{-1}(y))$$
$$= 1 - P(X \le t^{-1}(y));$$

therefore

$$H_Y(y) = 1 - H_X(t^{-1}(y)). \tag{3.21}$$

Equations (3.20) and (3.21) show how the distribution function H_Y of the output variate can be derived from the distribution function H_X of the input variate and the transformation t.

3.8.4 Derived Density Function

Again, two cases must be considered (see Figure 3.7). (i) When the transformation t is strictly increasing, the derivative of expression (3.20) with respect to y is

$$\frac{dH_Y(y)}{dy} = \frac{dH_X(t^{-1}(y))}{dy}$$
$$= \frac{dx}{dy} \frac{dH_X(x)}{dx}, \qquad x = t^{-1}(y),$$

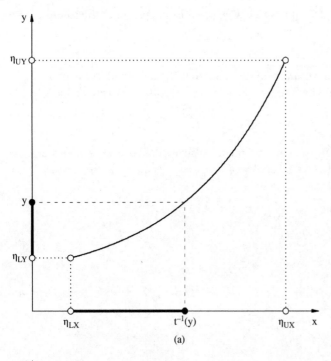

(a)

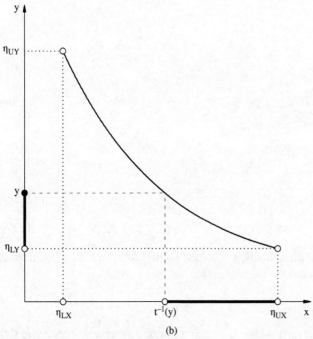

(b)

Figure 3.7 A transformation *t* between variate *X* having sample space $\mathcal{X} = (\eta_{LX}, \eta_{UX})$ and variate *Y* having sample space $\mathcal{Y} = (\eta_{LY}, \eta_{UY})$: (a) strictly increasing *t*; (b) strictly decreasing *t*.

which is equivalent to

$$h_Y(y) = \frac{dx}{dy} h_X(x);$$ (3.22)

note that $dx/dy > 0$ because t^{-1} is strictly increasing. (ii) When the transformation t is strictly decreasing, the derivative of expression (3.21) with respect to y is

$$\frac{dH_Y(y)}{dy} = -\frac{dH_X(t^{-1}(y))}{dy}$$

$$= -\frac{dx}{dy}\frac{dH_X(x)}{dx}, \qquad x = t^{-1}(y),$$

which is equivalent to

$$h_Y(y) = -\frac{dx}{dy} h_X(x);$$ (3.23)

note that $dx/dy < 0$ because t^{-1} is strictly decreasing; thus, $-dx/dy > 0$.

Equations (3.22) and (3.23) can be combined into a single equation:

$$h_Y(y) = \left|\frac{dx}{dy}\right| h_X(x).$$ (3.24)

When the inverse transformation, $x = t^{-1}(y)$, is introduced explicitly, the equation takes the form

$$h_Y(y) = \left|\frac{dt^{-1}(y)}{dy}\right| h_X(t^{-1}(y)).$$ (3.25)

The derivative of the input with respect to the output is called the *Jacobian of the transformation*:

$$\frac{dx}{dy} = \frac{dt^{-1}(y)}{dy}.$$ (3.26)

In summary, the density function of the output variate is equal to the product of the density function of the input variate and the absolute value of the Jacobian of the transformation (with the substitution $x = t^{-1}(y)$).

Dimensional analysis. Whenever a density function is derived according to equation (3.24) or (3.25), it is prudent to check the result via dimensional analysis. Recall from Section 2.3.2 that density $h_X(x)$ has dimension $1/[x]$, which is the reciprocal of the dimension of x. The dimension of the Jacobian (3.26) is $[x]/[y]$. Hence the dimensional form of equation (3.24) is

$$\frac{1}{[y]} = \frac{[x]}{[y]}\frac{1}{[x]},$$

which confirms that density $h_Y(y)$ must have dimension $1/[y]$.

3.8.5 Examples of Power Transformation

Example 3.6 *(Volume of a drop — Example 3.4 continued)*
Suppose the diameter of the drop X has the sample space $\mathcal{X} = (\eta_L, \eta_U)$, with $0 < \eta_L < \eta_U$. Given the transformation $y = \pi x^3/6$, the volume of the drop Y has the sample space $\mathcal{Y} = (\pi\eta_L^3/6, \pi\eta_U^3/6)$. Next suppose X has the power type II distribution, $X \sim \text{P2}(\beta, \eta_L, \eta_U)$ (see Section C.4.2). Thus the distribution function H_X is given by

$$H_X(x) = 1 - \left(\frac{\eta_U - x}{\eta_U - \eta_L}\right)^\beta.$$

With the transformation $t(x) = \pi x^3/6$ being a strictly increasing function of x and having the inverse

$$t^{-1}(y) = \left(\frac{6y}{\pi}\right)^{1/3},$$

the distribution function H_Y can be derived via equation (3.20):

$$H_Y(y) = H_X((6y/\pi)^{1/3})$$

$$= 1 - \left(\frac{\eta_U - (6y/\pi)^{1/3}}{\eta_U - \eta_L}\right)^{\beta}.$$

The density function h_X is given by

$$h_X(x) = \frac{\beta}{\eta_U - \eta_L}\left(\frac{\eta_U - x}{\eta_U - \eta_L}\right)^{\beta-1}.$$

With the inverse transformation t^{-1} as above, the Jacobian of the transformation can be found via equation (3.26):

$$\frac{dt^{-1}(y)}{dy} = \left(\frac{2}{9\pi}\right)^{1/3}\frac{1}{y^{2/3}}.$$

Finally, the density function h_Y can be derived via equation (3.25):

$$h_Y(y) = \left(\frac{2}{9\pi}\right)^{1/3}\frac{1}{y^{2/3}}h_X((6y/\pi)^{1/3})$$

$$= \left(\frac{2}{9\pi}\right)^{1/3}\frac{\beta}{(\eta_U - \eta_L)y^{2/3}}\left(\frac{\eta_U - (6y/\pi)^{1/3}}{\eta_U - \eta_L}\right)^{\beta-1}.$$

For the dimensional analysis of this result, let us suppose that the diameter x is measured in millimeters (mm); then the shift parameters η_L and η_U must be in mm as well, and the volume y is in mm^3. Hence the units inside the large parentheses on the right side cancel out, and the only remaining dimensional quantity is $1/[(\eta_U - \eta_L)y^{2/3}]$, whose unit is $1/[(\text{mm})(\text{mm}^3)^{2/3}] = 1/\text{mm}^3$. Correctly, the reciprocal of the unit of y becomes the unit of the density $h_Y(y)$.

Example 3.7 *(Weibull distribution — Example 3.5 continued)*
The specified distribution function H_X is

$$H_X(x) = 1 - \exp(-x).$$

The inverse transformation takes the form

$$t^{-1}(y) = \left(\frac{y - \eta}{\alpha}\right)^{\beta},$$

which is a strictly increasing function. Thus the distribution function H_Y can be derived via equation (3.20):

$$H_Y(y) = H_X\left(\left(\frac{y - \eta}{\alpha}\right)^{\beta}\right)$$

$$= 1 - \exp\left[-\left(\frac{y - \eta}{\alpha}\right)^{\beta}\right],$$

which is a Weibull distribution function of variate Y, with scale parameter α, shape parameter β, and shift parameter η (see Section C.3.2).

The specified density function h_X is

$$h_X(x) = \exp(-x).$$

With the inverse transformation t^{-1} as above, the Jacobian of the transformation can be found via equation (3.26):

$$\frac{dt^{-1}(y)}{dy} = \frac{\beta}{\alpha}\left(\frac{y - \eta}{\alpha}\right)^{\beta-1}.$$

Finally, the density function h_Y can be derived via equation (3.25):

$$h_Y(y) = \frac{\beta}{\alpha}\left(\frac{y-\eta}{\alpha}\right)^{\beta-1} h_X\left(\left(\frac{y-\eta}{\alpha}\right)^{\beta}\right)$$

$$= \frac{\beta}{\alpha}\left(\frac{y-\eta}{\alpha}\right)^{\beta-1} \exp\left[-\left(\frac{y-\eta}{\alpha}\right)^{\beta}\right],$$

which is a Weibull density function of variate Y, with scale parameter α, shape parameter β, and shift parameter η (see Section C.3.2).

3.8.6 Example of Log-Ratio Transformation

Section C.4 defines a *class* of log-ratio distributions obtained via the derived distribution theory, and lists the expressions from which the functions h_Y, H_Y, and H_Y^{-1} can be constructed for any particular distribution in this class. To illustrate this approach to modeling, we construct the *log-ratio logistic distribution*. Its mnemonic, LR1(η_L, η_U)-LG(β, α), means: the log-ratio type I transformation of a variate on the bounded open interval (η_L, η_U) has a logistic distribution with parameters (β, α).

The variates involved are: the original variate Y having the sample space $\mathcal{Y} = (\eta_L, \eta_U)$ is treated as the "output"; the transformed variate X having the sample space $\mathcal{X} = (-\infty, \infty)$ is treated as the "input". For any $y \in \mathcal{Y}$,

$$x = t^{-1}(y) = \ln\frac{y-\eta_L}{\eta_U - y}, \qquad x \in \mathcal{X},$$

which is the log-ratio type I transformation.

Density function
The *density function* h_Y is constructed thusly.

(i) Get from Section C.4.3 the expression for a log-ratio density function of Y:

$$h_Y(y) = \frac{\eta_U - \eta_L}{(y-\eta_L)(\eta_U - y)} h_X\left(\ln\frac{y-\eta_L}{\eta_U - y}\right).$$

(ii) Extract the expression for the density function of X; here $X \sim$ LG(β, α), and thus from Section C.2.1:

$$h_X(x) = \frac{1}{\alpha}\exp\left(-\frac{x-\beta}{\alpha}\right)\left[1 + \exp\left(-\frac{x-\beta}{\alpha}\right)\right]^{-2}.$$

(iii) Use this h_X in the expression for h_Y.

When the explicit expression for h_Y is not needed, or is overly complicated, the numerical calculation of $h_Y(y)$ may be accomplished thusly. Given $y \in \mathcal{Y}$, calculate the argument $x = t^{-1}(y)$ and the Jacobian $J = (\eta_U - \eta_L)/[(y-\eta_L)(\eta_U - y)]$. Next, calculate $h_X(x)$. Finally, let $h_Y(y) = Jh_X(x)$.

Distribution function
The *distribution function* H_Y is constructed thusly.

(i) Get from Section C.4.3 the expression for a log-ratio distribution function of Y:

$$H_Y(y) = H_X\left(\ln\frac{y-\eta_L}{\eta_U - y}\right).$$

(ii) Extract the expression for the distribution function of X; here $X \sim$ LG(β, α), and thus from Section C.2.1:

$$H_X(x) = \left[1 + \exp\left(-\frac{x-\beta}{\alpha}\right)\right]^{-1}.$$

(iii) Use this H_X in the expression for H_Y:

$$H_Y(y) = \left[1 + \exp\left(-\frac{\ln\frac{y-\eta_L}{\eta_U-y} - \beta}{\alpha}\right)\right]^{-1}.$$

This is the explicit expression for the log-ratio logistic distribution function of Y; in shorthand, $Y \sim \mathrm{LR1}$ $(\eta_L, \eta_U)\text{-LG}(\beta, \alpha)$.

Quantile function

The *quantile function* H_Y^{-1} is constructed thusly.

(i) Get from Section C.4.3 the expression for a log-ratio quantile function of Y:

$$H_Y^{-1}(p) = \eta_U - \frac{\eta_U - \eta_L}{\exp\left[H_X^{-1}(p)\right] + 1}.$$

(ii) Extract the expression for the quantile function of X; here $X \sim \mathrm{LG}(\beta, \alpha)$, and thus from Section C.2.1:

$$H_X^{-1}(p) = \beta + \alpha \ln\left(\frac{p}{1-p}\right).$$

(iii) Use this H_X^{-1} in the expression for H_Y^{-1} :

$$H_Y^{-1}(p) = \eta_U - \frac{\eta_U - \eta_L}{\exp\left[\beta + \alpha \ln(p/(1-p))\right] + 1}.$$

This is the explicit expression for the log-ratio logistic quantile function of Y.

3.8.7 Example of Reflection Transformation

Section C.5 presents a procedure for constructing a distribution of variate Y on a bounded-above interval from (i) any distribution of variate X on a bounded-below interval and (ii) the reflection transformation $X = -Y$. Here is an illustration of this procedure.

For the original variate Y, given are the sample, the specified sample space, and the hypothesized parametric model:

$$\{y\} = \{1.5, -1.9, 2.3, 0.2, 2.8\},$$

$$\mathcal{Y} = (-\infty, -\eta) = (-\infty, 3),$$

$$\mathrm{EX}(\alpha, \eta, -1) = \mathrm{EX}(\alpha, -3, -1);$$

the specified shift parameter is $-\eta = 3$; hence $\eta = -3$. For the transformed variate X, the same three objects are:

$$\{x\} = \{-1.5, 1.9, -2.3, -0.2, -2.8\},$$

$$\mathcal{X} = (\eta, \infty) = (-3, \infty),$$

$$\mathrm{EX}(\alpha, \eta) = \mathrm{EX}(\alpha, -3).$$

The least squares method (Section B.1) applied to the sample $\{x\}$ yields an estimate of the scale parameter: $\alpha = 2.2$ (see Table 3.5 for details). The expressions for the exponential distribution $\mathrm{EX}(2.2, -3)$ of variate X (Section C.3.1) yield:

$$h_X(x) = \frac{1}{2.2} \exp\left(-\frac{x+3}{2.2}\right), \qquad x > -3,$$

$$H_X(x) = 1 - \exp\left(-\frac{x+3}{2.2}\right), \qquad x > -3,$$

$$H_X^{-1}(p) = -2.2 \ln(1-p) - 3, \qquad 0 < p < 1.$$

When these expressions are inserted into the equations given in Section C.5.1, one obtains the expressions for the *reflected* exponential distribution EX(2.2, −3, −1) of variate Y:

$$h_Y(y) = \frac{1}{2.2} \exp\left(-\frac{3-y}{2.2}\right), \qquad y < 3,$$

$$H_Y(y) = \exp\left(-\frac{3-y}{2.2}\right), \qquad y < 3,$$

$$H_Y^{-1}(p) = 2.2 \ln(p) + 3, \qquad 0 < p < 1.$$

An evaluation of the goodness of fit in terms of the MAD, or in terms of the K-S statistic, yields the same result for H_X and for H_Y. Here, MAD = 0.0322, and its calculation is shown for H_X (Table 3.5) as well as for H_Y (Table 3.6). Note how the reflection transformation $X = -Y$ modifies all the numbers, except the ordinates of the empirical distribution function. Figure 3.8 shows the graphs of H_X and H_Y, together with the points of the corresponding empirical distribution functions, $\{(x_{(n)}, p_n)\}$ and $\{(y_{(n)}, p_n)\}$.

Table 3.5 Estimation, via the least squares method, of the scale parameter α of the exponential distribution EX(α, −3) of variate X.

	n	1	2	3	4	5
Ordered sample	$x_{(n)}$	−2.8	−2.3	−1.5	−0.2	1.9
Empirical DF	p_n	0.0756	0.2538	0.5	0.7462	0.9244
Transformed set	v_n	−1.6094	−0.3567	0.4055	1.0296	1.5892
of points	u_n	−2.5433	−1.2284	−0.3665	0.3157	0.9487
Hypothesized DF	$H_X(x_{(n)})$	0.0869	0.2725	0.4943	0.7199	0.8922
MAD	$\left\|p_n - H_X(x_{(n)})\right\|$	0.0113	0.0187	0.0057	0.0263	**0.0322**

$v_n = \ln(x_{(n)} - \eta)$ $\bar{v} = 0.2116$ $\hat{a} = 0.7864$ MAD = 0.0322
$u_n = \ln[-\ln(1 - p_n)]$ $\bar{u} = -0.5748$ $\hat{a} = 2.20$

Table 3.6 Evaluation of the goodness of fit of the *reflected* exponential distribution EX(2.2, −3, −1) of variate Y.

	n	1	2	3	4	5
Ordered sample	$y_{(n)}$	−1.9	0.2	1.5	2.3	2.8
Empirical DF	p_n	0.0756	0.2538	0.5	0.7462	0.9244
Hypothesized DF	$H_Y(y_{(n)})$	0.1078	0.2801	0.5057	0.7275	0.9131
MAD	$\left\|p_n - H_Y(y_{(n)})\right\|$	**0.0322**	0.0263	0.0057	0.0187	0.0113

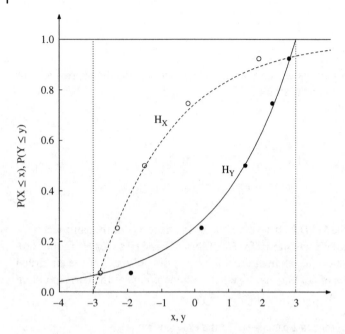

Figure 3.8 The reflected exponential distribution function H_Y of variate $Y \sim \text{EX}(2.2, -3, -1)$ on the sample space $\mathcal{Y} = (-\infty, 3)$, derived from the exponential distribution function H_X of variate $X \sim \text{EX}(2.2, -3)$ on the sample space $\mathcal{X} = (-3, \infty)$; the dots and circles are the points of the corresponding empirical distribution functions.

Exercises

1 **Probability integral transform**. Let X be a continuous variate having distribution function H, and let U be another continuous variate. The *probability integral transform* (PIT) of X is

$$U = H(X).$$

That is, the distribution function H serves also as a transformation. Prove that U has the uniform distribution $\text{UN}(0, 1)$.

Note. The inverse PIT, written for realizations as $x = H^{-1}(u)$, is the centerpiece of Monte Carlo simulation: a realization x is generated by evaluating the quantile function H^{-1} at a realization u, which comes from a random number generator.

2 **Normal quantile transform**. Let X be a continuous variate having distribution function H; let Z be another continuous variate; and let Q^{-1} be the standard normal quantile function. The *normal quantile transform* (NQT) of X is

$$Z = Q^{-1}(H(X)).$$

That is, the composition $Q^{-1} \circ H$ serves as a transformation. Prove that Z has the standard normal distribution $\text{NM}(0, 1)$.

Note. The NQT, written for realizations as $z = Q^{-1}(H(x))$, is used in multivariate data analyses: each X may have a different H; its sample is transformed into a sample of Z. Then the transformed data are analyzed in the multivariate standard normal space, where the assumptions of a linear model are more likely to hold.

3 **Calibration of plotting positions**. Read Exercises 1 and 2 above. Now, replace H by the empirical distribution function $\check{H}$ defined by equation (3.2). The result is the *empirical* NQT:

$$z_{(n)} = Q^{-1}(\check{H}(x_{(n)})) = Q^{-1}(p_n), \qquad n = 1, \dots, N.$$

It transforms the ordered sample $\{x_{(n)} : n = 1, \dots, N\}$ of variate X into the ordered sample $\{z_{(n)} : n = 1, \dots, N\}$ of the standard normal variate $Z \sim \text{NM}(0, 1)$. Most significantly, this sample of Z does not depend upon the sample of X, but only upon the plotting positions $\{p_n : n = 1, \dots, N\}$. Hence the following property.

Definition A plotting position formula is said to be *well calibrated* against the standard normal variate Z, up to its second moment, if the sample mean of Z equals 0 and the sample variance of Z equals 1 for every sample size $N \geq 2$.

Test numerically whether or not this definition is satisfied. Toward this end, for each of the three plotting position formulae, (3.4), (3.5), (3.6), and for each of the three sample sizes specified in the option: (i) Calculate the plotting positions $\{p_n : n = 1, \dots, N\}$. (ii) Calculate the realizations $\{z_{(n)} : n = 1, \dots, N\}$ using the rational function approximation to Q^{-1} (Table 2.3). (iii) Calculate the sample mean and the sample variance (see Section 2.4).

3.1 Summarize all sample moments in a table.

3.2 Compare the results and draw conclusions, allowing for the fact that the numerical calculations are approximate.

Hint. For the standard plotting positions, replace $p_N = 1$ (which would yield $z_{(N)} = \infty$) by $p_N = 0.9999$.

Options for sample sizes:

  (5) $N = 5, 15, 25$.
  (7) $N = 7, 17, 27$.
  (9) $N = 9, 19, 29$.
  (11) $N = 11, 21, 31$.
  (13) $N = 13, 23, 33$.

4 **Distribution modeling**. Apply the distribution modeling methodology to the variate which is specified in the option. Hypothesize at least two parametric models. Estimate the parameters via the LS method. Document all steps of the methodology; in particular, report the following items.

4.1 Assessed bounds on the sample space (if any), hypothesized parametric models, estimates of the parameters, the MAD for each hypothesized model, your rationale for choosing a model, the results of the K-S goodness-of-fit test of the chosen model.

4.2 A figure showing the graph of the empirical distribution function and the superposed graph of the chosen parametric distribution function. A figure showing the graph of the corresponding parametric density function. (*Reminder*: label the axes and show the units.)

Requirement. The graph of a parametric function should be continuous, and should extend beyond the range of the sample so that each tail is well depicted.

Options for the variate:

  (E*j*)   Score received by a student on exam $j, j \in \{1, 2, 3, 4\}$; Table 3.7.
  (HP)   Horsepower of a car; Table 3.8.
  (FM)   Fuel mileage of a car; Table 3.8.

5 **Area of a square**. In the process of manufacturing motorcycle batteries, a square plate is produced whose side is x. But what determines the electrical capacity of the plate is its area $y = x^2$. As part of quality control, a

Table 3.7 Scores from four final examinations received by students in the course taught from a draft of this textbook in four consecutive years. The score ranges from 0 to 100 in increments of 1.

Exam 1			Exam 2			Exam 3			Exam 4		
55	77	66	85	62	72	95	82	89	67	50	84
60	71	49	66	72	81	70	85	83	63	81	78
85	83	75	76	71	75	84	99	84	66	87	73
73	42	61	71	72	76	93	97	87	70	84	76
71	68	91	57	74	76	88	92	92	77	92	74
60	75	66	84	60	83	93	77	83	60	89	78
92	65	81	71	75	87	91	96	82	55	90	81
69	75		72	59	76	74	94	94	81	79	74
68	57		85	74	83	88	95	96	61	87	73
53	87		89	50	76	94	73	85	70	82	82
68	71		81	59	80	92	94	93	61	88	91
72	63		84	70	81	95	83	78	92	96	82
68	63		78	83	83	88	73	96	94	88	58
71	74		85	70	72	86	98	84	94	90	72
69	65		72	65	82	90	97	66	92	70	94
91	55		86	65	85	81	88	87	84	81	83
78	46		70	74	92	90	99	80	91	74	85
74	64		81	88	91	89	84	91	92	79	71
73	63		74	67	86	84	93	96	72	72	80
73	53		93	63	80	84	76		82	74	69
69	90		88	63		95	86		92	71	67
60	54		92	70		91	83		82	66	80
83	59		87	79		91	79		84	66	52
69	73		92	52		90	96		88	77	72
73	78		93	81		87	89		85	82	44
60	71		72	89		92	91		82	74	
78	80		69	91		86	95		75	66	
70	48		62	89		97	90		78	82	
57	74		85	97		85	77		66	78	
72	68		92	87		88	93		63	87	
75	84		94	87		89	87		89	88	
63	37		90	92		81	83		89	84	
66	74		95	80		89	89		84	91	
63	71		89	85		93	92		80	96	
55	70		84	80		97	81		98	93	
65	59		84	87		87	94		84	63	
82	75		97	86		94	90		78	82	
53	78		91	73		79	92		83	75	
59	83		88	83		95	92		85	85	
76	76		87	85		90	90		89	82	

Table 3.8 Fuel mileage [mpg] of passenger cars developing between 220 and 330 horsepower [hp], sold in the USA, and tested by *Road & Track* (March 2009). Each mpg measurement was made "largely during urban driving" and falls between city and highway estimates of the US Environmental Protection Agency.

hp	mpg	hp	mpg	hp	mpg	hp	mpg
305	18.0	306	18.0	258	22.3	260	20.0
265	19.1	330	21.0	291	15.9	265	22.2
300	23.9	303	20.7	237	17.3	305	16.0
302	18.5	220	23.7	270	18.7	268	26.3
260	20.4	240	19.0	290	18.8	250	18.2
252	18.0	263	20.4	306	18.9	227	22.8
285	22.0	272	22.5	260	23.0	311	16.3
315	16.0	232	16.6	295	18.9	235	21.5
268	16.0	268	21.0	280	17.3		
237	23.4	263	19.6	252	19.0		

sensor measures the side; the measurement varies from plate to plate due to production tolerances. Thus, the side X is random, and so is the area Y. Based on a sample of side measurements, a parametric distribution function of X was estimated and is specified in the option.

5.1 Write the expressions for the distribution function H_X and the density function h_X; derive the expressions for the distribution function H_Y and the density function h_Y. (*Reminder*: identify the sample spaces of X and Y.) Perform dimensional analysis of the equation for $h_Y(y)$.

5.2 Using the parameter values specified in the option, graph all four functions. (*Reminder*: label the axes and show the units.) Characterize in words the distinctions between the shapes of H_X and H_Y, and between the shapes of h_X and h_Y. Are the distributions of X and Y of the same type?

Options for the distribution:

(EX)	$X \sim \mathrm{EX}(\alpha, \eta)$;	$\alpha = 2.04$ mm,		$\eta = 110$ mm.
(WB)	$X \sim \mathrm{WB}(\alpha, \beta, \eta)$;	$\alpha = 2.26$ mm,	$\beta = 2.5$,	$\eta = 110$ mm.
(IW)	$X \sim \mathrm{IW}(\alpha, \beta, \eta)$;	$\alpha = 1.18$ mm,	$\beta = 0.5$,	$\eta = 110$ mm.
(LW)	$X \sim \mathrm{LW}(\alpha, \beta, \eta)$;	$\alpha = 0.76$ mm,	$\beta = 5.5$,	$\eta = 110$ mm.
(LL)	$X \sim \mathrm{LL}(\alpha, \beta, \eta)$;	$\alpha = 1.08$ mm,	$\beta = 1.5$,	$\eta = 110$ mm.

6 **Volume of a drop.** The problem is as described in Example 3.4 and as solved in Section 3.8.5 where $X \sim \mathrm{P2}(\beta, \eta_L, \eta_U)$. Here, X has the distribution specified in the option.

6.1 Write the expressions for the distribution function H_X and the density function h_X; derive the expressions for the distribution function H_Y and the density function h_Y. (*Reminder*: identify the sample spaces of X and Y.) Perform dimensional analysis of the equation for $h_Y(y)$.

6.2 Using the parameter values specified in the option, graph all four functions. (*Reminder*: label the axes and show the units.) Characterize in words the distinctions between the shapes of H_X and H_Y, and between the shapes of h_X and h_Y. Are the distributions of X and Y of the same type?

Options for the distribution:

(EX) $X \sim \text{EX}(\alpha, \eta)$; $\alpha = 1.02$ mm, $\eta = 0.9$ mm.

(WB) $X \sim \text{WB}(\alpha, \beta, \eta)$; $\alpha = 1.13$ mm, $\beta = 2.0$, $\eta = 0.9$ mm.

(IW) $X \sim \text{IW}(\alpha, \beta, \eta)$; $\alpha = 0.59$ mm, $\beta = 0.5$, $\eta = 0.9$ mm.

(LW) $X \sim \text{LW}(\alpha, \beta, \eta)$; $\alpha = 0.38$ mm, $\beta = 6.0$, $\eta = 0.9$ mm.

(LL) $X \sim \text{LL}(\alpha, \beta, \eta)$; $\alpha = 0.54$ mm, $\beta = 1.5$, $\eta = 0.9$ mm.

7 Project cost. A consulting firm charges a fixed fee of $5000 per project, plus $200 per work-hour of engineering staff. For the next project, the firm estimated the median number of work-hours to be 1800. From past experience, the firm knows that the actual number of work-hours per project is never smaller than $1/2$ of the median, is larger than $4/3$ of the median in 10% of the projects, and has a distribution type which is specified in the option.

7.1 Formulate an equation for the cost of the next project as a function of work-hours. Derive the inverse and the Jacobian of this transformation.

7.2 Write the expressions for the distribution function and the density function of the work-hours to be spent on the next project; determine the parameter values; graph both functions.

7.3 Derive the expressions for the distribution function and the density function of the cost of the next project; graph both functions. (*Reminder*: label the axes and show the units.) Are the distributions of the input and output variates of the same type?

7.4 A client is concerned with the project cost and asks for three quantiles whose nonexceedance probabilities are 0.55, 0.85, 0.95. Calculate these quantiles and indicate them on the graph of the distribution function.

Hint. To determine the parameter values, use the quantile function to set a system of two equations with two unknowns.

Options for the distribution:

(WB) $X \sim \text{WB}(\alpha, \beta, \eta)$.

(IW) $X \sim \text{IW}(\alpha, \beta, \eta)$.

(LW) $X \sim \text{LW}(\alpha, \beta, \eta)$.

(LL) $X \sim \text{LL}(\alpha, \beta, \eta)$.

8 Construction cost. A contractor must submit an estimate of the total cost of building a custom-designed house. Based on the architect's drawings and specs, she calculated the cost of all materials to be $388 000. She knows the wages that will be paid to workers (backhoe operator, masons, carpenters, roofers, electricians, plumber, plasterer, painter). But she worries about the time needed to build the house. There may be rocks below the topsoil that will complicate the excavation for the foundations. There are custom features such as porticos, balconies, tourets, and solatubes, which her crew rarely makes. Then there are elegant finishes — crown mouldings, mantels, corbels, medallions — whose fitting to perfection may consume extra time. Based on the records of previous projects and her experience, she quantified the uncertainties by assessing judgmentally (as you will learn in Chapter 9) the distribution function of the number of days, X, needed to build the house; then she fitted a parametric model, which is specified in the option.

Assumptions. A work day has 8 hours; the construction crew each day has four men, each costing $70 per hour in wages and benefits. The contractor's overhead (which includes her wages and company's profit) equals 25% of the cost of materials and labor.

8.1 Formulate an equation for the total construction cost as a function of the time needed to build this house. Derive the inverse and the Jacobian of this transformation.

8.2 Write the expressions for the distribution function and the density function of the number of days to be spent on this construction; graph both functions.

8.3 Derive the expressions for the distribution function and the density function of the total construction cost; graph both functions. (*Reminder*: label the axes and show the units.) Are the distributions of the input and output variates of the same type?

8.4 The contractor has a decision rule: she submits to the client the cost estimate such that the probability of the actual cost exceeding this estimate (and thus reducing her profit or even causing a loss) is q. Calculate such a cost estimate for $q = 0.25, 0.15, 0.05$.

Options for the distribution:

(EX)	$X \sim \text{EX}(47, 160)$.
(WB)	$X \sim \text{WB}(53, 2, 160)$.
(IW)	$X \sim \text{IW}(22, 1.5, 160)$.
(LW)	$X \sim \text{LW}(3.8, 4, 160)$.
(LL)	$X \sim \text{LL}(42, 3, 160)$.
(LR-LG)	$X \sim \text{LR1}(160, 230)\text{-LG}(-1.3, 2)$.
(LR-LP)	$X \sim \text{LR1}(160, 230)\text{-LP}(1.2, 3)$.
(LR-GB)	$X \sim \text{LR1}(160, 230)\text{-GB}(-1.5, 2)$.
(LR-RG)	$X \sim \text{LR1}(160, 230)\text{-RG}(1.4, 3)$.

9 **Power load: demand for cooling**. Every day the manager of an electric power plant must decide the maximum power to be generated on the following day. Typically, the forecast of short-term (24 hours or less) maximum power load is bifurcated into a weather-insensitive component and a weather-sensitive component. The most significant predictor of the latter is temperature: the maximum daily temperature determines the maximum power load due to cooling of buildings; the minimum daily temperature determines the maximum power load due to heating of buildings. The transformation between the temperature x [°F] and the power load y [MW] is a two-branch logistic function:

$$y = t(x) = \begin{cases} m\left[1 + \exp\left(\dfrac{\theta_c - x}{\xi_c}\right)\right]^{-1}, & \eta_c < x; \\[4mm] m\left[1 + \exp\left(\dfrac{x - \theta_h}{\xi_h}\right)\right]^{-1}, & x < \eta_h; \end{cases}$$

where m is the maximum weather-sensitive load [MW], $(\xi_c, \theta_c, \eta_c)$ are parameters for the cooling branch, and $(\xi_h, \theta_h, \eta_h)$ are parameters for the heating branch. The values of the parameters depend upon the region, the season, and the day of the week; they are specified in the option. Your task: use the function t to transform a probabilistic forecast of the uncertain temperature X into a probabilistic forecast of the uncertain power load Y. The forecast of X is in the form of a distribution specified in the option.

Note. With $\eta_c < \eta_h$, the two branches of t overlap over the interval (η_c, η_h) of "comfortable" temperatures, which require neither cooling nor heating of buildings. Being an approximation to the transformations actually used, the function t yields the power load $t(x)$ that is slightly greater than zero, and slightly different in each branch, at every $x \in (\eta_c, \eta_h)$. The forecast of X specifies a distribution either on $\mathcal{X}_c = (\eta_c, \infty)$, or on $\mathcal{X}_h = (-\infty, \eta_h)$.

9.1 Identify the sample space $\mathcal{Y}$ of the power load Y due to the demand for cooling. (*Hint*: the calculated lower bound should be rounded down.) Graph the transformation t on $\mathcal{X}_c$.

9.2 Write the expressions for the distribution function H_X and the density function h_X on the sample space $\mathcal{X}_c$.

9.3 Derive the expressions for the distribution function H_Y and the density function h_Y on the sample space $\mathcal{Y}$. Perform dimensional analysis of the equation for $h_Y(y)$.

9.4 Using the parameter values specified in the option, graph all four functions. (*Reminder*: label the axes and show the units.) Characterize in words the distinctions between the shapes of H_X and H_Y, and between the shapes of h_X and h_Y.

Options for the transformation:

(A) $m = 300$ MW, $\xi_c = 3.7°$F, $\theta_c = 80°$F, $\eta_c = 50°$F.
(B) $m = 400$ MW, $\xi_c = 3.6°$F, $\theta_c = 78°$F, $\eta_c = 48°$F.
(C) $m = 500$ MW, $\xi_c = 3.5°$F, $\theta_c = 82°$F, $\eta_c = 52°$F.

Options for the distribution:

(WB) $X \sim$ WB $(51, 3, \eta_c)$.
(IW) $X \sim$ IW $(20, 3, \eta_c)$.
(LW) $X \sim$ LW $(2.7, 3, \eta_c)$.
(LL) $X \sim$ LL $(19, 3, \eta_c)$.

10 **Power load: demand for heating**. Do Exercise 9 with the parameter values for t and the distribution of X specified in the options below, and with the following modifications.

10.1 Identify the sample space $\mathcal{Y}$ of the power load Y due to the demand for heating. (*Hint*: the calculated lower bound should be rounded down.) Graph the transformation t on $\mathcal{X}_h$.

10.2 Study Sections C.5 and 3.8.7 before performing these tasks. (i) Derive the expressions for the density function and the distribution function of a variate whose sample space is an interval bounded above $(-\infty, -\eta)$. (ii) Adapt these expressions to obtain h_X and H_X of the temperature X on the sample space $\mathcal{X}_h$.

10.3 Same as 9.3.

10.4 Same as 9.4.

Options for the transformation:

(A) $m = 300$ MW, $\xi_h = 4.6°$F, $\theta_h = 35°$F, $\eta_h = 70°$F.
(B) $m = 400$ MW, $\xi_h = 4.5°$F, $\theta_h = 33°$F, $\eta_h = 68°$F.
(C) $m = 500$ MW, $\xi_h = 4.4°$F, $\theta_h = 37°$F, $\eta_h = 72°$F.

Options for the distribution:

(R-WB) $X \sim$ WB $(52, 3, -\eta_h, -1)$.
(R-IW) $X \sim$ IW $(21, 3, -\eta_h, -1)$.
(R-LW) $X \sim$ LW $(2.8, 3, -\eta_h, -1)$.
(R-LL) $X \sim$ LL $(20, 3, -\eta_h, -1)$.

Mini-Projects

11 **Fuel efficiency**. The efficiency of a car with respect to fuel consumption is characterized in two ways. Americans measure X — the distance traveled (output) per unit of fuel consumed (input), expressed in miles per gallon [mpg]. Europeans measure Y — the fuel consumed (input) per unit of distance traveled (output), expressed in liters per 100 kilometers [l/100 km]; the American unit of Y is gallons per 100 miles [gpc], where "c" comes from Latin *centum* (hundred).

11.1 Given the sample in Table 3.8, apply the distribution modeling methodology to obtain a parametric model for the distribution of X. Estimate the parameters via the method specified in the option. Document all steps of the methodology (see Exercise 4 for particulars). Write the expressions for the distribution function H_X and the density function h_X.

11.2 Formulate the transformation between X [mpg] and Y [gpc], and graph it.

11.3 Derive the expressions for the distribution function H_Y and the density function h_Y.

11.4 Graph all four functions: H_X, h_X; H_Y, h_Y. (*Reminder*: label the axes and show the units.) Next, characterize in words the distinctions between the shapes of H_X and H_Y, and between the shapes of h_X and h_Y.

11.5 Consider the improvements of fuel efficiency in two cars: (A) from 16 to 17.4 mpg; (B) from 40 to 50 mpg. (i) Which improvement saves significantly more fuel (and money) on a 1000-mile trip? (ii) Express the two improvements in gpc (rounding to two decimals).

11.6 Consider a car that makes 25 mpg in the city and 40 mpg on the highway. In a year, it is driven 9000 miles in the city and 5000 miles on the highway. What is the average fuel efficiency experienced by the owner after a year? Express it in mpg and gpc.

11.7 Which measure, X [mpg] or Y [gpc], makes the comparison, and the averaging, of fuel efficiencies intuitively easier and why?

Options for the estimation method:

 (LS) Least squares method.

 (UD) Uniform distance method (using DFit or other software).

Note. Critiques of the X [mpg] measure have been published, for example, in *Road & Track* (May 2006, March 2009), *Science* (vol. 320, June 2008), and *The Wall Street Journal* (22–23 April 2017).

12 **Timber log**. The trunk of a fallen tree is cut into logs. For analyses, a log is approximated by a cylinder having height h and diameter x measured at $h/2$. Only logs having diameter larger than 40 cm are hauled from a forest to the sawmill. The diameters of 25 randomly selected logs were measured (Table 3.9).

12.1 Suppose the diameter X [cm] is a variate with the sample space $\mathcal{X} = (\eta, \infty)$ and a distribution function H_X belonging to one of the two families: the log-Weibull or the log-logistic. Find the best parametric model for H_X (the distribution family and the parameter values). Toward this end: (i) construct the empirical distribution function $\check{H}_X$ of X; (ii) estimate the parameters of each hypothesized model for H_X via the LS method; (iii) choose the best model; (iv) perform the K-S goodness-of-fit test of the chosen model; (v) graph the empirical distribution function of X, the chosen parametric distribution function H_X, and the corresponding density function h_X.

Table 3.9 The diameter [cm] of a log.

95	49	78	84	55
64	57	52	61	58
71	74	53	69	60
56	67	62	54	66
47	51	59	55	63

12.2 A dispatcher is concerned with the weight Y [in metric tons, t] of the load to be hauled by a truck. Nominally, the truck hauls n logs. A cylindrical log having diameter x, height h, and mass density c (of the wood from a particular tree species) has the weight $0.25\pi x^2 hc$. Thus

$$Y = 0.25\pi hcnX^2.$$

Let $\pi = 3.14$, $h = 9$ m, $c = 1.12$ t/m^3 (white oak), $n = 12$. Given the distribution function H_X and the density function h_X of the diameter X found in Exercise 12.1: (i) derive the distribution function H_Y and the density function h_Y of Y; (ii) graph H_Y and h_Y; (iii) calculate the probability of the load exceeding the capacity y of the truck, when $y = 45$ t and $y = 95$ t.

Note. White oak grows from central Minnesota to eastern Texas, from Maine to northern Florida; its wood is used in making furniture, yachts, and barrels for whiskey and sherry.

13   **Highway flooding**. State Highway 250 near Dailey, West Virginia, lies in the floodplain of the Tygart Valley River, some 4 meters above the datum of the nearby river gauge. A civil engineer must decide whether or not to elevate the highway in order to reduce the frequency of inundations and, thereby, the cost of repairs and other flood-related losses. To begin the analysis, the engineer obtained from the National Weather Service flood crests recorded by the river gauge (Table 3.10).

13.1 Estimate the probability g of the event: at least one flood occurs in a year.

Table 3.10   Flood crests recorded by a river gauge on the Tygart Valley River in Dailey, West Virginia, from 1932 to 1996.

Date [day.month.year]	Crest [m]	Date [day.month.year]	Crest [m]
04.02.1932	5.24	26.01.1978	4.51
17.03.1936	4.30	21.12.1978	3.35
03.02.1939	4.33	21.01.1979	3.19
24.08.1942	3.46	16.02.1979	3.87
30.07.1943	3.87	10.10.1979	3.11
23.02.1944	4.07	21.05.1980	3.22
26.12.1944	3.56	06.06.1981	3.35
14.02.1948	3.84	20.03.1982	4.44
20.03.1963	3.72	14.02.1984	3.24
05.03.1964	3.52	13.08.1984	3.60
07.03.1967	4.00	12.03.1985	3.06
31.12.1970	3.98	05.11.1985	5.06
22.12.1971	3.91	24.03.1993	3.27
26.02.1972	3.98	16.04.1994	4.06
09.12.1973	3.83	08.05.1994	4.11
02.06.1974	4.18	19.01.1996	4.17
09.10.1976	3.59	17.05.1996	4.75
25.02.1977	3.38	31.07.1996	4.44

13.2 Apply the distribution modeling methodology to the flood crest variate X. Hypothesize at least two parametric models on a bounded-below interval $\mathcal{X}$. Estimate the parameters of each model via the method specified in the option. Document all steps of the methodology (see Exercise 4 for particulars).

13.3 Let F denote the distribution function of X obtained above. Calculate $F(4)$, $F(5)$; interpret each value for the engineer.

Hint. The sample of X from which F was estimated includes all floods observed in 65 years; there were years with more than one flood, and years with no flood. Reflect; then complete each sentence: (i) Distribution function F of X applies only to years .... (ii) Distribution function F of X is conditional on ....

13.4 Given probability g and the conditional distribution function F, the unconditional distribution function H of X, in any year, is

$$H(x) = (1 - g) + F(x)g, \qquad x \in \mathcal{X}.$$

(i) Apply the total probability law to justify this equation. (ii) Graph the functions F and H. (iii) Write the expression for the conditional quantile function F^{-1}. (iv) Derive the expression for the unconditional quantile function H^{-1}.

13.5 When deciding the elevation of a highway, civil engineers employ the concept of a *T-year flood*, where T is a positive integer. For instance, a highway is said to be *designed against* the 50-year flood crest if, each year, the probability of the flood crest exceeding the design elevation is $1/T = 1/50 = 0.02$. In short, this design elevation equals the unconditional 0.98-*probability quantile* of X:

$$P(X > x_{0.98}) = 1 - P(X \le x_{0.98}) = 1 - 0.98 = 0.02.$$

The integer T is called the *return period* of the flood because the expected number of years between two consecutive T-year floods is T. (To justify this fact, view the occurrence of a T-year flood in any year as the success of a Bernoulli trial with probability $q = 1/T$; then the number of Bernoulli trials (years) between two consecutive successes has a geometric distribution, and its expectation equals $1/q = T$.) (i) Calculate the 50-year flood crest and the 100-year flood crest for State Highway 250 near Dailey. (ii) Indicate these crests on the graph of the distribution function H.

Options: (LS) Least squares method.

(UD) Uniform distance method (using DFit or other software).

14 **Stock performance**. Consider two long-term investment alternatives: (S) buy the stock of a company, or (M) buy an index of the overall stock market. (A major US stock-market index, such as Dow Jones Industrial Average, S&P 500, or Russell 3000, can be traded on an exchange as a portfolio fund or as a futures contract.) What is the probability of a stock outperforming the index, say by 50% over 35 years? Specifically, let X denote the difference between the return rate from a stock and the return rate from Russell 3000 index, each expressed as the percentage. Thus, $X = 0$ means equal return rates (i.e., equal performance of each investment alternative); $X < 0$ means lower return rate from the stock than from the index (i.e., the stock underperforms the market, the stock is a percentage "loser"); $X > 0$ means a higher return rate from the stock than from the index (i.e., the stock outperforms the market, the stock is a percentage "gainer"). A sample of X from the period of 35 years (1980–2014) for 11 250 stocks was compiled as follows (Table 3.11): the sample space was partitioned into the intervals of 50% width, and the number of stocks whose realizations x fell within each interval was counted. Your task is to obtain a parametric model for the distribution of X.

Assumptions. (i) Variate X is continuous. (ii) Variates representing different stocks form a random sample of X. (iii) The observed realizations of X are the midpoints of the intervals. (iv) The number of stocks per interval is interpreted as the number of identical realizations of X.

Table 3.11 Return rate from a stock, as the percentage of the return rate from Russell 3000 index, for 11 250 stocks over the period of 35 years (1980–2014). (The numbers are estimated from a graph in *The Wall Street Journal*, 4 November 2014.)

Return rate [%]	Number of stocks	Return rate [%]	Number of stocks	Return rate [%]	Number of stocks	Return rate [%]	Number of stocks
(−450, −400]	36	(200, 250]	225	(850, 900]	19	(1450, 1500]	14
(−400, −350]	56	(250, 300]	150	(900, 950]	21	(1500, 1550]	8
(−350, −300]	309	(300, 350]	141	(950, 1000]	21	(1550, 1600]	6
(−300, −250]	413	(350, 400]	122	(1000, 1050]	28	(1600, 1650]	10
(−250, −200]	553	(400, 450]	94	(1050, 1100]	19	(1650, 1700]	12
(−200, −150]	984	(450, 500]	66	(1100, 1150]	19	(1700, 1750]	7
(−150, −100]	1472	(500, 550]	56	(1150, 1200]	21	(1750, 1800]	3
(−100, −50]	2053	(550, 600]	52	(1200, 1250]	17	(1800, 1850]	6
(−50, 0]	1519	(600, 650]	52	(1250, 1300]	9	(1850, 1900]	5
(0, 50]	1003	(650, 700]	36	(1300, 1350]	15	(1900, 1950]	4
(50, 100]	684	(700, 750]	36	(1350, 1400]	10	(1950, 2000]	5
(100, 150]	459	(750, 800]	38	(1400, 1450]	10	(2000, 2050]	6
(150, 200]	328	(800, 850]	18				

14.1 Apply the distribution modeling methodology to variate X. Hypothesize three parametric models: Laplace, log-ratio Laplace, a model of your choice. Estimate the parameters of each model via the method specified in the option. Document all steps of the methodology (see Exercise 4 for particulars).

14.2 Based on its graph, characterize the shape of the parametric density function of X.

14.3 Calculate: (i) the quantiles of X corresponding to probabilities 0.25, 0.5, 0.75; (ii) the probabilities of seven events, $\{X < -250\}$, $\{X < -50\}$, $\{X < 0\}$, $\{X > 0\}$, $\{X > 50\}$, $\{X > 250\}$, $\{X > 500\}$.

14.4 Interpret the quantiles and four selected probabilities for an investor who is considering the two alternatives: S, M.

14.5 Suppose an investor has neither time nor ability to forecast the returns from stocks of individual companies. Still, he can read the current market news, select a company intuitively, and decide to buy its stock or to buy the index. (i) State the advice you would give to such an investor. (ii) State the rationale behind your advice.

Options: (LS) Least squares method.
  (UD) Uniform distance method (using DFit or other software).

15 **Agricultural yield: bounded interval**. The yield of a crop, such as soybean, corn, or wheat, varies from year to year due to the variability of weather (mostly rainfall and temperature) during the growing season. The density function of the crop yield is often *platykurtic* (has short tails) and asymmetrical: the right tail is short because there exists an upper bound on yield (which depends on climate, soil, and the level of agronomic technology); the left tail may be somewhat longer because small yield is possible, but is naturally bounded by zero. When these properties are discernible in the empirical distribution function, it is logical to specify the sample space as either a bounded open interval or a bounded-above interval (recall Section 3.3). Let us analyze the extent to which this general characterization applies to the regional average yield of soybeans in the state of Mato Grosso, Brazil, in the years 1998–2017 (Table 3.12).

Table 3.12 Regional average yield of soybeans in the state of Mato Grosso, Brazil, in bags/hectare.

Year	Yield	Year	Yield	Year	Yield	Year	Yield
1998	45.03	2003	48.07	2008	52.42	2013	49.32
1999	47.30	2004	45.97	2009	51.33	2014	51.27
2000	50.37	2005	48.42	2010	50.28	2015	51.83
2001	50.92	2006	44.70	2011	53.72	2016	48.28
2002	51.00	2007	50.32	2012	52.15	2017	55.03

15.1 Apply the distribution modeling methodology to obtain a parametric model for the distribution of the soybean yield. Estimate the parameters via the method specified in the option. Document all steps of the methodology.

15.2 Graph the parametric distribution function and the corresponding density function. Characterize in words the shape of each function. (*Reminder*: label the axes and show the units.)

15.3 Calculate the quantiles of the yield corresponding to the probabilities 0.1, 0.25, 0.5, 0.75, 0.9.

15.4 For soybeans, the Brazilian customary unit of bags/hectare equals 0.891 bushels/acre, the US customary unit; and one bushel/acre equals 67.32 kilograms/hectare, the metric unit.

(i) Derive the expressions for the distribution function, the density function, and the quantile function of the soybean yield in metric tons/hectare. (ii) Recalculate the quantiles from Exercise 15.3.

Note. Brazil and the United States are the largest producers of soybeans, with the global market share of 36% and 35%, respectively, in 2017/2018. The production cycle, from planting to harvesting, lasts approximately 8 months: October–May in Brazil, April–November in the USA.

Options for the estimation method:

    (LS)   Least squares method.

    (UD)   Uniform distance method (using DFit or other software).

Options for the distribution:

    (LR-LG)   Log-ratio logistic.

    (LR-LP)   Log-ratio Laplace.

    (LR-GB)   Log-ratio Gumbel.

    (LR-RG)   Log-ratio reflected Gumbel.

16 **Agricultural yield: bounded-above interval**. Do Exercise 15 with the distribution type selected from the options below, and with the following modifications.

16.1 Study Sections C.5 and 3.8.7 before performing these tasks. (i) Derive the expressions for the density function, the distribution function, and the quantile function of the soybean yield whose sample space is an interval bounded above $(-\infty, -\eta)$. (ii) Apply the distribution modeling methodology to obtain a parametric model for the distribution of the soybean yield. Estimate the parameters via the method specified in the option. (*Reminder*: Section C.5.3 is now a part of the estimation method.) Document all steps of the methodology.

16.2 Same as 15.2.

16.3 Same as 15.3.

16.4 Same as 15.4.

Options for the estimation method:

(LS)   Least squares method.

(UD) Uniform distance method (using DFit or other software).

Options for the distribution:

(R-WB)   Reflected Weibull.

(R-IW)   Reflected inverted Weibull.

(R-LW)   Reflected log-Weibull.

(R-LL)   Reflected log-logistic.

Part II

Discrete Models

4

Judgmental Forecasting

Expert judgment plays an important role in sciences and engineering. In particular, experts can gather predictive information, identify sources of uncertainty in a decision problem, and quantify uncertainty about an event or a hypothesis. This chapter formalizes the judgmental task of quantification of uncertainty. It introduces probability as a measure of the degree of uncertainty, prescribes the procedure for assessing probability judgmentally, and explores two particular problems: forecasting fraction of events, and revising probability sequentially.

4.1 A Perspective on Probability

The mathematical theory of probability defines probability formally as a number that satisfies certain axioms. It is assumed herein that all probabilities conform to a mathematical definition. In applications, two basic questions arise: How to interpret a probability? How to determine a probability?

4.1.1 Interpretation of Probability

Let A denote an event or a hypothesis, and P stand for probability. The expression $P(A)$ reads: "the probability of event A" or "the probability of hypothesis A". The terms "event" and "hypothesis" are interchangeable and their usage is dictated by the context. One usually speaks of the probability of occurrence of an event (e.g., a tornado strikes Kansas City in May) or the probability of the truth of a hypothesis (e.g., the defendant in a court trial is guilty). The term "probability" admits either or both of the following interpretations.

Frequency interpretation. The probability $P(A)$ is the *relative frequency* with which event A occurs (hypothesis A is true) in a large number of repetitions of the process that generates the event (validates the hypothesis).

Subjective interpretation. The probability $P(A)$ is the numerical measure of the *degree of certainty* about the occurrence of event A (the truth of hypothesis A).

The frequency interpretation of probability presumes it is feasible to observe a large number of repetitions of the process that randomly generates the event. This interpretation is natural for *repetitive events*. Here are some examples of repetitive events: a computer chip exiting the production line is defective; an airplane on a transatlantic flight arrives late at its destination; a rain gauge in Savannah, Georgia, records rain during a 24-h period beginning at 0600 Eastern Time in August. In each example, the number of repetitions of the process that randomly generates the event is large: a company manufactures millions of computer chips per year; there are thousands of transatlantic flights per year; the National Centers for Environmental Information in Asheville, North Carolina — the world's largest archive of climatic data — stores hourly rainfall measurements made at

Savannah International Airport for over 65 years, which represent over 2015 (= 65 years × 31 days) repetitions of the daily weather process in the month of August.

The subjective interpretation of probability makes no presumption about the nature of the event. This may be a *repetitive event*, as explained above; a *rare event* for which no reliable estimate of the relative frequency can be obtained because the number of observations is small; a *unique event* to which the notion of repetition does not apply; or a *factual event* (hypothesis, statement) which is either true or false, but one is uncertain.

An example of a rare event: a tornado will strike Kansas City next May. An example of a unique event: you will be offered a job by the company of your choice at least 3 months before graduation. An example of a factual event: the computer code you wrote, but have not yet tested, performs calculations correctly. Though the relative frequency of an event cannot be estimated, one may still want to forecast the event. This can be done by expressing the forecast in terms of probability interpretable as a measure of the *degree of certainty* (the *degree of certitude*, the *degree of belief*, the *strength of belief*) about the occurrence of an event (the truth of a hypothesis or a statement).

In summary, the subjective interpretation of probability is the most general because it is valid regardless of the nature of the forecasted event (repetitive, rare, unique, or factual) and regardless of the method by which the forecast probability is determined.

4.1.2 Determination of Probability

While many methods are used in practice to determine probabilities, there are four kinds of methods which might be classified as "pure methods". Accordingly, there are four kinds of probabilities.

1. *Statistical probability* is estimated as a relative frequency of an event in a finite sample of realizations. The statistical probability is not unique because its value is a function of the sample. For instance, a larger sample may reveal a different relative frequency than does a smaller sample. In general, the statistical probability is subject to *sampling uncertainty*; this uncertainty becomes practically insignificant only if the sample is sufficiently large.
2. *Mathematical probability* is calculated from a mathematical model of a phenomenon, for instance, a queueing model, a macroeconomic model, an atmospheric model. The mathematical probability is *not unique* because a model always embodies subjective assumptions of its developer. (The term *physical probability* appears sometimes in analyses of reliability of materials, machines, or structures.)
3. *Logical probability* is deduced as, for example, one deduces that obtaining heads twice in two tosses of a fair coin is 1/4. In general, the logical probability is *unique* in that it is universally agreed upon.
4. *Judgmental probability* is assessed by an individual or a group. It is *not unique*, but depends upon several factors, which we shall study. (Synonyms: *subjective probability*, *intuitive probability*.)

Once probabilities have been estimated statistically, calculated mathematically, deduced logically, or assessed judgmentally, the same calculus of probability applies to all of them. In real-world applications, probabilities assigned to events are seldom of one kind. Often, one uses a combination of samples, models, logic, and judgment to arrive at probability values. This effort at *combining* (or aggregating, or fusing) information from different sources is one of the highlights of judgmental forecasting based on Bayesian theory, which we study in Section 4.4.

Example 4.1   *(Herd immunity prediction)*
In February 2021, an expert predicted that Americans will soon acquire herd immunity against Covid-19 — "the inevitable result of viral spread and vaccination". As he explained: "My prediction that Covid-19 will be mostly gone by April is based on laboratory data, mathematical data, published literature and conversations with experts … also based on direct observation". *Post factum*: By May, with 47% of Americans immunized (10% post-infection

plus 37% vaccinated), many states became nearly free of Covid-19 and lifted the mandates (on face masks, public gatherings, school operations, etc.).

4.1.3 Probability as Logic

The judgmental probability theory derives from axioms (see Appendix A). As such, it is a mathematical theory. But its authors had a practical purpose in mind: to create the logic that would formalize (i) the reasoning about uncertainty (**probabilistic reasoning**) and (ii) the quantification of judgment about uncertainty (in terms of numerical probability). This purpose dates to Cicero (106–43 BC) — Roman statesman, orator, and author — quoted by Good (1950): "Probability is the very guide of life."

Two millennia later, Sherlock Holmes — the great fictional detective of the Victorian era — defended his investigative method against the suggestion of it being guesswork: "we balance probabilities and choose the most likely. It is the scientific use of the imagination, but we have always some material basis on which to start" (from *The Hound of the Baskervilles*).

In 2008, the world economy descended into a devastating financial crisis, which caught the major investment banks and financial institutions unawares because their models failed to predict it. Allan Greenspan, former chairman of the Federal Reserve, reflected afterwards on the economic forecasting and decision making in the public arena: "It's a tough job. We can't see over the horizon, but since we live in the future, we have no choice but to try to make forecasting judgments" (*The Wall Street Journal*, 4 February 2013).

So here we are: the expert judgment remains the last line of defense against the risk of relying on precise but erroneous forecasts outputted by deterministic models. Such models, however complex, may not capture every possible cause of a future event and do not reveal the resultant uncertainty.

At this point, we assume the reader is familiar with elementary methods of the probability calculus and statistical analysis. Our objective is to present (i) methods for assessing judgmental probabilities and (ii) methods for revising (updating) probabilities.

4.1.4 Judgmental Task

This is the first of several judgmental "warm-up" tasks in this book. A task is to be performed intuitively, sans any preparation and aid. The purpose is twofold: (i) to focus your mind on a problem before studying the scientific approach; and (ii) to compare your intuitive response with the normative answer, if it exists, and thereby to learn about the performance of a human as an intuitive forecaster or an intuitive decider.

Task 4.1 (*Judgmental probability*).   Consider three events:
  R — The next Thanksgiving day in Charlottesville, Virginia, will be rainy.
  G — Your grade from the midterm exam in this course will be B or higher.
  D — The invasion of Normandy by allied forces began before 6 April 1944.
Assess judgmentally your probability of each event.
Use only probability values from the set:
    $\{0, 0.1, 0.2, 0.3, 0.4, 0.5, 0.6, 0.7, 0.8, 0.9, 1\}$.
Responses: $P(R) = $ _____
              $P(G) = $ _____
              $P(D) = $ _____

An analysis of the task is presented in Section 4.5. It may be read now, or at the conclusion of this chapter.

4.2   Judgmental Probability

4.2.1   Definition of Judgmental Probability

A human can be viewed as an information processor (Figure 4.1). At a given time, he possesses knowledge (K) and experience (E) within him, and receives information (I) from the world outside. He then processes (I, K, E) into a *degree of certainty* about the occurrence of event A, or the truth of hypothesis A. The degree of certainty (the degree of certitude, the degree of belief, the intensity of belief) is a primitive notion — undefined but assumed to be formable in one's mind. The judgmental task is to quantify the degree of certainty by assigning a probability number $P(A)$ to event A. This assignment should satisfy two principles.

> **Conditioning principle**. Because your degree of certainty depends on (I, K, E), the probability you assign to A is *conditional on* (I, K, E); this can be indicated by writing $P(A|I, K, E)$. As your information, knowledge, and experience change, so may your degree of certainty and its numerical measure $P(A|I, K, E)$.

> **Internal coherence principle**. Consider event A and some other event B; if your belief that A will occur is stronger than your belief that B will occur, then the probability $P(A|I, K, E)$ you assign to event A should be greater than probability $P(B|I, K, E)$ you assign to event B. Thus a coherent assignment of probability to an event requires a *comparative judgment* of at least two events.

Henceforth, either the abbreviated notation or the complete notation will be employed, assuming that

$$P(A) = P(A|I, K, E),$$

and remembering that judgmental probability is always conditional.

Example 4.2   *(Medical diagnosis)*
A physician possesses medical knowledge K learned in school and expanded through reading journals and attending conferences. Her experience E, acquired through residency training and years of practice, has two dimensions: (i) the *diagnostic skill* in applying medical knowledge to identify an illness, to treat it, and to evaluate the outcome; and (ii) the *metadiagnostic skill* in applying probabilistic reasoning to judge the degree of certainty that a diagnosis (a hypothesized illness) is correct, and to quantify this degree of certainty in terms of the probability. During the visit of a patient, the physician gathers information I, such as medical history and symptoms; hypothesizes an illness A; judges the degree to which she is certain that illness A afflicts the patient; and finally, expresses this degree of certainty in terms of probability $P(A|I, K, E)$.

Example 4.3   *(Crime investigation)*
It is now noon on the first day of the final examinations, and the police investigate a theft of books from a college library. One relevant event, A, is "fewer than 3 persons were in the reading room of the college library at 0700

Figure 4.1   Judgmental task of assessing probability.

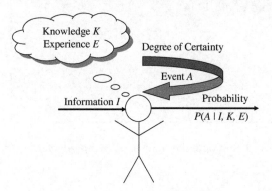

today". To an individual who has never been to a college library, this event is highly uncertain, as information I_1 and experience E_1 pertinent to it are lacking; hence, only general knowledge K_1 of libraries and students can condition his assignment of probability, say $P(A|K_1) = 0.6$. To an individual who attended another college of similar size and spent many mornings in its library reading room, the experience E_2, in addition to general knowledge K_2, helps to assign probability $P(A|K_2, E_2) = 0.9$. To a librarian who walks through the reading room at 0800 every morning, there is available general knowledge K_3, experience E_3 from the past, and information I_3 from this morning, as she recalls seeing 5 persons, so she assigns probability $P(A|I_3, K_3, E_3) = 0.7$. But to a student who entered the reading room at 0700 this morning and counted 8 persons already at the desks, this first-hand information I_4 is sufficient to declare event A impossible and to assign $P(A|I_4) = 0$.

4.2.2 Assessment of Judgmental Probability

To assess a probability judgmentally, the forecaster should follow a five-step procedure.

Step 1. Define event A precisely.

Step 2. Gather, organize, and analyze all predictive information.

Step 3. Judge the **veracity** of information; judge the **predictive strength** of information; form an overall judgment of the degree of certainty about the occurrence of event A, given information. (Information is veracious if it can be taken at its face value; for instance, a witness reports seeing 5 wolves; were there exactly 5, and not 4 or 6? Predictive strength refers to the degree of stochastic dependence between information and event; for instance, a cumulonimbus is a predictor of convective wind, but a cirrus is not.)

Step 4. Assign probability $P(A)$ to the event. This can be accomplished in one of three ways, by assessing:

    4.1 *Probability number*. Assess directly a real number p, $0 < p < 1$, such that in your judgment "the probability of event A is p". For example, $P(A) = 0.8$.

    4.2 *Real number odds*. First, judge whether an event or its complement is more likely; say, A is more likely than A^c. Second, assess a real number $r > 1$ such that in your judgment "event A is r times more likely than its complement A^c". This gives the *odds on* event A:

$$\frac{P(A)}{P(A^c)} = \frac{r}{1}; \qquad \text{(read: } r \text{ to one)}$$

and because $P(A^c) = 1 - P(A)$, you can write

$$\frac{P(A)}{1 - P(A)} = \frac{r}{1},$$

wherefrom

$$P(A) = \frac{r}{r+1}.$$

For example, when $r = 1.5$, the odds on event A are $1.5/1$, and the probability of event A is $P(A) = 1.5/2.5 = 0.6$. (The *odds against* event A are defined as $P(A^c)/P(A) = 1/r$.)

    4.3 *Rational number odds*. Assess two positive integers, $m > 0$ and $n > 0$, such that in your judgment "the *odds on* event A are m/n". Thus, you can write

$$\frac{P(A)}{P(A^c)} = \frac{m}{n}, \qquad \text{(read: } m \text{ to } n)$$

wherefrom

$$P(A) = \frac{m}{m+n}.$$

For example, $m/n = 3/5$ yields $P(A) = 3/8 = 0.375$. (Equivalently, you may state the *odds against* event A, which are defined as $P(A^c)/P(A) = n/m$. Assessing the odds against an event is the favorite method of uncertainty quantification by bookmakers. A bookmaker usually writes 5–3 to mean the odds against an event are 5 to 3. We write 5/3 because, mathematically, the odds is a quotient.)

Step 5. Validate the internal coherence of this assignment. Toward this end:

5.1 Identify a *reference event R* whose probability $P(R)$ is a logical probability and is equal to $P(A)$:

$$P(A) = P(R).$$

5.2 Compare your degree of certainty about event A with your degree of certainty about event R. Are they about equal? They should be. If they are not, then (i) adjust $P(A)$ upward (when A appears more certain than R) or downward (when A appears less certain than R), and (ii) return to Step 5.1 (to validate the adjusted $P(A)$).

A reference event can be constructed in many ways. For instance, by tossing a coin once, twice, thrice, etc., one can generate reference events having probabilities 1/2, 1/4, 1/8, etc. By throwing a die, one can generate events having probabilities 1/6, 1/3, 1/2, etc. A particularly versatile device is the *probability wheel* (Figure 4.2). Suppose it is a white wheel with a red sector R whose area makes up a fraction $P(R)$ of the wheel's area, $0 < P(R) < 1$. The wheel is spun. When it stops, event R occurs if the red sector faces a fixed pointer. The probability of this event is obviously $P(R)$. To apply this fact to a particular assessment, adjust the sector's size so that event R appears to you as likely as event A. Then determine the fraction $P(R)$ and set $P(A) = P(R)$. The premise behind this procedure is that the probability of an event can be visualized as a *proportion*. And given an opportunity to win, say $500, by betting either on the occurrence of event A or on red in the "wheel of fortune", you should be indifferent because you judged both events to be equally likely.

4.2.3   Allowable Forecast Probabilities

In principle, the probability assessed by a forecaster can be any number from the closed interval $\mathcal{Y} = [0,1]$. In practice, it is often desirable to restrict the set of allowable forecast probabilities to a finite set of numbers,

$$\mathcal{Y} = \{y_0, y_1, \dots, y_I\},$$

where $0 \le y_0 < y_1 < \cdots < y_I \le 1$. The choice of the numbers depends on the events being forecasted. For instance, when the National Weather Service forecasts the occurrence of precipitation, each forecast probability is rounded to the nearest number in the set

$$\{0, 0.02, 0.05, 0.1, 0.2, 0.3, 0.4, 0.5, 0.6, 0.7, 0.8, 0.9, 1\}.$$

The advantage of a finite set $\mathcal{Y}$ is twofold. First, the precision with which a human can quantify uncertainty is limited. Thus a finite scale is often sufficient to convey the judgmental estimates. In fact, a finite scale may facilitate the judgmental task of assessing a probability. Second, when judgmental forecasts are made repetitively, a finite set $\mathcal{Y}$ facilitates the recording and verification of forecasts, and the conveyance of feedback to the forecaster — the topics of Chapter 6.

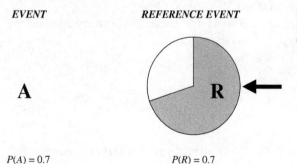

EVENT **A** $P(A) = 0.7$

REFERENCE EVENT **R** $P(R) = 0.7$

Figure 4.2   Validation of internal coherence: if events A and R are judged equally likely, then the assessed probability $P(A)$ of event A should equal the known probability $P(R)$ of reference event R generated by the probability wheel.

4.2.4 Cromwell's Rule

Special consideration should be given to the endpoints of the allowable probability scale. Should $y_0 = 0$ and $y_1 = 1$ always? Should the forecaster ever assign probabilities 0 or 1? In either case, the forecaster states that there is absolute certainty about the event. The mathematical implication of making such a statement is strong: the forecast probability which is either 0 or 1 can never be revised. To illustrate, suppose the forecaster initially assigned probability $P(A)$. Subsequently, new information J was received. To keep notation from expanding, let J also denote the event that generates the new information. Suppose event J is possible under the hypothesis that A is true as well as under the complementary hypothesis; mathematically this means that $P(J|A) > 0$ and $P(J|A^c) > 0$; and the larger the ratio $P(J|A) / P(J|A^c)$, the stronger the evidence in favor of hypothesis A. Given new information J, the initial probability $P(A)$ should be revised. Bayes' theorem provides the mathematical framework for this task:

$$P(A|J) = \frac{P(J|A)P(A)}{P(J|A^c)P(A^c) + P(J|A)P(A)},$$

where $P(A^c) = 1 - P(A)$. An implication of Bayes' theorem is that

$$P(A) = 0 \implies P(A|J) = 0,$$
$$P(A) = 1 \implies P(A|J) = 1.$$

In other words, probability 0 or 1 can never be revised, no matter how strong evidence against it is brought to light by new information. Hence the following.

> **Advice** (Cromwell's rule). Never assign probability 0 or 1 to any event A, unless you can prove, using rules of logic or mathematics, that information you have implies event A or its complement A^c.

When evidence is strong but a proof is lacking, leave some room for doubt: assign a very small or a very large probability $P(A)$, but such that $0 < P(A) < 1$, so that a revision remains mathematically possible.

4.3 Forecasting Fraction of Events

Let $A_1, \ldots, A_N$ be events stochastically independent of each other. To each event A_n the forecaster assigned probability $p_n = P(A_n)$, for $n = 1, \ldots, N$. Denoting the number of events that will occur by X ($0 \leq X \leq N$), the quotient X/N represents the fraction of events that are forecasted to occur.

Theorem (Forecast of fraction) *The expectation of the number X of events to occur equals the sum of forecast probabilities,*

$$E(X) = p_1 + \cdots + p_N; \tag{4.1}$$

the expectation of the fraction X/N of events to occur equals the mean of forecast probabilities,

$$E\left(\frac{X}{N}\right) = \frac{1}{N}(p_1 + \cdots + p_N). \tag{4.2}$$

The variances are:

$$Var(X) = \sum_{n=1}^{N} p_n(1 - p_n), \tag{4.3}$$

$$Var\left(\frac{X}{N}\right) = \frac{1}{N^2} \sum_{n=1}^{N} p_n(1 - p_n). \tag{4.4}$$

Exercise 10 directs you to prove this theorem, which has many applications, particularly in forecasting unique events by groups of experts. One such application arises from Task 4.1 and is discussed in Section 4.5. Another one is in sports.

Example 4.4   *(Olympic medals)*

Before the Olympic Games, a popular contest is forecasting the gold medals to be won by athletes from a particular country. Various forecasting methods are used, but one judgmental method could proceed thusly. Suppose a country will participate in N sports. For each sport $n \in \{1, \dots, N\}$, an expert in that sport gathers, organizes, and analyzes predictive information (e.g., results of recent athletics meetings, investments in sports, demographic statistics of the country), and on this basis assigns probability $p_n = P(G_n)$ to the event G_n — an athlete from that country will win the gold medal in sport n. Then the sum of N forecast probabilities, equation (4.1), gives the expected number of gold medals, and the mean of N forecast probabilities, equation (4.2), gives the expected fraction of sports in which the country will win the gold medals. For a generalization of this method to many countries, or teams, see Exercise 23.

Example 4.5   *(London Olympics)*

For the London Olympics in 2012, *The Wall Street Journal* prepared a forecast of the gold medal counts and the total medal counts by countries. The forecast was published before the Olympics on 23 July, and the verification was reported one day after the closing ceremony, on 13 August. The forecasting system was partly subjective and partly statistical; it included data analysis and computer simulation; it used information such as "interviews with sports experts" and "recent performances of athletes". But the gist of the system was like the method of this section: a probability was assigned to each medal contender, and then the probabilities were mapped into the estimates of medal counts. Table 4.1 reproduces a subset of the verification data. Examine the data and judge the goodness of the forecast. (Overall, did the Journal's forecast based on probabilities capture the differences in the strengths of the ten countries?)

Table 4.1   A subset of the London Olympics data published by *The Wall Street Journal* on 13 August 2012: forecasted and actual gold medal counts for the top 10 gold-medal-winning countries, and the corresponding rankings.

Country	Gold medal count			Country rank	
	Forecast	Actual	Error	Forecast	Actual
USA	40	46	6	1	1
China	38	38	0	2	2
Russia	29	24	−5	3	4
Great Britain	22	29	7	4	3
Germany	16	11	−5	5	6–7
Japan	13	7	−6	6	10[a]
France	9	11	2	7–8	6–7
Italy	9	8	−1	7–8	8–9
South Korea	7	13	6	9	5[a]
Hungary	5	8	3	10	8–9

a)   Two largest reversals of the rank.

4.4 Revising Probability Sequentially

4.4.1 Revision Paradigms

The Bayesian theory of sequential revision of probability not only supports mathematical models and judgmental procedures but also deepens our understanding of probabilistic forecasting — the essence of which is the **conditioning** of probability on information. To highlight this fact, the abbreviated notation retains I (and omits K, E, which are assumed to remain invariant during the revision process):

$$P(A|I) = P(A|I, K, E).$$

Information I is now defined as a set that comprises $N + 1$ subsets; in vector notation $I = (I_0, I_1, \ldots, I_N)$; equivalently, borrowing from sentential logic the *conjunction symbol* $\wedge$, which reads "and",

$$I = I_0 \wedge I_1 \wedge \cdots \wedge I_N.$$

No two subsets can be identical; that is, I_n should include something not already included in $I_0, \ldots, I_{n-1}$. Behind each subset I_n $(n = 0, 1, \ldots, N)$ lies a quantitative or qualitative description; it may be a datum, fact, evidence, intelligence, pattern of a time series, observation of a phenomenon, or inference based on a model (conceptual or mathematical); it may either strengthen or weaken the degree of certainty about the occurrence of event A. This *structure* of I is encountered in two paradigms, explained below for $N = 2$.

Decomposition of information set. To facilitate the mental processing of the huge amount of information contained in set I, you identified three subsets of logically cohesive pieces of information. Now you want to *decompose* the probability assessment into three stages, whereby at each stage your mental analysis concentrates primarily on information in one subset. Your tasks are (i) to enlarge predictive information sequentially, $I_0, I_0 \wedge I_1, I_0 \wedge I_1 \wedge I_2$, and (ii) to assess a sequence of probabilities,

$$P(A|I_0), \ P(A|I_0, I_1), \ P(A|I_0, I_1, I_2) = P(A|I).$$

Can it be done? Will it facilitate the judgmental process? Will it yield a higher-quality forecast of A?

Sequence of information subsets. At the time you had to forecast event A, with lead time LT_0, you received information I_0, and thus you assessed the probability $P(A|I_0)$. Now, at lead time $LT_1 < LT_0$, you receive additional information, I_1. Do you have to discard $P(A|I_0)$ and to repeat the assessment procedure with information $I_0 \wedge I_1$ in order to obtain $P(A|I_0, I_1)$? Or, could you just concentrate your mental analysis primarily on the new information I_1 and, based on it, *revise $P(A|I_0)$ into $P(A|I_0, I_1)$*? These questions will arise again at lead time $LT_2 < LT_1$, when you anticipate receiving additional information I_2, with the task of producing $P(A|I_0, I_1, I_2) = P(A|I)$.

Example 4.6 *(Economic recession)*
During four months covering the pandemic outbreak of Covid-19, an economist made three forecasts of a recession in 2020 (event A). Here are the approximate forecast times, the assigned probabilities, and the available information subsets.

- 30 December 2019: $P(A|I_0) = 0.15$. The 11-year bull market continues: US stock prices have been rising, with the Dow Jones Industrial Average (DJIA) hitting record highs above 28 500, and the unemployment rate reaching new lows below 3%.
- 15 February 2020: $P(A|I_0, I_1) = 0.45$. On 10 January, China reported the first death from coronavirus, which begun spreading rapidly throughout the Asia-Pacific region. Countries responded by ordering nationwide quarantines, closing borders, limiting travel, and restricting public life. The DJIA jittered, but then resumed its climb.
- 11 March 2020: $P(A|I_0, I_1, I_2) = 0.9$, with a hint of another revision soon. Coronavirus infections surged around the world, with an exponential growth in Italy and the USA. The DJIA reached an all-time high of 29 551 on 12 February, and then declined. On 10 March, it suffered its worst decline since 2008: it lost 2013 points and closed at 23 851, some 19.3% below the peak.

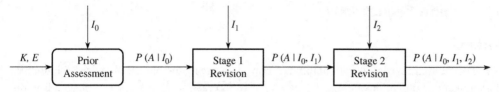

Figure 4.3 Sequential Bayesian revision of the prior probability $P(A|I_0)$ of event A (or hypothesis A) — assessed by an expert based on information I_0, knowledge K, and experience E — into the sequence of posterior probabilities $P(A|I_0, I_1)$, $P(A|I_0, I_1, I_2)$, conditional on I_0 and on new information subsets I_1, I_2.

4.4.2 Bayesian Revision Theory

The process of revising probability sequentially (in stages: $n = 1, \ldots, N$) is depicted in Figure 4.3 for $N = 2$. The Bayesian theory underlying this process involves (i) the quantification of stochastic dependence among $(A, A^c; I_0, I_1, I_2)$ in terms of input probabilities and (ii) the derivation of the revised (posterior) probabilities; all probabilities are assumed to obey Cromwell's rule. The process begins with

$P(A|I_0)$ — *prior probability* of event A occurring (of hypothesis A being true) assessed or estimated based on I_0; when the prior information I_0 is unspecific, it should be omitted, here and in all subsequent probabilities.

Stage 1 revision
Input probabilities may be of any kind (recall Section 4.1.2). Their interpretation is independent of the method through which they are determined.

$P(I_1|A, I_0)$ — *probability* of receiving information I_1, **conditional on the hypothesis** that A occurs (is true), and given prior information I_0 (which had already been utilized).
— *likelihood* of A, **conditional on** I_1, and given I_0 (which had already been utilized); this interpretation reverses the conditioning to reflect the reality of forecasting: I_1 is known (has been received), while A is uncertain (must be forecasted).
$P(I_1|A^c, I_0)$ — *probability* of receiving information I_1, **conditional on the hypothesis** that A does not occur (is false), and given prior information I_0 (which had already been utilized).
— likelihood of A^c, **conditional on** I_1, and given I_0.

Conditional expected probability of I_1, given I_0 — from the *total probability law*, with $P(A^c|I_0) = 1 - P(A|I_0)$:

$$P(I_1|I_0) = P(I_1|A^c, I_0)P(A^c|I_0) + P(I_1|A, I_0)P(A|I_0). \tag{4.5}$$

Posterior probability of A, given (I_0, I_1) — from *Bayes' theorem*, which revises the prior probability $P(A|I_0)$:

$$P(A|I_0, I_1) = \frac{P(I_1|A, I_0)P(A|I_0)}{P(I_1|I_0)}. \tag{4.6}$$

Stage 2 revision
Input probabilities are conditional on an additional information subset, but otherwise have the same structure as those in the first stage; therefore their definitions are abbreviated.

$P(I_2|A, I_0, I_1)$ — probability of I_2, conditional on A, and given (I_0, I_1).
— likelihood of A, conditional on I_2, and given (I_0, I_1).
$P(I_2|A^c, I_0, I_1)$ — probability of I_2, conditional on A^c, and given (I_0, I_1).
— likelihood of A^c, conditional on I_2, and given (I_0, I_1).

Conditional expected probability of I_2, given (I_0, I_1) — from the total probability law, with $P(A^c|I_0, I_1) = 1 - P(A|I_0, I_1)$:

$$P(I_2|I_0, I_1) = P(I_2|A^c, I_0, I_1)P(A^c|I_0, I_1) + P(I_2|A, I_0, I_1)P(A|I_0, I_1). \tag{4.7}$$

Posterior probability of A, given (I_0, I_1, I_2) — from Bayes' theorem, which revises the prior probability $P(A|I_0)$:

$$P(A|I_0, I_1, I_2) = \frac{P(I_1, I_2|A, I_0)P(A|I_0)}{P(I_1, I_2|I_0)}$$

$$= \frac{P(I_2|A, I_0, I_1)}{P(I_2|I_0, I_1)} \frac{P(I_1|A, I_0)P(A|I_0)}{P(I_1|I_0)}$$

$$= \frac{P(I_2|A, I_0, I_1)P(A|I_0, I_1)}{P(I_2|I_0, I_1)}. \tag{4.8}$$

Each conditional joint probability of (I_1, I_2), in the numerator and in the denominator of the first equality, is factorized in the second equality; then the second quotient is recognized as the right side of equation (4.6); this leads to its substitution in the third equality.

Exploiting the pattern of equations (4.6) and (4.8) for revision stages 1 and 2, one can write equations for revision stages n and $n + 1$ (for any integer $n \geq 2$) and prove by induction the following principle.

> **Sequential revision principle**. In the sequential Bayesian revision of a prior probability $P(A|I_0)$ of event A based on information subsets $I_1, I_2, \ldots, I_N$, the *posterior probability* $P(A|I_0, I_1, \ldots, I_n)$ from one revision stage, n ($1 \leq n < N$), becomes the prior probability for the next revision stage, $n + 1$, as depicted in Figure 4.3.

4.4.3 Conditional Stochastic Independence

It is apparent that the input probabilities expand their conditioning by one information subset at each revision stage; this makes their statistical estimation, or judgment assessment, impracticable for large N. A trade-off between the exactness and the practicableness of the theory is achieved via simplifying assumptions. Their background is this definition.

Definition   Event A is *stochastically independent* of event B if

$$P(A|B) = P(A).$$

Theorem (*Stochastic independence*)   *The following equalities are equivalent (i.e., one holds if and only if all others hold):*

$$P(A|B) = P(A), \qquad P(B|A) = P(B), \qquad P(A, B) = P(A)P(B);$$
$$P(A|B^c) = P(A), \qquad P(B|A^c) = P(B), \qquad P(A^c|B) = P(A^c);$$
$$P(A^c|B^c) = P(A^c), \qquad P(B^c|A^c) = P(B^c), \qquad P(B^c|A) = P(B^c).$$

Exercise 12 asks you to prove this theorem for a subset of the equalities. They illustrate the elementary simplification of stochastic dependence. Now we are ready to simplify the dependence structure for the sequential Bayesian revision.

Definition

(i) **Conditional on** event A (on event A^c), information I_1 is *stochastically independent* of information I_0 if

$$P(I_1|A, I_0) = P(I_1|A),$$
$$P(I_1|A^c, I_0) = P(I_1|A^c).$$

(ii) **Conditional on** event A (on event A^c), information I_2 is *stochastically independent* of information (I_0, I_1) if

$$P(I_2|A, I_0, I_1) = P(I_2|A),$$
$$P(I_2|A^c, I_0, I_1) = P(I_2|A^c).$$

In essence, definition (i) says that the probability of receiving information I_1 depends solely on the event occurring (A) or not occurring (A^c), and is unaffected by the prior information I_0. Both equalities must hold because one does not imply the other. In short, we say that information subsets I_0, I_1 are *conditionally stochastically independent relative to* events A, A^c. The reader should interpret definition (ii) similarly. It can be generalized to the N-stage revision process.

Definition **Conditional on** A (on A^c), I_n is *stochastically independent* of $(I_0, I_1, \dots, I_{n-1})$ if

$$P(I_n|A, I_0, I_1, \dots, I_{n-1}) = P(I_n|A),$$
$$P(I_n|A^c, I_0, I_1, \dots, I_{n-1}) = P(I_n|A^c).$$

When this definition holds for every $n \in \{1, \dots, N\}$, we say, in short, that information subsets $I_0, I_1, \dots, I_N$ are *conditionally stochastically independent relative to* events A, A^c. The practical implication is enormous: every input probability, being now conditional on only A or A^c, is feasible to estimate statistically or to assess judgmentally.

4.4.4 Bayesian Revision Model

Embedding the definition of conditional stochastic independence in equations (4.5)–(4.8) yields the following model.

Stage 1 revision

$$P(I_1|I_0) = P(I_1|A^c)P(A^c|I_0) + P(I_1|A)P(A|I_0), \tag{4.9}$$

$$P(A|I_0, I_1) = \frac{P(I_1|A)P(A|I_0)}{P(I_1|I_0)}. \tag{4.10}$$

Stage 2 revision

$$P(I_2|I_0, I_1) = P(I_2|A^c)P(A^c|I_0, I_1) + P(I_2|A)P(A|I_0, I_1), \tag{4.11}$$

$$P(A|I_0, I_1, I_2) = \frac{P(I_2|A)P(A|I_0, I_1)}{P(I_2|I_0, I_1)}. \tag{4.12}$$

These equations for revision stages $n = 1, 2$ establish a pattern that can be readily continued when writing equations for revision stages $n = 3, \dots, N$.

Example 4.7 *(Bayesian revisions)*

Suppose the assessed prior probability is $P(A|I_0) = 0.3$, and the input probabilities to two revision stages are:

$$P(I_1|A^c) = 0.2, \qquad P(I_1|A) = 0.9;$$
$$P(I_2|A^c) = 0.5, \qquad P(I_2|A) = 0.2.$$

They imply that information I_1 is $0.9/0.2 = 4.5$ times more probable to exist when A is true than otherwise, while information I_2 is $0.5/0.2 = 2.5$ more probable to exist when A is false than otherwise. The results of the calculations, rounded to four decimal places, are:

$$P(I_1|I_0) = 0.2 \times 0.7 + 0.9 \times 0.3 = 0.41,$$
$$P(A|I_0, I_1) = 0.9 \times 0.3/0.41 = 0.6585,$$
$$P(I_2|I_0, I_1) = 0.5 \times 0.3415 + 0.2 \times 0.6585 = 0.3025,$$
$$P(A|I_0, I_1, I_2) = 0.2 \times 0.6585/0.3025 = 0.4354.$$

As one may have anticipated, the prior probability $P(A|I_0)$ is revised upward based on I_1; subsequently, the probability is revised downward based on I_2. Is the value of the posterior probability $P(A|I_0, I_1, I_2)$ affected by the order of information subsets, (I_1, I_2) or (I_2, I_1)? To verify your intuition, do Exercise 13.

Equation (4.9) suggests that I_1 is not *stochastically independent* (SI) of I_0, even though it is *conditionally stochastically independent* (CSI). Figure 4.4 explains this visually using an arrow to depict the *stochastic dependence* (SD). The arrow from I_0 to A (A^c) depicts the prior SD; the arrow from A (A^c) to I_1 depicts the likelihood SD. The total probability law (4.9) creates an arrow from I_0 to I_1; essentially, this law is a form of *stochastic transitivity*:

If I_1 is SD on A (on A^c), and A (A^c) is SD on I_0, then I_1 is SD on I_0.

Likewise, equation (4.11) suggests that I_2 is not SI of (I_0, I_1), even though it is CSI. Figure 4.4 explains this visually as a consequence of stochastic transitivity. Continuing this pattern to revision stages $n = 3, \ldots, N$ yields $P(I_n|I_0, I_1, \ldots, I_{n-1})$, which means that, in general, information subset I_n is stochastically dependent on all preceding information subsets, even though it is CSI of them.

Figure 4.4 Stochastic dependence (SD) structures in the sequential Bayesian revision of the prior probability $P(A|I_0)$ of event A (or hypothesis A) based on new information subsets I_1, I_2, which are conditionally stochastically independent (CSI) relative to events A, A^c. By stochastic transitivity, I_1 is SD on I_0, and I_2 is SD on (I_0, I_1).

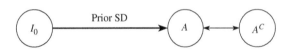

Prior Assessment

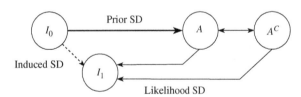

Stage 1 Revision

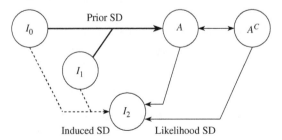

Stage 2 Revision

Example 4.8 *(CSI does not imply SI)*

Suppose the unconditional prior probability, assessed sans specific information, is $P(A) = 0.4$, and the input probabilities to Stage 1 revision are the same as in Example 4.7. Then

$$P(I_1) = P(I_1|A^c)P(A^c) + P(I_1|A)P(A)$$
$$= 0.2 \times 0.6 + 0.9 \times 0.4 = 0.48,$$
$$P(I_1|I_0) = P(I_1|A^c)P(A^c|I_0) + P(I_1|A)P(A|I_0)$$
$$= 0.2 \times 0.7 + 0.9 \times 0.3 = 0.41.$$

Because $P(I_1|I_0) \neq P(I_1)$, information subsets I_0, I_1 are not SI.

Equations (4.9)–(4.12) also suggest that the magnitude of probability revision based on new information subset I_n ($n = 1, \ldots, N$) depends on the strength of *stochastic dependence* of I_n on A (on A^c). There are two limiting information subsets.

Definition

(i) Subset I_n is said to be *perfect* for event A if

$$P(I_n|A) = 1 \quad \text{and} \quad P(I_n|A^c) = 0,$$

or

$$P(I_n|A) = 0 \quad \text{and} \quad P(I_n|A^c) = 1.$$

(ii) Subset I_n is said to be *uninformative* for event A if

$$P(I_n|A) = P(I_n).$$

(iii) Otherwise, subset I_n is said to be *informative* for event A.

Theorem *(Limiting posterior probabilities)* *The posterior probability resulting from revision based on limiting information subset I_n is determined thusly:*

$$\text{perfect } I_n \quad \implies P(A|I_0, I_1, \ldots, I_n) = 1 \text{ or } 0;$$
$$\text{uninformative } I_n \implies P(A|I_0, I_1, \ldots, I_n) = P(A|I_0, I_1, \ldots, I_{n-1}).$$

The proof is deferred to Exercises 14 and 15. The plain meaning is this: Perfect I_n eliminates the uncertainty about A. Uninformative I_n does not change the prior probability of A. Furthermore,

$$I_n \text{ is perfect for A} \quad \iff I_n \text{ is perfect for } A^c,$$
$$I_n \text{ is uninformative for } A \iff I_n \text{ is uninformative for } A^c,$$
$$I_n \text{ is informative for } A \quad \iff I_n \text{ is informative for } A^c.$$

Example 4.9 *(Uninformativeness implies SI)*

Suppose that in Example 4.8, $P(A|I_0) = 0.4$; this implies that subset I_0 is uninformative for event A because it does not change the unconditional prior probability $P(A) = 0.4$. Now,

$$P(I_1|I_0) = 0.2 \times 0.6 + 0.9 \times 0.4 = 0.48.$$

Because $P(I_1|I_0) = P(I_1)$, information subsets I_0, I_1 are SI. This is a complementary illustration of the following theorem.

Theorem (*Informativeness precludes SI*) *If I_0, I_1 are CSI relative to A, A^c, and each is informative for A, then I_0 and I_1 cannot be SI.*

Proof: Information subsets I_0, I_1 are SI if $P(I_1|I_0) = P(I_1)$. The two equations in Example 4.8 imply that this equality holds if and only if $P(A|I_0) = P(A)$, which means that I_0 is uninformative for A; this contradicts the assumption. By switching the positions of I_0 and I_1 in the above equations, one can reach the same conclusion for I_1. QED.

The theorem extends to $I_0, I_1, \dots, I_N$. The theorem clarifies casual claims regarding the desirability of "independent" information subsets for forecasting. What is desirable is their CSI because it simplifies the modeling; but what is desirable formost is their informativeness — which precludes their SI

4.4.5 Probabilistic Reasoning

With the Bayesian revision model (4.9)–(4.12) as a mathematical guide, let us highlight the fundamental distinctions between probabilistic reasoning and deterministic reasoning in the context of forecasting.

Deterministic reasoning
In this mode of reasoning, one usually strives to identify *logical dependence* or *physical dependence* (essentially a causal relationship) between information and event. For instance, a flaming match thrown into a puddle of gasoline (I_0) causes fire (A). In the formal language of logic: if I_0, then A; symbolically, $I_0 \implies A$. In the language of probability: given information I_0, event A is certain to occur; $P(A|I_0) = 1$; there is no need for probability theory.

Probabilistic reasoning
In this mode of reasoning, one must judge *stochastic dependence*, which may be one of three kinds.

Direct SD arises when logical dependence between information and event can be determined *a posteriori* (after the event occurred), but can be judged only possible at the forecast time. For instance, after a forest fire, lightning is determined to be the cause. But when forecasting forest fire (A) during July, lightning (I_0), while almost certain to be observed during that month, is only a possible cause because not every instance of lightning ignites forest fire. Hence, the forecaster should assign a probability such that $0 < P(A|I_0) < 1$. In Example 4.7, this is the prior probability $P(A|I_0) = 0.3$.

Indirect SD arises when information alters the judgment of circumstances under which the forecasted event will, or will not, occur. For instance, prolonged drought (I_1) has preceded the July thunderstorm season. Upon learning this, the forecaster should revise the probability so that $P(A|I_0, I_1) > P(A|I_0)$. In Example 4.7, this is the posterior probability $P(A|I_0, I_1) = 0.659$.

Contextual SD arises always between information subsets (which are CSI and informative for forecasting a particular event), but does not imply the causality. For instance, the conditional expected probability, $P(I_1|I_0) = 0.41$ in Example 4.7, does not mean lightning (I_0) may cause drought (I_1) that preceded it! It means that both information subsets may exist at the forecast time because each is informative for predicting forest fire (A). Consequently, I_0 and I_1 become SD **in the context of forecasting a particular event** A. This is the implication of stochastic transitivity induced by the total probability law (recall Section 4.4.4 and Figure 4.4).

To exploit Example 4.7 to its end, suppose heavy rainfall (I_2) will follow lightning, possibly suppressing any ignited forest fire. Given this information, the forecaster revised the posterior probability to $P(A|I_0, I_1, I_2) = 0.435$. The conditional expected probability is now $P(I_2|I_0, I_1) = 0.302$. What does this not mean? What does it mean?

4.4.6 Adjustment Factors for Judgmental Revision

Following equations (4.9)–(4.12) in arithmetic calculations is trivial. But following them in judgmental assessments is another matter. Hence our next objective: to restructure these equations into a form that offers normative guidance for the judgmental revision procedure (to be formulated in Section 4.4.7).

Stage 1 revision

$$P(A|I_0, I_1) = \frac{P(I_1|A)P(A|I_0)}{P(I_1|I_0)}$$

$$= \frac{P(I_1)}{P(I_1|I_0)} \frac{P(I_1|A)}{P(I_1)} P(A|I_0).$$

This is equivalent to

$$P(A|I_0, I_1) = \alpha_1 P(A|I_0),$$ (4.13)

with the following substitutions and interpretations.

Revision factor:

$$\alpha_1 = \frac{\lambda_1}{\gamma_1}.$$ (4.14)

Per equation (4.13), α_1 is the factor by which the prior probability is increased if $\alpha_1 > 1$, or decreased if $\alpha_1 < 1$, or unchanged if $\alpha_1 = 1$. It has three properties. (i) The constraints $0 < P(A|I_0, I_1) < 1$ impose the bounds

$$0 < \alpha_1 < \frac{1}{P(A|I_0)}.$$

For example, $0 < \alpha_1 < 5$ when $P(A|I_0) = 0.2$, and $0 < \alpha_1 < 1.25$ when $P(A|I_0) = 0.8$. (ii) In the limit, as $P(A|I_0) \to 1$, the bounds are $0 < \alpha_1 \leq 1$. Thus, when event A is near certain *a priori*, new information I_1 can only decrease its probability (if $\alpha_1 < 1$) or have no influence (if $\alpha_1 = 1$). (iii) Per equation (4.14), the revision factor is *decomposable* into two factors, which quantify two separate attributes of information I_1.

Information factor:

$$\lambda_1 = \frac{P(I_1|A)}{P(I_1)}.$$ (4.15)

This measures the *predictive strength* of I_1 with respect to A — specifically, the sign and the relative strength of SD between I_1 and A. The more λ_1 departs from 1, the stronger this SD:

$$P(I_1|A) > P(I_1) \iff \lambda_1 > 1 \iff I_1 \text{ and } A \text{ are } positively\ dependent;$$
$$I_1 \text{ increases the prior probability of } A.$$

$$P(I_1|A) < P(I_1) \iff \lambda_1 < 1 \iff I_1 \text{ and } A \text{ are } negatively\ dependent;$$
$$I_1 \text{ decreases the prior probability of } A.$$

$$P(I_1|A) = P(I) \iff \lambda_1 = 1 \iff I_1 \text{ and } A \text{ are } independent;$$
$$I_1 \text{ is } uninformative \text{ for } A.$$

Dependence factor:

$$\gamma_1 = \frac{P(I_1|I_0)}{P(I_1)}.$$ (4.16)

This measures the *stochastic dependence* of I_1 on I_0 — specifically, the sign and the relative strength of SD between I_1 and I_0. The more γ_1 departs from 1, the stronger this SD:

$$P(I_1|I_0) > P(I_1) \iff \gamma_1 > 1 \iff I_1 \text{ and } I_0 \text{ are } positively\ dependent;$$
$$\text{this decreases the influence of } I_1 \text{ on probability revision.}$$

$$P(I_1|I_0) < P(I_1) \iff \gamma_1 < 1 \iff I_1 \text{ and } I_0 \text{ are } negatively\ dependent;$$
$$\text{this increases the influence of } I_1 \text{ on probability revision.}$$

$$P(I_1|I_0) = P(I_1) \iff \gamma_1 = 1 \iff I_1 \text{ and } I_0 \text{ are } independent;$$
$$I_1 \text{ is } uninformative \text{ for } A.$$

Table 4.2 The sign rules of interaction between the information factor λ_1 and the dependence factor γ_1, which create the revision factor $\alpha_1 = \lambda_1/\gamma_1$.

Predictive		Stochastic dependence γ_1		
		negative	none	positive
strength	λ_1	$\frac{1}{2}$	1	2
positive	2	4		1
none	1		1	
negative	$\frac{1}{2}$	1		$\frac{1}{4}$

A numerical example in matrix form (Table 4.2) illustrates the *sign rules* of interaction between the information factor λ_1 and the dependence factor γ_1, which create the revision factor α_1 per equation (4.14). Firstly, the most influential adjustments occur under the opposite signs of dependence: positively dependent $(I_1|A)$ and negatively dependent $(I_1|I_0)$, which yield high $\alpha_1 > 1$, or negatively dependent $(I_1|A)$ and positively dependent $(I_1|I_0)$, which yield low $\alpha_1 < 1$. Secondly, along the main diagonal of the matrix, where $(I_1|A)$ and $(I_1|I_0)$ are both positively dependent, or negatively dependent, or independent, the two factors cancel out, resulting in $\alpha_1 = 1$, which leaves the prior probability unchanged. Thirdly, four cells of the matrix are empty because an informative I_1 cannot be SI of I_0, and an uninformative I_1 cannot be stochastically dependent on I_0.

Stage 2 revision

$$P(A|I_0, I_1, I_2) = \frac{P(I_2|A)P(A|I_0, I_1)}{P(I_2|I_0, I_1)}$$

$$= \frac{P(I_2)}{P(I_2|I_0, I_1)}\frac{P(I_2|A)}{P(I_2)}P(A|I_0, I_1).$$

This is equivalent to

$$P(A|I_0, I_1, I_2) = \alpha_2 P(A|I_0, I_1), \tag{4.17}$$

with the following substitutions and interpretations.

Revision factor:

$$\alpha_2 = \frac{\lambda_2}{\gamma_2}; \tag{4.18}$$

with the bounds

$$0 < \alpha_2 < \frac{1}{P(A|I_0, I_1)}.$$

Information factor:

$$\lambda_2 = \frac{P(I_2|A)}{P(I_2)}. \tag{4.19}$$

This measures the *predictive strength* of I_2 with respect to A. Its value admits interpretation parallel to that of λ_1.

Dependence factor:

$$\gamma_2 = \frac{P(I_2|I_0, I_1)}{P(I_2)}. \tag{4.20}$$

This measures the *stochastic dependence* of I_2 on (I_0, I_1) — specifically, the sign and the relative strength of SD between I_2 and (I_0, I_1). Its value admits interpretation parallel to that of γ_1.

Prior Assessment

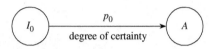

Figure 4.5 Stochastic dependence (SD) structures between event *A* (or hypothesis *A*) and information subsets I_0, I_1, I_2 under the CSI assumption, quantified in terms of the prior probability p_0, the information factors λ_1, λ_2, and the dependence factors γ_1, γ_2.

Stage 1 Revision

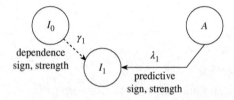

Stage 2 Revision

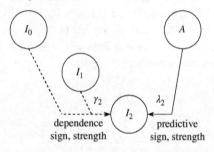

Conceptual revision algorithm

Figure 4.5 depicts the quantifications involved in revision stages $n = 1, 2$. Exploiting the pattern of equations, the sequential Bayesian revision model, with the CSI assumption about information subsets $(I_0, I_1, \ldots, I_N)$ and the decomposable revision factors, can be condensed into an algorithm. It is framed so that its steps can be mapped into a judgmental revision procedure. With the notation abbreviated to

$$p_0 = P(A|I_0),$$
$$p_n = P(A|I_0, I_1, \ldots, I_n), \qquad n = 1, \ldots, N,$$

every revision stage n $(n = 1, \ldots, N)$ involves these steps.

0. Input: p_{n-1}, I_n.
1. Determine information factor: $\lambda_n = P(I_n|A)/P(I_n)$.
2. Adjust the prior probability: $\hat{p}_n = \lambda_n\, p_{n-1}$.
3. Determine dependence factor: $\gamma_n = P(I_n|I_0, I_1, \ldots, I_{n-1})/P(I_n)$.
4. Readjust the prior probability: $p_n = \hat{p}_n/\gamma_n$.
5. Determine the revision factor: $\alpha_n = \lambda_n/\gamma_n$.
6. Output: p_n, α_n.

The algorithm can be executed numerically to obtain normative values of the factors $\lambda_n, \gamma_n, \alpha_n$. For this purpose, an additional input is the unconditional prior probability $P(A)$, assessed sans specific information. It is needed to calculate

$$P(I_n) = P(I_n|A^c)P(A^c) + P(I_n|A)P(A), \qquad n = 1, \ldots, N, \tag{4.21}$$

Table 4.3 Results for Example 4.10 of sequential revision of the prior probability $p_0 = P(A|I_0) = 0.3$ in two stages ($n = 1, 2$), through information factors λ_n and dependence factors γ_n, which create the revision factors $\alpha_n = \lambda_n/\gamma_n$.

n	1	(sign)	2	(sign)
$P(I_n|A)$	0.9		0.2	
$P(I_n)$	0.48		0.38	
$P(I_n|I_0, \dots, I_{n-1})$	0.41		0.3025	
λ_n	1.8750	(+)	0.5263	(−)
$\hat{p}_n$	0.5625		0.3466	
γ_n	0.8542	(−)	0.7961	(−)
p_n	0.6585		0.4354	
α_n	2.1951		0.6612	

which, in turn, is needed to calculate the factors λ_n and γ_n; however, equations (4.14)–(4.16) make it clear that $P(I_n)$ serves only the internal reasoning and has no influence on the numerical output: p_n, α_n. But there is a caveat: $\hat{p}_n > 1$ is possible! In such a case, λ_n should be decreased so that $\hat{p}_n = 1$, and γ_n should be decreased so that α_n remains unchanged.

Example 4.10 *(Revisions through factors)*

The input probabilities come from Example 4.7. In addition, $P(A) = 0.4$, as in Example 4.8. Results are reported in Table 4.3. In every revision stage, there are two adjustment steps. In Stage 1, the prior probability of 0.3 is first adjusted up to 0.563 based on the positive predictive strength of I_1, and is next readjusted up to 0.659 based on the negative dependence of I_1 on I_0. In Stage 2, the prior probability of 0.659 is first adjusted down to 0.347 based on the negative predictive strength of I_2, and is next readjusted up to 0.435 based on the negative dependence of I_2 on (I_0, I_1). The total adjustment yields the posterior probability

$$p_2 = \alpha_2\alpha_1 p_0 = 0.6612 \times 2.1951 \times 0.3$$
$$= 1.4514 \times 0.3 = 0.4354.$$

While this probability is the same as that calculated in Example 4.7, the reasoning that accompanies the calculations is different.

Compounding effect of evidence

Generalizing the last expression, the Bayesian revision model with the CSI information subsets takes the product form:

$$p_N = \alpha_N \cdots \alpha_2\alpha_1 p_0. \tag{4.22}$$

This simple mathematical expression embodies an ancient principle of jurisprudence:

> **Newman's principle.** When a hypothesis (e.g., the guilt of an accused person) is initially doubtful (p_0 close to 0), but becomes supported by many pieces of different circumstantial evidence ($\alpha_n > 1$, $n = 1, \dots, N$), then even if each piece of evidence is by itself insufficient to raise the prior belief decisively, their compounding effect may produce strong posterior belief (p_N close to 1).

Bayesian Revision Model

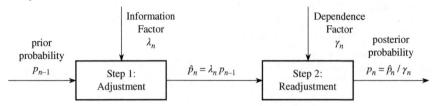

Judgmental Revision Procedure

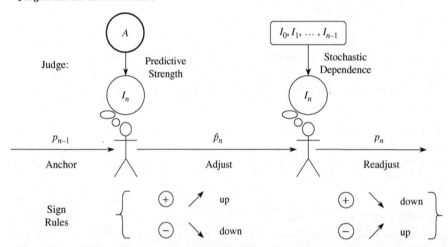

Figure 4.6 Schemata of two-step revision of the prior probability $p_{n-1} = P(A|I_0, I_1, \ldots, I_{n-1})$ into the posterior probability $p_n = P(A|I_0, I_1, \ldots, I_n)$ of event A, given new information I_n obtained at stage n of the sequential revision process: (a) Bayesian revision model that provides the normative guidance; (b) judgmental revision procedure that follows the guidance.

4.4.7 Judgmental Revision Procedure

When assessing quantities, individuals tend to follow heuristic rules, which simplify and structure the task. One such a heuristic is *anchoring and adjustment*. Remarkably, this heuristic is congruous with the Bayesian model: at revision stage n, the "anchor" (the initial estimate) is the prior probability p_{n-1}, the "adjustment" is the factor α_n, and the output is the posterior probability $p_n = \alpha_n p_{n-1}$. However, our model contains an innovation. By defining $\alpha_n = \lambda_n / \gamma_n$, we decompose the adjustment into two steps and offer normative guidance, schematized in Figure 4.6.

> **Probability adjustment principle**. At revision stage n, the prior probability $P(A|I_0, I_1, \ldots, I_{n-1})$ can be adjusted in two steps that reflect two separate influences. (i) The adjustment that reflects the *predictive strength* of new information I_n with respect to event A (factor λ_n): by itself, new information may suggest an upward (+) or downward (−) revision of the prior probability. (ii) The readjustment that reflects the *stochastic dependence* of new information I_n on the prior information $(I_0, I_1, \ldots, I_{n-1})$, which has already been utilized (factor γ_n): this dependence may be positive (+) or negative (−), suggesting a downward or upward readjustment, respectively.

To revise a prior probability judgmentally in accordance with the above principle at revision stage n ($n = 1, \ldots, N$), the forecaster should follow a six-step procedure.

Step 1. Review the prior probability p_{n-1} of event A, and the prior information $I_0, I_1, \ldots, I_{n-1}$.

Step 2. Gather, organize, and analyze the new information I_n. Judge its veracity.

Step 3. Judge the *predictive strength* of new information; form an overall judgment of the influence of new information on the degree of certainty about the occurrence of event A. (i) Does I_n increase or decrease p_{n-1}? (ii) By how much?

Step 4. Adjust the prior probability p_{n-1} to the new number $\hat{p}_n$. For example, $p_{n-1} = 0.7$ is adjusted upward to $\hat{p}_n = 0.8$.

Step 5. Analyze the *stochastic dependence* of the new information I_n on the prior information $I_0, I_1, \ldots, I_{n-1}$, which has already been utilized. Judge the strength and the sign $(+, -)$ of this dependence by posing questions. For example, given the prior information: Is the new information similar $(+)$ or dissimilar $(-)$? Is the new information unsurprising $(+)$ or surprising $(-)$? Is the new information more probable $(+)$ or less probable $(-)$ to exist than it would if prior information did not exist?

Step 6. Readjust the probability $\hat{p}_n$ to the new number p_n, based on the strength and the sign of the dependence:

decrease $\hat{p}_n$ if I_n is positively dependent $(+)$ on the prior information;

increase $\hat{p}_n$ if I_n is negatively dependent $(-)$ on the prior information.

For example, $\hat{p}_n = 0.8$ is readjusted downward to $p_n = 0.75$ when the dependence is positive $(+)$; it is readjusted upward to $p_n = 0.85$ when the dependence is negative $(-)$.

Step 7. Output the posterior probability p_n of event A.

Example 4.11   *(Medical diagnosis — Examples 4.7, 4.8, 4.10 continued)*

To illustrate a practical interpretation of the numbers and a probabilistic reasoning congruous with the Bayesian model, let us add context: imagine a novel disease A affecting some residents of a city.

Example 4.12   *(Bayesian revisions — Example 4.7 continued)*

Physicians observed that symptom I_0 preceded the onset of disease A in some of their patients, and on that basis assessed that, given symptom I_0, the patient had a 30% chance of being eventually diagnosed with disease A. Thereafter, another symptom I_1 was identified: it had been reported by 90% of patients who eventually got sick, and by 20% of patients who never got sick. When expressed as probabilities, these are the three inputs to Stage 1 revision. The two outputs inform us that: (i) 41% of patients who first report symptom I_0, later report symptom I_1; and (ii) 66% of patients complaining of both symptoms eventually are diagnosed to have the disease. *Hint*: Interpret similarly the inputs to, and the outputs from, Stage 2 revision.

Example 4.13   *(CSI does not imply SI — Example 4.8 continued)*

Disease A has spread, and now 40% of the city population carry it; thus the probability of a randomly met resident having the disease is $P(A) = 0.4$. The calculated marginal probability $P(I_1) = 0.48$ informs us that if the entire city population were surveyed, 48% would report experiencing symptom I_1. Interestingly, symptom I_1 occurs more frequently than disease A. Moreover, the two symptoms, I_0 and I_1, are not SI of each other because $P(I_1|I_0) = 0.41 \neq 0.48$. Whether or not the two symptoms are medically dependent (through some biochemical reactions) cannot be inferred from probabilities alone. More generally: stochastic dependence does not imply causation.

Example 4.14   *(Revisions through factors — Example 4.10 continued)*

The numbers from Table 4.3 are interpreted and connected with the steps of the judgmental revision procedure. (i) *Steps 3–4.* Per information factor λ_1, formula (4.15), symptom I_1 is more probable, (by a factor of 1.875) to occur in a person having disease A than in a randomly met resident. This positive dependence $(+)$ of I_1 on A justifies adjusting the prior probability of A upward, from 0.3 to 0.563. (ii) *Steps 5–6.* Per dependence factor γ_1, formula (4.16), symptom I_1 is less probable (by a factor of 0.854) to occur in a person experiencing symptom I_0 than in a randomly met resident. This negative dependence $(-)$ of I_1 on I_0 justifies readjusting the probability of A upward, from 0.563 to 0.659. Altogether, the Stage 1 revision illustrates the working of the sign rules (Table 4.2): the opposite signs, here $(+)$ and $(-)$, enhance the upward revision of the prior probability from 0.3 to 0.563 to 0.659. *Hint*: Interpret similarly the numbers from Table 4.3 for Stage 2 revision.

4.4.8 Likelihood Ratios for Judgmental Revision

A different judgmental revision procedure, utilizing the likelihood ratios, is sketched for the sake of completeness. Consider Stage 1 revision, equation (4.10) for the posterior probability of event A, and write a parallel equation for the posterior probability of the complementary event A^c:

$$P(A^c|I_0, I_1) = \frac{P(I_1|A^c)P(A^c|I_0)}{P(I_1|I_0)}. \tag{4.23}$$

Now, form the quotient of equations (4.10) and (4.23):

$$\frac{P(A|I_0, I_1)}{P(A^c|I_0, I_1)} = \frac{P(I_1|A)}{P(I_1|A^c)} \frac{P(A|I_0)}{P(A^c|I_0)}.$$

In the abbreviated form,

$$r_1 = l_1 r_0, \tag{4.24}$$

with the interpretations that extend the concepts from Section 4.2.2. To wit, the *prior odds on* event A are

$$r_0 = \frac{P(A|I_0)}{P(A^c|I_0)};$$

the *posterior odds on* event A are

$$r_1 = \frac{P(A|I_0, I_1)}{P(A^c|I_0, I_1)};$$

and the *likelihood ratio on* event A, given information I_1, is

$$l_1 = \frac{P(I_1|A)}{P(I_1|A^c)}.$$

The generalization to the N-stage revision process employs the abbreviated notation introduced in Section 4.4.6,

$$r_0 = p_0/(1 - p_0),$$
$$r_n = p_n/(1 - p_n),$$
$$l_n = P(I_n|A)/P(I_n|A^c),$$

and yields

$$r_n = l_n r_{n-1}, \qquad n = 1, \dots, N. \tag{4.25}$$

The posterior odds r_n can be transformed into the posterior probability p_n via an equation from Section 4.2.2.

In a judgmental revision procedure, the task of the forecaster at each revision stage n is to assess the likelihood ratio l_n in a manner parallel to the assessment of real number odds (Section 4.2.2, Step 4.2). First, judge whether information I_n is more likely to exist when the hypothesis is true (event A) or when the hypothesis is false (event A^c); say when A. Second, assess a real number $l_n > 1$ such that in your judgment "information I_n is l_n times more likely to exist when A is true than when A^c is true".

Example 4.15 *(Revisions through likelihood ratios)*

A human–computer system was designed for probabilistic forecasting of events (hypotheses regarding military actions by nations playing significant political roles on the world stage in 1975). Experts were tasked with

(i) interpreting information arriving sequentially from three sources (the radar system for detecting ballistic missiles, the intelligence system, the reconnaissance satellite system), and (ii) assessing the likelihood ratios $l_1, \dots, l_N$, with $N = 60$. A computer was coded to display information, to elicit assessments, and to calculate posterior probabilities $p_1, \dots, p_N$ (Edwards et al., 1968).

4.5   Analysis of Judgmental Task

4.5.1   Analysis of Events

We now return to Task 4.1.

Event R — rain (*repetitive event*). For any lead time greater than 14 days (the limit of weather predictability), it is reasonable to forecast R in terms of the *statistical probability*, which equals the relative frequency of rain on Thanksgiving day in Charlottesville. Estimated from a multi-year record of weather observations, $P(R) = 0.3$; meteorologists call it the *climatic probability*. Not knowing this, you may assess the *judgmental probability* $P(R)$ based on your recollection of geography.

Event G — grade (*unique event*). You will observe event G or its complement G^c only once. You do have partial control over G. You also possess some information about this course; and you possess experience from previous courses. Based on such considerations, you may assess probability $P(G)$ which reflects your current degree of uncertainty about the grade.

Event D — date (*factual event*). The first day of invasion, known as D-Day, was 6 June 1944. Thus the statement is false. If you know this, you should assign probability $P(D) = 0$. If you are uncertain, assess probability $0 < P(D) < 1$, reflecting the degree to which you believe the statement is true based on your recollection of history.

4.5.2   Analysis of Responses

Task 4.1 was performed by 159 students who took this course at the University of Virginia over 2 years. Results are summarized in Table 4.4.

Event R — rain (*repetitive event*). In each year, the range of the judgmental probabilities is wide, but the *group median probability* is equal to, or close to, the statistical probability $P(R) = 0.3$. This suggests that *combining* individually assessed probabilities into a group probability improves the forecast relative to the forecast prepared by a randomly selected group member. This phenomenon has been observed in many different forecasting experiments.

Table 4.4 Statistics of the judgmental probabilities assessed in Task 4.1, repeated twice in 2 years with groups of N students.

| Event | Year | Judgmental probability | | | | Actual fraction (ϕ) | N |
		Min	Median	Mean (p)	Max		
R	1	0.1	0.3	0.29	0.6		113
	2	0.1	0.4	0.39	0.7		46
G	1	0.2	0.5	0.55	0.9	0.42	113
	2	0.0	0.6	0.62	1.0	0.56	46
D	1	0.0	0.1	0.19	1.0		113
	2	0.0	0.2	0.35	1.0		46

Event G — grade (*unique event*). The assessed probability $P(G)$ reflects solely the individual judgment; its range is wide. Yet, the *group mean probabilities p* differ by only 0.07. And based on equation (4.2), p may be interpreted as the expected fraction of students whose grades will be B or higher. Subsequently, the grades from the midterm yielded the actual fraction ϕ. In each year, $p > \phi$, which means that the group was overconfident. But in year 2, ϕ differs from p by only 0.06, which means that the group was approximately *well calibrated*. A more refined definition of calibration is introduced in Chapter 6.

Event D — date (*factual event*). Out of all 159 students, 44% assigned probability $P(D) = 0$, thereby expressing their certainty of knowing the truth — correctly. Another 16% assigned probability 0.1, expressing near certainty. Again, in each year, the *group median probability*, 0.1 or 0.2, offers a better forecast of the truth than do probabilities $P(D) > 0.2$ assessed by 32% of all students. Moreover, 4% of all students assigned $P(D) = 1$, thereby claiming to know the truth with certainty — falsely. This exemplifies a violation of Cromwell's rule (Section 4.2.4). One should learn to avoid it — leave some room for doubt!

Historical Notes

Ramsey (1926), an English academic prodigy associated with the University of Cambridge (Misak, 2020), laid down philosophical and mathematical arguments for the theses that (i) probability theory is the logic of partial beliefs, and (ii) the degree of belief is measurable in terms of probability, which is the basis for decision.

Judgmental probabilities attached to some weather forecasts appeared in England and France around 1800 (Murphy, 1998). A system producing routinely nationwide probabilistic forecasts of events was deployed in the United States by the National Weather Service in 1965 (Murphy, 1981). In its latest version, dynamical–statistical models (run centrally every 6 hours) output probability of precipitation (PoP) occurrence in time intervals covering 10 days at grid points covering the contiguous United States. Then meteorologists at some 120 weather offices review and revise (if they judge it necessary) these PoPs based on local and recent information (not utilized in the models).

Bibliographical Notes

Cromwell's rule is stated after Dennis V. Lindley (1985); it bears the name of Oliver Cromwell (1599–1658), English general and statesman, who counseled: "think it possible you may be mistaken". Newman's principle is stated after Bruno de Finetti (1974); it bears the name of John Henry Newman (1801–1890), English cardinal and writer. The use of information (evidence) in probabilistic reasoning guided by Bayes' theorem is the subject of a comprehensive treatise by Schum (1994). The anchoring and adjustment heuristic comes from Tversky and Kahneman (1982). The expert in Example 4.1 is Marty Makary, professor at the Johns Hopkins School of Medicine (*The Wall Street Journal*, 19 February 2021). Example 4.6 is based on the article by Alan S. Blinder, former vice chairman of the Federal Reserve (*The Wall Street Journal*, 11 March 2020). Besides the models in Sections 4.4.6 and 4.4.8, there are other models that may aid in intuitive interpretation of the Bayesian revision of probability; they involve the "weights of evidence" (Good, 1950) and the "Bayes factors" (Bernado and Smith, 1994).

Exercises

1 **Degree of certainty** (geologists). A geologist interprets lithospheric maps and seismic records and then assigns probability to the hypothesis that a well to be drilled at a specified location will yield oil; the probabilities assigned to five locations are 0.5, 0.6, 0.4, 0.3, 0.7. This task is next repeated by another geologist having about the same knowledge but different experience; the probabilities assigned to the same five locations are

0.6, 0.7, 0.1, 0.2, 0.9. Which geologist, the first or the second, expressed a higher degree of certainty? Justify your answer.

2 **Degree of certainty** (radiologists). A radiologist examines the magnetic resonance imaging (MRI) of the chest of a patient suspected of having lung cancer. When he detects features that are atypical, such as lesions or nodules, but indeterminate, he assigns a probability that they will develop into cancerous ones. Then another radiologist performs an independent interpretation and probability assignment. For five such indeterminate MRIs, the probabilities assigned by the first radiologist are 0.7, 0.4, 0.2, 0.1, 0.9, whereas the probabilities assigned by the second radiologist are 0.6, 0.5, 0.8, 0.3, 0.4. Which radiologist, the first or the second, expressed a higher degree of certainty? Justify your answer.

3 **Odds on event** (radiologist). Each time a physician suspects a condition whose diagnosis could be aided by an x-ray image, he refers the patient to a radiologist. The radiologist supplies an x-ray image, his diagnosis of the condition, and his judgmental probability that the diagnosis is correct. For six different patients these probabilities are 0.9, 0.5, 0.8, 0.99, 0.75, 0.95. Transform each probability into the *odds on* the diagnosis being correct.

4 **Odds on event** (runner). Before each 10-km race, a runner conceives winning it — an uncertain event due to a combination of uncontrollable circumstances (weather, initial pace, race tactics of opponents, etc.). He judges that for every m combinations favoring him, there are n combinations favoring other runners. Then he transforms this judgment into the probability of winning. For six consecutive races these probabilities are 0.375, 0.7, 0.45, 0.65, 0.8, 0.875. For each probability, uncover the *odds on* winning, m/n, as judged originally by the runner.

5 **Odds against event** (cities). During competition for rights to host the 2020 Summer Olympic Games, a bookmaker assigned *odds against* each city being the winner (*The Wall Street Journal*, 17 February 2012): Doha, Qatar, at 6/1; Baku, Azerbaijan, at 8/1; Tokyo, Japan, at 6/4; Madrid, Spain, at 9/4; Istanbul, Turkey, at 4/1.
   5.1 Transform each odds into the probability of a city being the winner.
   5.2 Did the bookmaker provide a valid quantification of uncertainty? (Specifically, do the probabilities calculated from the odds satisfy the axioms of probability?) Justify your answer.
   Note. A bookmaker usually writes 6–4 to mean the odds against an event are 6 to 4. We write 6/4 because, mathematically, the odds is a quotient.

6 **Odds against event** (horses). Before his final career race, the $1\frac{1}{4}$-mile Breeder's Cup Classic in October 2015, the Triple Crown winner, American Pharoah, was the favorite, with *odds against* 4/5 (see note below Exercise 5). Other horses to run were Keen Ice at 8/1, Honor Code at 5/1, and Tonalist at 4/1 (*The Wall Street Journal*, 30 October 2015).
   6.1 Transform each odds into the probability of a horse being the winner.
   6.2 Did the bookmaker provide a valid quantification of uncertainty? (Specifically, do the probabilities calculated from the odds satisfy the axioms of probability?) Justify your answer.

7 **Repetitive events**. In each of the following events, the day means the 24-h period beginning at 0600 Eastern Time. Event M: the Sunday of the next Memorial Day weekend in Indianapolis, Indiana, will be without rainfall. Event T: the next Thanksgiving Day in Miami, Florida, will be rainy. Event N: the next New Year's Day in Buffalo, New York, will be marked by snowfall.

7.1 Assess judgmentally your probability of each event. Rely on your general knowledge; do not gather information.

7.2 Do not read this paragraph until you complete Exercise 7.1. The probability of each event was estimated as the relative frequency of that event in a climatic sample. This sample consisted of precipitation measurements during 7 years, 1997–2003, and included all days from the month in which the event day falls; there were missing measurements. The estimates are:

$$P(M) = 75/(75 + 74),$$
$$P(T) = 44/(93 + 44),$$
$$P(N) = 98/(94 + 98).$$

For each event, compare this "climatic probability" with your judgmental probability. Draw conclusions.

8 Unique events. Consider four events defined by the following statement. Your grade from this course will be ...
A minus or higher (event A).
B minus or higher (event B).
C minus or higher (event C).
D minus or higher (event D).

8.1 State the inequalities which the probabilities of these events must satisfy.

8.2 Assess judgmentally your probability of each event.

9 Probabilistic test. The table specified in the option lists 16 hypotheses. A hypothesis is either true or false.

9.1 To each hypothesis assign a probability that reflects your degree of certainty that the hypothesis is true. Use only probabilities from the set $\{0.1, 0.3, 0.5, 0.7, 0.9\}$. Rely on your general knowledge; do not gather information.

9.2 Do not read this paragraph until you complete Exercise 9.1. For each hypothesis, find out whether it is true or false. The code is as follows: in the sequence of symbols

$$T, F, T, T, T, F, F, T, F, F, T, T, F, T, F, F,$$

T denotes true and F denotes false; the corresponding sequence of hypotheses, identified by their numbers, and beginning with hypothesis $n = \binom{5}{2} + 1$, is

$$n, n + 1, \ldots, 16, 1, \ldots, n - 1.$$

9.3 Summarize the results in a contingency table. (The format of a contingency table is shown in Section 6.1.4. The indicator w should be interpreted thusly: $w = 0 \Leftrightarrow$ the hypothesis is false; $w = 1 \Leftrightarrow$ the hypothesis is true.) Store the table for the purpose of verification of your assessments (Exercise 1 in Chapter 6).

Options: (A) Table 4.5.
(B) Table 4.6.

10 Forecast of fraction. Prove the theorem from Section 4.3. Here are a few hints. For each $n \in \{1, \ldots, N\}$, introduce a Bernoulli variate (an indicator variate) such that $W_n = 1$ if A_n occurs, and $W_n = 0$ if A_n^c occurs.

10.1 Write and justify the expressions for $E(W_n)$ and $Var(W_n)$.

10.2 Express X as a function of $W_1, \ldots, W_N$. Next, write and justify the expressions for $E(X)$ and $Var(X)$.

10.3 Define the fraction $Z = X/N$, and derive its mean $E(Z)$ and variance $Var(Z)$.

Table 4.5 General knowledge hypotheses.

1. Vilnius is the capital of Norway.
2. Herodotus visited Egypt in the fifth century BC.
3. A viola is smaller than a violin.
4. The opera *The Barber of Seville* was composed by Verdi.
5. Venus is closer to the Sun than Earth is.
6. The first probabilistic forecast of rain was made in Australia in 1905.
7. Malta is a peninsula in the Adriatic Sea.
8. There is a mission church built by Spaniards in Tucson, Arizona.
9. A whiffletree is a bush growing in eastern Ontario, Canada.
10. Venice, Italy, lies south of Chicago, Illinois.
11. Excelsior was the name of an American motorcycle.
12. Rio de Janeiro, Brazil, lies west of New York City, USA.
13. Stradivari was an Italian violin maker.
14. Nicolaus Copernicus was born in the second half of the fifteenth century.
15. Nikola Tesla was an American electrician and inventor.
16. A gargoyle is an alpine flower.

Table 4.6 General knowledge hypotheses.

1. The Mesa Verde National Park lies in Utah.
2. An American woman won the 2017 New York City Marathon.
3. Police radar cannot clock a car's speed from behind.
4. Achilles' heel means speed and agility.
5. The first Thanksgiving in America was observed at Berkeley, Virginia.
6. Europeans learned to cultivate and eat potatoes before North Americans.
7. A bathyscaphe is a research vehicle for exploring planets.
8. A siesta is customary in the hot climate of Latin America.
9. The largest city in Texas is Dallas.
10. Massachusetts lies north of Montana.
11. Flappers is a nickname of young women in the 1920s who defied convention.
12. Potato famine in Ireland happened in the 1740s.
13. One of the first stars of rock 'n' roll was Elvis Presley.
14. Aphrodite was a Greek goddess whose Roman name is Venus.
15. A kayak is styled after an Eskimo canoe.
16. A puffin is a type of Australian kangaroo that puffs while leaping.

11 **Verification of forecasts**. A high school senior applied to 9 colleges, and then asked a counselor to assess his chance of being admitted to each. The counselor assigned the probabilities: 0.2, 0.8, 0.7, 0.3, 0.4, 0.9, 0.6, 0.5, 0.8. The admission decisions (A—admitted; R—rejected), in the order of probabilities, were R, A, A, R, R, A, R, A, A. The number of forecasts (just 9) is too small to verify them comprehensively according to the Bayesian verification theory of Chapter 6.

   11.1 Can the theorem of Section 4.3 be applied to verify the goodness of these forecasts in-the-large, that is, in terms of expectations? If so, then present your rationale.

   11.2 Apply the theorem, report the results, and explain their meaning to the counselor.

   11.3 Judge the goodness of the counselor's forecasts. Overall, did these forecasts capture the uncertainty associated with the admission decisions?

12 **Stochastic independence**. Prove that the three equalities specified in the option are equivalent (i.e., one holds if and only if the other two hold).

   Options: (A) $P(A|B) = P(A)$, $\quad P(B|A) = P(B)$, $\quad P(A|B^c) = P(A)$.

   (B) $P(A|B^c) = P(A)$, $\quad P(B^c|A) = P(B^c)$, $\quad P(B^c|A^c) = P(B^c)$.

   (C) $P(A^c|B) = P(A^c)$, $\quad P(B|A^c) = P(B)$, $\quad P(A^c|B^c) = P(A^c)$.

   (D) $P(A^c|B^c) = P(A^c)$, $\quad P(B^c|A^c) = P(B^c)$, $\quad P(A^c|B) = P(A^c)$.

13 **Order of information subsets**. Revisit Example 4.7. Reverse the order in which the information subsets are utilized in the two revision stages. Calculate all output probabilities. Compare them with those calculated in the example. Draw a conclusion.

14 **Limiting information subsets** (stage 1). Consider the revision of a prior probability $P(A|I_0)$ under the assumption that information subsets (I_0, I_1) are CSI relative to events A, A^c. Prove the following statements.

   14.1 If subset I_1 is perfect for event A, then

   $$P(I_1|I_0) = P(A|I_0) \text{ or } P(A^c|I_0),$$
   $$P(A|I_0, I_1) = 1 \text{ or } 0.$$

   14.2 If subset I_1 is uninformative for event A, then it is uninformative for A^c, and

   $$P(A|I_0, I_1) = P(A|I_0).$$

15 **Limiting information subsets** (stage 2). Consider the revision of a prior probability $P(A|I_0, I_1)$ under the assumption that information subsets (I_0, I_1, I_2) are CSI relative to events A, A^c. Prove the following statements.

   15.1 If subset I_2 is perfect for event A, then

   $$P(I_2|I_0, I_1) = P(A|I_0, I_1) \text{ or } P(A^c|I_0, I_1),$$
   $$P(A|I_0, I_1, I_2) = 1 \text{ or } 0.$$

   15.2 If subset I_2 is uninformative for event A, then it is uninformative for A^c, and

   $$P(A|I_0, I_1, I_2) = P(A|I_0, I_1).$$

16 **Revision factors** (interest rate). The Federal Reserve sets the benchmark interest rate, which propagates through the economy and ultimately affects individuals via the prices of consumer goods and the interest rates the banks charge on loans and pay on deposits. Futures markets forecast the decision that the Federal Open Market Committee (FOMC) will make at its next meeting. Before the meeting scheduled for 29–30 October 2019, the Chicago Mercantile Exchange Group assessed daily the probability of a quarter-percentage-point rate cut (the complementary event being no rate cut). Listed below are forecast probabilities on the dates of

turning points (transitions from a downward trend to an upward trend, or vice versa) of the daily sequence of probabilities published in *The Wall Street Journal* on 30 October 2019.

Forecasts:

24 Sept. 0.63,	3 Oct. 0.88,	8 Oct. 0.82,	16 Oct. 0.89,
30 Sept. 0.40,	7 Oct. 0.75,	11 Oct. 0.67,	25 Oct. 0.93.

16.1 Infer the values of the revision factors, starting from the unconditional prior probability equal to 0.5, valid in mid-September.

16.2 Graph the sequence of the revision factors and the sequence of the forecast probabilities. Comment on their behavior as functions of the decreasing lead time (as if you were explaining these graphs to a client).

Note: The FOMC did vote, 8 to 2, to lower the benchmark interest rate by a quarter point to the range between 1.5% and 1.75%.

17 **Revision factors** (economic recession). Revisit Example 4.6 with three forecasts of the economic recession, and analyze it as follows.

17.1 Infer the values of the revision factors.

17.2 Focus on the available information subsets. (i) Judge the sign of the stochastic dependence between I_1 and I_0, and between I_2 and (I_0, I_1). Is this dependence positive, or negative, or null (the subsets are stochastically independent)? (ii) Assess judgmentally the dependence factors. Rely on the descriptions of I_0, I_1, I_2 and on your general knowledge.

17.3 Based on the above results, infer the values of the information factors.

17.4 Graph the sequences of the three factors (information, dependence, revision) and of the forecast probabilities p_0 and $\{(\hat{p}_n, p_n) : n = 1, 2\}$. Comment on their behavior (as if you were explaining these graphs to a client).

18 **Revision factors** (drug trials). A new drug developed by a pharmaceutical company must pass a sequence of clinical trials before it can be approved by the US Food and Drug Administration (FDA). A perfect forecast of a drug's success would benefit the company, the investors, and the patients. In reality, there exists huge uncertainty: it evolves in stages and over many years, causing volatility of the company's stock price and posing risk to the investors.

Stage 0 The company collects preclinical information about the new drug, including results of studies on models and animals, then applies to the FDA for permission to conduct clinical trials.

Stage 1 The goal is to demonstrate safety, usually on a small cohort of sick and healthy individuals. Of all drugs that enter this stage, 70% pass it.

Stage 2 The goal is to establish efficacy and identify side effects. The trials are usually randomized, involve several hundred individuals, and last up to 2 years. Of all drugs that enter this stage, 33% pass it.

Stage 3 The goal is to validate efficacy and monitor side effects in a large cohort of patients over several years. Of all drugs that enter this stage, 26% pass it. (Data source: *The Wall Street Journal*, 26 December 2019.)

18.1 Let event A be passing all three stages of the clinical trials. Calculate the prior probability of A known at stage 0, and the two posterior probabilities of A, after passing stage 1 and after passing stage 2.

18.2 Calculate the two revision factors.

18.3 Suppose a company contemplates an investment in preclinical testing in order to increase the posterior probability of A before entering stage 3 trial (the longest and costliest) to p_2^*. Assuming the same revision factors as calculated above, what is the smallest value of the prior probability of A necessary to reach $p_2^* = 0.55$, $p_2^* = 0.8$?

19 **Risk of heart attack**. In a population of adults 65–74 years old, comprising 40% of men (M) and 60% of women (W), 2 out of 100 men experience a heart attack within a year (event A); the chance of A for a woman is half that for a man. A normal product of protein metabolism in the body is homocysteine, but its elevated level (information H) is associated with A: it is found in 27% of adults who experience A, but only in 12% of adults who do not experience A.

19.1 What is the probability of A for an adult in that population?

19.2 What is the probability of A for (i) a man with H, (ii) a woman with H?

19.3 What is the value of the revision factor that updates the probability: (i) from $P(A|M)$ to $P(A|M,H)$; (ii) from $P(A|W)$ to $P(A|W,H)$?

20 **Detective work**. Apply the Bayesian revision model (define notation, write the equations, perform the calculations) to aid a detective in the following case. A jewelry store had been robbed by a man. Based on initial investigation, the detective identified three suspects, each equally likely to be the thief.

20.1 The court approved three search warrants, but only two were executed because suspect A was traveling. Of the two suspects searched and interrogated, one revealed nothing and one produced an alibi, confirmed to be true. What is the probability of suspect A being the thief, given information gathered thus far?

20.2 The apartment of suspect A was searched upon his return, and pieces of jewelry were found. The detective's assessments: (i) conditional on the hypothesis that A is the thief, the probability of finding this evidence is $7/8$; (ii) conditional on the hypothesis that A is not the thief, the probability of finding these jewelry pieces, which are not unusual, in the apartment he occupies with his wife is $1/4$. What is the probability of suspect A being the thief, given all information gathered till now?

21 **Default on debt**. A bond is an interest-bearing certificate of indebtedness, the holder of which owns the debt of the issuer. A bond with a put gives the holder an option to sell it back to the issuer at par (at face value) before or on a specified date. Imagine working as an analyst for an investment fund that owns B-rated bonds of a large retailer, RRR, with puts that can be exercised in 7 months, on 31 January. Your principals are considering this option because they begin to doubt the ability of RRR to repay the debt. You are tasked with assessing the probability of RRR defaulting on the bonds (event A). The historical annual rate of default on B-rated bonds is known. The mid-year financial report (I_0) just released by RRR suggests a strain on their cash flow. On that basis, you assessed $P(A|I_0)$. Next you determined the information that may help you revise this probability.

Information I_1 By the end of August, you will learn from suppliers the sizes of orders placed for the holiday season; very large orders will suggest that RRR may be overspending, thereby increasing the likelihood of default.

Information I_2 By the end of November, you will learn whether RRR's Black Friday sales were successful; if not, then the default will become more likely.

Information I_3 By early January, you will track the volume of the returned merchandise and judge its impact on RRR's revenue from holiday sales; huge return volume will make the default more likely.

Given the input probabilities specified in the option, do the following.

21.1 Calculate the sequence of the posterior probabilities from equations of the Bayesian revision model (total probability law, Bayes' theorem).

21.2 Calculate and graph: (i) the sequences of the three factors (information, dependence, revision); (ii) the sequence of the probabilities (prior, adjusted, readjusted).

21.3 Comment on the behavior of each graph (as if you were explaining the graph to your principals). *Hint*: Explain separately (i) the evolution of uncertainty in time, and (ii) the influence of information subsets on probability revisions.

Options for the input:

(A) $P(A) = 0.10$, $P(A|I_0) = 0.3$, $P(I_1|A) = 0.6$, $P(I_1|A^c) = 0.3$;
$P(I_2|A) = 0.2$, $P(I_2|A^c) = 0.5$;
$P(I_3|A) = 0.7$, $P(I_3|A^c) = 0.4$.

(B) $P(A) = 0.05$, $P(A|I_0) = 0.5$, $P(I_1|A) = 0.6$, $P(I_1|A^c) = 0.5$;
$P(I_2|A) = 0.8$, $P(I_2|A^c) = 0.5$;
$P(I_3|A) = 0.6$, $P(I_3|A^c) = 0.4$.

(C) $P(A) = 0.07$, $P(A|I_0) = 0.4$, $P(I_1|A) = 0.6$, $P(I_1|A^c) = 0.7$;
$P(I_2|A) = 0.8$, $P(I_2|A^c) = 0.4$;
$P(I_3|A) = 0.5$, $P(I_3|A^c) = 0.3$.

22 **Choice of information**. Suppose you assessed $P(A|I_0)$ of some event A with the lead time of 3 months. To update this probability, you plan to acquire additional information after each passing month. Two sources are available but you can afford only one; the first one can supply I_1 and I_3; the second one can supply I_2 and I_4. The content of each subset is unknown until it is received, but based on experience with the past performance of each source, you assumed that, conditional on A (on A^c), each subset I_n, $n \in \{1, 2, 3, 4\}$, is stochastically independent of the other three subsets. Perform the analyses whose results would aid you to choose the source. Justify your analyses and your choice of the source.

Options for the conditional probabilities:

(A) $P(A|I_0) = 0.3$;
$P(I_1|A) = 0.6$, $P(I_1|A^c) = 0.5$; $P(I_3|A) = 0.1$, $P(I_3|A^c) = 0.4$;
$P(I_2|A) = 0.7$, $P(I_2|A^c) = 0.2$; $P(I_4|A) = 0.9$, $P(I_4|A^c) = 0.8$.

(B) $P(A|I_0) = 0.7$;
$P(I_1|A) = 0.5$, $P(I_1|A^c) = 0.4$; $P(I_3|A) = 0.9$, $P(I_3|A^c) = 0.6$;
$P(I_2|A) = 0.8$, $P(I_2|A^c) = 0.3$; $P(I_4|A) = 0.2$, $P(I_4|A^c) = 0.1$.

(C) $P(A|I_0) = 0.5$;
$P(I_1|A) = 0.2$, $P(I_1|A^c) = 0.4$; $P(I_3|A) = 0.9$, $P(I_3|A^c) = 0.3$;
$P(I_2|A) = 0.7$, $P(I_2|A^c) = 0.1$; $P(I_4|A) = 0.8$, $P(I_4|A^c) = 0.6$.

Mini-Projects

23 **Track meet**. Before the track meet, a sports expert $m \in \{1, 2, 3, 4\}$ issues a judgmental forecast in terms of the probability $p_{jn}(m)$ of the event: a runner from college $j \in \{1, \ldots, 6\}$ will win the race over distance $n \in \{1, \ldots, 7\}$; the distances are: 100 m, 200 m, 400 m, 1000 m, 3 km, 5 km, 10 km. The forecasting procedure is this: for each distance n, the expert identifies the most likely winner j^*, assigns to it probability q, and then distributes probability $1 - q$ uniformly among the rest of the competitors (Table 4.7). The mean of probabilities $p_{jn}(m)$ assigned by the participating experts is taken as the group probability p_{jn}. The group probabilities

Table 4.7 Probabilistic forecast for a track meet by expert $m \in \{1, 2, 3, 4\}$: college $j^* \in \{1, \dots, 6\}$ whose runner is most likely to win over distance $n \in \{1, \dots, 7\}$, and the probability q of that win.

m	Forecast element	n						
		1	2	3	4	5	6	7
1	j^*	2	6	3	5	2	5	4
	q	0.6	0.7	0.5	0.8	0.6	0.4	0.9
2	j^*	3	5	3	2	3	2	3
	q	0.7	0.6	0.6	0.7	0.5	0.5	0.7
3	j^*	3	1	5	5	3	2	6
	q	0.5	0.6	0.4	0.6	0.7	0.6	0.8
4	j^*	3	6	3	5	2	2	6
	q	0.4	0.8	0.7	0.8	0.7	0.7	0.4

are used to estimate the number X_j of races, and the fraction Z_j of races, which will be won by college j. An option for the meet specifies the participating experts, the competing colleges, and the distances to be run. *Options for the meet*:

(A) $m = 1, 2$; $j = 2, 3, 4, 5$; $n = 1, 2, 3, 4, 5, 6$.
(B) $m = 3, 4$; $j = 2, 3, 5, 6$; $n = 1, 3, 4, 5, 6, 7$.
(C) $m = 1, 2, 3$; $j = 1, 2, 3, 4, 5$; $n = 2, 3, 4, 5, 6$.
(D) $m = 2, 3, 4$; $j = 1, 2, 3, 4, 6$; $n = 2, 3, 4, 6, 7$.
(E) $m = 1, 2, 4$; $j = 1, 2, 3, 5, 6$; $n = 1, 2, 4, 5, 6$.
(F) $m = 1, 3, 4$; $j = 2, 3, 4, 5, 6$; $n = 1, 3, 4, 5, 7$.
(G) $m = 1, 2, 3, 4$; $j = 1, 2, 3, 4, 5, 6$; $n = 1, 2, 3, 4, 5, 6, 7$.

23.1 Based on the specification for the meet, construct an array $\{p_{jn}(m)\}$ with forecast probabilities assigned by the experts, and an array $\{p_{jn}\}$ with the group probabilities.

23.2 For each competing college j, calculate the expectation and the standard deviation of X_j and of Z_j. Rank the colleges based on the expected number of wins.

23.3 Suppose the meet will consist of two races over every specified distance, and the forecast probabilities are identical for each race. Repeat Exercise 23.2.

23.4 Compare the uncertainty about X_j (the number of races to be won) and the uncertainty about Z_j (the fraction of races to be won) in the two forecasts from Exercises 23.2 and 23.3. Explain which predictand is less uncertain and why.

24 **Ticket lottery.** A college distributes gratis half of the tickets to the last home basketball game through a lottery. The campus paper reported the history and statistics of the lottery. First, it reported that of all people who apply for tickets only 10% win them, but a student (I_0) has a 30% chance of winning. Second, it reported that senior students (I_1) constitute 60% of all applicants getting the tickets, and 20% of all applicants not getting the tickets. Third, it reported that 50% of all applicants getting tickets are varsity members (I_2); among applicants not getting tickets, 10% are varsity members. Four friends have applied for the tickets, identifying themselves as (I_0, I_1, I_2), (I_0, I_1^c, I_2), (I_0, I_1, I_2^c), (I_0, I_1^c, I_2^c), where the last one is a student, but neither a senior nor a varsity member.

24.1 For each student, calculate the sequence of the posterior probabilities from equations of the Bayesian revision model (total probability law, Bayes' theorem).

24.2 For each student, calculate and graph: (i) the sequences of the three factors (information, dependence, revision); (ii) the sequence of the probabilities (prior, adjusted, readjusted).

24.3 Select the information set, either (I_0, I_1^c, I_2) or (I_0, I_1, I_2^c), and interpret every number in the set $\{(\lambda_n, \gamma_n, \alpha_n; \hat{p}_n, p_n) : n = 1, 2\}$, as if you were explaining it to the student.

24.4 Calculate the probabilities of events: a combination of X friends out of four will win the tickets, $X \in \{0, 1, 2, 3, 4\}$. Calculate the expected number of friends who will win the tickets. Calculate the expected fraction of friends who will win the tickets.

25 **Activity success**. A company conducting research and development (R&D) evaluates each candidate project before deciding whether or not to pursue it. The evaluation includes a forecast of the technical success of the R&D project. Success means achieving precisely defined technical outcomes. The probability of success is assessed judgmentally by a team (e.g., the originator of the project, the R&D manager, an expert not involved in the project). In this vein, forecast your academic success this semester. Specifically, set yourself a target grade T, such that if you complete each course with a grade of at least T, then you will be content and declare the semester successful (event A).

25.1 Gather, organize, and analyze information about each course you are taking — information that is pertinent to predicting your grade. Select 3 most difficult (challenging) courses, and order them: the easiest (I_1), the more difficult (I_2), the most difficult (I_3). Denote by I_0 information about the other courses, and any additional information pertinent to predicting your grades.

25.2 Based on your knowledge of yourself, your experience in the college, and information subset I_0, assess the probability $P(A|I_0)$. *Hint*: The conditioning on I_0 means, *inter alia*, that you are taking only the other courses.

25.3 Follow the six-step procedure to revise your prior probability $P(A|I_0)$ sequentially, incorporating one information subset I_n at each stage $n \in \{1, 2, 3\}$. Your final posterior probability should be $P(A|I_0, I_1, I_2, I_3)$.

25.4 Graph the sequence of the probabilities (prior, adjusted, readjusted). Calculate and graph the sequences of the three factors (information, dependence, revision). Comment on the behavior of each graph (as if you were explaining its meaning to your peer).

26 **Sequential revision problem**. Come up with a sequential forecasting problem to which the Bayesian revision model with $N \geq 3$ is applicable.

26.1 Describe the problem; define the forecasted event A, the required forecast lead times, and the available information subsets $I_0, I_1, \ldots, I_N$.

26.2 Assign plausible numbers to all input probabilities, assuming that the information subsets are conditionally stochastically independent, relative to events A, A^c. Comment on the reasonableness of this assumption.

26.3 Calculate the sequence of the posterior probabilities from equations of the Bayesian revision model (total probability law, Bayes' theorem).

26.4 Calculate and graph: (i) the sequences of the three factors (information, dependence, revision); (ii) the sequence of the probabilities (prior, adjusted, readjusted). Comment on the behavior of each graph (as if you were explaining the graph to a client).

Hint: To ensure the coherence between these calculations and those in Exercise 26.3, assign a number to $P(A)$, and use the total probability law to obtain $P(I_n)$, $n = 1, \ldots, N$.

26.5 Which form of calculations, those in Exercise 26.3 or those in Exercise 26.4, would you present to a client in order to explain (i) the evolution of uncertainty, and (ii) your rationale behind the assessed probabilities? *Note*: Assume that the client has some knowledge of probability, but is unfamiliar with the Bayesian theory of forecasting.

5

Statistical Forecasting

A *repeatable event* (such as occurrence of rain, presence of a defect in a manufacture, decline of a commodity price) is to be forecasted with a specified *lead time*. A record of past occurrences of the event exists; a record of past values of a *predictor* — a variate containing some predictive information about the event — exists also. How can these records be used to build a statistical model for forecasting (i.e., for quantifying the uncertainty about) future occurrences of the event?

This chapter lays down a Bayesian theory for statistical forecasting of a binary predictand. First, the theoretic structure of the *Bayesian forecaster* is derived. Subsequently, procedures are presented for model development, parameter estimation, and predictor selection.

5.1 Bayesian Forecaster

5.1.1 Variates

Let W be the *predictand* — a variate whose realization must be forecasted. It is a binary variate having the sample space $\{0, 1\}$; it serves as the indicator of an event of interest, with $W = 1$ if and only if the event occurs, and $W = 0$ otherwise. Its realization is denoted w, where $w \in \{0, 1\}$.

Let X be the *predictor* — a variate whose realization provides information that reduces uncertainty about the event of interest. Suppose X is a continuous variate having the sample space $\mathcal{X}$; its realization is denoted x, where $x \in \mathcal{X}$.

5.1.2 Input Elements

With P denoting a probability and p denoting a generic density function, the inputs into the Bayesian forecaster are defined as follows.

$g = P(W = 1)$ is the *prior probability* of event $W = 1$. The prior probability g quantifies the uncertainty about the predictand W that exists before a realization of the predictor X is available. Equivalently, it characterizes the natural variability of the predictand.

$f_w(x) = p(x|W = w)$ for all $x \in \mathcal{X}$ and for $w = 0, 1$. For a fixed $w \in \{0, 1\}$, the object f_w is the *density function* of predictor X, **conditional on the hypothesis** that the event is $W = w$. For a fixed realization $X = x$ of the predictor, the object $f_w(x)$ is called the *likelihood* of event $W = w$. The pair of values $(f_0(x), f_1(x))$ constitutes the *likelihood function* of the predictand W, with $f_0(x)$ being the likelihood of event $W = 0$, and $f_1(x)$ being the likelihood of event $W = 1$. All possible realizations x of the predictor (and there are infinitely many of them), generate a *family of likelihood functions*, denoted by (f_0, f_1). The family of likelihood functions quantifies the stochastic dependence between predictor X and predictand W. Equivalently, it characterizes the informativeness of the predictor with respect to the predictand. The term "informativeness" is defined precisely in Section 5.5.

Probabilistic Forecasts and Optimal Decisions, First Edition. Roman Krzysztofowicz.
© 2025 John Wiley & Sons Ltd. Published 2025 by John Wiley & Sons Ltd.
Companion website: www.wiley.com/go/ProbabilisticForecastsandOptimalDecisions1e

5.1.3 Output Elements

The outputs from the Bayesian forecaster are defined as follows.

$\kappa(x) = p(x)$ for all $x \in \mathcal{X}$. The function κ is the *expected density function* of predictor X; it may also be called the marginal density function of X.

$\pi(x) = P(W = 1|X = x)$ is the *posterior probability* of event $W = 1$, **conditional on a realization** of the predictor $X = x$, where $x \in \mathcal{X}$. The function $\pi : \mathcal{X} \rightarrow (0, 1)$, which maps the sample space $\mathcal{X}$ into the probability interval $(0, 1)$, constitutes the solution to this forecasting problem. Once π is obtained, W may be forecasted repeatedly: given realization $x \in \mathcal{X}$ of the predictor, the posterior probability of event $W = 1$ is simply calculated as $y = \pi(x)$, yielding $y \in (0, 1)$.

5.1.4 Theoretic Structure

The prior probability g and the family of likelihood functions (f_0, f_1) carry information about the prior uncertainty and the informativeness of the predictor into the Bayesian forecaster — the name of the general solution for function π. It is derived in two steps. First, the *expected density function* of the predictor X is obtained from the *total probability law*:

$$p(x) = p(x|W = 0)P(W = 0) + p(x|W = 1)P(W = 1);$$

in the abbreviated notation,

$$\kappa(x) = f_0(x)(1 - g) + f_1(x)g, \qquad x \in \mathcal{X}. \tag{5.1}$$

Second, the *posterior probability* of event $W = 1$, conditional on a realization of the predictor $X = x$, is obtained from *Bayes' theorem*:

$$P(W = 1|X = x) = \frac{p(x|W = 1)P(W = 1)}{p(x)};$$

in the abbreviated notation,

$$\pi(x) = \frac{f_1(x)g}{\kappa(x)}, \qquad x \in \mathcal{X}, \kappa(x) > 0. \tag{5.2}$$

When expression (5.1) is inserted into equation (5.2), one obtains

$$\pi(x) = \frac{f_1(x)g}{f_0(x)(1 - g) + f_1(x)g}.$$

When both the numerator and the denominator are divided by $f_1(x)g$, one obtains an alternative equation, called the *Bayesian forecaster*:

$$\pi(x) = \left[1 + \frac{1 - g}{g} \frac{f_0(x)}{f_1(x)} \right]^{-1}. \tag{5.3}$$

The posterior probability, $y = \pi(x) = P(W = 1|X = x)$, constitutes a probabilistic forecast of the event of interest, given a realization of the predictor. Equation (5.3) reveals that the posterior probability is determined by the product of two quotients. The first quotient,

$$\frac{1 - g}{g} = \frac{P(W = 0)}{P(W = 1)}, \tag{5.4}$$

is called the *prior odds against* event $W = 1$; alternatively, it may be called the *prior odds on* event $W = 0$. The larger these prior odds, the smaller the posterior probability y of event $W = 1$. The second quotient,

$$\frac{f_0(x)}{f_1(x)} = \frac{p(x|W = 0)}{p(x|W = 1)},\tag{5.5}$$

is called the *likelihood ratio against* event $W = 1$, **conditional on a realization** of the predictor $X = x$; alternatively, it may called the *likelihood ratio on* event $W = 0$. The larger this likelihood ratio, the smaller the posterior probability y of event $W = 1$.

Equation (5.3) defines the theoretic structure of the Bayesian forecaster for a binary predictand. This structure has four properties that are advantageous in applications.

5.1.5 Structural Properties

1. *Generality.* (i) Equation (5.3) is the general solution for π because it arises, via Bayes' theorem and the total probability law, directly from the axioms of probability theory, sans any assumptions. (ii) The shape of function π is unrestricted; it is dictated solely by the likelihood ratio function f_0/f_1. In turn, the shape of each f_w ($w = 0, 1$) can be molded by data — as we shall learn in Section 5.3. (iii) By changing the expressions for f_0 and f_1, a variety of functional relationships π between x and y can be created.

2. *Decomposition.* The process of applying the Bayesian forecaster to a particular problem may be decomposed into two independent tasks. (i) Assessment or estimation of the prior probability g. (ii) Modeling and estimation of the conditional density functions (f_0, f_1). In each of these tasks, we may use the same or different data sets. These possibilities give rise to yet another property: a Bayesian forecaster can *fuse information* from different sources. Figure 5.1 shows a schema for this concept.

3. *Prior conditionality.* Suppose that the stochastic dependence between predictor X and predictand W is characterized by (f_0, f_1) for each subject in a population, but event $W = 1$ has the prior probability g_A for a subject in subpopulation A, and g_B for a subject in subpopulation B. Then by inserting either g_A or g_B into equation (5.3),

Figure 5.1 A schema of the Bayesian forecaster of binary predictand W using continuous predictor X.

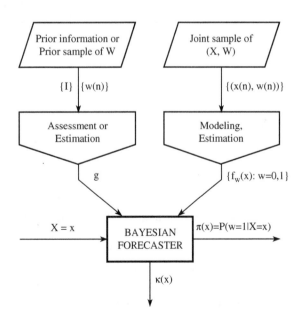

the posterior probability, either $\pi_A(x)$ or $\pi_B(x)$, can be calculated for any subject from each subpopulation. In general, the prior probability itself may be a *conditional probability*.

4. *Marginal calibration*. The prior probability g plays an additional role — it calibrates the posterior probability in the following sense.

Definition The forecaster is said to be *well calibrated in-the-margin* if the expectation of the conditional probability $P(W = 1|X = x)$, viewed as a function of $x \in \mathcal{X}$, equals the marginal probability $P(W = 1)$; that is, if

$$E[P(W = 1|X)] = P(W = 1).$$

Theorem *(Calibration in-the-margin)* *The Bayesian forecaster (5.3) is well calibrated in-the-margin.*

Proof: Take the expectation with respect to X on both sides of equation (5.2):

$$
\begin{aligned}
E[\pi(X)] &= \int_{\mathcal{X}} \frac{f_1(x)g}{\kappa(x)} \kappa(x)\, dx \\
&= g \int_{\mathcal{X}} f_1(x)\, dx \\
&= g;
\end{aligned}
\tag{5.6}
$$

the last equality holds because the integral of density function f_1 over sample space $\mathcal{X}$ equals one. QED.

The equality between the expected posterior probability and the prior probability is a necessary as well as a desirable property. For it means that after making many forecasts, the average conditional probability with which event $W = 1$ has been forecasted equals the marginal probability of the event. Hence, if g represents the relative frequency of event $W = 1$ occurring in "nature", then the Bayesian forecaster simulates "nature" faithfully: it predicts event $W = 1$ with the same relative frequency g, on average.

Note on calibration. This property has two versions: calibration in-the-margin is a *weak* version; a *strong* version is presented in Section 6.6.

5.2 Samples and Examples

5.2.1 Samples

For modeling and estimation, data records should be organized into two samples. (i) A *prior sample* of the predictand W is a set

$$\{w(n) : n = 1, \ldots, N'\} \tag{5.7}$$

which consist of N' realizations of W, indexed by n. (ii) A *joint sample* of the predictor X and predictand W is a set

$$\{(x(n), w(n)) : n = 1, \ldots, N\} \tag{5.8}$$

which consists of N joint realizations of (X, W), indexed by n. Table 5.1 provides an example.

It is possible that the sample sizes are equal, $N = N'$, and that the prior sample is just the marginal sample of the joint sample. But in general, the prior sample is allowed to be larger than the joint sample, $N < N'$. This is called *sample asymmetry*. A situation in which the two samples are asymmetrical and disjoint is also possible and allowed.

Table 5.1 Structure of the data: a long prior sample of predictand *W*; a short joint sample of predictor *X* and predictand *W*. (Example: *W*, precipitation occurrence indicator; *X*, mean relative humidity [%]; $N' = 620, N = 124$.)

Index of realization *n*	Realizations	
	Predictor *x(n)*	Predictand *w(n)*
N'		1
⋮		⋮
$N + 3$		0
$N + 2$		0
$N + 1$		1
N	43	1
⋮	⋮	⋮
3	54	0
2	73	1
1	27	0

Classical statistical techniques, such as the logistic regression method, cannot deal with this *sample asymmetry*; they simply ignore the large prior sample. In effect, these methods throw away vast amounts of information about the predictand. In contrast, the Bayesian forecaster can use both samples effectively. It extracts information from each sample and then optimally fuses information according to the laws of probability: the total probability law and Bayes' theorem.

5.2.2 Examples

Each of the two examples that follow describes a different context in which the samples are collected, and illustrates a different situation in which it is advantageous, or even necessary, to fuse information from different sources.

Example 5.1 *(Weather forecasting)*
Suppose the event of interest is the occurrence of precipitation at a given location during the 6-h period 1200–1800 local time in October. Precipitation occurrence is defined as the accumulation of at least 0.254 mm (0.01 inch) of water in a rain gauge. The forecast is to be made based on the output from a numerical weather prediction model that uses data collected up to 1800 the preceding day. One of the most often utilized predictors is the mean relative humidity calculated by the model for the specified location and time. To develop a Bayesian forecaster, two samples are collected, as depicted in Table 5.1. First, there is a prior sample of predictand *W*. To collect it, it is plausible to assume that, in the absence of any other information, the probability of precipitation occurrence is identical every day throughout the month of October. Suppose a homogeneous record of precipitation observations over 20 years exists. Thus one can extract a prior sample of size $N' = 20$ years $\times$ 31 days = 620 realizations. From this sample the prior probability *g* of precipitation occurrence can be estimated. Second, there is a joint sample of the predictor and predictand (X, W). Because the numerical prediction model is modified every few years in order to incorporate new results of meteorological research, the strength of stochastic dependence between *X* and *W* may change after every significant modification. This effectively limits the possible size of

the joint sample. Suppose a homogeneous record of the model output over the last 4 years exists, so that one can form a joint sample of size $N = 4$ years $\times 31$ days $= 124$ realizations. From this sample, the two conditional density functions, f_0 and f_1, of predictor X can be estimated. Note that this joint sample is much shorter than the prior sample: $N < N'$. Such sample asymmetry is typical of forecasting problems in environmental sciences (meteorology, hydrology, climatology, geology, oceanography), where long climatic records covering decades or even centuries exist, but forecasting models are relatively young, and are updated frequently, so that the joint samples of model output and predictand realizations are relatively short.

Example 5.2 *(Medical diagnosis)*
Suppose a blood test had been invented to diagnose a particular type of cancer. It measures the concentration of a protein, X. The premise is that the more elevated the concentration, the higher the probability of cancer being present. A joint sample, abbreviated $\{(x, w)\}$, of protein measurement $X = x$ and patient state $W = w$ was collected from $N = N_0 + N_1 = 220$ individuals, $N_0 = 99$ having no cancer ($W = 0$) and $N_1 = 121$ having cancer ($W = 1$). (The state $W = 1$ was first suspected based on magnetic resonance imaging and then diagnosed via a biopsy of the suspicious tissue — a definitive but complicated and costly procedure because tissue samples must be collected by a surgeon in an operating room under anesthesia and then examined by a pathologist under a microscope. If this procedure could be replaced by a blood test, or performed only when the test result $X = x$ yields a posterior probability $\pi(x)$ of cancer high enough, the benefits could be enormous.) From the joint sample, the two conditional density functions, f_0 and f_1, can be estimated. But the relative frequency of cancer in the joint sample, $N_1/N = 121/220 = 0.55$, would be a wrong estimate of the prior probability $g = P(W = 1)$ of cancer occurrence in the population of potential patients. This is because the 220 individuals were selected to represent the two states, $W = 0$ and $W = 1$, about equally — to ensure that f_0 and f_1 are estimated equally well. The proper source of g then is a prior sample, abbreviated $\{w\}$, of states $W = w$ of as many individuals in a population as feasible, say N'. A possible source is epidemiological data collected by government agencies and health organizations. Again, the sample asymmetry, $N < N'$, appears inevitable; and the fusion of information from the two samples is necessary.

5.3 Modeling and Estimation

The forecasting equation (5.3) is built of three elements: the prior probability g of the event of interest, $W = 1$; the density function f_0 of the predictor X, conditional on the hypothesis that the event is $W = 0$; and the density function f_1 of the predictor X, conditional on the hypothesis that the event is $W = 1$. To be able to implement the Bayesian forecaster, we must learn how to model and estimate the three elements.

5.3.1 Prior Probability

The prior probability g of event $W = 1$ may be obtained in one of four ways. It may be assessed judgmentally, as described in Chapter 4, based on information that does not include the predictor X. It may be estimated from a *prior sample* $\{w(n) : n = 1, \ldots, N'\}$ whenever such a sample is available. It may be estimated from the *marginal sample* $\{w(n) : n = 1, \ldots, N\}$ of the joint sample when a prior sample is not available. It may be obtained through a mixed procedure that utilizes a sample, a model, and a judgment.

5.3.2 Conditional Distribution Functions

The two conditional density functions f_0, f_1 should be derived from the corresponding conditional distribution functions F_0, F_1. The latter are defined as follows.

$F_w(x) = P(X \leq x | W = w)$ for all $x \in \mathcal{X}$ and for $w = 0, 1$. For a fixed $w \in \{0, 1\}$, the function F_w is the *distribution function* of predictor X, **conditional on the hypothesis** that the event is $W = w$.

Our approach is to choose a parametric model for each conditional distribution function, F_0 and F_1, to estimate the parameters of the two models from the joint sample, and finally to derive the corresponding parametric model for each conditional density function, f_0 and f_1. The procedure for modeling and estimation of F_0 and F_1 is as follows.

Step 0. Given is a *joint sample* that consists of N joint realizations of the predictor X and the predictand W:

$$\{(x(n), w(n))\} = \{(x(n), w(n)) : n = 1, \ldots, N\}.$$

Step 1. Stratify this joint sample into two *subsamples*:

$$\{(x(n), 0)\} \quad \text{and} \quad \{(x(n), 1)\}.$$

Let N_0 and N_1 denote the sizes of these subsamples, so that $N = N_0 + N_1$.

Step 2. Construct two empirical distribution functions, $\check{F}_0$ and $\check{F}_1$, by applying the procedure detailed in Section 3.2 to each of the subsamples.

Step 3. Specify the sample space $\mathcal{X}$ of predictor X. This is done subjectively based on (i) the definition of X, (ii) the extreme realizations of X, and (iii) the tails of the empirical distribution functions $\check{F}_0$ and $\check{F}_1$; Section 3.3 provides additional explanations. Next repeat Steps 4–8 twice, for $w = 0, 1$.

Step 4. Hypothesize a parametric model for F_w. To judge whether or not a parametric model is suitable as a hypothesis, consider (i) the sample space $\mathcal{X}$ of the predictor, (ii) the shape of the empirical distribution function $\check{F}_w$, (iii) the analytic convenience of the parametric model, and (iv) the empirical precedents. Section 3.4 provides additional explanations. Appendix C contains a catalogue of parametric models.

Step 5. Estimate the parameters of the hypothesized model. This task is explained in Section 3.5.

Step 6. Evaluate the goodness of fit of the hypothesized parametric model for F_w. The measure to be used herein is the *maximum absolute difference* (MAD) between the empirical distribution function $\check{F}_w$ and the hypothesized distribution function F_w; details are given in Section 3.6. If the fit is poor, then return to Step 4 and hypothesize another parametric model. If the fit is good, or if all suitable hypotheses have been evaluated, then proceed to the next step.

Step 7. Choose the best parametric model for F_w. When two or several parametric models have been evaluated, follow the procedure detailed in Section 3.6.

Step 8. Perform a statistical goodness-of-fit test of the chosen parametric model for F_w. This should be a test that is appropriate for continuous distribution functions. Details are provided in Section 3.6.

5.3.3 Conditional Density Functions

The chosen parametric model provides an expression for the conditional distribution function F_w. The expression for the corresponding conditional density function f_w follows from the relationship

$$f_w(x) = \frac{dF_w(x)}{dx}, \qquad w = 0, 1.$$

The catalogue of parametric models in Appendix C lists both expressions, for the distribution function and for the density function. Therefore, the expression for f_w can be written immediately; and f_w inherits the parameter estimates from F_w.

5.3.4 Monotone Likelihood Ratio Function

In many applications, logic dictates that function π should be strictly monotone on $\mathcal{X}$; hence the likelihood ratio function f_0/f_1 should also be strictly monotone on $\mathcal{X}$. This must be validated for two reasons. (i) The parametric distribution function F_w ($w = 0, 1$) is seldom a perfect model, even if it passes a goodness-of-fit test. Then F_w is differentiated to obtain f_w ($w = 0, 1$). And finally, f_0 is divided by f_1. Obviously, any errors in

F_0 and F_1 are propagated into f_0/f_1. (ii) A gross error in f_0/f_1 occurs most often in a region of $\mathcal{X}$ where both f_0 and f_1 have tails; equivalently, where both the numerator and the denominator of the ratio $f_0(x)/f_1(x)$ are close to zero.

To detect, and to cope with, such potential errors, the following procedure for validation and rectification of the monotonicity of the likelihood ratio is recommended.

Step 1. *Monotonicity requirement.* Analyze qualitatively any laws (e.g., physical, economic, behavioral) that imply a stochastic dependence between the predictor X and the predictand W, and infer a *logical requirement* for the monotonicity of the likelihood ratio function f_0/f_1 on $\mathcal{X}$. For instance, when X is a relative humidity, expressed in percent, and W is an indicator of precipitation occurrence, it is logical for the posterior probability $\pi(x)$ to strictly increase with the predictor value x, for all $x \in \mathcal{X} = (0, 100)$; hence the logical monotonicity requirement: f_0/f_1 should be strictly decreasing on $\mathcal{X}$.

Step 2. *Monotonicity validation.* Given the models for f_0, f_1, validate the monotonicity requirement by graphing the likelihood ratio function f_0/f_1 (or the function π) on $\mathcal{X}$.

Step 3. *Monotonicity rectification.* There are several ways of rectifying the monotonicity error; here are two of them. (i) When the monotonicity requirement is violated in the tails of f_0 or f_1, perhaps restricting π to an open subinterval $(x_L, x_U) \subset \mathcal{X}$ on which f_0/f_1 is strictly monotone is acceptable for operational forecasting. (ii) When $\mathcal{X}$ is a bounded open interval, adjusting slightly one bound or both bounds of $\mathcal{X}$, and re-estimating the parameters of F_0 and F_1 may rectify the monotonicity of f_0/f_1, at least on a subinterval $(x_L, x_U) \subset \mathcal{X}$ that is acceptable for operational forecasting.

5.3.5 Conditional Sample Spaces

Suppose predictor X has sample space $\mathcal{X}$, but conditional on $W = w$, predictor X has sample space $\mathcal{X}_w \subset \mathcal{X}$ ($w = 0, 1$). This situation may arise in one of two ways.

1. It may be advantageous to specify the sample space $\mathcal{X}_w$ independently for each w ($w = 0, 1$), and end up with $\mathcal{X}_0 \neq \mathcal{X}_1$ due to different bounds (Step 3 of the algorithm in Section 5.3.2), because this may improve the goodness of fit of a parametric model for F_w to the empirical distribution function $\check{F}_w$.
2. It may be logical. For instance, suppose $X \in \mathcal{X}$, but it is also known, as a fact, that $X \in \mathcal{X}_0$ if $W = 0$, and $X \in \mathcal{X}_1$ if $W = 1$, with $\mathcal{X}_0 \neq \mathcal{X}_1$, and

$$\mathcal{X}_0 \cap \mathcal{X}_1 \neq \varnothing, \qquad \mathcal{X}_0 \cup \mathcal{X}_1 = \mathcal{X}.$$

Regardless of the genesis, the situation with conditional sample spaces requires special care when modeling the likelihood ratio function f_0/f_1, as illustrated below.

Example 5.3 *(Signal detection: Weibull model)*
First, read Exercise 6 to become familiar with radar. Next, consider the following modification thereof. Let X represent the voltage of the received signal, with the sample space $\mathcal{X} = (-\infty, \infty)$ for convenience of modeling. Let W be a binary indicator of the presence ($W = 1$) or absence ($W = 0$) of a target. When $W = 0$, the signal X equals the noise voltage, and always $X < \eta_0$. When $W = 1$, the signal X equals the target voltage plus the noise voltage, and always $X > \eta_1$, with $\eta_1 < \eta_0$. Hence the conditional sample spaces are $\mathcal{X}_0 = (-\infty, \eta_0)$ and $\mathcal{X}_1 = (\eta_1, \infty)$. The parametric models for the conditional density functions are as follows.

f_0 on $\mathcal{X}_0$ is reflected Weibull: $\quad$ WB$(\alpha, \beta, -\eta_0, -1)$,

f_1 on $\mathcal{X}_1$ is Weibull: $\quad$ WB(α, β, η_1).

The expression for f_1 is extracted directly from Section C.3.2. The expression for f_0 is obtained by applying the formula from Section C.5.1 for the density function of a reflected variate. The resultant expressions are

$$f_0(x) = \frac{\beta}{\alpha}\left(\frac{\eta_0 - x}{\alpha}\right)^{\beta-1} \exp\left[-\left(\frac{\eta_0 - x}{\alpha}\right)^{\beta}\right], \qquad x < \eta_0,$$

$$f_1(x) = \frac{\beta}{\alpha}\left(\frac{x - \eta_1}{\alpha}\right)^{\beta-1} \exp\left[-\left(\frac{x - \eta_1}{\alpha}\right)^{\beta}\right], \qquad x > \eta_1.$$

The expression for the likelihood ratio function, after simplification, is

$$L(x) = \frac{f_0(x)}{f_1(x)} = \left(\frac{\eta_0 - x}{x - \eta_1}\right)^{\beta-1} \exp\left[\left(\frac{x - \eta_1}{\alpha}\right)^{\beta} - \left(\frac{\eta_0 - x}{\alpha}\right)^{\beta}\right], \qquad \eta_1 < x < \eta_0.$$

Two observations are vital. (i) The domain of L is the intersection of the conditional sample spaces on which f_0 and f_1 are defined — an open interval: $\mathcal{X}_0 \cap \mathcal{X}_1 = (\eta_1, \eta_0)$. (ii) As is always understood, each density function takes on value zero at every point outside the sample space: $f_0(x) = 0$ at any $x \geq \eta_0$, and $f_1(x) = 0$ at any $x \leq \eta_1$. It follows that when the sample space $\mathcal{X} = \mathcal{X}_0 \cup \mathcal{X}_1$ is considered, the likelihood ratio is specified by

$$\frac{f_0(x)}{f_1(x)} = \begin{cases} \infty & \text{if } x \leq \eta_1, \\ L(x) & \text{if } \eta_1 < x < \eta_0, \\ 0 & \text{if } \eta_0 \leq x; \end{cases} \tag{5.9}$$

and consequently, via equation (5.3), the posterior probability is specified by

$$y = \begin{cases} 0 & \text{if } x \leq \eta_1, \\ \pi(x) & \text{if } \eta_1 < x < \eta_0, \\ 1 & \text{if } \eta_0 \leq x. \end{cases} \tag{5.10}$$

Advice (Conditional sample spaces). (i) If bounds on the sample space are to be specified for advantage in modeling, and so that $\mathcal{X}_0 \neq \mathcal{X}_1$, then they must be set far enough in order to avoid issuing forecast probability 0 or 1, which violates Cromwell's rule. (ii) If the predictor indeed behaves like the one in Example 5.3 above, then $\mathcal{X}_0 \neq \mathcal{X}_1$ is logically justified, and is a way of modeling a *partially-perfect predictor*: when its realization is either very small ($x \leq \eta_1$) or very large ($\eta_0 \leq x$), it eliminates the uncertainty about W.

5.4 An Application

5.4.1 Predictand and Predictor

The event ($W = 1$) to be forecasted is the occurrence of precipitation in Buffalo, New York, during the 6-h period 1200–1800 Eastern Time. This event is said to occur if at least 0.254 mm (0.01 inch) of water accumulates in the rain gauge at Buffalo-Niagara International Airport. The forecast is to be made on each day of the cool season (October–March). The predictor X is the relative vorticity on the isobaric surface of 850 millibars above the rain gauge at 1200. The value of the predictor is output from a numerical weather prediction model run by the National Weather Service at 1800, 42 h before the beginning of the forecast period. Thus the *lead time* of the forecast is 42–48 h.

(The particular numerical weather prediction model that supplies the predictor value is the Global Spectral Model. Vorticity is a measure of the spin of horizontally flowing air parcels about a vertical axis. It is positive when air spins counterclockwise; it is negative when air spins clockwise. Its unit is the number of revolutions per second (s^{-1}). Relative vorticity means vorticity relative to the earth surface. As the earth spins, it too has vorticity — the earth's vorticity.)

5.4.2 Samples

The prior sample of the predictand was extracted from the database of the National Centers for Environmental Information in Asheville, North Carolina. This is a 7-year-long sample extending from 1 January 1996 through 31 March 2003. The size of the sample for the cool season (October–March) is $N' = 1143$; the precipitation occurred $N_1' = 267$ times. Thus the prior probability of precipitation occurrence is $g = 267/1143 = 0.234$. This prior probability remains the same for each day of the cool season.

The joint sample of the predictor and the predictand came from the database of the National Weather Service. This is a 4-year-long sample extending from 1 April 1997 through 31 March 2001. The size of the sample for the cool season (October–March) is $N = 718$. When this sample is stratified into two subsamples according to the realization of the predictor, $w = 0$ or $w = 1$, the sizes of the subsamples are $N_0 = 560$ and $N_1 = 158$. (Note that if the joint sample were used to estimate the prior probability, the result would be $g = 158/718 = 0.220$.)

5.4.3 Conditional Distribution Functions

Figure 5.2 shows the empirical conditional distribution functions of the predictor X. The function $\check{F}_0$ is estimated from the subsample of size $N_0 = 560$, and the function $\check{F}_1$ is estimated from the subsample of size $N_1 = 158$. The fact that these two conditional distribution functions are separated implies that the predictor X is informative for forecasting the predictand W.

Because the relative vorticity X has no natural lower bound or upper bound, Figure 5.2 is examined visually to decide whether or not bounds should be specified. Evidently, each empirical distribution function, $\check{F}_0$ and $\check{F}_1$, is

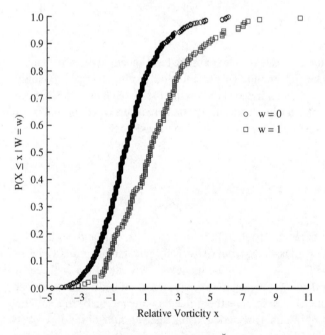

Figure 5.2 Empirical distribution functions $\check{F}_w$ of the relative vorticity X, conditional on event $W = w$ (precipitation nonoccurrence, $w = 0$; and precipitation occurrence, $w = 1$); subsample sizes $N_0 = 560$ and $N_1 = 158$; 6-h forecast period 1200–1800 Eastern Time, beginning 42 h after the model run; cool season; Buffalo, New York.

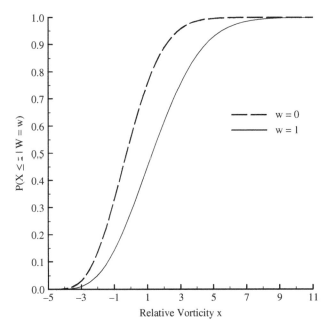

Figure 5.3 Parametric distribution functions F_w of the relative vorticity X, conditional on event $W = w$ (precipitation nonoccurrence, $w = 0$; and precipitation occurrence, $w = 1$); function F_0 is log-Weibull, LW(1.847, 6.651, −5); function F_1 is log-Weibull, LW(2.093, 7.123, −5); 6-h forecast period 1200–1800 Eastern Time, beginning 42 h after the model run; cool season; Buffalo, New York.

skew to the right — has a relatively short left tail and a relatively long right tail. Thus the decision is to bound the sample space from below so that every parametric model listed in Section C.3 can be hypothesized for F_0 and F_1. (Another decision could have been to bound the sample space from below and from above so that every parametric model listed in Section C.4 could be hypothesized for F_0 and F_1.)

The sample space $\mathcal{X}$ of the predictor is a bounded-below interval $\mathcal{X} = (-5, \infty)$. The parametric models ultimately chosen for the conditional distribution functions, F_0 and F_1, are as follows.

F_0 is log-Weibull: LW(α_0, β_0, η)

$\qquad \alpha_0 = 1.847, \quad \beta_0 = 6.651, \eta = -5.$

F_1 is log-Weibull: LW(α_1, β_1, η)

$\qquad \alpha_1 = 2.093, \quad \beta_1 = 7.123, \eta = -5.$

The log-Weibull distribution is characterized in Section C.3.4. The expression for the distribution function F_w of predictor X, conditional on event $W = w$ ($w = 0, 1$), is

$$F_w(x) = 1 - \exp\left[-\left(\frac{\ln(x - \eta + 1)}{\alpha_w}\right)^{\beta_w}\right].$$

Figure 5.3 shows the two conditional distribution functions, F_0 and F_1. Figure 5.4 shows each of these functions overlaid on the empirical conditional distribution function shown earlier in Figure 5.2; one can judge the goodness of fit visually.

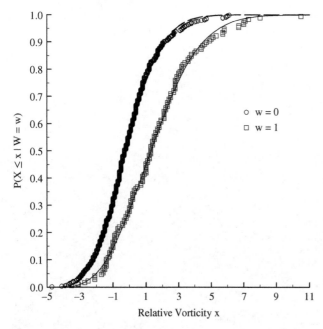

Figure 5.4 Parametric conditional distribution functions F_w ($w = 0, 1$) from Figure 5.3 overlaid on empirical conditional distribution functions $\check{F}_w$ ($w = 0, 1$) from Figure 5.2.

5.4.4 Conditional Density Functions

The expression for the density function f_w of predictor X, conditional on event $W = w$ ($w = 0, 1$), is

$$f_w(x) = \frac{\beta_w}{\alpha_w(x - \eta + 1)} \left(\frac{\ln(x - \eta + 1)}{\alpha_w} \right)^{\beta_w - 1} \cdot \exp\left[-\left(\frac{\ln(x - \eta + 1)}{\alpha_w} \right)^{\beta_w} \right].$$

Figure 5.5 shows the two conditional density functions, f_0 and f_1. The fact that these two conditional density functions are separated implies that the predictor X is informative for forecasting the predictand W. Figure 5.6 shows the graph of the likelihood ratio $f_0(x)/f_1(x)$ versus the predictor value x; this likelihood ratio (against the occurrence of precipitation) is a strictly decreasing function of the relative vorticity.

5.4.5 Posterior Probability

Given the prior probability g and the two conditional density functions, f_0 and f_1, the posterior probability $y = \pi(x)$ can be graphed as a function of the predictor value x according to equation (5.3). Figure 5.7 shows three graphs, each for a different value of g; this demonstrates the sensitivity of y to g.

Regardless of the value of g, the posterior probability $\pi(x)$ of precipitation occurrence is a strictly increasing, nonlinear, irreflexive function of the relative vorticity x. The shape of this function π depends on the shape of the likelihood ratio function, f_0/f_1, shown in Figure 5.6, as well as the value of the prior probability g.

5.4.6 Real-Time Forecasting

The Bayesian forecaster thus developed is valid on each day of the cool season (October–March) because both samples, the prior sample and the joint sample, contain realizations from all days of that season. The first day on

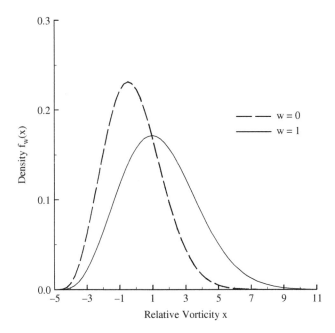

Figure 5.5 Conditional density functions f_w ($w = 0, 1$) corresponding to the conditional distribution functions F_w ($w = 0, 1$) shown in Figure 5.3.

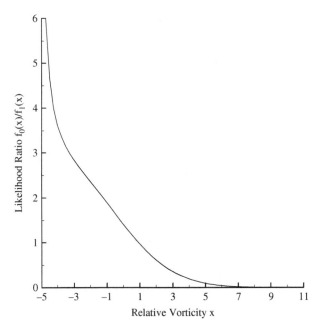

Figure 5.6 Likelihood ratio $f_0(x)/f_1(x)$ against the occurrence of precipitation as a function of the relative vorticity x; the conditional density functions f_w ($w = 0, 1$) are those shown in Figure 5.5.

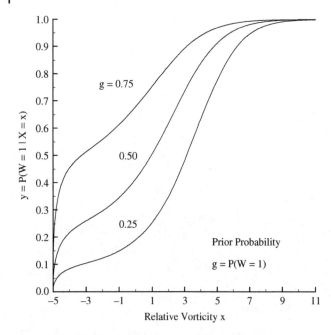

Figure 5.7 Posterior probability $y = \pi(x)$ of precipitation occurrence as a function of the relative vorticity x, given three values of the prior probability g; the likelihood ratio function f_0/f_1 is that shown in Figure 5.6; 6-h forecast period 1200–1800 Eastern Time, beginning 42 h after the model run; cool season; Buffalo, New York.

Table 5.2 Example of forecasts on three days.

Day	x	$f_0(x)$	$f_1(x)$	$f_0(x)/f_1(x)$	y
1	5	0.0049	0.0511	0.0959	0.761
2	−3	0.0618	0.0217	2.8479	0.097
3	3	0.0447	0.1239	0.3608	0.459

which this forecaster could have been used is 1 October 2003 because the latest realizations in the sample came from 31 March 2003.

To make a forecast on any day of the cool season, the value of the relative vorticity x output from the numerical weather prediction model at 1800 must be obtained. Then the posterior probability y of precipitation occurrence in Buffalo during the 6-h period 1200–1800 on the second day (42 h after the model run) can be calculated from equation (5.3), or read off a graph like the one shown in Figure 5.7. For instance, with the prior probability $g = 0.234$ estimated in Section 5.4.2, which gives the prior odds $(1 - g)/g = 3.2735$, the forecasts on three consecutive days could be calculated as shown in Table 5.2.

5.4.7 Nonstationary Prior Probability

The third structural property of the Bayesian forecaster (Section 5.1.5) offers this practical advantage: even though the conditional density functions f_0, f_1 remain fixed during a season (here the cool season), the prior probability g can change within a season. Such a setup is appropriate when the predictive performance of the numerical

Table 5.3 Sample sizes and estimates of the prior probability g of precipitation occurrence; 6-h forecast period 1200–1800 Eastern Time; Buffalo, New York.

	\multicolumn{6}{c}{Month}	Cool Season
	Oct	Nov	Dec	Jan	Feb	Mar	
N'	182	183	203	213	184	178	1143
N_1'	32	48	52	54	43	38	267
g	0.176	0.262	0.256	0.254	0.234	0.213	0.234

weather prediction model (which is quantified by f_0, f_1) is stationary within the season, but the prior uncertainty about the precipitation occurrence (which is quantified by g) is nonstationary within the season. Table 5.3 reveals that g is indeed *nonstationary*: when the prior sample for the cool season is partitioned into six subsamples, one for each month, and the prior probability g is estimated from each subsample, it is found that g varies from month to month. Thus, it is appropriate to use $g = 0.176$ for each day in October, $g = 0.262$ for each day in November, and so on. This setup ensures that the posterior probability is well calibrated, in the sense of equation (5.6), against the climatic probability for each month, which is obviously advantageous. Another instance of a nonstationary prior probability occurs in quality control — see Exercises 8 and 9.

5.5 Informativeness of Predictor

5.5.1 The Concept

Suppose there are two candidate predictors, X^A and X^B, each of which could be employed in the Bayesian forecaster, and we wish to select one. This selection problem may be framed from the viewpoint of rational deciders — those who will use forecasts in decision models developed according to the principles of Bayesian theory. This theory, which is presented in Chapter 7, prescribes the criterion: select a predictor that maximizes the economic value of the forecaster for every rational decider who will use the forecasts. Hence the following definition.

Definition Predictor X^A is said to be *more informative than* predictor X^B with respect to predictand W if, for every rational decider, the economic value of the forecaster using predictor X^A is **at least as high as** the economic value of the forecaster using predictor X^B.

5.5.2 Limiting Predictors

For any decision problem, the economic value of a forecaster is bounded below by zero (the value of an uninformative forecaster — who does not reduce the prior uncertainty, and thus is useless, so to speak), and bounded above by a maximum (the value of a perfect forecaster — the imaginary clairvoyant, who eliminates the prior uncertainty). Each of these limits may be characterized in terms of the conditional density functions (f_0, f_1) of the predictor.

Definition The variate X, having the sample space $\mathcal{X}$, is said to be an *uninformative predictor* for the predictand W if $f_0(x) = f_1(x)$ at every $x \in \mathcal{X}$.

Definition The variate X, having the sample space $\mathcal{X}$, is said to be a *perfect predictor* for the predictand W if there exist subsets $\mathcal{X}_0 = \{x : f_0(x) > 0\}$ and $\mathcal{X}_1 = \{x : f_1(x) > 0\}$ such that

$$\mathcal{X}_0 \cap \mathcal{X}_1 = \emptyset, \qquad \mathcal{X}_0 \cup \mathcal{X}_1 = \mathcal{X}.$$

Let us elaborate. (i) An uninformative predictor is one that yields the likelihood ratio $f_0(x)/f_1(x) = 1$ at every $x \in \mathcal{X}$. Consequently, the posterior probability equals the prior probability, $\pi(x) = g$, regardless of the predictor value $x \in \mathcal{X}$. Hence no reduction of prior uncertainty. (ii) A perfect predictor is one whose sample space $\mathcal{X}$ may be dichotomized into a subset $\mathcal{X}_0$ of x values which occur only if $W = 0$ (so that $f_0(x) > 0$ and $f_1(x) = 0$), and a subset $\mathcal{X}_1$ of x values which occur only if $W = 1$ (so that $f_0(x) = 0$ and $f_1(x) > 0$). Consequently,

$$\frac{f_0(x)}{f_1(x)} = \begin{cases} \infty & \text{if } x \in \mathcal{X}_0, \\ 0 & \text{if } x \in \mathcal{X}_1; \end{cases} \tag{5.11}$$

and

$$\pi(x) = \begin{cases} 0 & \text{if } x \in \mathcal{X}_0, \\ 1 & \text{if } x \in \mathcal{X}_1. \end{cases} \tag{5.12}$$

Hence the elimination of prior uncertainty. (The probability 0 or 1 is assigned by a logical rule (5.12) — an exception permitted by Cromwell's rule (Section 4.2.4).)

5.5.3 Receiver Operating Characteristic

One of the deepest yet practical results of Bayesian theory is this: the two conditional density functions, f_0 and f_1, of predictor X are sufficient to construct a measure of informativeness of X, dubbed the *receiver operating characteristic* (ROC) by electrical engineers. Then any two predictors, X^A and X^B, can be compared in terms of their ROCs to determine whether or not one is more informative than the other. Remarkably, such an inference is possible sans calculating the economic value of each predictor for every rational decider. The theory underlying this sufficient comparison can be fully explained only after the forecaster and the decider are coupled into an F–D system, in Chapters 6 and 7. This section only defines the ROC, with minimal explanations, under a restrictive and simplifying assumption.

1. *Assumption.* The likelihood ratio function f_0/f_1 is strictly monotone and continuous on $\mathcal{X}$; so is π. (Thus an uninformative predictor and a perfect predictor are excluded.)
2. *Detection problem.* This involves making a binary decision a, such that

$$a = \begin{cases} 0 & \Longleftrightarrow \text{ action is not taken (e.g., against a target),} \\ 1 & \Longleftrightarrow \text{ action is taken (e.g., against a target),} \end{cases}$$

which is followed by event $W = w$, such that

$$w = \begin{cases} 0 & \Longleftrightarrow \text{ event does not occur (e.g., target is absent),} \\ 1 & \Longleftrightarrow \text{ event occurs (e.g., target is present).} \end{cases}$$

Conditional on $W = w$ ($w = 0, 1$), the decision $a = 1$ leads to one of the two states:

$$\text{Detection,} \quad D \Longleftrightarrow (a = 1 | W = 1),$$
$$\text{False alarm,} \quad F \Longleftrightarrow (a = 1 | W = 0).$$

To find the probability of each state, one must know the decision rule. Suppose it is

$$a = \begin{cases} 0 & \text{if } \pi(x) \leq p^*, \\ 1 & \text{if } \pi(x) > p^*, \end{cases}$$

where p^* is a threshold probability specified by the decider. Because π is a strictly monotone and continuous function, it has an inverse which yields a unique threshold $x^* = \pi^{-1}(p^*)$, a point in $\mathcal{X}$.

3. *Thresholds.* Inasmuch as every decider may have different p^* (and x^*), and the definition of informativeness requires that every rational decider be considered, all thresholds $x^* \in \mathcal{X}$ must be considered. Equivalently, all points $x \in \mathcal{X}$ must be considered. Next, two cases of the monotonicity assumption must be distinguished.

4. *Probabilities: case SI.* The function π is strictly increasing, so that $\pi(X) > p^* \iff X > x^*$. Hence for any $x \in \mathcal{X}$,

$$P(D|x) = P(a = 1|W = 1) = P(X > x|W = 1)$$
$$= 1 - F_1(x); \tag{5.13}$$

$$P(F|x) = P(a = 1|W = 0) = P(X > x|W = 0)$$
$$= 1 - F_0(x). \tag{5.14}$$

5. *Probabilities: case SD.* The function π is strictly decreasing, so that $\pi(X) > p^* \iff X < x^*$. Hence for any $x \in \mathcal{X}$,

$$P(D|x) = P(a = 1|W = 1) = P(X < x|W = 1)$$
$$= F_1(x); \tag{5.15}$$

$$P(F|x) = P(a = 1|W = 0) = P(X < x|W = 0)$$
$$= F_0(x). \tag{5.16}$$

Definition A graph of the probability of detection $P(D|x)$ versus the probability of false alarm $P(F|x)$ for all predictor values $x \in \mathcal{X}$ is called the *receiver operating characteristic* (ROC) of predictor X.

5.5.4 ROC Construction and Usage

Construction and properties

Under the monotonicity and continuity assumption, the ROC of predictor X can be constructed numerically from its two conditional distribution functions, F_0 and F_1.

Step 1. From the sample space $\mathcal{X}$, select I points, uniformly and closely spaced: $\{x_i : i = 1, \dots, I\}$. If $\mathcal{X} = \mathcal{X}_0 \cup \mathcal{X}_1$ and $\mathcal{X}_0 \cap \mathcal{X}_1 = (\eta_1, \eta_0)$, the case described in Section 5.3.5, then include η_1 and η_0 among the selected points.

Step 2. At each point x_i, calculate $v_i = P(D|x_i)$ and $u_i = P(F|x_i)$ using F_0 and F_1 either in equations (5.13)–(5.14) or in equations (5.15)–(5.16), as dictated by the monotonicity of π (or f_0/f_1).

Step 3. Use the set of points $\{(u_i, v_i) : i = 1, \dots, I\}$ to graph a smooth curve (by interpolation) on the unit square $[0, 1] \times [0, 1]$, beginning at point $(0, 0)$ and ending at point $(1, 1)$.

Figure 5.8 shows the ROC constructed from F_0 and F_1 obtained in Section 5.4.3 and graphed in Figure 5.3. Because f_0/f_1 is a strictly decreasing function (Figure 5.6), and hence π is a strictly increasing function (Figure 5.7), equations (5.13)–(5.14) were used in the construction.

The ROC has four main properties that should guide its interpretation.

1. The ROC displays all *trade-offs* between the probability of detection $P(D|x)$ and the probability of false alarm $P(F|x)$ that a given predictor offers to all deciders.
2. The ROC of a *perfect predictor* consists of one operating point, $(0, 1)$.
3. The ROC of an *uninformative predictor* consists of two operating points, $(0, 0)$ and $(1, 1)$.
4. The ROC of an informative predictor is a strictly concave function between point $(0, 0)$ and point $(1, 1)$. The ROC of a *partially-perfect predictor* (defined in Section 5.3.5) begins with a vertical line segment, or ends with a horizontal line segment, or both.

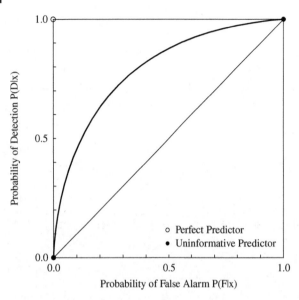

Figure 5.8 Receiver operating characteristic (ROC) of the relative vorticity X, constructed from the conditional distribution functions F_w ($w = 0, 1$) shown in Figure 5.3.

The ROC of predictor X in Figure 5.8 clearly possesses property 4. Furthermore, this predictor (i) is informative, as its ROC lies decisively above the diagonal line (which joins the operating points of an uninformative predictor — property 3), but (ii) is not perfect, as its ROC passes far from the upper left corner (the operating point of a perfect predictor — property 2). A heuristic rule for intuitive interpretation is this: the more concave the ROC, the more informative the predictor.

Informativeness relation

Most importantly, ROC fulfills its purpose, which is the comparison of predictors in terms of the *informativeness relation*.

Theorem (*Comparison of predictors*) *Predictor X^A is more informative than predictor X^B if and only if the ROC of X^A is superior to the ROC of X^B.*

Section 7.6.3 gives the proof. Here are examples of possible inferences based on Figure 5.9. (i) Predictor X^A is more informative than predictor X^B, as the ROC of X^A is superior to the ROC of X^B. (Superiority means that at every probability of false alarm, X^A offers probability of detection at least as high as X^B offers.) (ii) Predictor X^C is more informative than predictor X^B. (iii) Neither is X^A more informative than X^C, nor is X^C more informative than X^A; this is so because their ROCs cross each other (neither is superior to the other).

A general conclusion: The order of predictors in terms of the informativeness relation may be *incomplete*. Above, X^B is ordered third, but X^A and X^C cannot be ordered. Specifically, X^B is the least valuable economically for all deciders; but X^A may be more valuable for some deciders, whereas X^C may be more valuable for other deciders.

If the forecasting model being developed will serve a single decider (or a class of deciders facing identical decision problems), then it should be coupled at this point with the decision model from which the economic value of the forecaster (using a particular predictor) can be calculated — the subject of Chapter 7. Thereby, the most valuable economically predictor, for that decision problem, can be selected. If the forecasting model will serve many different deciders, then the selection of X^A or X^C is left to the discretion of the analyst.

Area under ROC

The discretionary selection of the predictor is sometimes guided by the area under the ROC, denoted by ξ — the larger the better. For a perfect predictor, $\xi = 1$. For an uninformative predictor, $\xi = 1/2$. Thus for any informative predictor, $1/2 < \xi < 1$. The interpretation of ξ arises thusly. Let X_w denote the variate having distribution function F_w. Imagine a realization x_w of X_w is generated (as in Monte Carlo simulation) from F_w, for $w = 0, 1$. Because

Figure 5.9 Receiver operating characteristics (ROCs) of three predictors X^A, X^B, X^C.

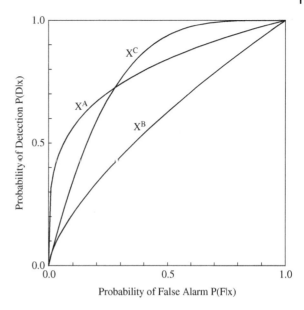

the two variates are continuous, the order of their realizations is complete: $x_0 < x_1$ or $x_0 > x_1$. Now consider the problem of predicting the inequality before x_0 and x_1 are generated. The prediction amounts to determining the probability $P(X_0 < X_1)$; then $P(X_0 > X_1) = 1 - P(X_0 < X_1)$.

Theorem *(Area under ROC)* *Given the ROC of predictor X defined in Section 5.5.3,*

$$P(X_0 < X_1) = \xi \qquad \textit{if } \pi \textit{ is strictly increasing,}$$
$$P(X_0 > X_1) = \xi \qquad \textit{if } \pi \textit{ is strictly decreasing.}$$

Corollary 5.1 *(Informativeness implies area)* *If predictor X^A is more informative than predictor X^B, then $\xi^A \geq \xi^B$.*

The theorem asserts that the area under the ROC, ξ, equals the probability of X_0 being smaller or larger than X_1. The corollary follows directly from the comparison of predictors theorem: the superior ROC has area beneath it at least as large as the inferior ROC. But when two ROCs cross each other, as they do in Figure 5.9, the order of ξ^A and ξ^C does not imply the order of predictors in terms of their informativeness. In other words, the area under the ROC is a necessary (but not sufficient) measure of informativeness.

Bibliographical Notes

The concept of informativeness was originated by Blackwell (1951, 1953) in the context of sufficient comparisons of statistical experiments, and was later operationalized by electrical engineers in the context of signal detection. Its recasting for the purpose of evaluating probabilistic forecasters of binary predictands was initiated by DeGroot and Fienberg (1982, 1983) and continued by Krzysztofowicz and Long (1990a, 1991). That a Bayesian forecaster is well calibrated asymptotically was argued and proven by Dawid (1982).

Exercises

1 **Normal model**. Suppose the density function f_w of predictor X, conditional on the hypothesis that the event is $W = w$, is normal with mean $E(X|W = w) = \mu_w$ and variance $Var(X|W = w) = \sigma_w^2$ ($w = 0, 1$).

1.1 Write the expression for the likelihood ratio and simplify it to the form

$$\frac{f_0(x)}{f_1(x)} = \frac{\sigma_1}{\sigma_0} \exp(-dx^2 - ax - c), \quad -\infty < x < \infty,$$

where d, a, c are parameters; express these parameters in terms of the parameters $\mu_0, \sigma_0; \mu_1, \sigma_1$.

1.2 Take the values of $(g; \mu_0, \sigma_0; \mu_1, \sigma_1)$ specified in the option. Graph the conditional density functions. Graph the posterior probability y versus the predictor value x; characterize in words the shape of this relationship. Calculate the values of y at $x = 14, 16, 18$.

Options: (A) $g = 0.35$; $\mu_0 = 15$, $\sigma_0 = 3$; $\mu_1 = 17$, $\sigma_1 = 4$.

(B) $g = 0.65$; $\mu_0 = 15$, $\sigma_0 = 4$; $\mu_1 = 17$, $\sigma_1 = 3$.

(C) $g = 0.35$; $\mu_0 = 17$, $\sigma_0 = 3$; $\mu_1 = 15$, $\sigma_1 = 4$.

(D) $g = 0.65$; $\mu_0 = 17$, $\sigma_0 = 4$; $\mu_1 = 15$, $\sigma_1 = 3$.

2 Normal model: logistic regression. Suppose the density function f_w of predictor X, conditional on the hypothesis that the event is $W = w$, is normal with mean $E(X|W = w) = \mu_w$ and variance $Var(X|W = w) = \sigma^2$ $(w = 0, 1)$. (This is a special case of the normal model from Exercise 1, which arises when the conditional variances of the predictor are equal: $\sigma_0^2 = \sigma_1^2 = \sigma^2$.)

2.1 Write the expression for the likelihood ratio and simplify it to the form

$$\frac{f_0(x)}{f_1(x)} = \exp(-ax - c), \quad -\infty < x < \infty,$$

where a, c are parameters; express these parameters in terms of the parameters μ_0, μ_1, σ.

2.2 Take the values of $(g; \mu_0, \mu_1, \sigma)$ specified in the option. Graph the conditional density functions. Graph the posterior probability y versus the predictor value x; characterize in words the shape of this relationship. Calculate the values of y at $x = 15, 17, 19$.

2.3 Show that expression (5.3) for the posterior probability can be written in the form

$$y = \left[1 + \exp(-ax - b)\right]^{-1}, \quad -\infty < x < \infty,$$

where a, b are parameters; express these parameters in terms of the parameters μ_0, μ_1, σ of the conditional density functions f_0, f_1, and the prior probability g of event $W = 1$.

2.4 Derive the expression for x_g such that $f_0(x_g)/f_1(x_g) = 1$; prove that $\pi(x_g) = g$; calculate the value of x_g. Derive the expression for $x_{0.5}$ such that $\pi(x_{0.5}) = 0.5$; calculate the value of $x_{0.5}$.

Note. The above form of the relationship $y = \pi(x)$ is known as the *logistic regression*; its parameters a, b are usually estimated directly and solely from a joint sample of (X, W). Thus, the logistic regression is a special (and very restrictive) case of the Bayesian forecaster. It arises also from an exponential model — see Exercise 3.

Options: (A) $g = 0.25$; $\mu_0 = 16$, $\mu_1 = 18$, $\sigma = 3$.

(B) $g = 0.65$; $\mu_0 = 18$, $\mu_1 = 16$, $\sigma = 3$.

(C) $g = 0.25$; $\mu_0 = 15$, $\mu_1 = 19$, $\sigma = 4$.

(D) $g = 0.65$; $\mu_0 = 19$, $\mu_1 = 15$, $\sigma = 4$.

3 Exponential model: logistic regression. Suppose the density function f_w of predictor X, conditional on the hypothesis that the event is $W = w$, is exponential with scale parameter α_w and shift parameter η $(w = 0, 1)$.

3.1 Write the expression for the likelihood ratio $f_0(x)/f_1(x)$ and simplify it.

3.2 Take the values of $(g; \alpha_0, \alpha_1, \eta)$ specified in the option. Graph the conditional density functions. Graph the posterior probability y versus the predictor value x; characterize in words the shape of this relationship; identify the range of y. Calculate the values of y at $x = 6.5, 7, 7.5$.

3.3 Show that expression (5.3) for the posterior probability can be written in the form

$$y = \left[1 + \exp(-ax - b)\right]^{-1}, \qquad \eta < x,$$

where a, b are parameters; express these parameters in terms of the parameters α_0, α_1, η of the conditional density functions f_0, f_1, and the prior probability g of event $W = 1$.

3.4 Derive the expression for x_g such that $f_0(x_g)/f_1(x_g) = 1$; prove that $\pi(x_g) = g$; calculate the value of x_g. Derive the expression for $x_{0.5}$ such that $\pi(x_{0.5}) = 0.5$, calculate the value of $x_{0.5}$.

Note. The above form of the relationship $y = \pi(x)$ is known as the *logistic regression*; its parameters a, b are usually estimated directly and solely from a joint sample of (X, W). Thus, the logistic regression is a special (and very restrictive) case of the Bayesian forecaster. It arises also from a normal model — see Exercise 2.

    Options: (A) $g = 0.3$; $\alpha_0 = 0.2$, $\alpha_1 = 0.8$, $\eta = 5.7$.

            (B) $g = 0.6$; $\alpha_0 = 0.2$, $\alpha_1 = 0.8$, $\eta = 5.7$.

            (C) $g = 0.3$; $\alpha_0 = 0.9$, $\alpha_1 = 0.3$, $\eta = 5.8$.

            (D) $g = 0.6$; $\alpha_0 = 0.9$, $\alpha_1 = 0.3$, $\eta = 5.8$.

4 **Weibull models**. Suppose the conditional density functions f_w $(w = 0, 1)$ of predictor X are of the same type (WB, or IW, or LW, as designated by the option), but differ in terms of the scale parameter; f_0 has parameters (α_0, β, η) and f_1 has parameters (α_1, β, η).

4.1 Write the expression for f_w $(w = 0, 1)$. Write the expression for the likelihood ratio $f_0(x)/f_1(x)$ and simplify it. Identify the domain and the range of the likelihood ratio function f_0/f_1.

4.2 Take the parameter values specified in the option. Graph the conditional density functions. Graph the likelihood ratio function; characterize in words its shape. Graph the ROC.

4.3 For each of the three values of the prior probability g ($g = 0.3, 0.5, 0.7$), graph the posterior probability y versus the predictor value x; identify the range of y. Characterize in words the shape of the relationship $y = \pi(x)$, and the impact of g.

    Options for distribution type: (WB), (IW), (LW).

    Options for parameter values:

        (A) $\alpha_0 = 1.1$, $\alpha_1 = 1.6$, $\beta = 2.5$, $\eta = 5$.

        (B) $\alpha_0 = 2.1$, $\alpha_1 = 2.6$, $\beta = 3.0$, $\eta = 4$.

        (C) $\alpha_0 = 1.6$, $\alpha_1 = 1.1$, $\beta = 2.5$, $\eta = 5$.

        (D) $\alpha_0 = 2.6$, $\alpha_1 = 2.1$, $\beta = 3.0$, $\eta = 4$.

5 **Power model**. Suppose the conditional density functions f_w $(w = 0, 1)$ of predictor X are modeled thusly: f_0 is $P2(\beta_0, \eta_L, \eta_U)$ and f_1 is $P1(\beta_1, \eta_L, \eta_U)$.

5.1 Write the expression for the likelihood ratio $f_0(x)/f_1(x)$ and simplify it.

5.2 Sketch and characterize in words (i) the shape of f_0 when $\beta_0 = 1$ and when $\beta_0 = 4$, and (ii) the shape of f_1 when $\beta_1 = 1$ and when $\beta_1 = 4$.

5.3 Write the expression for the posterior probability y as a function of the predictor value x.

5.4 Let $\beta_0 = 6.5$, $\beta_1 = 3.5$, $\eta_L = 3$, $\eta_U = 9$, and $g = 0.65$. Graph the conditional density functions. Graph the ROC of the predictor. Graph the posterior probability y versus the predictor value x; characterize in words the shape of this relationship. Calculate the values of y at $x = 5, 6, 7$.

5.5 Consider four special cases of the parameter values: $\beta_0 \neq 1$ and $\beta_1 \neq 1$, $\beta_0 = 1$ and $\beta_1 \neq 1$, $\beta_0 \neq 1$ and $\beta_1 = 1$, $\beta_0 = \beta_1 = 1$. In each case, determine the limits of the likelihood ratio $f_0(x)/f_1(x)$: when $x \to \eta_L$, when $x \to \eta_U$.

5.6 Sketch and characterize in words the shape of the relationship $y = \pi(x)$ in each of the four cases investigated in Exercise 5.5. In each case, identify the range of y; and if this range is a proper subset of the interval $(0, 1)$, then explain the implication.

Mini-Projects

6 **Signal detection: normal model.** A radar (*radio detecting and ranging*) system is built of a transmitter and a receiver. The transmitter sends out a signal (ultrahigh-frequency radio waves) in the direction of the hypothesized target that, if present, reflects the signal. The receiver captures a signal, determines if it contains a reflected version of the transmitted signal, and then analyzes it for some purpose, for instance, to estimate the distance to, or the velocity of, a target. When the purpose is to decide whether a target (invisible to a human observer because of the distance, darkness, fog, or clouds) is present ($W = 1$) or not ($W = 0$), the strength of the received signal serves as the predictor of W. Here is a basic model. The prior probability $g = P(W = 1)$ quantifies the uncertainty about the presence of the target at a given location and time — the uncertainty that exists before any signal is received. Variate X represents the voltage of the *received signal*. Variate $\Theta \sim NM(\mu, \sigma)$ represents the voltage of the *noise*. (Its sources are many, e.g., the signal reflected from microscopic particles of dust and vapor ever present in the air.) The constant $s > 0$ represents the voltage of the *reflected signal* from the target. (Its value depends upon the target range (the closest target returns the strongest signal), the type of surface (e.g., metallic or plastic, flat or curvy), the frontal area, etc. For example, police radar can detect the vertical metallic back of a tractor-trailer from over 7 km, but a bicycle from only a few hundred meters.) Conditional on the target being absent, the signal voltage equals the noise voltage: $X = \Theta$. Conditional on the target being present, the signal voltage equals the target voltage plus the noise voltage: $X = s + \Theta$.

6.1 Write the expressions for the conditional means $E(X|W = w)$, the conditional variances $Var(X|W = w)$, and the conditional density functions f_w of the signal voltage X ($w = 0, 1$).

6.2 Write the expression for the likelihood ratio $f_0(x)/f_1(x)$ and simplify it to the form that shows the *signal-to-noise ratio, s/σ*; this is the ratio of the reflected signal strength to the standard deviation of the noise. Next, analyze two special signals. (i) Derive the expression for signal x_g such that $f_0(x_g)/f_1(x_g) = 1$; prove that $\pi(x_g) = g$; interpret this result. (ii) Derive the expression for signal $x_{0.5}$ such that $\pi(x_{0.5}) = 0.5$; interpret this result.

6.3 Let $\mu = 5$, $\sigma = 3$, $s = 6$, $g = 0.4$. (i) Graph the conditional density functions. (ii) Graph the likelihood ratio function. Is it strictly decreasing? (iii) Graph the ROC of the predictor. Is it concave? (iv) Graph the posterior probability y versus the predictor value x; characterize in words the shape of this relationship; report the values of y at $x = 4, 6, 8, 10, 12, 14$. (v) Calculate the values of x_g and $x_{0.5}$; show them on the graph of y versus x.

6.4 Increase the reflected signal strength to $s = 10$, and repeat Exercise 6.3.

6.5 Synthesize the results. (i) Compare the two radar systems (with $s = 6, 10$) in terms of their ROCs; draw a conclusion. (ii) Compare the two radar systems in terms of their functions π; draw a conclusion. (iii) Analyze the shape of function π in the limits: when $s \to 0$, when $s \to 8\sigma$; sketch the limiting π in each case. (iv) Characterize in words the influence of the reflected signal strength s on the predictability of W. *Hint*: interpret s as a measure of *separation* between f_0 and f_1; consider the signal-to-noise ratio.

7 **Fault diagnosis: Rayleigh model.** Upon leaving an assembly line, each drone has its vision, navigation, and control systems tested in a flight to a remote target point T. The drone may miss point T and land at

Table 5.4 Randomly generated joint sample of the coordinates (Y_1, Y_2), in meters, of the actual landing point A, and the state W of the drone ($W = 0$, faultless; $W = 1$, faulty).

y_1	y_2	w	y_1	y_2	w	y_1	y_2	w	y_1	y_2	w
−0.39	0.41	0	1.30	−1.90	1	−2.86	−2.18	0	−1.38	−5.92	1
−1.29	−0.41	0	2.13	1.36	0	−4.14	−0.01	1	2.71	6.16	1
0.15	1.56	1	−1.11	2.46	0	2.79	−3.12	0	−5.85	4.40	1
1.13	−1.11	0	−2.38	2.41	1	3.39	−3.48	1	6.99	−4.11	1
0.29	1.81	0									

point A due to various random factors: variability of wind speed and direction, fluctuation of skill or mis-judgment of the drone pilot, tolerances of the navigation and control systems, and so on. While the *distance* X between point A and point T varies randomly from flight to flight, it tends to be larger when the drone has a fault. Whether the drone is faulty or faultless can be diagnosed through laboratory tests, but they are costly. Therefore, the factory manager wants to quantify the uncertainty about the state of the drone after the test flight. Specifically, the measured distance $X = x$ is to be used to calculate the probability of the drone being faulty. (Then, based on this probability, a decision can be made whether or not to perform the laboratory tests in order to avoid the possibility of shipping a faulty drone to a customer — see Exercise 19 in Chapter 7.) In the pilot production run, 150 drones underwent laboratory tests, and 18 were found faulty. Selected drones were next tested in flight; the results are compiled in Table 5.4.

Model of random distance. (i) To model the distance X between the actual landing point A and the target point T, the Cartesian coordinates are set with the origin at point T and the axes oriented east and north. When point A has the coordinates (Y_1, Y_2), the distance is $X = (Y_1^2 + Y_2^2)^{1/2}$. (ii) To model the randomness of point A, its coordinates Y_1 and Y_2 are assumed to be independent and identical variates, each having a normal distribution NM$(0, \tau)$. Then X has the *Rayleigh* distribution with scale parameter $\alpha = \sqrt{2\tau}$, a special case of the Weibull distribution: WB$(\alpha, 2, 0)$.

7.1 Formulate a Bayesian forecaster. (Define notation for variates, sample spaces, prior probability, conditional density functions, and posterior probability.)

7.2 Estimate the prior probability of the drone being faulty.

7.3 Model each conditional distribution function of the distance using the Rayleigh model; follow the designated option (MS or LS). Graph the estimated conditional distribution function and the empirical conditional distribution function; evaluate the goodness of fit in terms of the MAD.

7.4 Graph the conditional distribution functions of the distance, the corresponding conditional density functions, the likelihood ratio function, and the ROC of the predictor.

7.5 Consider three values of the prior probability g : $g_0, 2g_0, 3g_0$, where g_0 is the estimate from Exercise 7.2. For each g, graph the posterior probability as a function of the distance.

7.6 Calculate the posterior probability of the drone being faulty, given the distance between the landing point and the target point measured during a test flight. Perform this calculation nine times, given each of the distances $(2, 4, 6)$ m, and given each value of the prior probability $(g_0, 2g_0, 3g_0)$.

7.7 Given $g = g_0$, find the values of the distance at which the posterior probability equals $0.1, 0.5, 0.9$.

7.8 Interpret results and draw conclusions.

Options for the estimation method:

(MS) Method of moments. Estimate the scale parameter of the Rayleigh model using the method of moments thusly. (i) Take the appropriate subsample of (Y_1, Y_2) and calculate from it the conditional sample variances of Y_1 and Y_2. Do they satisfy, at least approximately, the model assumption? (ii) Take their geometric mean as the final estimate, and calculate the scale parameter. (iii) Is this a reasonable procedure? Is the geometric mean preferable to the algebraic mean? Justify your answers.

(LS) Method of least squares. Estimate the scale parameter of the Rayleigh model using the method of least squares thusly. (i) Take the appropriate subsample of (Y_1, Y_2) and transform it into a subsample of X. (ii) Construct an empirical conditional distribution function of X. (iii) Take the Weibull model, and in its linearized quantile function set $b = 1/\beta = 1/2$ and $\eta = 0$; then let $v' = \ln x$ and $u' = bu = u/2$. (iv) The linearized quantile function is now $v' = u' + a$; it has only one parameter a. Apply the appropriate version of the LS method.

8 Quality control: exponential model. A rolling mill produces 22 steel sheets per completed shift, but occasionally gets out of tune in the course of production and must be stopped. A review of service records found that, on average, the mill gets out of tune once per 146 shifts while outputting the first steel sheet in a shift, and once per 20 shifts while outputting the last steel sheet in a shift; the denominator of this relative frequency varies linearly from the first sheet to the last sheet in a shift. To detect the out-of-tune state during the shift, each steel sheet has its thickness measured. Then the absolute difference between the measurement and a specification is calculated; this is called the *departure*. The departure varies randomly from sheet to sheet, but tends to be larger when the mill is out of tune. Based on a sample of measurements, it was estimated that when the mill is in tune, the mean of the departure is 5 μm (micrometers) and the variance of the departure is 27 μm^2; and when the mill is out of tune, the two statistics are 17 μm and 285 μm^2, respectively. The factory manager wants to quantify the uncertainty about the state of the rolling mill in the course of production. Specifically, the departure observed in the steel sheet just outputted is to be used to calculate the probability of the mill being out of tune. (Then, based on this probability, a decision can be made whether or not to stop the mill immediately in order to avoid the possibility of outputting defective steel sheets — see Exercise 20 in Chapter 7.)

Hint. A sequence of variates is called a *stochastic process*. With variate W_j denoting the state of the rolling mill while outputting the jth steel sheet in a shift ($j = 1, \ldots, 22$), the predictands $\{W_1, \ldots, W_j, \ldots, W_{22}\}$ form a stochastic process. This particular stochastic process is *nonstationary* because the prior probability is nonstationary; that is, $g_j = P(W_j = 1)$ varies with j. Consider every other element of the Bayesian forecaster. Is it nonstationary (and thus needing subscript j) or *stationary* (i.e., independent of j, and thus not needing subscript j)?

8.1 Formulate a Bayesian forecaster. (Define notation for variates, sample spaces, prior probability, conditional density functions, posterior probability.)

8.2 Estimate the prior probability of the rolling mill being out of tune while outputting the jth steel sheet in a shift ($j = 1, \ldots, 22$). Graph the prior probability as a function of j. What might be an explanation (from the manufacturing perspective) behind the shape of this graph?

8.3 Model each conditional distribution function of the departure as follows. (i) Hypothesize that it is an exponential distribution. (ii) Use the method of moments to obtain two estimates of the scale parameter: one from the sample mean and one from the sample variance; take their geometric mean as the final estimate. (iii) Is this a reasonable procedure? Justify your answer.

8.4 Graph the conditional distribution functions of the departure and the corresponding conditional density functions. Graph the likelihood ratio function. Graph the ROC of the predictor.

8.5 Consider three events: the rolling mill is out of tune while outputting the jth steel sheet in a shift, where $j = 2, 12, 22$. For each event, graph the posterior probability as a function of the departure.

8.6 Calculate the posterior probability of the rolling mill being out of tune given a departure observed during a shift. Perform this calculation nine times, given each of the departures $(7, 19, 49)$ μm, and given that these departures were observed (i) in the 2nd sheet, (ii) in the 12th sheet, (iii) in the 22nd sheet.

8.7 Interpret results and draw conclusions.

9 **Quality control: power model**. Perform Exercise 8 with the following modifications to data, models, and tasks.

Frequency of out-of-tune state: once per 130 shifts while outputting the first sheet, once per 25 shifts while outputting the last sheet.

Sample statistics when in tune: 9 μm, 56 μm^2.

Sample statistics when out of tune: 37 μm, 99 μm^2.

9.1 Same as 8.1.

9.2 Same as 8.2.

9.3 Model the conditional distribution functions of the departure as follows. Hypothesize that conditional on the mill being in tune, the departure has the power type II distribution, and that, conditional on the mill being out of tune, the departure has the power type I distribution. Suppose that the departure never exceeds 50 μm. Then for each of the hypothesized models: (i) estimate the shape parameter from the sample mean; (ii) using this parameter estimate, calculate the conditional variance of the departure under the model; (iii) compare the calculated variance with the sample variance; and (iv) draw a conclusion.

9.4 Same as 8.4.

9.5 Same as 8.5.

9.6 Same as 8.6, except the observed departures are $(5, 23, 41)$ μm.

9.7 Same as 8.7.

10 *Challenger* **accident forecast**. The National Aeronautics and Space Administration (NASA) operated a fleet of six space shuttles from 1981 to 2011. During the launch from Kennedy Space Center at Cape Canaveral, Florida, the shuttle sat atop the external liquid-fuel tank to which two solid rocket motors were attached. Each rocket had three primary O-rings, whose purpose was to seal the joints between its four segments; an O-ring was 7.1 mm thick and 11.43 m in diameter, and was made of a rubbery compound. On 28 January 1986 the space shuttle *Challenger* disintegrated 73 seconds into the flight; all seven astronauts perished. The cause of the accident, as established by the Presidential Commission, was an O-ring failure: the combustion gas — hot and under high pressure — leaked through the aft joint of the right rocket and "eventually weakened and/or penetrated the external tank, initiating vehicle structural breakup". The near certainty of such an event had been predicted by five engineers at Morton Tiokol — the maker of the solid rocket motors. On the eve of the ill-fated launch, they argued emphatically for a delay because the weather forecast called for a near-freezing temperature at which an O-ring might stiffen and not seal the joint. These engineers had knowledge of material science, experience in rocket design, and the courage to speak up. But their judgmental forecast was not supported convincingly by a display of data on the O-ring thermal distress from 23 previous shuttle flights. At the conclusion of contentious discussions, Morton Tiokol managers asserted that "temperature data [are] not conclusive on predicting primary O-ring blowby". And the NASA administrators were determined to launch the *Challenger* on schedule. "It's going to blow up", predicted Bob Ebeling, one of the five engineers. Let us turn the clock back to the eve of 28 January 1986, and imagine being asked by Morton Tiokol managers to apply a Bayesian forecaster to their data in order to aid them in deciding: to launch or not to launch. The data (Table 5.5) came from 23 previous shuttle flights. At the time of launch, the temperature, X, of the primary joint was recorded. After the launch, the rockets were recovered from the ocean, each O-ring was examined, and its state, W, was classified as not damaged ($W = 0$) or damaged ($W = 1$) by blowby or erosion. Your main objective is to calculate the posterior probability of O-ring damage, conditional on the temperature value. On the morning of 28 January 1986, the estimate of this temperature was 31°F.

Assumption. Conditional on the temperature, the postlaunch states of the six O-rings are stochastically independent of each other and identically distributed. Thus the data from Table 5.5 may be arranged into

Table 5.5 Record of the joint temperature (*T*) in degrees Fahrenheit, and the number (*N*) of primary O-rings damaged (due to blowby or erosion) in 23 space shuttle launches (between April 1981 and January 1986) before the *Challenger* accident.

Flight	T	N	Flight	T	N	Flight	T	N	Flight	T	N
1	66		7	73		13	67		19	76	
2	70	1	8	70		14	53	2	20	79	
3	69		9	57	1	15	67		21	75	2
4	68		10	63	1	16	75		22	76	
5	67		11	70	1	17	70		23	58	1
6	72		12	78		18	81				

a joint sample of $23 \times 6 = 138$ realizations of predictor X and predictand W, with each of the 23 flights contributing 6 realizations.

10.1 Display the data in a scatterplot showing one point per flight: temperature as the abscissa, number of damaged O-rings as the ordinate. Sketch intuitively a trend between the temperature and the expected number of O-rings damaged during the launch. Extrapolate the trend to 31°F and make a prediction.

Note. This was, basically, the reasoning of Morton Tiokol experts, but it was derailed by two mistakes, one leading to the next. First, their display excluded the 16 points having ordinate zero. Second, the U-configuration of the seven points having ordinate 1 or 2 led to the wrong inference that "temperature data [are] not conclusive on predicting" the O-ring damage.

10.2 Formulate a Bayesian forecaster (define notation for variates, sample spaces, prior probability, conditional density functions, posterior probability). To obtain a parametric model for each conditional distribution function of the predictor, hypothesize at least two models; estimate the parameters of each model; choose the models that best fit the empirical distribution functions and that yield a strictly monotone likelihood ratio function on the interval $[30, 82]$. Document all steps of the development; report all parameter estimates; report evaluations of the goodness of fit. Then graph the following functions on the domain $[30, 82]$: the empirical conditional distribution functions, the chosen parametric conditional distribution functions, the corresponding conditional density functions, the likelihood ratio function. Graph the ROC of the predictor.

Requirement. As one of the hypothesized parametric models, consider $LW(\alpha_0, \beta_0, \eta_0, -1)$ for F_0 on $\mathcal{X}_0 = (-\infty, 90)$, and $WB(\alpha_1, \beta_1, \eta_1, -1)$ for F_1 on $\mathcal{X}_1 = (-\infty, 84)$.

10.3 Estimate the prior probability of the O-ring damage from the joint sample.

10.4 Graph the posterior probability of the O-ring damage as a function of the temperature on the domain $[30, 82]$. Calculate the posterior probability at $X = 31°F$, the temperature estimated for the planned launch; at $X = 53°F$, the lowest temperature recorded in all previous launches; and at $X = 71°F$.

Note. The crux of this forecasting problem is the enormous extrapolation that must be performed in the sample space of the predictor: relative to the interval $[53, 81]°F$, which comprises all recorded temperatures, the value of 31°F is an outlier (22°F below an interval of width 28°F)! A sensible extrapolation requires (i) a forecasting model with the correct theoretic structure (which the Bayesian forecaster provides), and (ii) an expert judgment (which the five engineers at Morton Tiokol did provide, albeit not formally).

10.5 Given the posterior probability $\pi(x)$, the expected number $R(x)$ of damaged O-rings during a launch at temperature $X = x$ can be calculated. Perform the calculation at sufficiently many points x to construct a graph of function R on the domain $[30, 82]$. Overlay this graph on the display of data from Exercise 10.1.

Compare the graph of *R* with the trend you sketched intuitively based on raw data. How close did your intuitive sketch come to the theoretic regression *R* derived through the Bayesian forecaster?

10.6 Perform a sensitivity analysis. To wit, repeat Exercise 10.4 twice, each time replacing the prior probability by a different value: 0.01, 0.15. (i) Place all three graphs of π on one figure. (ii) List all nine calculated values $\pi(x)$ in one table. (iii) Discuss the sensitivity of $\pi(x)$ to *x* and *g*.

10.7 The two solid rocket motors have six O-rings. Consider two events: (i) At least one O-ring becomes damaged during the launch. (ii) All six O-rings become damaged during the launch. Calculate the prior probability of each event, given *g* estimated in Exercise 10.3. Calculate the posterior probability (obtained from that *g*) of each event, when $X = 31, 53, 71$.

10.8 Summarize all results, draw conclusions, and make a recommendation to Morton Tiokol managers.

    Bibliographical note. The story of the *Challenger's* last launch is based on the report of Presidential Commission (1986), the article by Maranzano and Krzysztofowicz (2008), and the interviews with Bob Ebeling (*Washington Post*, 28 January 2016, 22 March 2016). Some technical details and the data in Table 5.5 come from Dalal et al. (1989).

11   **Language proficiency forecasting**. A foreign applicant to a graduate program at an American university must submit the scores from the Test of English as a Foreign Language Internet-based Test (TOEFL iBT) taken before the application deadline in January. There are four scores — for reading, listening, speaking, and writing — each ranging from 0 to 30 in increments of 1. An applicant who is admitted and is offered a graduate teaching assistantship (GTA) must also pass the Speaking Proficiency English Assessment Kit (SPEAK) test upon arriving on campus in August before the semester begins. The score ranges from 20 to 60 in increments of 5; a score of at least 55 is required to qualify for GTA duties. The uncertainty about the SPEAK test outcome creates a risk for the graduate program director at admission time because failure entails consequences: the prospective graduate teaching assistant must complete English training courses before assuming teaching duties, which may take several semesters; in the meantime, another student qualified for the suddenly vacated GTA position must be found, hired, and prepared for duties at short notice. Let us aid the program director to decide whether or not to offer a GTA to a foreign applicant by quantifying the uncertainty! The predictand is a binary indicator of passing ($W = 0$) or failing ($W = 1$) the SPEAK test in August. The predictor (assumed to be a continuous variate) is the speaking score from the TOEFL iBT taken in January or earlier. Thus the *lead time* of the forecast will be at least 7 months. Given an applicant's TOEFL iBT speaking score, the forecast should state the probability of failing the SPEAK test at a specified threshold. A record of scores from the tests taken by 65 past students is compiled in Table 5.6; some TOEFL scores are missing.

Options for threshold:

(55) The minimum SPEAK test score required for GTA duties is 55.

(50) The minimum SPEAK test score required for supporting duties (e.g., grading homework assignments and written examinations, which do not require oral communication with students) is 50.

11.1 Formulate a Bayesian forecaster (define notation for variates, sample spaces, prior probability, conditional density functions, posterior probability). To estimate its parameters, use all the scores recorded in Table 5.6. For each conditional distribution function of the predictor, hypothesize at least two parametric models; estimate the parameters of each model; choose the models that best fit the empirical distribution functions and that yield a strictly monotone likelihood ratio function on an interval which contains all (or almost all) the predictor realizations from Table 5.6. Document all steps of the development; report all parameter estimates; report evaluations of the goodness of fit; justify the choice of models. Then graph the following elements: the empirical conditional distribution functions, the chosen parametric conditional distribution functions, the corresponding conditional density functions, the likelihood ratio function, the ROC of the predictor, and the posterior probability as a function of the predictor value.

Table 5.6 Record of scores from two tests of spoken English taken by 65 foreign applicants to a graduate program at the University of Virginia in the years 2009–2014. (A blank signifies a missing score.)

TOEFL iBT speaking score	SPEAK test score	TOEFL iBT speaking score	SPEAK test score	TOEFL iBT speaking score	SPEAK test score
	50	19	45	20	45
18	45	22	50	20	45
27	55	19	45	19	40
19	45		55	22	50
	55	19	45	23	50
23	55	22	50	22	45
	40	24	55	20	45
	50		55	19	40
	60	22	45	20	50
22	40	22	55	22	45
27	50	23	40	22	45
	40	20	45	24	50
27	45	22	50	19	50
	50	26	55	23	40
	55	27	55	22	45
20	45	24	50	24	55
	45	20	50	18	35
	50	19	50	22	45
	50	23	60	22	40
	50	23	45	22	45
	55		50	22	45
17	40		55		

Requirement. As one of the hypothesized parametric models for each conditional distribution function, consider $\text{WB}(\alpha, \beta, \eta)$ with $\eta = 16$.

11.2 Validate the Bayesian forecaster. For this purpose use the equation developed for the posterior probability to calculate a forecast for each applicant whose TOEFL iBT score is recorded. (These forecasts are made retrospectively — they are *hindcasts*.) Next, store the joint sample $\{(y, w)\}$ of the forecast probability y and the actual value w of the predictand for the purpose of verification (Exercise 13 in Chapter 6).

11.3 Adapt the Bayesian forecaster. You learn that a university-wide sample of SPEAK test scores of 803 students (including the 65 students whose scores are recorded in Table 5.6) yields a probability of failing the SPEAK test equal to 0.65 when the threshold for passing is 55, and equal to 0.43 when the threshold for passing is 50. Adapt the forecaster you developed in Exercise 11.1 to this new information. Graph the posterior probability as a function of the predictor value. Compare this graph with the one from Exercise 11.1; explain the impact of the prior probability on the shape of the graph.

12 Snow forecasting. Buffalo, New York, is known for snowstorms. The city lies on the mean trajectory of the polar front. The interplay of the incoming Arctic air and the warming effects of the Lake Erie (until it freezes completely, usually by February) causes frequent and heavy snowfalls. In January 1997, the city recorded new snowfall on each of 22 consecutive days. Let us forecast the occurrence of snow in Buffalo! The snow, or more generally the precipitation, is said to occur during a specified period if at least 0.254 mm (0.01 inch) of water accumulates in the rain gauge at Buffalo-Niagara International Airport. The forecast is to be made on each day at 1800 Eastern Time for the 24-h period beginning at 0600 the next day; thus the *lead time* of the forecast is 12–36 h. The predictor to be used is the mean relative humidity of the air column in the location of the rain gauge at the beginning of the forecast period (0600 Eastern Time). The value of the predictor is output from a numerical weather prediction model run by the National Weather Service at 1800. (The particular model that supplied the predictor values is the Global Spectral Model.) Table 5.7 reports a sample of the predictand from the years 1998–1999; Table 5.8 reports a joint sample of the predictand and the predictor from the years 2000–2001. (For the dates not listed in the tables, the realizations are missing.)

Options: (J) Use samples from January.

(F) Use samples from February.

(B) Use samples from January–February.

12.1 Formulate a Bayesian forecaster (define notation for variates, sample spaces, prior probability, conditional density functions, posterior probability). Estimate its parameters using samples from the years 1998–2000. (Imagine it is the summer of 2000, and you are developing a forecaster for the winter of 2001.) To obtain a parametric model for each conditional distribution function of the predictor, hypothesize at least two models on a bounded open interval; assess the bounds; estimate the parameters of each model; choose the models that best fit the empirical distribution functions and that yield a monotone likelihood ratio function on the interval (η_L, η_U). Document all steps of the development; report all parameter estimates; report evaluations of the goodness of fit. Then graph the following elements: the empirical conditional distribution functions, the chosen parametric conditional distribution functions, the corresponding conditional density functions, the likelihood ratio function, the ROC of the predictor, and the posterior probability as a function of the predictor value.

Requirement. As one of the hypothesized parametric models, consider LR1(η_L, η_U)-LG(β, α) with $\eta_L = 7$ and $\eta_U = 100$.

12.2 Validate the Bayesian forecaster. For this purpose use the equation developed for the posterior probability to produce a forecast from a given value of the predictor. Produce two sets of forecasts: (i) using the predictor values from the joint sample from the year 2000; this is the *estimation sample* (the forecasts are made retrospectively — they are *hindcasts*); and (ii) using the predictor values from the joint sample from the year 2001; this is the *validation sample* (imagine the forecasts are made in real-time, one forecast each day, as the winter unfolds). Next, store each set of forecasts and the actual values of the predictand, as a joint sample $\{(y, w)\}$, for the purpose of verification (Exercise 14 in Chapter 6).

13 **Diabetes forecasting**. Type 2 diabetes is a disease in which body cells gradually diminish their response to insulin (a hormone released into the bloodstream by the pancreas to stimulate the transport of glucose from the blood into the cell). This *insulin resistance*, in short, deprives the body cells of energy from glucose, while glucose builds up in the blood above the normal level (between 65 and 120 milligrams per deciliter, mg/dL). Consequences include increased likelihood of stroke, heart attack, kidney failure, and vision problems (up to blindness). According to studies by the National Institute of Diabetes and Digestive and Kidney Diseases, the highest rates of occurrence of type 2 diabetes are found among Pima Indians of central Arizona. For example, among 768 Pima Indian women, 268 had type 2 diabetes. Their medical records constitute the *Pima Indians Diabetes Database*, part of which is reproduced in Tables 5.9 and 5.10. There are two joint samples collected from 52 and 75 patients; each sample includes realizations of four predictors of the diabetes and the predictand. Predictors were measured during a medical examination at the time the patient was nondiabetic and of age at least 21. Predictand is a binary indicator of absence ($W = 0$) or presence ($W = 1$) of diabetes within 1–5 years of the examination. Because patients diagnosed with diabetes within 1 year of the examination were excluded from the database, the *lead time* of any forecast will be 1–5 years. Your task is to forecast the onset of diabetes in a patient within 1–5 years of the medical examination based on a realization of a single predictor specified in the option.

Options for predictor:
- (A) Age [years].
- (S) Skin [mm] — triceps skin fold thickness.
- (G) Glucose [mg/dL] — plasma glucose concentration.
- (I) Insulin [μU/mL] — serum insulin concentration.

Each predictor can be measured during a medical examination of the patient. Measurements of predictors G and I are obtained through a *glucose tolerance test*: the concentrations of plasma glucose and serum insulin are measured in specimens collected 2 hours after the patient takes a 75 g glucose load orally while fasting.

Options for samples:
- (a) For estimation — sample I (Table 5.9).
   For validation — sample II (Table 5.10).
- (b) For estimation — sample II (Table 5.10).
   For validation — sample I (Table 5.9).
- (c) For estimation — sample I.
   For validation — union of samples I and II.
- (d) For estimation — sample II.
   For validation — union of samples I and II.

13.1 Formulate a Bayesian forecaster (define notation for variates, sample spaces, prior probability, conditional density functions, posterior probability). Use the predictor designated by the option. Estimate the prior probability as a relative frequency in the total sample of 768 patients. Estimate all other parameters from the sample designated by the option. To obtain a parametric model for each conditional distribution function of the predictor, hypothesize at least two models and choose the one that best fits the empirical distribution function. Document all steps of the development; report all parameter estimates; report evaluations of the goodness of fit. Then graph the following elements: the empirical conditional distribution functions, the chosen parametric conditional distribution functions, the corresponding conditional density functions, the likelihood ratio function, the ROC of the predictor, and the posterior probability as a function of the predictor value.

Requirement. As one of the hypothesized parametric models, consider the model designated for your predictor and the conditional distribution function F_w ($w = 0, 1$):
- (A) F_0 is EX($\alpha_0, 20$),
   F_1 is P2($\beta_1, 20, 85$);
- (S) F_0 is WB($\alpha_0, \beta_0, 0$),
   F_1 is LL($\alpha_1, \beta_1, 0$);
- (G) F_0 is LR1(55, 245)-LG(β_0, α_0),
   F_1 is LR1(55, 245)-LG(β_1, α_1);
- (I) F_0 is LL($\alpha_0, \beta_0, 0$),
   F_1 is LL($\alpha_1, \beta_1, 0$).

13.2 Validate the Bayesian forecaster. For this purpose use the equation developed for the posterior probability to produce a forecast from a given value of the predictor. Produce two sets of forecasts: (i) using the predictor values from the *estimation sample* (the forecasts are made retrospectively — they are *hindcasts*); and (ii) using the predictor values from the *validation sample* (imagine the forecasts are made in real-time, one forecast per patient, as they come for their annual medical examination). Next, store each set of forecasts and the actual values of the predictand, as a joint sample $\{(y, w)\}$, for the purpose of verification (Exercise 15 in Chapter 6).

14 Diabetes forecasting — *continued*: **comparison of predictors**. If Exercise 13.1 was performed two or more times, each time using a different predictor but the same samples, then the predictors can be compared in terms of their informativeness.

14.1 Graph in one figure the ROCs of all predictors. Compare the predictors in terms of the trade-offs they offer.

14.2 If only one predictor could be available in real-time forecasting, what would be your preference order? Justify your recommendation.

Bibliographical note. The Pima Indians Diabetes Database, part of which is reproduced in Tables 5.9 and 5.10, was obtained on 1 February 2012 from an archive available online at that time.

Table 5.7 A prior sample at Buffalo from January–February 1998–1999; first column: date (year, month, day); second column: *P*, precipitation amount [cm] observed during the 24-h period beginning at 0600 Eastern Time.

Date			*P*	Date			*P*	Date			*P*	Date			*P*
98	1	1	0.00	98	2	1	0.00	99	1	1	0.00	99	2	1	0.46
		2	0.00			2	0.00			4	1.02			2	0.25
		3	0.71			3	0.03			5	0.05			3	0.00
		4	0.25			4	0.00			7	0.00			4	0.05
		5	0.69			5	0.00			8	1.37			5	0.10
		6	0.13			6	0.00			9	0.28			6	0.03
		7	5.61			7	0.00			10	0.25			7	0.41
		8	1.42			8	0.00			11	1.07			8	0.00
		9	0.30			9	0.00			12	1.37			9	0.00
		10	0.10			10	0.00			13	0.00			10	0.00
		11	0.00			11	0.23			14	1.35			13	0.05
		12	0.30			12	0.53			15	0.15			14	0.00
		13	0.10			13	0.00			16	0.00			15	0.00
		14	0.03			14	0.00			17	0.36			17	0.03
		16	0.00			15	0.00			19	0.00			18	0.00
		17	0.00			16	1.12			20	0.00			19	0.00
		19	0.10			17	1.96			21	0.00			20	0.05
		20	0.00			18	0.28			22	0.71			21	0.00
		21	0.00			19	0.08			23	0.30			22	0.00
		22	0.38			20	0.18			24	0.00			23	0.00
		23	0.79			21	0.00			25	0.33			24	0.00
		24	0.00			22	0.00			26	0.03			25	0.08
		25	0.03			23	0.05			27	0.00			26	0.00
		26	0.03			24	0.43			28	0.15			27	0.05
		27	0.00			25	0.00			29	0.00			28	0.53
		28	0.00			26	0.00			30	0.00				
		29	1.09			27	0.05			31	0.00				
		30	0.25			28	1.04								
		31	0.00												

Table 5.8 A joint sample at Buffalo from January–February 2000–2001; first column: date (year, month, day); second column: P, precipitation amount [cm] observed during the 24-h period beginning at 0600 Eastern Time; third column: H, mean relative humidity [%] at 0600 output from a numerical weather prediction model.

Date			P	H	Date			P	H	Date			P	H	Date			P	H
00	1	1	0.00	40	00	2	1	0.00	87	01	1	1	0.05	50	01	2	2	0.48	90
		2	0.30	79			2	0.00	35			3	0.33	36			3	0.00	46
		3	1.04	59			3	0.20	87			4	0.10	66			4	0.00	84
		4	0.71	92			4	0.00	83			5	0.53	97			5	0.15	87
		5	0.03	63			5	0.13	50			6	0.00	67			6	0.33	84
		6	0.15	58			6	0.00	38			7	0.10	68			7	0.00	57
		7	0.05	84			7	0.00	63			8	0.00	68			8	0.99	82
		8	0.15	52			8	0.00	18			9	0.53	49			9	1.04	90
		9	0.30	71			9	0.00	55			10	0.00	45			10	0.00	38
		12	0.76	50			10	0.53	69			11	0.00	41			11	0.03	37
		13	0.36	89			11	0.00	88			12	0.00	26			12	0.00	42
		14	0.00	52			12	0.00	24			13	0.00	49			13	0.00	45
		15	0.00	62			13	0.38	62			14	0.10	49			14	1.37	89
		16	0.05	67			15	0.03	36			15	0.46	80			15	0.00	50
		17	0.00	24			17	0.00	27			16	0.28	89			16	0.00	52
		20	0.18	99			18	0.86	84			17	0.10	74			17	0.00	54
		21	0.00	56			19	0.15	73			18	0.00	60			18	0.00	42
		22	0.00	39			20	0.10	61			19	0.00	85			19	0.00	58
		23	0.00	83			21	0.00	55			20	0.00	40			20	0.00	76
		24	0.00	56			22	0.00	51			21	0.00	47			21	0.00	50
		26	0.00	75			23	0.00	74			22	0.00	36			22	0.15	41
		27	0.00	34			24	0.51	72			23	0.00	36			23	0.03	55
		28	0.00	34			25	0.03	85			24	0.10	48			24	0.91	36
		29	0.00	22			26	0.00	61			25	0.00	64			25	0.20	94
		30	0.25	50			27	0.08	65			26	0.36	25			26	0.10	45
		31	0.89	93			28	0.00	68			27	0.05	97			27	0.08	71
							29	0.00	28			28	0.23	48			28	0.00	42
												29	0.18	35					
												30	1.35	98					
												31	0.69	90					

Table 5.9 Joint sample I of size $N = 52$ from Pima Indians Diabetes Database: A, age [years]; S, skin [mm]; G, glucose [mg/dL]; I, insulin [μU/mL]; W, indicator of diabetes.

A	S	G	I	W	A	S	G	I	W
21	41	139	480	0	33	41	128	58	1
25	26	129	205	0	23	18	119	92	0
59	23	189	846	1	48	40	88	54	0
35	30	158	328	1	23	13	68	15	0
37	15	133	155	0	28	49	172	579	1
43	25	95	180	1	42	29	100	196	0
21	46	102	78	0	39	37	153	140	0
47	41	163	114	1	35	39	93	72	0
53	45	197	543	1	45	18	125	122	1
29	52	86	65	0	31	47	118	230	1
22	21	127	335	0	22	26	108	63	0
25	63	180	14	1	26	44	109	99	1
23	13	82	95	0	21	19	97	82	0
24	26	140	130	1	34	33	124	402	0
22	37	104	64	1	46	37	134	370	1
25	15	81	76	0	22	35	123	240	0
24	42	135	250	1	21	8	109	182	0
26	17	96	49	0	57	29	196	280	1
21	29	78	40	0	24	21	157	168	0
29	39	119	220	1	36	22	174	194	1
25	33	144	135	1	32	30	125	120	0
58	32	173	265	0	30	32	122	156	1
33	34	152	171	1	28	23	84	115	0
22	28	106	135	0	51	28	155	150	1
22	36	179	159	1	27	27	146	100	0
27	23	117	106	0	24	12	90	43	0

Table 5.10 Joint sample II of size $N = 75$ from Pima Indians Diabetes Database: A, age [years]; S, skin [mm]; G, glucose [mg/dL]; I, insulin [μU/mL]; W, indicator of diabetes.

A	S	G	I	W	A	S	G	I	W
23	42	153	485	0	27	20	110	100	0
28	29	105	325	0	33	14	173	168	1
52	24	124	600	1	24	40	102	90	0
26	44	181	510	1	22	15	98	84	0
30	25	147	293	0	53	34	145	165	1
46	26	155	495	1	25	35	139	160	1
22	10	108	278	0	21	18	94	76	0
26	43	165	255	0	41	18	104	156	1
34	33	187	392	1	28	39	121	74	0
29	33	189	325	1	36	21	95	73	0
34	45	123	230	0	32	49	129	155	1
25	32	91	210	0	48	37	120	150	1
22	31	117	188	0	24	28	86	71	0
58	34	176	300	1	45	28	133	140	1
22	42	181	293	1	43	18	96	67	0
31	39	184	277	1	21	22	84	66	0
23	13	111	182	0	47	23	161	132	1
38	40	151	271	1	42	35	136	130	1
28	21	131	166	0	25	47	119	63	0
29	36	196	249	1	22	11	80	60	0
22	37	106	148	0	33	24	75	55	0
51	48	148	237	1	25	41	109	129	1
26	35	139	140	0	54	29	150	126	1
51	26	136	135	0	25	16	87	52	0
43	46	167	231	1	23	36	102	120	1
31	51	122	220	1	28	29	129	115	1
24	45	112	132	0	43	32	136	110	1
29	41	158	210	1	21	26	100	50	0
26	33	124	130	0	25	25	92	41	0
21	27	126	120	0	25	56	162	100	1
24	45	128	194	1	32	30	115	96	1
27	25	128	190	1	35	26	180	90	1
39	34	96	115	0	28	20	100	90	1
57	2	145	110	0	23	34	89	37	0
29	37	120	105	0	21	30	99	18	0
22	14	188	185	1	27	31	80	70	1
46	46	144	180	1	24	25	95	36	1
36	32	112	175	1					

6

Verification of Forecasts

After a forecaster assigned probability $y = 0.01$ to event A and the event did occur, can one say the forecaster was wrong? The forecaster did not state that A will not occur. Rather he stated the *degree of certainty* about the occurrence of A that existed, in his judgment, *at the forecast time*. And as long as the forecaster allows for uncertainty by assigning probability y which is neither 0 nor 1, adjectives "right", "wrong", "correct", "incorrect", "accurate", "inaccurate" are inappropriate for characterizing the forecaster's performance. More generally, a single probabilistic forecast cannot be verified. Only a sample of probabilistic forecasts can be verified.

This chapter presents methods for verifying probabilistic forecasts according to the Bayesian verification theory. The theory prescribes two kinds of verification measures: measures of *calibration* and measures of *informativeness*. These verification measures characterize the necessary and sufficient attributes of a probabilistic forecast from the viewpoint of a rational decider.

6.1 Data and Inputs

6.1.1 Variates and Samples

Let W be the *predictand* — a variate whose realization w is forecasted. It is a Bernoulli variate with the sample space $\mathcal{W} = \{w : w = 0, 1\}$, and serves as the *indicator* of event A of interest:

$$W = 0 \iff \text{event } A \text{ does not occur,}$$
$$W = 1 \iff \text{event } A \text{ occurs.}$$

Let Y be the *probability variate* — a variate whose realization y is the probability assigned to the event $W = 1$ by the forecaster, and whose sample space is $\mathcal{Y}$. When the probability is assessed judgmentally, $y = P(W = 1|I, K, E)$ as explained in Chapter 4. When the probability is produced statistically, $y = P(W = 1|X = x)$ as explained in Chapter 5.

Probabilistic forecasts can be verified only after the forecaster has assigned probabilities to many events whose realizations have been recorded. Then one can form a *sample of joint realizations* of the probability variate Y and the predictand W:

$$\{(y(n), w(n)) : n = 1, \dots, N\}, \tag{6.1}$$

where n is the index of realizations (events), N is the *sample size*, $y(n)$ is the forecast probability assigned to the nth event, and $w(n)$ is the indicator of the nth event, with $y(n) \in \mathcal{Y}$ and $w(n) \in \mathcal{W}$ for $n = 1, \dots, N$. The joint sample can arise in many ways.

Example 6.1   *(Forecast of rain)*

In the evening of each day n ($n = 1, \ldots, 92$) in July, August, and September, a meteorologist on duty in the Weather Forecast Office in Phoenix, Arizona, assigns a probability $y(n)$ to the occurrence of rain in the city during the 24-h period beginning at 0400 Mountain Time. Afterwards, if the rain gauge at Phoenix Sky Harbor International Airport reports an accumulation of 0.25 mm or more, then the event indicator $w(n)$ is set to 1; else $w(n)$ is set to 0.

The verification method presented in this chapter assumes that Y is a discrete variate with the sample space

$$\mathcal{Y} = \{y : y = y_0, y_1, \ldots, y_I\}, \tag{6.2}$$

where $0 \leq y_0 < y_1 < \cdots < y_I \leq 1$. When forecasts have been made judgmentally, $\mathcal{Y}$ is the set of probabilities the expert has used. When forecasts have been made statistically, each probability is a real number from the interval $[0, 1]$, and the set $\mathcal{Y}$ must be determined via the discretization algorithm of Section 6.1.3. (*Note*: there does exist a verification method that treats Y as a continuous variate, but it is involved.)

6.1.2   Necessary Sample Properties

In order to verify the forecasts, it must be validated, at least approximately, that each vector $(y(n), w(n))$ is an independent realization of the same random vector (Y, W). In practice, this validation is usually performed intuitively by reasoning using the following formal argument. Let us back up to a time **before** the forecasting began. At that time, the joint realizations $(y(n), w(n))$ for $n = 1, \ldots, N$ were yet unknown. Thus, they had to be viewed as future realizations of random vectors $(Y(n), W(n))$ for $n = 1, \ldots, N$, where $Y(n)$ is the probability variate and $W(n)$ is the predictand defined for the nth event to be forecasted.

Definition   If the N random vectors $(Y(1), W(1)), \ldots, (Y(N), W(N))$ are stochastically independent and identically distributed, then they are said to form a *random sample*.

Because every random vector $(Y(n), W(n))$ for $n = 1, \ldots, N$ has an identical distribution, one can introduce a random vector (Y, W) that has the common distribution. And because the N random vectors are stochastically independent, the probability of any realization of the random sample can be factorized into N bivariate probabilities:

$$\prod_{n=1}^{N} P(Y = y(n), W = w(n)).$$

Next, each of the bivariate probabilities can be factorized into a conditional probability and a marginal probability:

$$P(Y = y(n), W = w(n)) = P(Y = y(n) | W = w(n)) P(W = w(n)).$$

From this relationship, one can infer two necessary conditions for a sample to be random:

1. Each forecasted event must have *a priori* the same probability of occurrence $P(W = 1)$.
2. For fixed $y \in \mathcal{Y}$ and $w \in \mathcal{W}$, the conditional probability $P(Y = y | W = w)$ must be the same for each forecasted event.

The joint sample (6.1) that meets these necessary conditions is called *homogeneous*. The first condition, that the prior probabilities must be identical, is straightforward. The second condition can be loosely interpreted as saying that the performance of the forecaster across all events must be stationary. This might be expected when the forecaster has employed the same thought process, has used the same type of information, and has not undergone any training that would substantially increase his knowledge or experience. The stationarity of performance might also be expected when the same model has been used to produce forecasts based on the realizations of the same predictor whose veracity has not changed across the forecasted events.

Example 6.2 *(Homogeneity of sample)*
Suppose a joint sample includes (i) forecasts of the occurrence of a tornado in Kansas City within a specified time interval and (ii) forecasts of the occurrence of rain in San Francisco within the same time interval. The sample is obviously nonhomogeneous, if only because information, knowledge, and experience required to forecast tornadoes are different than those required to forecast rain. On the other hand, a joint sample that includes (i) forecasts of the occurrence of rain in Kansas City on any day in July and (ii) forecasts of the occurrence of rain in St. Louis on any day in July is possibly homogeneous because the events are generated by the same physical phenomenon occurring in the same climate. Consequently, the prior probabilities of the events are about the same, and the performance of the forecaster across the events is most likely stationary.

In summary, the joint sample (6.1) must be homogeneous. This property should be validated, at least intuitively. Realizations that do not meet the two necessary conditions for homogeneity should not be included in the sample.

6.1.3 Discretization Algorithm

The algorithm is applicable whenever the original probability variate Y is continuous and, therefore, the forecast probability y is a real number from the interval $[0, 1]$. Its purpose is to replace the interval $[0, 1]$ by a finite set (6.2), and to replace the original joint sample (6.1) by a discrete one.

Step 0. Given is the *original joint sample* (6.1). Extract from it the marginal sample of Y:

$$\{y(n) : y(n) \in [0, 1], n = 1, \ldots, N\}.$$

Step 1. Display this marginal sample as a dot diagram on the interval $[0,1]$; Figure 6.1 illustrates this and the next two steps.

Step 2. Partition the interval $[0, 1]$ into $I + 1$ $(I \geq 1)$ subintervals $\mathcal{Y}_0, \mathcal{Y}_1, \ldots, \mathcal{Y}_I$, such that:
- any discernible cluster of realizations $y(n)$ belongs to one subinterval;
- the number N'_i of realizations $y(n)$ in subinterval $\mathcal{Y}_i$ is sufficiently large, for every i $(i = 0, 1, \ldots, I)$. (By definition of a *partition*, the subintervals must be disjoint and their union must equal $[0, 1]$.)

Step 3. For $i = 0, 1, \ldots, I$, calculate the sample mean y_i of the realizations $y(n)$ falling into the subinterval $\mathcal{Y}_i$. Form the finite sample space (6.2).

Step 4. For $n = 1, \ldots, N$, if $y(n) \in \mathcal{Y}_i$, then replace $y(n)$ by $y_i, i \in \{0, 1, \ldots, I\}$. This creates the discrete marginal sample of Y:

$$\{y(n) : y(n) \in \mathcal{Y}, n = 1, \ldots, N\}.$$

Step 5. In the original joint sample (6.1), replace the marginal sample of Y by the discrete one. This creates the *discrete joint sample* of (Y, W), to be utilized in the verification of forecasts.

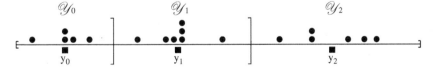

Figure 6.1 Discretization of the continuous probability variate Y: dot diagram of the marginal sample $\{y(n) : n = 1, \ldots, 18\}$, partition of the interval $[0, 1]$ into subintervals $\mathcal{Y}_0 \cup \mathcal{Y}_1 \cup \mathcal{Y}_2$; mean forecast probabilities y_0, y_1, y_2, which form the finite sample space of Y.

6.1.4 Contingency Table

The joint sample (6.1) can be summarized in a *contingency table* whose format is defined in Table 6.1. For $w = 0, 1$ and $i = 0, 1, \ldots, I$, the contingency table reports a count, N_{wi} — the number of forecasts that specified probability y_i and were followed by the indicator value w. Then for $w = 0, 1$,

$$N_w = \sum_{i=0}^{I} N_{wi}$$

is the number of events $W = w$ included in the joint sample; and for $i = 0, 1, \ldots, I$,

$$N_i' = N_{0i} + N_{1i}$$

is the number of forecasts that specified probability y_i. Finally, $N = N_0 + N_1$ is the sample size. Table 6.2 gives an artificial example.

If, for some $i \in \{0, 1, \ldots, I\}$, the sample size N_i' is small, then column i should be combined with one of the adjacent columns. For example, when columns i and $i + 1$ are combined, the newly created column retains index i, is assigned numbers $N_{0i} \equiv N_{0i} + N_{0,i+1}$ and $N_{1i} \equiv N_{1i} + N_{1,i+1}$, and represents forecast probability equal to the weighted average:

$$y_i \equiv \frac{y_i N_i' + y_{i+1} N_{i+1}'}{N_i' + N_{i+1}'}.$$

The contingency table supplies (i) part of the data (or all of the data in some circumstances) for the estimation of the prior probability, and (ii) all of the data for the estimation of the conditional probability functions — the two input elements needed for the verification of forecasts.

6.1.5 Prior Probability

The first input element needed for the verification of forecasts is the prior probability; it is defined as follows.

$g = P(W = 1)$ is the *marginal probability* of event $W = 1$. This probability characterizes the natural variability of

Table 6.1 Format of the contingency table for verification of probabilistic forecasts of a binary predictand.

		y					
		y_0	y_1	$\cdots$	y_i	$\cdots$	y_I
w	0	N_{00}	N_{01}	$\cdots$	N_{0i}	$\cdots$	N_{0I}	N_0
	1	N_{10}	N_{11}	$\cdots$	N_{1i}	$\cdots$	N_{1I}	N_1
		N_0'	N_1'	$\cdots$	N_i'	$\cdots$	N_I'	N

Table 6.2 Example of the contingency table for verification of probabilistic forecasts of a binary predictand.

		y				
		0.01	0.3	0.7	0.99	
w	0	60	45	30	15	150
	1	10	20	30	40	100
		70	65	60	55	250

the predictand. It is also called the *prior probability* of event $W = 1$. For if no forecast were available, one would use g to make a decision. In this sense, the prior probability g constitutes a reference against which the probabilistic forecasts are verified. It is, therefore, imperative to estimate g appropriately. Two cases arise.

Case 1 *(Joint sample)* When the only available realizations of W are those in the *joint sample*, as summarized in a contingency table (Table 6.1), the estimator of the prior probability is given by

$$g = \frac{N_1}{N}. \tag{6.3}$$

The data in Table 6.2 yield an estimate $g = 100/250 = 0.4$.

Case 2 *(Prior sample)* In many applications, the prior probability g should be estimated not from the joint sample, but from a prior sample. The *prior sample* is the longest available homogeneous sample of the predictand:

$$\{w(n) : n = 1, \ldots, M\}. \tag{6.4}$$

With M_w denoting the number of events $W = w$ included in the prior sample for $w = 0, 1$, and $M = M_0 + M_1$ being the sample size, the estimator of the prior probability is given by

$$g = \frac{M_1}{M}. \tag{6.5}$$

The estimate of g from the prior sample may be different than the estimate of g from the joint sample. There are two reasons.

First, the prior sample may be considerably larger than the joint sample: $M > N$. This happens when in addition to the N realizations of W in the joint sample, there exist $M - N$ realizations of W without the matching realizations of Y; these additional realizations of W could have been recorded either before or after the joint sample (Example 6.3).

Second, the joint sample may be unrepresentative of the predictand W either because it is too short (Example 6.3) or because the forecasting experiment that produced the joint sample was designed (for reasons of economy and efficiency) without considering the representativeness of the sample (Example 6.4). In the latter case, a prior sample that is *representative* of the predictand W must be collected independently of the joint sample.

Example 6.3 *(Forecast of rain: large prior sample — Example 6.1 continued)*
It is desired to verify daily forecasts of the occurrence of rain in Phoenix, Arizona, made during one year. For the summer season (July, August, September), the joint sample has size $31 + 31 + 30 = 92$. But from the archives of the National Centers for Environmental Information in Asheville, North Carolina, one could retrieve rainfall measurements made from July 1948 to the present. The record is incomplete; it is also not necessarily homogeneous due to changes in the rain gauge location and instrumentation and possibly due to nonstationarity of the climate. Let us suppose up to 10 years of daily rainfall measurements can be considered homogeneous, and let us estimate the prior probability g from several samples (Table 6.3). The estimate of g from the 2003 sample is nearly double the estimate of g from the 2002 sample; and the estimate from the 2-year sample is very different than the estimate from the 5-year sample. Clearly, a sample from one or two years is too small to provide a reliable estimate. On the other hand, the tiny difference between the estimates from the 5-year sample and the 10-year sample suggests that each of these samples may be large enough to provide a reliable estimate. In conclusion, the sample for the estimation of the prior probability g should be selected judiciously.

Example 6.4 *(Diagnosis of leukemia: representative prior sample)*
Leukemia is a cancer of the blood caused by growth of malignant white blood cells. To diagnose leukemia and its type (one of four), blood tests are performed. Suppose a new blood test has been developed that outputs the

Table 6.3 Estimates of the prior probability *g* obtained from different prior samples (Example 6.3).

Number of years	Years of record	Number of rainy days	Number of days with measurements	Estimate of *g*
1	2003	9	92	0.098
1	2002	5	92	0.054
2	2002–2003	14	184	0.076
5	1999–2003	51	460	0.111
10	1994–2003	101	858	0.118

probability of a patient having leukemia. To verify its performance, the test is tried on 200 patients, half of whom have the disease and half do not. The trial yields a joint sample from which the two conditional probability functions, f_0 and f_1, can be estimated (as detailed in the next section). But it would be incorrect to estimate the prior probability *g* from the joint sample (which yields $g = 0.5$) because the proportion of patients having the disease was chosen by an experimenter; hence, the joint sample is unrepresentative of the predictand. To find a representative prior sample, one could refer to epidemiological records of diseases diagnosed in the United States every year. For instance, during the 1990s, about 25,000 people were diagnosed with leukemia every year in the population of about 280,000,000; thus the prior probability of a person having leukemia was about 0.0001. A more refined estimate could be obtained by stratifying the prior sample according to age, as most frequently leukemia affects people over age 65.

6.1.6 Conditional Probability Functions

The second input element needed for the verification of forecasts is two conditional probability functions; they are defined as follows.

$f_w(y) = P(Y = y | W = w)$, for $y \in \mathcal{Y}$ and $w = 0, 1$. For a fixed $w \in \{0, 1\}$, the object f_w is the probability function of the probability variate *Y*, **conditional on the event** $W = w$. The two *conditional probability functions*, f_0 and f_1, quantify the stochastic dependence between the probability variate *Y* and the predictand *W*. Given the joint sample summarized in a contingency table (Table 6.1), the estimator of the conditional probability is given by

$$f_w(y_i) = \frac{N_{wi}}{N_w}, \qquad i = 0, 1, \ldots, I; \; w = 0, 1. \tag{6.6}$$

Table 6.4 shows an example. To interpret this table, let us consider two elements: $f_0(0.7) = 0.2$ means that on those occasions on which the event does not occur ($W = 0$), 20% of the time the forecast probability is $Y = 0.7$; and $f_1(0.7) = 0.3$ means that on those occasions on which the event occurs ($W = 1$), 30% of the time the forecast probability is $Y = 0.7$. The prior probability of the event occurring is $g = 0.4$.

The two conditional probability functions, f_0 and f_1, are graphed in Figure 6.2. The graph conveys the forecaster's *discrimination ability*: loosely speaking, the smaller the overlap between f_0 and f_1, the greater the ability to discriminate between the conditions that lead to event $W = 0$ and the conditions that lead to event $W = 1$. For a perfect forecaster (a clairvoyant), f_0 and f_1 do not overlap. For an uninformative forecaster (a useless forecaster), f_0 and f_1 overlap completely $(f_0 = f_1)$. A rigorous procedure that quantifies these notions is presented in Section 6.3.

There is a second interpretation of $f_w(y)$. For a fixed forecast probability $Y = y$, the object $f_w(y)$ is called the *likelihood* of event $W = w$. The pair of values $(f_0(y), f_1(y))$ constitutes the *likelihood function* of the predictand *W*, with $f_0(y)$ being the likelihood of event $W = 0$, and $f_1(y)$ being the likelihood of event $W = 1$. Applying this interpretation to the example in Table 6.4, one can say: when the forecast probability is $Y = 0.7$, the likelihood

Table 6.4 Example of the conditional probability functions f_0, f_1 of the probability variate Y, and the prior probability g of event $W = 1$, estimated from a joint sample summarized in Table 6.2.

		y				
		0.01	0.3	0.7	0.99	
w	0	0.4	0.3	0.2	0.1	0.6
	1	0.1	0.2	0.3	0.4	0.4

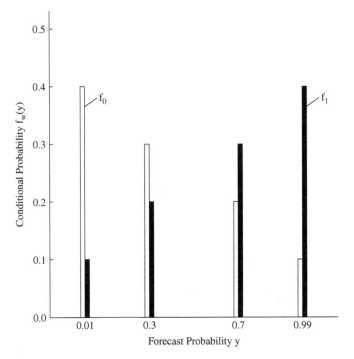

Figure 6.2 Conditional probability functions f_0 and f_1 of the probability variate Y; the conditioning is on events $W = 0$ and $W = 1$; the values of the ordinates come from Table 6.4.

of event $W = 0$ is 0.2, and the likelihood of event $W = 1$ is 0.3. There are as many likelihood functions of W as there are distinct forecast probabilities y in the sample space $\mathcal{Y}$. For this reason, the pair (f_0, f_1) is called the *family of likelihood functions*. For example, Table 6.4 specifies the family of four likelihood functions. The family of likelihood functions provides a complete characterization of the forecaster needed for forecast verification and for decision making.

6.2 Calibration

6.2.1 The Concept

Can the number y given by a forecaster be interpreted as a probability of event $W = 1$? Can the forecast probability y be used directly in a decision model?

In order that users may take a forecast probability at its face value, *normative interpretation* of the forecast must be maintained consistently over time or across events. Consistent interpretability requires *good calibration*. The concept is as follows. Suppose a forecaster assigned the same probability, say $y = 0.3$, to 1000 different but homogeneous events. It seems reasonable to anticipate that about 300 forecasts should be followed by the occurrence of events. In general, the probability should be interpretable as a *relative frequency*. The uniqueness of this interpretation makes the probability a means of communicating uncertainty from a forecaster to a decider, from an analyst to the public, from one individual to another. It is imperative, therefore, to have procedures for verifying the interpretation of probability. By their nature, logical probabilities (which are universally agreed upon) need not be verified; however, statistical probabilities, mathematical probabilities, and judgmental probabilities should be subjected to verification.

In a nutshell, the purpose of a calibration measure is to tell the forecaster and the users whether or not the normative interpretation has been maintained in a set of forecasts. This section presents a measure of calibration according to the Bayesian verification theory.

6.2.2 Bayesian Processor of Forecast

The Bayesian processor of forecast (BPF) takes as input the prior probability g, defined in Section 6.1.5, and the family of likelihood functions (f_0, f_1), defined in Section 6.1.6, and outputs two elements, defined as follows.

$\kappa(y) = P(Y = y)$, for $y \in \mathcal{Y}$. The function κ is the marginal probability function of the probability variate Y. The probability $\kappa(y)$ can be interpreted as the relative frequency with which the forecaster assigns probability y to event $W = 1$. The function κ is also called the *expected probability function* of Y — the name justified by its derivation. Because set $\mathcal{Y}$ contains only probabilities that have been used, $\kappa(y) > 0$ for every $y \in \mathcal{Y}$.

$\eta(y) = P(W = 1|Y = y)$, for $y \in \mathcal{Y}$. This is the probability of event $W = 1$, **conditional on forecast probability** $Y = y$. It is also called the *posterior probability* of event $W = 1$. It provides an answer to the following question posed by a user: When the forecaster issues probability $Y = y$, what is really the probability of event $W = 1$? To find the answer, the forecast probability $Y = y$ is treated as information; then, conditional on this information, probability $\eta(y)$ of event $W = 1$ is found.

The BPF applies two fundamental laws of probability. The first is the *total probability law*:

$$P(Y = y) = P(Y = y|W = 0)P(W = 0) + P(Y = y|W = 1)P(W = 1);$$

equivalently, employing the abbreviated notation,

$$\kappa(y) = f_0(y)(1 - g) + f_1(y)g, \qquad y \in \mathcal{Y}. \tag{6.7}$$

The second law is *Bayes' theorem*:

$$P(W = 1|Y = y) = \frac{P(Y = y|W = 1)P(W = 1)}{P(Y = y)};$$

equivalently, employing the abbreviated notation,

$$\eta(y) = \frac{f_1(y)g}{\kappa(y)}, \qquad y \in \mathcal{Y}. \tag{6.8}$$

Expressions (6.7) and (6.8) define the BPF.

Table 6.5 gives an example. To interpret this table, let us consider two elements: $\kappa(0.7) = 0.24$ means that the forecast probability $Y = 0.7$ is assigned to event $W = 1$ 24% of the time; $\eta(0.7) = 0.5$ means that on those occasions on which the forecast probability is $Y = 0.7$, the event occurs 50% of the time; in other words, given the forecast probability $Y = 0.7$, the probability of event $W = 1$ is 0.5.

Table 6.5 Example of the expected probability function κ of the probability variate Y and the posterior probabilities $\eta(y)$ of event $W = 1$ calculated from the BPF using f_0, f_1 listed in Table 6.4 and $g = 0.4$.

	y			
	0.01	0.3	0.7	0.99
$\kappa(y)$	0.28	0.26	0.24	0.22
$\eta(y)$	0.143	0.308	0.500	0.727

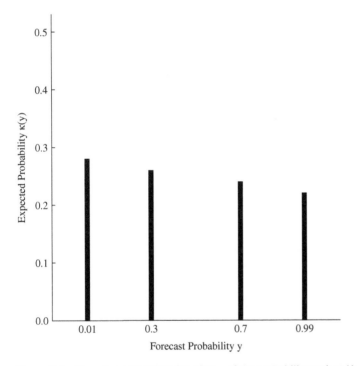

Figure 6.3 Expected probability function κ of the probability variate Y; the values of the ordinates come from Table 6.5.

The expected probability function κ of Y is graphed in Figure 6.3; it depicts the relative frequencies with which various forecast probabilities are used by the forecaster. The posterior probabilities $\eta(y)$ of event $W = 1$, conditional on forecast probabilities $y = 0.01, 0.3, 0.7, 0.99$, are graphed in Figure 6.4; before learning the purposes of this graph, let us examine the role of prior probability.

6.2.3 The Role of Prior Probability

First, look at expressions (6.7) and (6.8): the prior probability g impacts the expected probability $\kappa(y)$ linearly and the posterior probability $\eta(y)$ nonlinearly (because g enters both the numerator and the denominator, via $\kappa(y)$). To illustrate this impact, $g = 0.4$, which yields the functions κ and η listed in Table 6.5, is replaced by $g = 0.8$, functions κ and η are recalculated (using f_0, f_1 from Table 6.4), and the results are listed in Table 6.6. By comparing Table 6.6 with Table 6.5, we can judge the impact of g on κ and η. Clearly, this impact is significant. That is why it is imperative to estimate the prior probability g appropriately, as discussed in Section 6.1.5.

Table 6.6 Example of the expected probability function κ of the probability variate Y and the posterior probabilities $\eta(y)$ of event $W = 1$ calculated from the BPF using f_0, f_1 listed in Table 6.4 and $g = 0.8$.

	y			
	0.01	**0.3**	**0.7**	**0.99**
$\kappa(y)$	0.16	0.22	0.28	0.34
$\eta(y)$	0.500	0.727	0.857	0.941

Second, recall the concept of calibration introduced in Section 6.2.1. It is like its namesake in physics: each of the seven base physical quantities has defined a standard of measurement (e.g., the international prototype of the kilogram is the unit of mass), and every instrument for measuring a quantity should be calibrated against the standard (e.g., the balance in a grocery store should be calibrated as prescribed by a monitoring organization — in the United States, it is the National Institute of Standards and Technology). In forecasting, probability is the quantity, the forecaster is the instrument, and the standard against which the forecaster should be calibrated is the prior probability.

6.2.4 Probability Calibration Function

To characterize the performance of the forecaster with respect to his ability to maintain consistently the normative interpretation of the forecast probability, three terms are introduced.

Definition   A graph of the posterior probability $\eta(y)$ versus the forecast probability y, for all $y \in \mathcal{Y}$, is called the *probability calibration function* (PCF) of the forecaster.

Definition   The forecaster is said to be *well calibrated* **against** the prior probability g if for all $y \in \mathcal{Y}$,

$$\eta(y) = y.$$

Definition   The forecaster is said to be *semi-calibrated* if η is a strictly increasing function of y on $\mathcal{Y}$.

Let us apply these definitions to the example in Table 6.5. Figure 6.3 shows the expected probability function κ. From it, one can read the relative frequencies with which various forecast probabilities are used by the forecaster. Figure 6.4 shows the probability calibration function η. From it, one can infer that the forecaster is not well calibrated against the prior probability $g = 0.4$, but is semi-calibrated. When interpreting the PCF, one should consider also the sample sizes $N_0', N_1', \ldots, N_I'$ listed in the contingency table (Table 6.1). They help to judge the degree of statistical significance that one should attach to the points on the PCF: if for some $i \in \{0, 1, \ldots, I\}$, the sample size N_i' is small, then the statistical significance of the estimated conditional probability $\eta(y_i)$ is low. Thus good calibration, $\eta(y_i) = y_i$, or poor calibration, $\eta(y_i) \neq y_i$, at the particular forecast probability y_i may be an artifact of the sampling variability rather than an indication of the forecaster's skill.

The PCF serves two purposes. First, it answers the basic question of a decider: Can the forecast probability y be taken at its face value? If the forecaster is not well calibrated, then the answer is negative. In such a case, the PCF offers also the solution: having received the forecast probability y, the decider should use the posterior probability $\eta(y)$ instead of y. This process of mapping y into $\eta(y)$ is called *recalibration*.

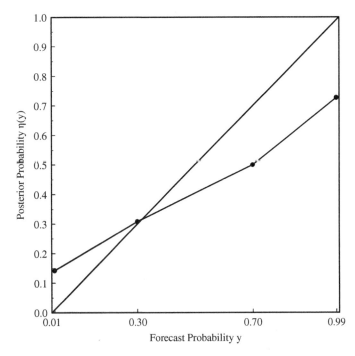

Figure 6.4 The probability calibration function (PCF) η of the forecaster when the set of used forecast probabilities is $\mathcal{Y} = \{y : y = 0.01, 0.3, 0.7, 0.99\}$; the values of the ordinates come from Table 6.5.

Second, the PCF provides feedback to the forecaster. If the forecaster is not well calibrated, then he may learn to adjust his judgments in the future. There is a strong empirical evidence that experts can improve their calibration through personal *feedback* and *training*. More on this topic is said in Section 6.5.

6.2.5 Generic Calibration Functions

When forecasts are produced judgmentally, the shape of the PCF reveals biases in judgments. It is instructive, therefore, to learn how to interpret the calibration functions for generic cases (Figure 6.5).

A) A *clairvoyant* needs only two probabilities, 0 and 1, to express her forecasts. Because she is always correct, the PCF consists of just two corner points, $\eta(0) = 0$ and $\eta(1) = 1$. We conclude that, in general, the PCF passing through the corners on the main diagonal is associated with good skill in detecting the conditions that lead to the occurrence or nonoccurrence of the event.

B) If on every occasion the forecaster assigns the same probability y, equal to the relative frequency g of the event, then the forecaster is well calibrated, and the PCF consists of just one point $\eta(g) = g$. Such a forecaster is called *naive* because to estimate probability g she needs only a sample of the predictand, but no information and knowledge about the phenomena that lead to a particular event and no skill in forecasting. We conclude that calibration is not a sufficient measure of forecaster's performance.

C) A PCF that is concave for $y < \frac{1}{2}$ and convex for $y > \frac{1}{2}$ is indicative of *overconfidence*. When the forecaster judges that event A is more likely to occur than not $(y > \frac{1}{2})$, she assigns probability y to A, which turns out to be larger than the conditional relative frequency of event A: $\eta(y) < y$. When the forecaster judges that the complementary event A^c is more likely to occur than not $(1 - y > \frac{1}{2})$, she assigns probability $1 - y$ to A^c, which turns out to be larger than the conditional relative frequency of event A^c: $1 - \eta(y) < 1 - y$, or $\eta(y) > y$.

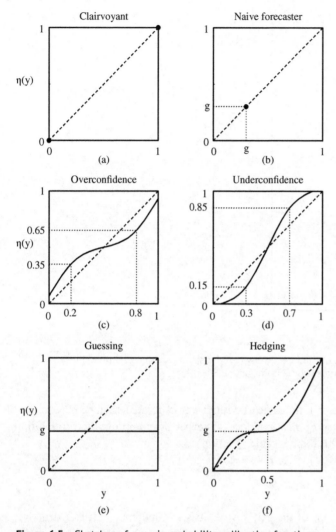

Figure 6.5 Sketches of generic probability calibration functions.

The extremes of overconfidence are manifested by the endpoints of the PCF not lying in the corners on the main diagonal: $\eta(0) > 0$ and $\eta(1) < 1$. The forecaster violated Cromwell's rule! (Recall Section 4.2.4.)

D) A PCF that is convex for $y < \frac{1}{2}$ and concave for $y > \frac{1}{2}$ is indicative of *underconfidence*. This is the opposite of overconfidence. At the extremes, the forecaster who possesses the skill of detecting with certainty event A on some occasions and event A^c on some other occasions, still assigns probabilities $y < 1$ and $y > 0$ on those occasions. Consequently, the endpoints of the PCF do not lie in the corners on the main diagonal but somewhere within the interval $(0, 1)$.

E) *Guessing* probability y amounts to generating y from a uniform probability function, independent of the event to occur, A or A^c. Mathematically, $f_0(y) = f_1(y) = 1/(I+1)$ for every $y \in \mathcal{Y}$. Consequently, equation (6.7) yields $\kappa(y) = 1/(I+1)$, and equation (6.8) yields $\eta(y) = g$ for every y. In other words, the conditional relative frequency $\eta(y)$ of event A is independent of the forecast probability y. We conclude that a constant PCF reflects the lack of forecasting skill.

Figure 6.6 Conceptual explanation of overconfidence in judgmental probability assessment.

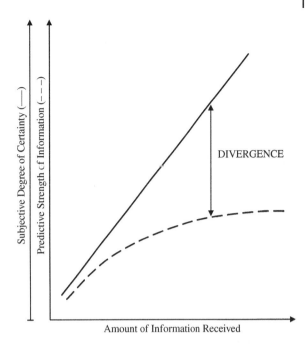

F) *Hedging* means assigning probability y closer to 0.5 whenever one has little or no predictive information. Normatively, on such occasions one should assign probability $y = g$ rather than $y = 0.5$. The PCF reveals this mistake by having a relatively flat section near $y = 0.5$ with $\eta(0.5) \approx g$. Another kind of hedging, premeditated and popular among economic, financial, and political analysts, is to assign probability $y = 0.4$ to any event A they have little clue about. Apparently, they perceive such "forecasts" as projecting expertise, while being defensible in hindsight, regardless of which event, A or A^c, occurs: if it is A, they note $P(A) = 0.4$ implied A was probable; if it is A^c, they note the implied odds were 6/4 against A (*The Wall Street Journal*, 27 February 2018). Again, the PCF can reveal this hedging tactic with $\eta(0.4) \approx g$.

6.2.6 Common Biases

The most common shape of the calibration function of an individual untrained in probabilistic reasoning and inexperienced in probability assessment is similar to the shape shown in graph C of Figure 6.5. A typical individual is mostly overconfident, hedges, and violates Cromwell's rule. Cognitive psychologists explain the tendency toward overconfidence (Figure 6.6): As the amount of information received and considered by an individual increases, the predictive strength of the total information increases initially, but eventually reaches a ceiling; however, the individual's degree of certainty increases with the amount of information received, regardless of its predictive strength. This divergence leads to overconfidence. In general, people appear to believe they know more than they actually do.

We can inject specificity to the above explanation by engaging the Bayesian theory of probability revision based on a sequence of information subsets (Section 4.4). Suppose an experiment is conducted in this paradigm, but the subjects are unaware of the probability adjustment principle (Section 4.4.7) and of the sign rules (Table 4.2). Consequently, if the predictive strength of each new information is positive (which is relatively easy to judge intuitively), and the stochastic dependence of the new information on the prior information is positive too, but is ignored (and is not easy to judge intuitively sans training), then the overestimation of probability y (when $y > \frac{1}{2}$)

is inevitable. Ergo, the PCF will reveal overconfidence — a predictable bias due to lack of training in probabilistic reasoning. (The bias of underconfidence can be explained similarly — see Exercise 9.)

6.3 Informativeness

6.3.1 The Concept

What is the forecaster's ability *to predict* the event $\{W = 1\}$? What is the forecaster's ability *to discriminate* between the event $\{W = 1\}$ and its complement $\{W = 0\}$? What is the forecaster's ability *to reduce uncertainty* about the predictand W? What is the forecaster's ability *to provide valuable information* for making decisions? These are equivalent questions that probe the second attribute of the forecaster, which is called *informativeness*.

The ultimate measure of informativeness is the economic value of the forecaster to rational deciders — those who make optimal decisions based on forecasts using decision models developed according to the principles of Bayesian decision theory, which we shall study in later chapters. When two or more forecasters provide forecasts of the same predictand, their economic values to a particular decider can be determined, and then the order of the forecasters in terms of their economic values can be established. However, the preference order of the forecasters may be different for every decider because the economic value of a forecaster depends upon elements of the decision model, in particular the prior distribution of the predictand and the utility function (or the loss function) which evaluates outcomes of decisions and events.

When a forecaster serves multiple users (as is the case with weather forecasters or market forecasters), it is infeasible to determine the economic value of the forecaster to every user. Likewise, when a forecast system is being developed, the users may still be unknown and thus modeling their decision problems may be infeasible. Still, the designer should like to select a forecast system that is the most valuable for every decider, and thus is preferred from a utilitarian point of view. Hence the following definition.

Definition Forecaster A is said to be *more informative than* forecaster B with respect to predictand W if, for every rational decider, the economic value of forecasts provided by A is **at least as high as** the economic value of forecasts provided by B.

The order of forecasters in terms of the *informativeness relation* does not always exist. Given two forecasters, it is possible that neither A is more informative than B, nor B is more informative than A. But if an order does exist, then it can be established via a statistical measure provided by the Bayesian verification theory; it is not necessary to determine the economic values of forecasters A and B for every rational decider!

In a nutshell, the purpose of an informativeness measure is to characterize the predictive skill of a forecaster, and to guarantee that the order of forecasters in terms of this measure corresponds to the order of forecasters in terms of the informativeness relation. This section presents such a measure.

6.3.2 Performance Probabilities

The informativeness measure is formulated from the viewpoint of a rational decider. Thus, it is necessary to sketch a decision problem. Suppose it is a detection problem: one must decide whether or not a target has been detected and an attack should be ordered; or one must decide whether or not an item exiting the production line is defective and should be rejected; or one must decide whether or not a hurricane warning should be issued and the population at risk should be evacuated. The elements of the detection problem needed now are as follows:

a — detection decision, $a \in \{0, 1\}$ with

$$a = 0 \quad \Longleftrightarrow \quad \text{action is not taken,}$$
$$a = 1 \quad \Longleftrightarrow \quad \text{action is taken;}$$

Table 6.7 Consequences of the detection decision, conditional on event.

		w	
		0	**1**
a	0	Q	M
	1	F	D

w — event indicator, $w \in \{0, 1\}$ with

$$w = 0 \quad \Longleftrightarrow \quad \text{event does not occur,}$$

$$w = 1 \quad \Longleftrightarrow \quad \text{event occurs.}$$

Conditional on event $W = w$ ($w = 0, 1$), each alternative decision ($a = 0, 1$) leads to different consequences. The totality of these consequences is given a label, as depicted in Table 6.7, and is termed as follows:

Q — Quiet state, $\qquad M$ — Missed event,

F — False alarm, $\qquad D$ — Detection.

The solution to the detection problem, to be derived in Chapter 7, provides an optimal decision rule which has the following structure. There exists a threshold t ($0 < t < \infty$) such that, given the forecast probability $y \in \mathcal{Y}$,

$$\text{if } \frac{f_1(y)}{f_0(y)} > t, \qquad \text{then } a = 1. \tag{6.9}$$

Functions f_1 and f_0 are the conditional probability functions of the probability variate Y defined in Section 6.1.6. The object $f_1(y)/f_0(y)$ is the *likelihood ratio on* event $W = 1$, given the forecast probability $Y = y$. The decision rule (6.9) states that it is optimal to take action when the likelihood ratio exceeds the threshold. Thus, given the set of used forecast probabilities

$$\mathcal{Y} = \{y : y = y_0, y_1, \ldots, y_I; f_0(y) + f_1(y) > 0\}, \tag{6.10}$$

one can construct a *detection set*

$$\mathcal{Y}_1(t) = \left\{ y : \frac{f_1(y)}{f_0(y)} > t, \quad y \in \mathcal{Y} \right\}, \tag{6.11}$$

such that $\mathcal{Y}_1(t) \subseteq \mathcal{Y}$. In other words, the detection set is a subset of the forecast probabilities used, each of which yields the likelihood ratio greater than the threshold t, and therefore implies taking action.

Finally, one can define the conditional probabilities of the consequences, given a value of the threshold t. These conditional probabilities provide a statistical characterization of the performance of the detection model which utilizes forecasts provided by a given forecaster: the *probability of detection* is

$$P(D|t) = P(Y \in \mathcal{Y}_1(t)|W = 1) = \sum_{y \in \mathcal{Y}_1(t)} f_1(y); \tag{6.12}$$

the *probability of a false alarm* is

$$P(F|t) = P(Y \in \mathcal{Y}_1(t)|W = 0) = \sum_{y \in \mathcal{Y}_1(t)} f_0(y); \tag{6.13}$$

the *probability of a missed event* is

$$P(M|t) = 1 - P(D|t); \tag{6.14}$$

the *probability of a quiet state* is

$$P(Q|t) = 1 - P(F|t). \tag{6.15}$$

6.3.3 The ROC Algorithm

The objective now is to calculate $P(D|t)$ and $P(F|t)$ for all values of t. The calculations are illustrated in Table 6.8 and proceed as follows.

Step 0 (Setup). Set up a table in which the first column is unlabeled and the remaining columns are labeled by $y = y_0, y_1, \ldots, y_I$. In the first two rows, list the conditional probabilities $f_1(y)$ and $f_0(y)$ for $y = y_0, y_1, \ldots, y_I$.

Step 1 (Likelihood ratios). For each $y = y_0, y_1, \ldots, y_I$, calculate the likelihood ratio $f_1(y)/f_0(y)$. Check that the likelihood ratios form an increasing sequence:

$$\frac{f_1(y_0)}{f_0(y_0)} \leq \frac{f_1(y_1)}{f_0(y_1)} \leq \cdots \leq \frac{f_1(y_I)}{f_0(y_I)}. \tag{6.16}$$

If this is not the case, then permute the columns of the table so that an increasing sequence is obtained. When two or more likelihood ratios are identical, the order of their columns is irrelevant.

Step 2 (Performance probabilities). Calculate $P(D|t)$ values using function f_1 and $P(F|t)$ values using function f_0 as follows. Place 0 in the last column. Next proceed leftward; in each column write the sum of the probabilities $f_w(y)$ listed to the right of the current column. Upon reaching the first column (the unlabeled column), the sum is 1. There are two special cases: (i) If two or more likelihood ratios are identical, then upon completing the calculations in the rightmost column with the common likelihood ratio, skip each subsequent column with the common likelihood ratio. (ii) If the likelihood ratio is ∞, then skip the column.

The two rows in Table 6.8 above Step 2 are not part of the algorithm but help to explain it. The objective of Step 2 is to calculate $P(D|t)$ and $P(F|t)$ for every $t \in (0, \infty)$. The likelihood ratios, calculated and ordered in Step 1, partition the space $(0, \infty)$ into $I + 2$ intervals. When inequalities (6.16) hold, with each being strict and none of the ratios being 0 or ∞, these intervals are

$$\left(0, \frac{f_1(y_0)}{f_0(y_0)}\right), \left[\frac{f_1(y_0)}{f_0(y_0)}, \frac{f_1(y_1)}{f_0(y_1)}\right), \cdots, \left[\frac{f_1(y_I)}{f_0(y_I)}, \infty\right). \tag{6.17}$$

Table 6.8 ROC tableau: calculation of the probabilities of detection $P(D|t)$ and the probabilities of false alarm $P(F|t)$ for all values of the threshold $t \in (0, \infty)$, when the set of used forecast probabilities is $\mathcal{Y} = \{y_0, y_1, y_2, y_3\}$.

	y	$\bullet$	y_0	y_1	y_2	y_3
Step 0	$f_1(y)$		0.1	0.2	0.3	0.4
	$f_0(y)$		0.4	0.3	0.2	0.1
Step 1	$\dfrac{f_1(y)}{f_0(y)}$		0.25	0.67	1.5	4
Interval of t		$(0, 0.25)$	$[0.25, 0.67)$	$[0.67, 1.5)$	$[1.5, 4)$	$[4, \infty)$
Detection set $\mathcal{Y}_1(t)$		$\mathcal{Y}$	$\{y_1, y_2, y_3\}$	$\{y_2, y_3\}$	$\{y_3\}$	$\varnothing$
Step 2	$P(D|t)$	1	0.9	0.7	0.4	0
	$P(F|t)$	1	0.6	0.3	0.1	0

Next, let us consider one interval at a time, beginning with the first. When the threshold t belongs to this interval, the detection set is $\mathcal{Y}_1(t) = \mathcal{Y}$; this follows from definition (6.11). When t belongs to the second interval, the detection set is $\mathcal{Y}_1(t) = \{y_1, \dots, y_I\}$. And so on. When t belongs to the last interval, the detection set is empty: $\mathcal{Y}_1(t) = \varnothing$. Thus to each interval of t, there corresponds a detection set. The $I + 2$ detection sets form a nested sequence. Given intervals (6.17), this sequence is

$$\mathcal{Y} \supset \{y_1, \dots, y_I\} \supset \{y_2, \dots, y_I\} \supset \dots \supset \{y_I\} \supset \varnothing. \tag{6.18}$$

It follows that as t increases from 0 to ∞, there are only $I + 2$ distinct detection sets $\mathcal{Y}_1(t)$. Hence each probability, $P(D|t)$ and $P(F|t)$, can take on only $I + 2$ distinct values. These values are calculated via equations (6.12)–(6.13) in Step 2.

6.3.4 Receiver Operating Characteristic

To characterize the performance of the forecaster with respect to his ability to predict the event, two constructs are introduced.

Definition   A graph of the probability of detection $P(D|t)$ versus the probability of false alarm $P(F|t)$ for all values of the threshold $t \in (0, \infty)$ is called the *receiver operating characteristic* (ROC) of the forecaster.

Definition   A piecewise linear function connecting the points of the ROC is called the ROC *outline*.

Figure 6.7 shows the ROC and the ROC outline resulting from the calculations reported in Table 6.8. From the viewpoint of any rational decider, the most preferred point of the unit square is the upper-left corner, $(0, 1)$, where $P(F|t) = 0$ and $P(D|t) = 1$ for all $t \in (0, \infty)$. This is the ideal point. A perfect forecaster would have the ROC consisting of just this point. The ROC of the forecaster in our example consists of five points. These points display the *trade-offs* between the probability of detection and the probability of false alarm that are offered by the forecaster to all deciders. For deciders with the threshold t for taking action such that $t \in [4, \infty)$, this forecaster offers the operating point $(0, 0)$; that is, $P(F|t) = 0$ and $P(D|t) = 0$. For deciders with $t \in [1.5, 4)$, this forecaster offers the operating point $(0.1, 0.4)$; that is, $P(F|t) = 0.1$ and $P(D|t) = 0.4$. A move from the operating point $(0, 0)$ to the operating point $(0.1, 0.4)$ entails an increase of the probability of detection (which is desirable) and a simultaneous increase of the probability of false alarm (which is undesirable); hence a trade-off. The nature of this trade-off is that as $P(F|t)$ increases, $P(D|t)$ increases at a diminishing rate; in other words, the ROC outline is a concave function. Finally, it should be noted that the values of the forecast probabilities $y_0, y_1, \dots, y_I$ are not involved at all in the calculation of $P(D|t)$ and $P(F|t)$. Hence, the ROC is independent of the forecast probabilities. To sum up, the ROC has five main properties:

1. The ROC displays all *trade-offs* between the probability of detection $P(D|t)$ and the probability of false alarm $P(F|t)$ that a given forecaster offers to all deciders.
2. The ROC of a *perfect forecaster* (a clairvoyant) consists of one operating point, $(0, 1)$.
3. The ROC of an *uninformative forecaster* (a useless forecaster) consists of two operating points, $(0, 0)$ and $(1, 1)$.
4. The ROC is independent of the forecast probabilities (and hence of the PCF).
5. The ROC outline is a piecewise linear, monotone increasing, concave function.

We can now turn to the primary purpose of the ROC, which is the comparison of forecasters.

Theorem (*Comparison of forecasters*)   *Forecaster A is more informative than forecaster B if and only if the ROC outline of A is superior to the ROC outline of B.*

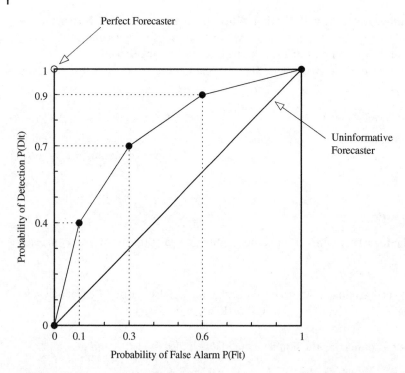

Figure 6.7 The receiver operating characteristic (ROC) of the forecaster who uses four probabilities, which yield distinct likelihood ratios. The ROC consists of five points; the ROC outline is a piecewise linear function connecting the ROC points; the coordinates of points come from Table 6.8.

Thus, the ROC is a sufficient and necessary measure for ordering two or more forecasters in terms of the *informativeness relation*. Section 7.6.3 gives the proof.

Figure 6.8 shows the ROC outlines of three forecasters: A, B, C. These are quite different forecasters if only because each uses a different number of probabilities (which can be inferred from the number of points of the ROC): A uses 4 probabilities, B uses 3 probabilities, and C uses 2 probabilities. Nevertheless, the theorem can be applied and the following inferences can be made: (i) Forecaster A is more informative than forecaster B. (ii) Forecaster C is more informative than forecaster B. (iii) Neither A is more informative than C, nor C is more informative than A; this is so because their ROC outlines cross each other (neither is superior to the other). It means that A may be more valuable for some deciders, whereas C may be more valuable for other deciders; however, each is preferred to B by all deciders. A general conclusion: The order of forecasters in terms of the informativeness relation may be *incomplete*.

6.3.5 Limiting Cases

The performance of any forecaster is, obviously, bounded by the performance of a perfect forecaster (a clairvoyant) and the performance of an uninformative forecaster (a useless forecaster). These two limiting cases, already mentioned in Section 6.3.4, are formally characterized in Table 6.9.

The *perfect forecaster* needs only two probabilities, $y_0 = 0$ and $y_1 = 1$, to express her forecasts; their conditional probabilities are $f_1(1) = f_0(0) = 1$ and $f_1(0) = f_0(1) = 0$. The two likelihood ratios are 0 and ∞. Consequently, regardless of their thresholds $t \in (0, \infty)$, all deciders take action if the forecast probability is 1 and do not take

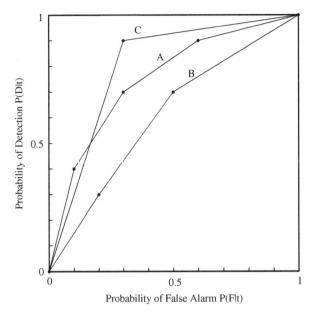

Figure 6.8 The ROC outlines of three forecasters: A, B, C.

action if the forecast probability is 0, so that $P(F|t) = 0$ and $P(D|t) = 1$ for all $t \in (0, \infty)$. Hence, the ROC consists of just one operating point, $(0, 1)$ — the ideal point. (Note that two intervals of t, $(0, 0)$ and $[\infty, \infty)$, are empty and, therefore, the first and the last column of the ROC tableau in Step 2 are left empty as well.)

The *uninformative forecaster* may use any number of probabilities $y_0, y_1, \ldots, y_I$ to express her forecasts; but every forecast y_i appears as if it were produced by *guessing*, or by generating y_i from a uniform probability function on the sample space $\mathcal{Y}$, independent of the event to occur, $W = w$ ($w = 0$ or $w = 1$). Thus the conditional probabilities are $f_0(y_i) = f_1(y_i) = 1/(I + 1)$ for every $i = 0, 1, \ldots, I$. Table 6.9 shows the case with $I = 1$. The two likelihood ratios,

Table 6.9 ROC tableau for a perfect forecaster and an uninformative forecaster, each employing only two probabilities from a set $\mathcal{Y} = \{y_0, y_1\}$.

	y	Perfect forecaster			Uninformative forecaster		
		•	y_0	y_1	•	y_0	y_1
Step 0	$f_1(y)$		0	1		0.5	0.5
	$f_0(y)$		1	0		0.5	0.5
Step 1	$\dfrac{f_1(y)}{f_0(y)}$		0	∞		1	1
Interval of t		$(0, 0)$	$[0, \infty)$	$[\infty, \infty)$	$(0, 1)$	$[1, 1)$	$[1, \infty)$
Detection set $\mathcal{Y}_1(t)$			$\{y_1\}$		$\mathcal{Y}$		$\varnothing$
Step 2	$P(D|t)$		1		1		0
	$P(F|t)$		0		1		0

1 and 1, are identical, which implies the lack of the discrimination ability. Consequently, all deciders with $t \in [1, \infty)$ never take action, so that $P(F|t) = 0$ and $P(D|t) = 0$; and all deciders with $t \in (0,1)$ always take action, so that $P(F|t) = 1$ and $P(D|t) = 1$. Hence, the ROC consists of two operating points, $(0,0)$ and $(1,1)$. (Note that one interval of t, $[1,1)$, is empty and, therefore, the middle column of the ROC tableau in Step 2 is left empty as well.)

6.3.6 Special Cases

In Step 1 of the ROC algorithm, one may encounter three special cases: (i) the likelihood ratios form a nonmonotone sequence and, thereby, violate condition (6.16); (ii) two or more likelihood ratios are identical; and (iii) one or more likelihood ratios are ∞. The instructions of the ROC algorithm handle each of these cases. Table 6.10 illustrates them; Figure 6.9 shows the resultant ROC.

First, the nonmonotone sequence of the likelihood ratios is identified in Step 1; the order of columns y_1 and y_2 is reversed; the permuted sequence of the likelihood ratios is increasing; and the algorithm is restarted.

Second, with the largest likelihood ratio being ∞, the interval $[\infty, \infty)$ of t is empty and, therefore, the last column of the ROC tableau in Step 2 is left empty as well. Consequently, the first operating point is $(0, 0.3)$; it implies a *partially perfect predictive skill*, which is available to all deciders with $t \in [1, \infty)$. Specifically, when the event does not occur ($W = 0$), the forecast probability is never y_3, as $f_0(y_3) = 0$; hence $P(F|t) = 0$ and y_3 implies the occurrence of event ($W = 1$). When the event does occur ($W = 1$), the forecast probability is y_3 with the relative frequency $f_1(y_3) = 0.3$; hence $P(D|t) = 0.3$.

Table 6.10 ROC tableau illustrating three special cases: a nonmonotone sequence of likelihood ratios, two identical likelihood ratios, and one infinite likelihood ratio.

	y	•	y_0	y_1	y_2	y_3
Step 0	$f_1(y)$		0.1	0.4	0.2	0.3
	$f_0(y)$		0.2	0.4	0.4	0.0
Step 1	$\dfrac{f_1(y)}{f_0(y)}$		0.5	1	0.5	∞

	Permutation of y	•	y_0	y_2	y_1	y_3
Step 0	$f_1(y)$		0.1	0.2	0.4	0.3
	$f_0(y)$		0.2	0.4	0.4	0.0
Step 1	$\dfrac{f_1(y)}{f_0(y)}$		0.5	0.5	1	∞
Interval of t		$(0, 0.5)$	$[0.5, 0.5)$	$[0.5, 1)$	$[1, \infty)$	$[\infty, \infty)$
Detection set $\mathcal{Y}_1(t)$		$\mathcal{Y}$		$\{y_1, y_3\}$	$\{y_3\}$	
Step 2	$P(D|t)$	1		0.7	0.3	
	$P(F|t)$	1		0.4	0.0	

Figure 6.9 The ROC calculated in Table 6.10; it consists of three operating points (dots); the operating point $(0, 0.3)$ implies a partially perfect predictive skill. Shown are also two redundant points (circles) resulting from permutations of two forecast probabilities which give identical likelihood ratios.

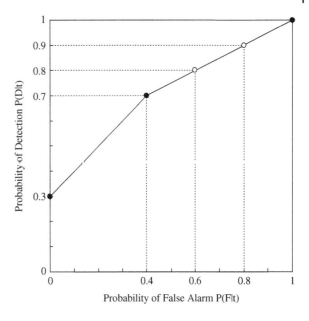

Third, with the likelihood ratios in columns y_0 and y_2 being identical, column y_0 is skipped in Step 2. Note that if the calculations were performed anyway, the resultant operating point would be $(0.8, 0.9)$; this point lies on the line segment with the endpoints $(0.4, 0.7)$ and $(1, 1)$; hence it does not alter the ROC outline (see Figure 6.9). If the order of columns y_0 and y_2 were reversed, then the column to be skipped in Step 2 would be y_2. If the calculations were performed anyway, the resultant operating point would be $(0.6, 0.8)$; this point lies on the line segment with the endpoints $(0.4, 0.7)$ and $(1, 1)$; hence it does not alter the ROC outline (see Figure 6.9). This illustrates why the order of columns having identical likelihood ratios is irrelevant.

6.4 Verification Scores

We already know that the forecaster's performance is characterized completely by two functions: a probability calibration function and a receiver operating characteristic, which serve different normative purposes, and are independent of each other.

For lesser purposes (e.g., for monitoring the forecaster's performance over time, on a monthly, or quarterly, or yearly time scale), it may be convenient to compress a function into a score and then to compare the scores. Of course, such a comparison is incomplete: a score is only a proxy measure of performance. With that word of caution, we present two scores that are consistent with the Bayesian verification theory: one score is a necessary (but not sufficient) measure of calibration, and another score, in two versions, is a necessary (but not sufficient) measure of informativeness. But first let us lay down the background.

6.4.1 Bernoulli Distribution

The predictand W is a Bernoulli variate. Given a prior probability $g = P(W = 1)$, the prior probability function (p.f.) of W is specified by

$$P(W = w) = g^w (1 - g)^{1-w}, \qquad w = 0, 1;$$

the prior mean of W is

$$E(W) = \sum_{w=0,1} wP(W = w)$$
$$= 0(1 - g) + 1g$$
$$= g;$$

and the prior variance of W is

$$Var(W) = E[(W - E(W))^2]$$
$$= \sum_{w=0,1} (w - g)^2 P(W = w)$$
$$= (0 - g)^2(1 - g) + (1 - g)^2 g$$
$$= g(1 - g).$$

Whereas the prior probability g remains fixed, the forecast probability y may be different on every occasion. Suppose the forecaster is well calibrated so that $\eta(y) = y$ for all $y \in \mathcal{Y}$, and thus probability y can be taken at its face value. Given $Y = y$, the Bernoulli variate W has

p.f.: $P(W = w|Y = y) = y^w(1 - y)^{1-w}, \qquad w = 0, 1;$

mean: $E(W|Y = y) = y;$

variance: $Var(W|Y = y) = y(1 - y).$

The mean is a linear function of y; the variance is a quadratic function of y that attains the minimum value of 0 at either $y = 0$ or $y = 1$, and the maximum value of 0.25 at $y = 0.5$. A clairvoyant would use only $y = 0$ or $y = 1$, and thus, the variance of W would be zero on every occasion — there would be no uncertainty about the occurrence of the event.

Now suppose the forecaster is not well calibrated. Therefore, the recalibrated probability $\eta(y)$ should be used instead of the probability y stated by the forecaster. Given $Y = y$, the Bernoulli variate W now has

p.f.: $P(W = w|Y = y) = [\eta(y)]^w[1 - \eta(y)]^{1-w}, \qquad w = 0, 1;$

mean: $E(W|Y = y) = \eta(y);$

variance: $Var(W|Y = y) = \eta(y)[1 - \eta(y)].$

6.4.2 Calibration Score

The *calibration score* is defined as the root-mean-square difference (a weighted Euclidean distance) between the probability calibration function and the line of perfect calibration:

$$CS = \{E[(E(W|Y) - Y)^2]\}^{1/2}$$
$$= \{E[(\eta(Y) - Y)^2]\}^{1/2}. \tag{6.19}$$

When the probability variate Y is discrete,

$$CS = \left\{ \sum_{i=0}^{I} (\eta(y_i) - y_i)^2 \kappa(y_i) \right\}^{1/2}. \tag{6.20}$$

The CS is bounded, $0 \leq CS \leq 1$, with $CS = 0$ implying perfect calibration and $CS = 1$ implying the worse possible calibration. Who could have $CS = 1$? A confused clairvoyant who states $y = 0$ when $W = 1$ and $y = 1$ when $W = 0$, so that $\eta(0) = 1$ and $\eta(1) = 0$.

6.4.3 Variance Score

The variance of the predictand W is a measure of uncertainty. Inasmuch as the purpose of a forecast is to reduce uncertainty, it seems sensible to consider the posterior variance of W as a proxy measure of informativeness.

The *variance score* is defined as the expected posterior variance of the predictand W:

$$VS = E[Var(W|Y)]$$
$$= E[\eta(Y)[1 - \eta(Y)]]. \tag{6.21}$$

When the probability variate Y is discrete,

$$VS = \sum_{i=0}^{I} \eta(y_i)[1 - \eta(y_i)]\kappa(y_i). \tag{6.22}$$

The VS is bounded, $0 \leq VS \leq 0.25$. The smallest score, $VS = 0$, implies that each recalibrated probability has value 0 or 1. Such a forecaster is equivalent to a clairvoyant. The largest score for the event occurring with relative frequency g is $VS = g(1 - g)$. This score can be achieved by either a naive forecaster, or by a forecaster who guesses, for whom $\eta(y_i) = g$ for all y_i. Such a forecaster is said to be *uninformative*. The upper bound, $VS = 0.25$, is possible only when the event occurs with relative frequency $g = 0.5$.

Besides the intuitive justification for the variance score, there is a theoretic justification. The variance score is a necessary (though not sufficient) measure of informativeness. To wit, in a comparison of two forecasters, A and B, whose scores are VS^A and VS^B, respectively, the following can be said.

Theorem *(Variance score)* *If forecaster A is more informative than forecaster B, then $VS^A < VS^B$.*

6.4.4 Uncertainty Score

Oftentimes it is desirable to compare the predictive performance of a forecaster (or several forecasters) across several predictands. For example, the National Weather Service may wish to compare the performance of its system for forecasting precipitation occurrence in different months. An obstacle arises because the prior probability g of precipitation occurrence varies from month to month, and so does the prior variance $g(1 - g)$ of the predictand. For instance, based on a 4-year sample (1997–2000) the estimates of g for the 6-h period 1200–1800 in Buffalo, New York, are as follows: $g = 0.098$ in October, which implies $0 \leq VS \leq 0.0884$; and $g = 0.268$ in December, which implies $0 \leq VS \leq 0.1962$. Because the range of the feasible VS values differs from month to month, a comparison of the variance scores for different months is meaningless. In general, the variance scores for different predictands are not comparable unless the prior probabilities of the predictands are identical. To overcome this obstacle, the variance score is rescaled.

The *uncertainty score* is defined as the ratio of the variances of the predictand W — the expected posterior variance divided by the prior variance:

$$US = \frac{E[Var(W|Y)]}{Var(W)}. \tag{6.23}$$

Given expression (6.22) and its properties, the uncertainty score equals the scaled variance score:

$$US = \frac{VS}{g(1 - g)}. \tag{6.24}$$

The US is bounded, $0 \leq US \leq 1$. The smallest score, $US = 0$, implies that the forecast eliminates the uncertainty; this is the case of a *perfect* forecaster. The largest score, $US = 1$, implies that the forecast does not reduce the uncertainty at all; this is the case of an *uninformative* forecaster. In general, the US provides a quantitative answer to the questions: By how much does the forecast reduce the variance (uncertainty) on average? What is the predictive skill of the forecaster? How much information is there in the forecast, on average?

Like the variance score, the uncertainty score is a necessary (though not sufficient) measure of informativeness. To wit, in a comparison of two forecasters, A and B, whose scores are US^A and US^B, respectively, the following can be said.

Theorem *(**Uncertainty score**)* *If forecaster A is more informative than forecaster B, then $US^A < US^B$.*

6.4.5 Quadratic Score

The quadratic score, also known as the Brier score, is defined by

$$QS = E[(W - Y)^2], \tag{6.25}$$

where the expectation is taken with respect to Y and W. In Section 6.6.4 it is proven that

$$QS = CS^2 + VS. \tag{6.26}$$

The QS is bounded, $0 \leq QS \leq 1$. The smallest score, $QS = 0$, is achievable only by a clairvoyant, for whom $CS = 0$ and $VS = 0$. The largest score, $QS = 1$, is achievable by a confused clairvoyant, for whom $CS = 1$ and $VS = 0$.

While popular, the quadratic score is dubious. It is a sum of two scores, each of which measures a different attribute of forecaster's performance. The calibration score, CS, measures the *calibration skill* — the ability to judge the degree of certainty. The variance score, VS, measures the *predictive skill* — the ability to reduce uncertainty. Except for the clairvoyant, the two skills need not be possessed to the same degree — they are independent. And possessing one skill is not a substitute for the lack of the other skill — they are not compensatory.

Furthermore, CS can be decreased mathematically via recalibration of the forecast probabilities (as the BPF does), whereas VS cannot be decreased by any external means, only by improving the forecaster — either enhancing the predictive skill of the expert (I, K, E in Chapter 4), or finding a more informative predictor (X in Chapter 5). Hence CS and VS should not be commingled. (Exercises 2, 3, and 4 illustrate the pitfall of ranking the forecasters in terms of the QS.)

Note on scores. There are many verification scores, in particular so-called proper scoring rules. To sift through them, Bayesian verification theory, which represents a coherent viewpoint of a rational decider, offers these guidelines. Firstly, no single score is sufficient for verification of probabilistic forecasts of events. Secondly, a score is potentially useful only if it constitutes a necessary measure of one attribute — either calibration or informativeness. Thirdly, a score must have a purpose — like those stated at the conclusions of Sections 6.2.4 and 6.3.4.

6.5 Forecast Attributes and Mental Processes

When probabilistic forecasts are made judgmentally, the two verification attributes — informativeness and calibration — can be connected with two mental processes — cognition and metacognition. This connection between statistics and psychology is not only interesting but also useful: it offers a framework for analyzing and improving one's judgmental skills.

6.5.1 Cognition and Metacognition

Cognition is the mental process of knowing, judging, deciding. *Metacognition* is the mental process of being aware of one's own mental processes in order to perform intellectual tasks better.

Example 6.5 *(Life in general)*

With each cognitive skill (CS) or ability, there is associated a metacognitive skill (MS) or ability.

CS: to learn new knowledge.
MS: to know that one knows.

Having answered a question on a test of factual knowledge, the student demonstrates this metacognitive skill by answering correctly the question: Is your answer to the original question correct?

CS: to recall a fact (a name) from one's memory.
MS: to know that a fact (a name) is stored in one's memory.

A manifestation of this metacognitive skill is the tip-of-the-tongue phenomenon: one is unable to recall a fact at the very instant the question is posed, especially if unexpectedly, but one is certain that one knows the fact and will recall it eventually.

CS: to read and comprehend.
MS: to know when one comprehended.

A manifestation of the lack of this metacognitive skill is a student surprised by failing to explain correctly a physical process during an examination, and offering as an excuse the effort he expended: "But I studied for 10 hours!" (A frequent reason for this failure: the student confuses knowledge with familiarity.)

CS: to possess an expertise in a domain.
MS: to assign the correct degree of certainty to one's opinion (hypothesis, diagnosis, prognosis).

This example brings us back to the main topic of our study — judgmental forecasting by an expert. Let us, therefore, consider in some detail how two experts, a physician and an investor, should reason.

Example 6.6 *(Family physician)*

A good physician possesses both a diagnostic skill (a cognitive skill) and a metadiagnostic skill (a metacognitive skill). (i) *Diagnostic skill* — the skill in identifying a set of diagnoses containing the correct one. Given patient information (e.g., medical history, current symptoms, test results), the physician identifies a set of diagnoses such that, in his judgment, each diagnosis is possible to be true and one is true, but presently he is uncertain which one. This judgment depends upon patient information, medical knowledge, and clinical experience. (ii) *Metadiagnostic skill* — the skill in assigning the correct degree of certainty to a hypothesized diagnosis. It depends upon the physician's clinical experience and training of intuition (in probabilistic reasoning).

Example 6.7 *(Commodity investor)*

A good investor possesses both a prognostic skill (a cognitive skill) and a metaprognostic skill (a metacognitive skill). (i) *Prognostic skill* — the skill in identifying a set of commodities containing at least one that will make a big move. Given market information (e.g., historical patterns of annual cycles of prices, current balances between supply and demand, possible scenarios of future economic, political, and weather events that may affect these balances), the investor identifies a set of commodities (e.g., crude oil, copper, soybeans, the Swiss franc) each of which may, in her judgment, make a big move (in terms of its price) within a specified time period (e.g., the next 6 months), and then predicts the direction of the move (up or down). This judgment depends upon market information, commodity knowledge, and trading experience. (ii) *Metaprognostic skill* — the skill in assigning the correct degree of certainty to a hypothesized move. It depends upon the investor's trading experience and training of intuition (in probabilistic reasoning).

6.5.2 Skill Measures

We conclude this chapter with a succinct integration of theoretical results and empirical findings that pertain to probabilistic forecasting.

Two connections. The Bayesian verification measures and the mental processes are connected thusly.

- Making well-calibrated probabilistic forecasts is a metacognitive skill. Therefore, measures of calibration are also measures of a metacognitive skill: PCF, *CS*.
- Making informative probabilistic forecasts is a cognitive skill. Therefore, measures of informativeness are also measures of a cognitive skill: ROC, *VS*, *US*.

Independence properties. As mathematical concepts, calibration and informativeness are independent attributes of the forecaster (human or statistical). As mental processes, metacognition and cognition are mostly independent of each other. In the context of probabilistic forecasting, the metacognitive skill is usually unrelated to information and knowledge; it can be mastered through training and experience (with personal verification feedback, of course). Thus, the metacognitive skill is potentially transferable from one domain to another.

The role of probability. The two connections between the verification measures and the mental processes shed light on the role of probability. In addition to being the logic that supports mathematical modeling and probabilistic reasoning (as discussed in Section 4.1.3):

- Probability is the language of uncertainty.
- It is a means of conveying to others how much one knows (about the event being forecasted).
- Assessing probability which is well calibrated, and therefore usable directly in decision making, requires mastering a metacognitive skill.

6.6 Concepts and Proofs

6.6.1 Calibration of Bayesian Forecaster

Suppose the forecast probability y of some repeatable event A will be calculated from a Bayesian forecaster developed according to the theory of Chapter 5. Can anything be said about the potential calibration of this forecaster? In Section 5.1.5, it was proven that the Bayesian forecaster is well calibrated in the margin — a weak version of calibration; a strong version was deferred to this section. Here it is.

Theorem *(Calibration of Bayesian forecaster)* *A Bayesian forecaster of binary predictand W, employing the prior probability $g = P(W = 1)$, is well calibrated against that prior probability.*

Proof: Recall Bayesian forecaster (5.3), which takes realization $x \in \mathcal{X}$ of a continuous predictor X, and outputs the posterior probability $\pi(x) = P(W = 1 | X = x)$:

$$y = \pi(x) = \left[1 + \frac{1-g}{g} \frac{f_0(x)}{f_1(x)} \right]^{-1}, \tag{6.27}$$

where $y \in (0, 1)$ is a real number. To keep the proof general and simple, the BPF from Section 6.2.2 is replaced by a continuous version, which requires no discretization of Y: (i) the conditional probability function f_w, defined in Section 6.1.6, is replaced by the conditional density function h_w, for $w = 0, 1$; (ii) equation (6.8) is rewritten in the form parallel to the above equation. The resultant BPF takes realization $y \in (0, 1)$ of the probability variate Y, and outputs the posterior probability $\eta(y) = P(W = 1 | Y = y)$:

$$\eta(y) = \left[1 + \frac{1-g}{g} \frac{h_0(y)}{h_1(y)} \right]^{-1}, \tag{6.28}$$

where $\eta(y) \in (0, 1)$ is a real number. Per definition, the forecaster is well calibrated against the prior probability g if $\eta(y) = y$ for all $y \in (0, 1)$. Replacing $\eta(y)$ by the right side of equation (6.28), and rearranging the terms, yields the equivalent definition: for all $y \in (0, 1)$,

$$\frac{h_0(y)}{h_1(y)} = \frac{g}{1-g} \frac{1-y}{y}.$$

Next, making use of the equality $y = \pi(x)$, replacing $\pi(x)$ by the right side of equation (6.27), and rearranging the terms, yields

$$\frac{h_0(y)}{h_1(y)} = \frac{g}{1-g} \frac{1-\pi(x)}{\pi(x)}$$

$$= \frac{g}{1-g} \frac{1-g}{g} \frac{f_0(x)}{f_1(x)}$$

$$= \frac{f_0(x)}{f_1(x)}.$$

That is, for every pair (x, y) such that $y = \pi(x)$, the two likelihood ratios against event $W = 1$ are equal; hence, the right sides of equations (6.28) and (6.27) are equal. Hence, the left sides are equal: $\eta(y) = y$. QED.

Of course, when the calibration is verified empirically, as described in this chapter, one should not be surprised that $\eta(y) = y$ does not hold exactly for all $y \in \mathcal{Y}$. But it is reassuring to know that the Bayesian forecaster is potentially well calibrated (and assuredly so under ideal conditions: when the models for f_0 and f_1 are perfect and the joint sample size N is infinite).

6.6.2 Calibration Measures

The PCF serves two normative purposes stated at the conclusion of Section 6.2.4. It is not the normative purpose of the PCF to order the alternative forecasters, or to choose the preferred one. Hence, there is no mathematical definition of a "better calibrated" forecaster. Of course, one may apply the term subjectively in some informal discussion.

The *CS* is a necessary (but not sufficient) measure of calibration in the logical sense: (i) given the PCF, one can calculate the *CS*, but not vice versa; (ii) given only the *CS*, one cannot accomplish the two normative purposes of the PCF.

6.6.3 Informativeness Measures

The proof that the ROC is a sufficient and necessary measure of informativeness (the theorem of Section 6.3.4) is presented in Chapter 7. The proofs that the *VS* and *US* are necessary measures of informativeness (the theorems of Sections 6.4.3 and 6.4.4) are beyond the mathematical scope of this book.

6.6.4 Decomposition of Quadratic Score

The proof of equation (6.26) proceeds from equation (6.25), in which the double expectation is replaced by the iterated expectation: $E_{Y,W} = E_Y \circ E_{W|Y}$. Specifically,

$$
\begin{aligned}
QS &= E_{Y,W}\left[(W - Y)^2\right] \\
&= E_Y\left[E_{W|Y}(Y^2 - 2YW + W^2)\right] \\
&= E\left[Y^2 - 2YE(W|Y) + E(W^2|Y)\right] \\
&= E\left[Y^2 - 2YE(W|Y) + E^2(W|Y) + Var(W|Y)\right] \\
&= E[(E(W|Y) - Y)^2] + E[Var(W|Y)] \\
&= CS^2 + VS.
\end{aligned}
$$

Between the third and fourth lines, use is made of an expression for the conditional variance:

$$
Var(W|Y) = E(W^2|Y) - E^2(W|Y).
$$

QED.

Bibliographical Notes

The Bayesian theory of probability calibration was formalized by DeGroot and Fienberg (1982, 1983), and by Lindley (1982). Early laboratory experiments on subjective probability assessments, empirical calibration functions, and inferences of judgmental biases were reviewed by Lichtenstein et al. (1982). That proper training and personal verification feedback can improve calibration of probability judgments was demonstrated in laboratory settings by Lichtenstein and Fischhoff (1980). In real-world settings, Allan H. Murphy, professor of atmospheric sciences at Oregon State University in Corvallis, directed numerous experiments on probabilistic forecasting of various weather elements. They demonstrated that such forecasting is feasible and that, with experience and verification feedback, professional weather forecasters become nearly perfectly calibrated (Murphy and Winkler, 1977, 1982; Murphy and Daan, 1984). Metacognition, as defined by Flavell (1976), has been studied extensively, especially by educators. The connection between metacognition and calibration was inspired by the works of Lichtenstein and Fischoff (1977), and Hosseini and Ferrell (1982), but diverges from them on the question of measures.

Exercises

1 **Probabilistic test** — Chapter 4, Exercise 9 *continued*. Retrieve the contingency table and complete the following exercises.
    1.1 Construct and graph your probability calibration function and your receiver operating characteristic.
    1.2 Calculate your calibration score and your uncertainty score.
    1.3 Interpret the results and draw conclusions concerning your performance on this "probabilistic test of general knowledge".

2 **Interpreting angiograms**. Two radiologists applied for a position in a hospital. To test their diagnostic skills, each was given the same set of 240 angiograms (x-ray pictures of veins or arteries into which iodine dye was injected), and was asked to assess, for every angiogram, the probability y of an abnormality (blockage or

widening in the blood vessels). The angiograms came from files of previous patients whose state (normal, $w = 0$; or abnormal, $w = 1$) was eventually diagnosed with certainty by other procedures. Each of the two radiologists, A and B, chose to use only two probability values. The contingency tables compiled after the assessments are as follows:

		Radiologist A					Radiologist B	
		y					*y*	
		0,1	0,9				0,4	0.6
w	0	72	48		*w*	0	60	60
	1	24	96			1	40	80

2.1 Determine which radiologist is better calibrated. (Construct the probability calibration functions, calculate the calibration scores, interpret them.)

2.2 Determine if one radiologist is more informative than the other. (Construct the receiver operating characteristics, calculate the uncertainty scores, interpret them.)

2.3 Overall, which candidate would you recommend for hiring? And what, if any, feedback would you give to each of the candidates?

3 **Interpreting angiograms**— *continued*: **training for calibration**. Having received the feedback, radiologist A contemplates what the verification of his diagnoses would have been had he used probability 0.3 in lieu of probability 0.1, and probability 0.7 in lieu of probability 0.9.

3.1 Produce a new probability calibration function, calibration score, receiver operating characteristic, and uncertainty score.

3.2 Compare the new verification results with the verification results obtained in Exercise 2; draw conclusions.

4 **Interpreting angiograms** — *continued*: **quadratic score**. A hospital administrator wants to boil down the comparison of the two radiologists to a comparison of two numbers — the quadratic scores.

4.1 Calculate the quadratic score of each radiologist using the verification results obtained in Exercise 2. Which candidate would be hired? Compare this decision with your recommendation in Exercise 2.3; draw a conclusion.

4.2 Calculate the quadratic score of radiologist A using the verification results obtained in Exercise 3. If this quadratic score of A were compared with the quadratic score of B, which candidate would be hired?

4.3 Argue for or against the use of the quadratic score as a performance measure of a radiologist (in particular) and a forecaster (in general).

5 **Probabilistic choice test**. Imagine a test with two-choice questions. On the deterministic version of the test, the student chooses an answer. A measure of the student's knowledge (equivalently, the student's cognitive ability to retain new knowledge) is the proportion of correct answers. On the probabilistic version of the test, the student chooses the answer he judges to be correct more likely than not, and assigns to it probability y ($\frac{1}{2} \leq y \leq 1$) to convey his degree of certainty about the answer being correct. During grading, each chosen answer gets label $w = 1$ if correct, and $w = 0$ if incorrect. Here are the contingency tables compiled after the test for two students:

		Student A					Student B		
			y					y	
		0.6	0.8	1.0			0.5	0.6	0.7
w	0	20	8	2	w	0	20	7	3
	1	30	32	8		1	25	28	17

5.1 Estimate the prior probability of event $W = 1$ for each student via the second viewpoint described in the hint below.

5.2 Construct and graph the PCF and ROC for each student.

5.3 Calculate the *CS*, *VS*, *US*, and *QS* for each student.

5.4 Interpret the results for each student and then compare them.

5.5 Design two rules for assigning a letter grade from the set {A, B, C, D, F}: (i) a rule based on (*CS*, *US*); (ii) a rule based on *QS*. Apply each rule to the two students. *Note*: the rule may include a parameter whose value is set by the teacher.

5.6 Which rule would you recommend to the teacher and why?

Hint regarding the prior probability. (i) One viewpoint presumes that *a priori* (before reading the test questions) the student is unable to know whether or not he can discriminate correct from incorrect answers; hence $g = P(W = 1) = \frac{1}{2}$. But students possessing metacognitive skill know before the test whether they are prepared well or not; hence for them $g \neq \frac{1}{2}$ is more sensible. (ii) Another viewpoint, therefore, recognizes g as a personal probability and allows its estimation from the joint sample. To wit, the student chooses the answer whose correctness he forecasts in terms of probability y; then this answer gets labeled true or false (the event chosen by the student occurs or not). Ergo, $g = P(W = 1)$ should equal the proportion of correct answers, which may vary from student to student. The PCF measures the calibration of the assigned probabilities y against the student's own g. In a sense, g is the prior probability but *revealed a posteriori*. (Conceivably, g could be assessed *a priori* by asking the student, before the start of the test, to assess the proportion of questions he believes he is prepared to answer correctly, or the proportion of the tested material he knows well. After the test, g could be compared with the proportion of correct answers, and thereby its calibration could be verified as well.)

6 **Typical choice experiment**. Subjects had received some training in probability assessment and then were asked to answer 500 general knowledge two-choice questions of the type: Which continent has larger population, Europe or Africa? Having chosen the answer, the subject assigned to it probability y ($\frac{1}{2} \leq y \leq 1$) to convey her degree of certainty about the answer being correct. Here is the contingency table for one subject:

				y			
		0.5	0.6	0.7	0.8	0.9	1.0
w	0	99	26	11	3	1	2
	1	190	52	31	19	28	38

6.1 Estimate the prior probability of event $W = 1$ via the second viewpoint described in the hint to Exercise 5.

6.2 Construct and graph the PCF and ROC.

6.3 Calculate *CS* and *US*.

6.4 Interpret the results. In particular, characterize the judgmental biases of this subject, per the guidelines of Section 6.2.5.

6.5 Repeat Exercises 6.2–6.4 with the prior probability $g = P(W = 1) = \frac{1}{2}$.

6.6 Compare the two sets of results obtained with different values of the prior probability. Draw conclusions. *Hint*. See the hint for Exercise 5.

Note. The contingency table was published by Lindley (1982), but it had come from Sarah Lichtenstein (see bibliographical notes at the end of Section 6.6).

7 **Rare event forecasts**. In 1884, before the tools of modern meteorology were invented (numerical weather prediction models, radars, satellites, supercomputers), a scientist made 2806 experimental forecasts of tornadoes occurring in the contiguous USA during the next 8 hours. The forecasts were deterministic, no or yes, and thus we interpret them as probabilities, 0 or 1. Having no complete contingency table, we suppose it was:

		y 0	*y* 1
w	0	2676	77
	1	21	32

The scientist reported only the *success rate* of 96.5% — the number of correct forecasts, $(0, 0)$ and $(1, 1)$, divided by the total number of forecasts. A critic suggested that a success rate of 98.1% could have been achieved if every forecast stated $y = 0$. Hence, the critic concluded, the scientist had no forecasting skill.

7.1 Verify these experimental forecasts: (i) construct and graph the PCF and ROC; (ii) calculate the *CS* and *US*; (iii) interpret the results.

7.2 Examine the dispute between the scientist and the critic. For the scientist: (i) show his calculation of the success rate; (ii) explain to him the fallacy of reporting only the success rate. For the critic: (iii) show his calculation of the success rate; (iv) construct the contingency table resulting from the forecasts he suggested; (v) verify these forecasts by showing the points of the PCF and of the ROC on the graphs constructed in Exercise 7.1, and by calculating *CS* and *US*; (vi) explain the fallacy of his suggestion and of his conclusion.

7.3 Draw overall conclusions. *Hint*: begin by characterizing the peculiarity of forecasting a rare event. *Note*. This exercise is a rendition of a story reported on the internet by Roger Pielke, Jr., professor of environmental studies at the University of Colorado in Boulder. It illustrates the pitfalls of *ad hoc* verification methods, which (i) focus on forecast "accuracy" ($w = y$), (ii) calculate a single score ("success rate"), and (iii) commingle the calibration "skill" with the predictive "skill".

8 **Sample homogeneity**. A group of primary care physicians uses the services of the same radiologist. Each time a physician suspects a condition whose diagnosis could be aided by an x-ray image, she sends the patient to the radiologist. The radiologist supplies an x-ray image, his diagnosis of the condition, and his judgmental probability that the diagnosis is correct. Eventually, the physician determines whether the condition suspected in a particular patient is true or false. The group of physicians specializes in detecting and treating 7 different conditions. The group has kept the records and is now asking you to evaluate the radiologist's performance.

What theoretical and practical considerations would guide your decision whether to combine all records into one sample and estimate a single set of performance measures (PCF, ROC; *CS*, *US*), or to stratify the records into several samples and estimate several sets of performance measures? Conclude with a list of factors that could serve as a basis for the stratification.

9 **Underconfidence bias**. Reread Section 6.2.6. Imagine an experiment was conducted in probability assessment based on a sequence of information subsets; the subjects were untrained in probabilistic reasoning. The PCF of each subject indicated underconfidence. Write a possible explanation of this bias, in parallel to the explanation of overconfidence.

Mini-Projects

10 **Precipitation occurrence**. During four years, July 1972 – June 1976, twice daily, a meteorologist on duty in the Weather Forecast Office in Chicago, Illinois, assessed judgmentally the probability of precipitation occurrence (an accumulation of 0.25 mm or more in the rain gauge) during the next 12-h period (day or night). Here are the contingency tables for two forecasters, A and B:

			0	0.02	0.05	0.1	0.2	0.3	0.4	0.5	0.6	0.7	0.8	0.9	1
									y						
A	w	0	208	48	660	285	495	112	99	140	15	84	36	12	1
		1	1	0	20	32	92	48	49	99	22	160	73	88	37
B	w	0	157	141	274	536	500	185	109	95	60	39	21	0	0
		1	4	5	8	39	89	72	63	108	87	120	61	38	9

10.1 Estimate and graph the conditional probability functions f_0 and f_1 of the probability variate Y; calculate and graph the expected probability function κ of the probability variate Y.

10.2 Construct and graph the probability calibration function of the forecaster; interpret it.

10.3 Construct and graph the receiver operating characteristic of the forecaster; interpret it.

10.4 Calculate the calibration score and the uncertainty score; interpret them.

10.5 What advice would you give to the forecaster? What advice would you give to the decider?

Options: (A) Forecaster A.
(B) Forecaster B.

Note. The contingency tables were obtained by digitizing Figures 2 and 3 from the article by Murphy and Winkler (1977).

11 **Snow removal**. Every day during the official "winter season", which lasts 100 days, the manager of a city's Highway Department must decide whether or not the night shift of the snow removal crew should remain on duty. To facilitate this decision, every afternoon a meteorologist prepares a forecast which states the probability of the event {snowfall during the night is at least 5 cm}. The manager has compiled the joint sample

of forecast probability and actual snowfall from the last two winter seasons (see the tables designated by the options below), and also counted the number of nights with snowfall of at least 5 cm during the last 13 winter seasons; the count was M_1.

11.1 Estimate and graph the conditional probability functions f_0 and f_1 of the probability variate Y; calculate and graph the expected probability function κ of the probability variate Y.

11.2 Construct and graph the probability calibration function of the forecaster; interpret it.

11.3 Construct and graph the receiver operating characteristic of the forecaster; interpret it.

11.4 Calculate the calibration score and the uncertainty score; interpret them.

11.5 What advice would you give to the forecaster? What advice would you give to the decider?

    Options: (A) Table 6.11, $M_1 = 585$.
                 (B) Table 6.12, $M_1 = 533$.

12 **Snow removal** — *continued*: **climate nonstationarity**. Under the hypothesis that climate is nonstationary, the mean number of nights with heavy snowfall (of at least 5 cm) per winter season is predicted to decrease by 20% in the next 7 years, relative to the mean in the last 13 years. No change in the forecaster's ability to discriminate between heavy snowfall and light snowfall is predicted for the next 7 years. The city manager wants to know the implication of these two predictions on the analysis performed in Exercise 11.

12.1 Which elements of this analysis must be re-estimated or recalculated?

12.2 Re-estimate or recalculate these elements for year 7.

12.3 Compare the new verification results under the predicted climate change with the verification results obtained in Exercise 11; draw conclusions.

13 **Language proficiency forecasting** — Chapter 5, Exercise 11.2 *continued*. Retrieve the joint sample $\{(y, w)\}$ from that exercise. Perform the complete verification of this set of forecasts. Report the PCF, the ROC, the calibration score, and the uncertainty score. Interpret results and draw conclusions.
Requirement. To construct the contingency table, apply the discretization algorithm of Section 6.1.3. Verify the calibration against the prior probability used in forecasting.

14 **Snow forecasting** — Chapter 5, Exercise 12.2 *continued*. Retrieve the two joint samples $\{(y, w)\}$ from that exercise. For each set of forecasts, those made using the *estimation sample* and those made using the *validation sample*, perform the complete verification. Report the PCF, the ROC, the calibration score, and the uncertainty score. Interpret results and draw conclusions.
Requirement. To construct the contingency table, apply the discretization algorithm of Section 6.1.3. Verify the calibration against the prior probability estimated from the sample from the years 1998–2000. (This is the prior sample available at the time the forecaster was developed.)

15 **Diabetes forecasting** — Chapter 5, Exercise 13.2 *continued*. Retrieve the two joint samples $\{(y, w)\}$ from that exercise. For each set of forecasts, those made using the *estimation sample* and those made using the *validation sample*, perform the complete verification. Report the PCF, the ROC, the calibration score, and the uncertainty score. Interpret results and draw conclusions.
Requirement. To construct the contingency table, apply the discretization algorithm of Section 6.1.3. Verify the calibration against the prior probability used in forecasting.

16 **Diabetes forecasting** — *continued*: **comparison of predictors**. If Exercise 13 from Chapter 5 was performed two or more times, each time using a different predictor but the same samples, then (i) repeat the above Exercise 15 for each predictor, and (ii) compare the predictors in terms of their performance.

Options: (E) Comparison on the estimation sample.

(V) Comparison on the validation sample.

(B) Comparison on both samples.

16.1 *Calibration of the forecaster*: Compare the predictors in terms of the probability calibration functions and the calibration scores. Which predictor enabled you to develop a better calibrated Bayesian forecaster?

16.2 *Informativeness of the forecaster*: Compare the predictors in terms of the receiver operating characteristics and the uncertainty scores. Which predictor enabled you to develop a more informative Bayesian forecaster?

16.3 *Order of predictors*: If only one predictor could be available in real-time forecasting, what would be your preference order? Justify your recommendation.

Table 6.11 Joint sample I of forecast probability y and actual snowfall x [cm] from two winter seasons, each lasting 100 days.

y	x	y	x	y	x	y	x	y	x
0.0	5.1	0.4	1.0	0.5	7.8	0.7	6.0	0.5	1.7
0.4	2.0	0.5	2.0	0.1	6.3	0.1	7.0	0.2	1.8
0.5	6.2	0.3	3.0	0.5	6.2	0.5	5.0	0.3	6.2
0.1	1.0	0.1	6.8	0.5	6.1	0.0	1.3	0.2	1.9
0.7	1.0	0.5	4.0	0.8	5.1	0.2	1.4	0.7	1.0
0.1	7.0	0.3	4.1	0.0	5.0	0.0	1.0	0.3	11.0
0.7	1.5	0.1	1.7	0.5	6.0	0.0	4.0	0.5	1.0
0.5	8.0	0.3	7.2	0.8	7.0	0.3	11.0	0.5	13.0
0.2	2.5	0.5	0.3	0.5	11.0	0.4	3.0	0.2	2.0
0.6	6.5	0.7	7.5	0.2	3.0	0.5	12.0	0.8	10.0
0.2	4.0	0.5	0.4	0.5	0.0	0.5	3.5	0.5	3.0
0.4	4.9	0.2	1.4	0.7	8.0	0.1	3.6	0.1	3.0
0.4	6.3	0.0	4.3	0.4	1.2	0.4	2.0	0.2	7.0
0.6	4.9	0.8	3.0	0.8	7.0	0.8	9.0	0.8	3.5
0.3	4.3	0.3	11.2	0.3	6.3	0.8	2.3	0.6	6.0
0.4	3.2	0.3	4.0	0.9	3.2	0.2	7.0	0.2	3.5
0.3	1.0	0.4	4.2	0.6	8.2	0.4	2.4	0.6	7.0
0.3	1.5	0.3	12.0	0.6	7.5	0.7	8.0	0.3	3.7
0.4	5.1	0.4	12.0	0.6	6.0	0.4	1.0	0.3	4.8
0.6	1.8	0.4	4.9	0.6	4.9	0.5	3.0	0.9	8.8
0.4	9.0	0.5	6.1	0.5	5.1	0.7	5.2	0.8	11.0
0.0	3.1	0.5	4.1	0.1	5.2	0.4	3.0	0.7	6.0
0.3	3.2	0.0	2.0	0.8	6.0	0.4	2.0	0.5	5.2
0.6	18.0	0.3	1.0	0.5	9.0	0.2	1.0	0.7	4.7
0.5	3.1	0.5	1.0	0.7	2.0	0.7	1.0	0.7	5.1
0.3	0.9	0.8	7.0	0.2	8.0	0.8	6.3	0.5	5.3
0.1	0.7	0.3	1.0	0.8	8.2	0.0	1.0	0.2	3.0
0.2	7.0	0.8	8.0	0.6	8.3	0.6	6.4	0.2	3.0
0.6	0.8	0.2	8.0	0.2	3.0	0.5	1.0	0.4	8.7
0.6	10.0	0.9	6.0	0.9	9.1	0.4	7.0	0.8	8.2
0.7	11.0	0.3	5.0	0.0	2.0	0.7	11.0	0.7	7.9
0.8	12.0	0.2	2.5	0.2	10.0	0.4	11.0	0.7	5.3
0.3	4.2	0.7	2.5	0.4	2.0	0.6	6.0	0.6	6.7
0.5	7.0	0.7	5.0	0.4	6.0	0.4	7.0	0.6	8.3
0.6	3.1	0.5	3.0	0.4	1.0	0.6	8.2	0.2	2.0
0.5	6.0	0.6	6.0	0.8	7.0	0.5	1.5	0.6	9.0
0.4	0.0	0.3	3.9	0.3	2.2	0.1	1.4	0.6	8.0
0.6	0.0	0.6	3.7	0.4	2.1	0.6	6.3	0.6	7.3
0.5	7.4	0.4	7.0	0.7	8.0	0.3	1.3	0.7	8.3
0.4	1.0	0.4	3.8	0.5	2.8	0.5	1.2	0.7	6.0

Table 6.12 Joint sample II of forecast probability *y* and actual snowfall *x* [cm] from two winter seasons, each lasting 100 days.

y	x	y	x	y	x	y	x	y	x
0.0	0.4	0.8	13.0	0.4	10.1	0.5	0.7	0.6	9.0
0.7	11.9	0.4	2.9	0.0	3.0	0.7	9.3	0.5	10.8
0.5	5.5	0.9	1.2	0.6	3.9	0.2	10.2	0.7	16.2
0.2	1.8	0.3	1.2	0.8	13.9	0.4	4.4	0.3	1.8
0.6	4.2	0.1	0.6	0.5	11.1	0.8	9.3	0.2	4.8
0.7	1.0	0.6	15.7	0.2	4.4	0.0	4.7	0.3	2.8
0.1	1.6	0.0	12.6	0.3	1.6	0.1	17.7	0.5	6.5
0.3	12.2	0.2	9.7	0.4	14.0	0.3	2.4	0.0	2.9
0.4	1.1	0.5	9.9	0.0	1.0	0.4	17.3	0.5	7.0
0.0	1.1	0.7	16.8	0.1	4.3	0.5	18.0	0.4	12.2
0.6	15.1	0.3	3.7	0.3	9.3	0.6	11.6	0.8	12.8
0.9	14.3	0.2	4.5	0.6	6.7	0.5	13.7	0.3	17.8
0.3	1.8	0.6	10.1	0.5	6.9	0.3	4.5	0.6	4.4
0.5	4.6	0.5	4.7	0.4	1.2	0.6	2.3	0.2	0.1
0.4	3.0	0.8	11.4	0.7	13.7	0.7	16.6	0.8	4.8
0.7	13.8	0.6	13.5	0.4	2.5	0.3	2.2	0.6	7.6
0.3	0.4	0.4	2.5	0.6	10.2	0.6	14.9	0.4	2.7
0.8	6.8	0.7	10.9	0.7	15.6	0.2	3.5	0.5	1.1
0.5	5.2	0.8	3.1	0.8	8.9	0.3	8.8	0.7	6.5
0.1	4.5	0.1	3.9	0.3	4.9	0.8	13.0	0.4	1.3
0.9	5.8	0.2	4.9	0.2	0.3	0.7	2.1	0.6	14.2
0.4	3.3	0.7	7.1	0.4	4.8	0.5	14.0	0.8	17.3
0.6	7.5	0.5	7.6	0.5	1.8	0.4	16.9	0.5	14.6
0.3	2.9	0.4	4.3	0.6	8.0	0.6	8.4	0.2	17.6
0.7	14.4	0.8	14.5	0.4	1.5	0.4	4.9	0.0	0.8
0.5	17.2	0.3	3.8	0.6	5.3	0.6	16.7	0.8	16.3
0.1	14.2	0.4	18.0	0.9	8.7	0.1	4.9	0.5	2.2
0.3	4.7	0.5	0.3	0.8	17.9	0.2	3.2	0.5	17.0
0.4	7.5	0.6	7.9	0.1	4.1	0.4	2.0	0.4	2.8
0.4	2.5	0.5	10.2	0.3	3.8	0.8	9.0	0.2	0.8
0.5	13.3	0.3	10.9	0.8	12.2	0.7	9.6	0.7	16.3
0.6	10.9	0.7	0.9	0.7	13.1	0.4	2.6	0.3	3.0
0.8	9.0	0.1	5.6	0.4	0.8	0.6	17.8	0.5	4.1
0.2	0.7	0.3	2.5	0.2	3.9	0.6	6.4	0.6	12.8
0.5	17.4	0.4	0.3	0.4	7.9	0.4	2.8	0.3	14.3
0.4	5.6	0.4	1.1	0.6	9.0	0.5	17.8	0.6	14.4
0.0	1.6	0.6	11.0	0.4	4.6	0.3	2.1	0.4	16.6
0.4	2.4	0.6	8.3	0.3	2.0	0.2	3.1	0.6	9.9
0.2	18.0	0.9	5.4	0.7	5.5	0.7	17.9	0.5	13.7
0.5	11.8	0.2	3.0	0.5	11.4	0.8	17.5	0.5	9.9

7

Detection-Decision Theory

There is a class of decision problems under uncertainty wherein the predictand is binary and the decision variable is binary. This class of problems includes hypothesis testing, target detection, quality control, medical diagnosis, rescue search, insurance purchasing, natural hazard warning, transport rerouting, activity rescheduling.

The basic decision model employs a prior probability of the uncertain event; the model prescribes the optimal decision and determines the (economic) value of perfect forecaster. The extended model employs a probabilistic forecast of the uncertain event; the model prescribes the optimal decision function for using the forecast probability and determines the (economic) value of the forecaster.

7.1 Prototypical Decision Problems

Below is a partial list of prototypical problems to which the detection-decision theory is applicable. Each problem is sketched by posing two questions: What is the uncertain event? What is the decision dilemma?

Hypothesis testing
  Is a hypothesis true or false?
  To accept or to reject the hypothesis?
Target detection
  Is a target present or not within a range?
  To shoot or not to shoot?
Quality control
  Is an item defective or not?
  To accept or to reject the item?
Medical diagnosis
  Is a disease present or not in a patient?
  To order or not to order a test?
  To medicate or not to medicate?
  To operate or not to operate?
Rescue search
  Is the missing person present or not in a sector?
  To search or not to search in the sector?
Insurance purchasing
  Will property damage occur or not within a year?
  To purchase or not to purchase insurance?

Probabilistic Forecasts and Optimal Decisions, First Edition. Roman Krzysztofowicz.
© 2025 John Wiley & Sons Ltd. Published 2025 by John Wiley & Sons Ltd.
Companion website: www.wiley.com/go/ProbabilisticForecastsandOptimalDecisions1e

Natural hazard warning for phenomena such as flash flood, flood, tornado, hurricane, tsunami

Will a flash flood occur or not within 3 hours?

To issue or not to issue a warning to the public?

Transport rerouting because of **severe weather** such as blizzard, lightning, hail, heavy rain, strong wind, fog, dust storm

Will severe weather occur or not along the route of a truck, a ship, an airplane?

To reroute or not to reroute?

Activity rescheduling because of **adverse weather** such as rain, wind, cold

Will adverse weather occur or not during the day?

To reschedule or not to reschedule a weather-sensitive activity such as road paving, concrete pouring, tower frame welding, high-rise window washing?

Some of these prototypical decision problems are detailed in examples or exercises.

7.2 Basic Decision Model

7.2.1 Elements

The basic model for detection decision is built of four elements as follows.

a — decision. This is a binary variable, $a \in \{0, 1\}$, such that

$$a = 0 \iff \text{action is not taken (the hypothesis is rejected),}$$

$$a = 1 \iff \text{action is taken (the hypothesis is accepted).}$$

W — predictand. This is a binary variate whose realization $w \in \{0, 1\}$ serves as the indicator of the event (or the hypothesis) of interest, such that

$$W = 0 \iff \text{event does not occur (the hypothesis is false),}$$

$$W = 1 \iff \text{event occurs (the hypothesis is true).}$$

g — probability of event occurring (probability of the hypothesis being true)

$$g = P(W = 1), \qquad 0 < g < 1.$$

d — disutility function that quantifies the strength of preference for outcomes of decision-event pairs; to wit,

$$d_{aw} = d(a, w)$$

is the disutility of outcome resulting from decision a and event $W = w$. This disutility is in a monetary unit when it represents cost and damage, or it is dimensionless when it represents a subjective valuation of the undesirability of an outcome.

Table 7.1 displays the four disutilities. They are assumed to be nonnegative, $d_{aw} \geq 0$. The natural preference order over decision-event pairs is such that

$$d_{00} < d_{10} \qquad \text{and} \qquad d_{11} < d_{01}.$$

This order ensures that the decision problem is nontrivial. For if $d_{00} < d_{10}$ and $d_{01} \leq d_{11}$, then the preferred decision would be $a = 0$, regardless of the event, $W = 0$ or $W = 1$; and if $d_{10} \leq d_{00}$ and $d_{11} < d_{01}$, then the preferred decision would be $a = 1$ regardless of the event, $W = 0$ or $W = 1$. In either case, there would be no need for further analysis.

Table 7.1 Disutilities of outcomes resulting from each pair of decision a and event $W = w$ in a detection problem.

		\multicolumn{2}{c}{w}	
		0	1
a	0	d_{00}	d_{01}
	1	d_{10}	d_{11}

7.2.2 Decision Tree

The detection-decision problem lends itself to a graphical representation in the form of a *decision tree* (Figure 7.1). Proceeding from left to right, we plot a *decision node* (a square) which represents the act of making a decision. Each branch emanating from the node represents an alternative decision. A decision branch leads to a *chance node* (a circle) which represents the process that generates an event. Each branch emanating from the node represents a possible event. An event branch leads to a disutility of the outcome.

Conditional on the event $W = w$ ($w = 0, 1$), there are two alternative decisions a ($a = 0, 1$), one of which is preferred. For convenience of communication, each of the four pairings is labeled as depicted in Table 7.2 and is termed as follows:

$(a = 0|W = 0)$ — *Quiet state* (Q).

$(a = 1|W = 0)$ — *False alarm* (F), as one issues a warning ($a = 1$) when the adverse event does not occur ($W = 0$); the term used in statistical testing in *Type II error*, as one accepts a hypothesis ($a = 1$) when it is false ($W = 0$); the term used in medical testing is *false positive* result, as one diagnoses an abnormality ($a = 1$) when it is absent ($W = 0$).

Figure 7.1 Decision tree of a detection problem.

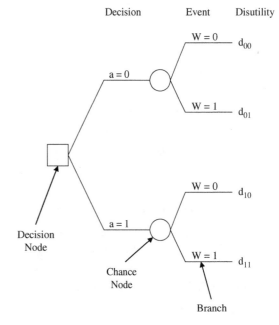

Table 7.2 Labels attached to decision *a*, conditional on event $W = w$: quiet state (Q), false alarm (F), missed event (M), detection (D).

		w	
		0	**1**
a	0	Q	M
	1	F	D

$(a = 0|W = 1)$ — *Missed event (M)*, as one does not issue a warning ($a = 0$) when the adverse event occurs ($W = 1$); the term used in statistical testing is *Type I error*, as one rejects a hypothesis ($a = 0$) when it is true ($W = 1$); the term used in medical testing is *false negative* result, as one diagnoses no abnormality ($a = 0$) when it is present ($W = 1$).

$(a = 1|W = 1)$ — *Detection (D)*.

7.2.3 Optimal Decision Procedure

First, each feasible decision *a* is evaluated in terms of its disutility. The *disutility of decision*, D_a, is defined as the *expected disutility of outcome* resulting from that decision:

$$D_a = E[d(a, W)]$$
$$= d(a, 0)P(W = 0) + d(a, 1)P(W = 1), \qquad a = 0, 1.$$

Using the abbreviated notation,

$$D_0 = d_{00}(1 - g) + d_{01}g,$$
$$D_1 = d_{10}(1 - g) + d_{11}g. \tag{7.1}$$

Next, the *minimum disutility of decision* (equivalently, the *minimum expected disutility of outcome*) is determined:

$$D^* = \min_a E[d(a, W)]$$
$$= \min\{D_0, D_1\}. \tag{7.2}$$

Finally, the *optimal decision* a^* is found. This is the decision that has the minimum disutility (equivalently, that minimizes the expected disutility of outcome); hence

$$a^* = \begin{cases} 0 & \text{if } D_0 \leq D_1, \\ 1 & \text{if } D_1 < D_0. \end{cases} \tag{7.3}$$

7.2.4 Optimality Condition

When $D_0 = D_1$, none of the decisions is preferred and the choice can be made arbitrarily. For convenience, it is assumed herein that action is taken, $a^* = 1$, only if the strict inequality holds, $D_1 < D_0$. The objective now is to express this condition for taking action in terms of the input elements, that is, the probability of an event occurring and the disutilities of outcomes. Using equation (7.1), the inequality $D_1 < D_0$ can be written

$$d_{10}(1 - g) + d_{11}g < d_{00}(1 - g) + d_{01}g;$$

that is,

$$\frac{1-g}{g} < \frac{d_{01} - d_{11}}{d_{10} - d_{00}}.$$

Equivalently,

$$\frac{1-g}{g} < r, \tag{7.4}$$

where the term on the right is the *ratio of disutility differences*,

$$r = \frac{d_{01} - d_{11}}{d_{10} - d_{00}}. \tag{7.5}$$

This ratio can be interpreted with the aid of Table 7.1 When the event occurs, $W = 1$, it is optimal to take action because $d_{11} < d_{01}$. Hence the difference $d_{01} - d_{11}$ can be interpreted as the *opportunity loss* from not taking action when needed; equivalently, one can say that this is the opportunity loss from the missed event (Type I error, false negative result). When the event does not occur, $W = 0$, it is optimal not to take action because $d_{00} < d_{10}$. Hence the difference $d_{10} - d_{00}$ can be interpreted as the *opportunity loss* from taking action when not needed; equivalently, one can say that this is the opportunity loss from the false alarm (Type II error, false positive result). In summary, r can be interpreted succinctly as the *ratio of the opportunity losses*:

$$r = \frac{\text{opportunity loss from missed event } (M)}{\text{opportunity loss from false alarm } (F)}.$$

Returning to condition (7.4), it can be rewritten in the form

$$g > \frac{1}{1+r},$$

with the term on the right being the *threshold probability for action*:

$$p^* = \frac{1}{1+r}. \tag{7.6}$$

Now the optimal decision rule (7.3) can be restated as follows:

$$a^* = \begin{cases} 0 & \text{if } g \leq p^*, \\ 1 & \text{if } g > p^*. \end{cases} \tag{7.7}$$

In conclusion, to find the optimal decision one needs to know only (i) the probability g of the event (a measure of uncertainty) and (ii) the ratio r of the opportunity losses (a measure of the relative undesirability of outcomes). The optimal decision is to take action whenever the probability of the event exceeds the threshold p^*, which depends solely on the ratio r.

7.2.5 Sensitivity Analysis

From equation (7.1), the disutilities of the two alternative decisions can be written in the form

$$D_0 = d_{00} + (d_{01} - d_{00})g,$$
$$D_1 = d_{10} + (d_{11} - d_{10})g. \tag{7.8}$$

Figure 7.2 illustrates the sensitivity of D_0 and D_1 to the probability g, given fixed disutilities of outcomes $(d_{00}, d_{01}, d_{10}, d_{11})$. The disutility of a decision is a linear function of the probability of the event. The opportunity losses from the missed event (M) and from the false alarm (F) appear as the vertical line segments between the two linear functions at $g = 1$ and $g = 0$, respectively. The abscissa at which the two linear functions intersect is the threshold probability for action p^*. The location of p^* is determined directly by the ratio of the opportunity

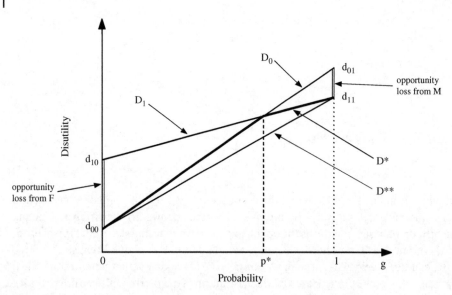

Figure 7.2 The disutility of decision D_a ($a = 0, 1$), the minimum disutility of decision D^* (equivalently, the minimum expected disutility of outcome), and the expected minimum disutility of outcome D^{**}, as functions of the probability g of event $W = 1$.

losses: as $r \to 0$, $p^* \to 1$; as $r \to \infty$, $p^* \to 0$. Finally, the minimum disutility of decision D^* (equivalently, the minimum expected disutility of outcome) is a two-piece linear, concave function of the probability g:

$$D^* = \min\{d_{00} + (d_{01} - d_{00})g, d_{10} + (d_{11} - d_{10})g\}. \tag{7.9}$$

(In statistics, D^* is often called the Bayes risk.)

7.2.6 Economic Estimation of Disutilities

Disutilities may be estimated via an economic analysis of the outcomes in two ways.

Total analysis estimates d_{aw} as the sum of all costs and damages resulting from decision-event (a, w), for $a, w = 0, 1$.

Incremental analysis takes the normal operating cost in the quiet state (Q) as the baseline and sets $d_{00} = 0$. Then, d_{10} represents the cost of action that turns out not needed (state F); d_{01} represents the damage caused by event not preceded by action (state M); d_{11} represents the cost of action plus the damage caused by the event preceded by action (state D).

7.2.7 Subjective Assessment of Disutilities

When there are no economic data for estimation of costs and damages, or when the outcomes include qualitative consequences (e.g., negative publicity, inconvenience, pain, injury, death), the decider may assess the disutilities subjectively by following a five-step procedure.

Step 1. Identify and describe the consequences which comprise the outcome of each decision–event pair (a, w); equivalently, the outcome of each system state.

Step 2. Order the four system states in terms of the undesirability of their outcomes, from the least undesirable (Q or D) to the most undesirable (F or M).

Figure 7.3 The scale for subjective assessment of the disutilities d_{aw} of outcomes resulting from decision a ($a = 0, 1$) and event $W = w$ ($w = 0, 1$).

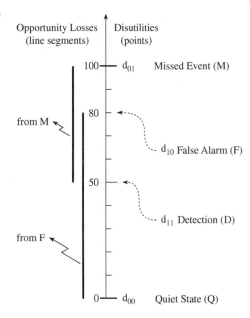

Step 3. *Disutility scale.* Assign disutility 0 to the least undesirable state, and disutility 100 to the most undesirable state. To explain the subsequent steps, let us suppose the preference order of the system states is Q, D, F, M. Thus

$$d_{00} = 0, \qquad d_{01} = 100,$$

define the disutility scale [0, 100], which can be displayed (Figure 7.3).

Step 4. *Assessment.* Guide your subjective valuations of consequences listed in Step 1 by asking two sets of questions.

- How undesirable is F relative to M and Q?
  Is F closer to M or to Q? How close?
  Locate d_{10} on the scale.
- How undesirable is D relative to F and Q?
  Is D closer to F or to Q? How close?
  Locate d_{11} on the scale.

Treat the assessed disutilities as initial, ask the questions reframed below, and adjust d_{10} and d_{11} as necessary to match your valuations.

- How undesirable is D relative to M and Q?
  Is D closer to M or to Q? How close?
  Adjust d_{11} as necessary.
- How undesirable is F relative to M and D?
  Is F closer to M or to D? How close?
  Adjust d_{10} as necessary.

Step 5. *Validate* the internal coherence of these assessments. Toward this end:

  5.1 Check the nontriviality conditions:

$$d_{00} < d_{10}, \qquad d_{11} < d_{01}.$$

5.2 Compare the opportunity losses:

$$\text{from state } M \ : \ l_M = d_{01} - d_{11},$$
$$\text{from state } F \ : \ l_F = d_{10} - d_{00}.$$

- Does l_M reflect the undesirability of M relative to D?
- Does l_F reflect the undesirability of F relative to Q?

5.3 Calculate the ratio of the opportunity losses:

$$r = \frac{l_M}{l_F}.$$

- If $r > 1$: Is the missed event r times more undesirable than the false alarm?
- If $r < 1$: Is the false alarm $1/r$ times more undesirable than the missed event?

Example 7.1 *(Flight cancellation)*

Severe weather (e.g., snowstorm, thunderstorm, strong wind, fog) is the most frequent cause of flight cancellations; for example, in the summer of 2017, Spirit Airlines cancelled 1441 flights. Place yourself in the position of an operations manager of a startup airline having no experience and no cost records. Yet you must decide to cancel ($a = 1$) or not to cancel ($a = 0$) a flight before its departure when there is uncertainty about the weather being severe ($W = 1$) or not severe ($W = 0$) at the destination airport during the scheduled arrival time of your flight; if the severe weather develops, then the airport will close, and the arriving airplane will be diverted to another airport. To prepare you for making this decision is the purpose of the disutility assessment procedure. (In fact, in 2017 the US Federal Aviation Administration initiated regular conference calls with airlines each afternoon to plan for severe weather that might affect flights the next day.) Suppose a manager applied the five-step assessment procedure, and summarized its results thusly.

Step 1.

$(0, 0)$: Quiet state (Q)
- normal operation

$(1, 1)$: Detection (D)
- cost of rebooking passengers and refunding some tickets
- cost of rescheduling the flight

$(1, 0)$: False alarm (F)
- cost of rebooking passengers and refunding some tickets
- cost of rescheduling the flight
- cost of transferring some passengers to competitors who fly
- displeased customers, negative publicity (may push future fliers to competitors)

$(1, 1)$: Missed event (M)
- cost of diverting the flying airplane to another airport
- cost of reimbursing passengers for hotels
- cost of transporting passengers to the original destination

Steps 2,3,4. A disutility scale $[0, 100]$ was set (Figure 7.3). The disutility d_{00} of the normal operation (Q) was attached to the bottom of the scale, and the disutility d_{01} of not cancelling the flight when the destination airport closed before arrival (M) was attached to the top of the scale. The other two disutilities were ordered, $d_{11} < d_{10}$, and then located on the scale to reflect the degree of the undesirability of state D (cancel the flight when the destination airport closes) and of state F (cancel the flight when the destination airport remains open), relative to each other, and relative to the states Q and M.

Step 5. The validation of internal coherence was performed. The ratio of the opportunity losses was found:

$$r = \frac{100 - 50}{80 - 0} = \frac{5}{8} = 0.625.$$

This prescribes the threshold probability (of severe weather at the destination airport) for cancelling the flight:

$$p^* = \frac{1}{1 + 0.625} = 0.6154 \approx 0.62.$$

Note on airline operations. The consequences depend also on the desired lead time of the decision: some airlines decide at the departure time, whereas other airlines prefer to cancel flights hours earlier to reduce the inconvenience for customers of coming to the airport in vain. Also the decisions vary across the airlines: for example, during the snowstorm on 11 February 2006, which affected flights into and out of three main New York airports (La Guardia, Kennedy, Newark), United Airlines cancelled 39% of its flights, but JetBlue cancelled only 3% of its flights. Such drastically different decisions in the same circumstances reflect individual airline preferences, which would be embedded in the ratio *r* of the detection-decision model. These facts and the example are based on articles by Scott McCartney (*The Wall Street Journal*, 21 February 2006, 6 September 2018), who for 20 years (2002–2021) wrote "The Middle Seat" column covering aviation and travel.

7.3 Decision with Perfect Forecast

7.3.1 Decision Tree with Perfect Forecast

The probability $g = P(W = 1)$ quantifies the uncertainty about the occurrence of an event (or the truth of a hypothesis) that exists at the time of making a decision. Obviously, the ideal situation would be one without uncertainty. Thus a question: How much is one losing, in terms of disutility, because of the presence of uncertainty? An alternative question: How much, at most, should one be willing to pay an expert for a perfect forecast of W? These two questions give rise to the concepts of the expected opportunity loss and the value of a perfect forecast. While a perfect forecast may be impossible to obtain, its value establishes an upper bound on the value of any forecast, or additional information, which may be obtained. As such, the value of a perfect forecast is useful because it provides a basis for deciding whether or not it is worthwhile to consider acquiring additional information before implementing the decision a^*.

To determine the value of a perfect forecast, we have to contemplate what decision would have been optimal if a clairvoyant revealed the true event. This decision problem can be represented in terms of a tree in which the order of nodes is reversed (Figure 7.4). Proceeding from left to right, the chance node (a circle) represents the clairvoyant;

Figure 7.4 Decision tree of a detection problem with the clairvoyant.

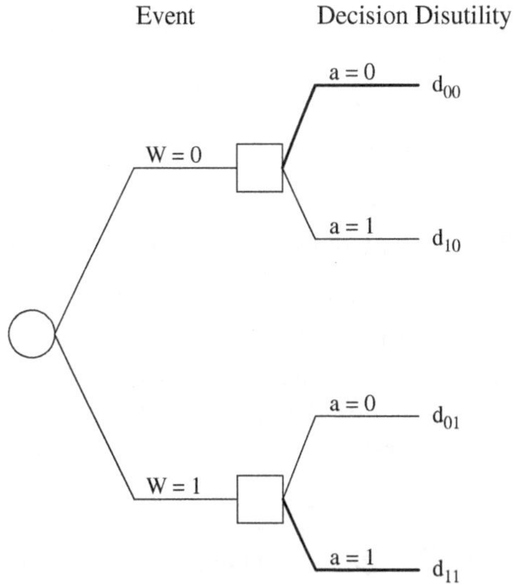

each branch emanating from this node represents an event that may be revealed as the true event. Given knowledge of the true event, the decision is made. This act is represented by a decision node (a square); each branch emanating from this node represents an alternative decision. A decision branch leads to the disutility of the outcome.

7.3.2 Decision Procedure with Perfect Forecast

Suppose the clairvoyant has revealed that $W = w$ is the true event ($w = 0$ or $w = 1$). Because the uncertainty has vanished, the decision can be made by minimizing the disutility of outcome, given $W = w$. Let a^{**} denote the optimal decision.

If $W = 0$, then

$$\min\{d_{00}, d_{10}\} = d_{00} \quad \text{and} \quad a^{**} = 0.$$

If $W = 1$, then

$$\min\{d_{01}, d_{11}\} = d_{11} \quad \text{and} \quad a^{**} = 1.$$

In a sense, this is a contingency plan: it specifies the minimum disutility and the optimal decision, given that $W = w$ is the true event. (The minimum disutility follows from the natural preference order, as discussed in Section 7.2.1.)

How to evaluate this plan? First, we determine the probability of the clairvoyant telling us that the true event is $W = w$. This probability is the same as the probability of the event itself: $P(W = 0) = 1 - g$, $P(W = 1) = g$. Second, we evaluate the decision procedure with a perfect forecast in terms of the *expected minimum disutility of outcome*:

$$
\begin{aligned}
D^{**} &= E[\min_a d(a, W)]. \\
&= \min\{d_{00}, d_{10}\}(1 - g) + \min\{d_{01}, d_{11}\}g \\
&= d_{00}(1 - g) + d_{11}g \\
&= d_{00} + (d_{11} - d_{00})g.
\end{aligned}
\tag{7.10}
$$

By comparing equation (7.10) with equation (7.2), we note that contemplating the clairvoyant amounts to reversing the order of operations (minimization and expectation), in parallel with reversing the order of nodes (decision and chance) in Figures 7.1 and 7.4.

By comparing the graphs of D^{**} and D^*, each as a function of the probability g (Figure 7.2), we note that D^{**} is a linear function which lies beneath D^*, for all $g \in (0, 1)$. This proves geometrically the following theorem.

Theorem *(Order of disutilities)* *For any disutility vector $(d_{00}, d_{01}, d_{10}, d_{11})$ and any probability g,*

$$0 \leq D^{**} \leq D^*.$$

7.3.3 Value of Perfect Forecaster

In general, the value of the perfect forecaster can be determined when the disutilities are in monetary units (representing costs and damages) as well as when the disutilities are dimensionless (representing subjective valuations of outcomes that are either monetary, or nonmonetary, or of both kinds). When the disutilities are in monetary units, the *value of the perfect forecaster* is defined as the difference between the minimum expected disutility of outcome and the expected minimum disutility of outcome:

$$VPF = D^* - D^{**},
\tag{7.11}$$

where D^* is given by equations (7.1)–(7.2) or by equation (7.9), and D^{**} is given by equation (7.10).

When D^* and D^{**} are in monetary units, VPF admits two useful interpretations. First, it is the maximum price the decider should be willing to pay for one perfect forecast. Second, it is an upper bound on the value of any imperfect forecast.

7.3.4 Properties of Value of Perfect Forecaster

It is instructive to derive an expression for VPF in terms of the input elements. Toward this end, two cases must be considered, in parallel to the two branches of the optimal decision rule (7.7).

If $g \leq p^*$, then $D^* = D_0$. Thus

$$VPF = D_0 - D^{**}$$
$$= [d_{00}(1 - g) + d_{01}g] - [d_{00}(1 - g) + d_{11}g]$$
$$= (d_{01} - d_{11})g.$$

If $g \geq p^*$, then $D^* = D_1$. Thus

$$VPF = D_1 - D^{**}$$
$$= [d_{10}(1 - g) + d_{11}g] - [d_{00}(1 - g) + d_{11}g]$$
$$= (d_{10} - d_{00})(1 - g).$$

Combining the above expressions into a single formula yields

$$VPF = \begin{cases} (d_{01} - d_{11})g & \text{if } g \leq p^*, \\ (d_{10} - d_{00})(1 - g) & \text{if } g \geq p^*. \end{cases} \tag{7.12}$$

The disutility differences that appear above are the same ones that appear in the ratio r from which the threshold probability for action p^* is calculated in Section 7.2.4. Namely, $d_{01} - d_{11}$ represents the opportunity loss from a missed event (M), and ($d_{10} - d_{00}$) represents the opportunity loss from a false alarm (F). Accordingly, expression (7.12) states that the value of the perfect forecaster equals the *minimum expected opportunity loss*.

Figure 7.5 illustrates the sensitivity of VPF to the probability g, given fixed disutilities of outcomes ($d_{00}, d_{01}, d_{10}, d_{11}$). From this figure and from expression (7.12), the following five properties of VPF for a detection-decision problem can be inferred:

1. The VPF equals the minimum expected opportunity loss.
2. The VPF is never negative. This follows from the fact that none of the opportunity losses can be negative, as $d_{01} - d_{11} > 0$ and $d_{10} - d_{00} > 0$.

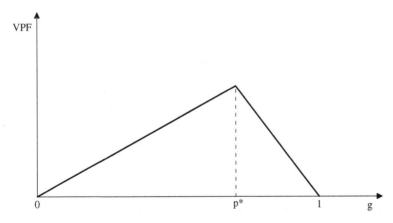

Figure 7.5 The value of a perfect forecaster VPF as a function of the probability g of event $W = 1$.

3. The *VPF* is a two-piece linear, concave function of the probability g.

4. The *VPF* attains the maximum at $g = p^*$; at this value of g, the choice between the alternative decisions is indeterminable because $D_0 = D_1$; and in a neighborhood of $g = p^*$, the difference between D_0 and D_1 is small. In other words, the perfect forecast has the highest value when it is difficult to ensure the optimality of the decision in the sense that a small change in the probability g can alter the choice.

5. The *VPF* decreases toward zero as either $g \to 0$ or $g \to 1$. In other words, the perfect forecast becomes worthless when the uncertainty about the occurrence of the event vanishes.

7.4 Decision Model with Forecasts

7.4.1 Repetitive Decisions with Forecasts

Consider a decision process consisting of a sequence of detection-decision problems which are essentially independent of each other and identical in all respects except for the posterior probability of the uncertain event. This probability varies from one decision time to the next.

Example 7.2 *(Construction scheduling)*
Each morning the manager of a construction company acquires a local weather forecast for the coming day. Then the probability of precipitation occurrence during the next 12-h period (0600–1800) is used to decide whether or not rain-sensitive activities such as laying of concrete, roofing, and paving should be scheduled during that day. *A priori*, for instance one month in advance, the probability of rain during the 12-h period is the same for each future day within a month (or season) and can be estimated from a climatic record of rain observations. But the forecast probability for the next 12-h period, calculated from real-time meteorological data collected up to the current day, varies from day to day.

Example 7.3 *(Quality control)*
Every time a batch of parts exits the production line, the quality inspector acquires the testing log for that batch (number of parts tested, types of tests, and results of tests). Then the probability of a part not being up to standard is used to decide whether the batch should be accepted or rejected. *A priori*, for instance before the start of a production shift, the probability of a part not being up to standard is the same for each future batch and can be estimated from a record of past tests. But the forecast probability for a part in the current batch, calculated from the results of tests performed on selected parts from that batch, varies from batch to batch.

Our objective is to adapt the basic model for the detection-decision problem to repetitive decisions with forecasts.

7.4.2 Input Elements

Let Y be the probability variate — a variate whose realization y is the probability assigned to the event $W = 1$ by the forecaster on a particular occasion. The forecast probability y may be either a judgmental forecast prepared according to the procedure described in Chapter 4, or a statistical forecast produced via some technique such as the Bayesian forecaster described in Chapter 5. It is assumed herein that Y is a discrete variate with the sample space

$$\mathcal{Y} = \{y : y = y_0, y_1, \ldots, y_I\}, \tag{7.13}$$

where $0 \le y_0 < y_1 < \cdots < y_I \le 1$. When the original forecast variate is continuous, it should be discretized in accordance with the discretization algorithm for the verification of forecasts (Section 6.1.3).

In order to couple the forecast system with the decision system, one needs the Bayesian processor of forecast (BPF) developed according to the theory presented in Chapter 6. The BPF outputs two elements which become input elements to the detection-decision model. These elements, defined and explained in Section 6.2.2, are recalled briefly here:

$\kappa(y) = P(Y = y)$ for $y \in \mathcal{Y}$, is the expected probability of the forecast being $Y = y$;

$\eta(y) = P(W = 1 | Y = y)$ for $y \in \mathcal{Y}$, is the posterior probability of event $W = 1$, conditional on forecast probability $Y = y$.

7.4.3 Optimal Decision Procedure

Inasmuch as the posterior probability $\eta(y)$ is a function of the forecast probability y, the optimal decision must be found for every $y \in \mathcal{Y}$. Specifically, what must be found is the *optimal decision function* α^* that maps the set of forecast probabilities $\mathcal{Y}$ into the set of decision indicators $\{0, 1\}$. Let $\alpha^*(y)$ denote the optimal decision, $\alpha^*(y) \in \{0, 1\}$, given the forecast probability $y \in \mathcal{Y}$. The optimal decision function α^* can be found by adapting the optimal decision rule (7.7):

$$\alpha^*(y) = \begin{cases} 0 & \text{if } \eta(y) \leq p^*, \\ 1 & \text{if } \eta(y) > p^*, \end{cases} \tag{7.14}$$

where p^* is the threshold probability for action given by equation (7.6).

Another way of characterizing the optimal decision is by constructing a *detection set*:

$$\mathcal{Y}_1 = \{y : \eta(y) > p^*, y \in \mathcal{Y}\}, \tag{7.15}$$

which is the set of forecast probabilities such that if $y \in \mathcal{Y}_1$, then it is optimal to take action, $\alpha^*(y) = 1$.

Figure 7.6 shows two examples of the detection set. In the first example, η is a strictly increasing function on $\mathcal{Y}$ (which implies that the forecaster is semi-calibrated), and consequently the detection set $\mathcal{Y}_1$ is **connected**. In the second example, η is a nonmonotone function on $\mathcal{Y}$ that crosses thrice the threshold probability for action p^*; consequently, the detection set is **disconnected**; to wit, $\mathcal{Y}_1$ is the union of two disjoint subsets of $\mathcal{Y}$.

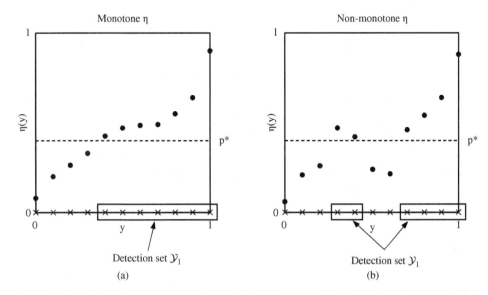

Figure 7.6 The detection set $\mathcal{Y}_1$ when the posterior probabilities $\eta(y)$ form (a) a monotone-increasing sequence and (b) a nonmonotone sequence that crosses thrice the threshold probability for action p^*.

7.4.4 Evaluation of Decision Procedure

Given the forecast probability $Y = y$, the *disutility of decision*, $D_a(y)$, is calculated as the expected disutility of outcome resulting from that decision:

$$D_a(y) = E[d(a, W)|Y = y]$$
$$= d(a, 0)P(W = 0|Y = y) + d(a, 1)P(W = 1|Y = y), \qquad a = 0, 1.$$

Using the abbreviated notation,

$$D_0(y) = d_{00}[1 - \eta(y)] + d_{01}\eta(y),$$
$$D_1(y) = d_{10}[1 - \eta(y)] + d_{11}\eta(y). \tag{7.16}$$

Given the forecast probability $Y = y$, the *minimum disutility of decision* is

$$D^*(y) = \min_a E[d(a, W)|Y = y]$$
$$= \min\{D_0(y), D_1(y)\}. \tag{7.17}$$

When these calculations are performed for every forecast probability $y \in \mathcal{Y}$, they produce function D^* on $\mathcal{Y}$.

The last equation offers the alternative to equation (7.14) for finding the *optimal decision function* α^*. For every forecast probability $y \in \mathcal{Y}$, the optimal decision $\alpha^*(y)$ is one that has the minimum disutility (equivalently, that minimizes the expected disutility of outcome); hence

$$\alpha^*(y) = \begin{cases} 0 & \text{if } D_0(y) \leq D_1(y), \\ 1 & \text{if } D_1(y) < D_0(y). \end{cases} \tag{7.18}$$

To obtain an overall evaluation of the forecast–decision system, the *integrated minimum disutility* is calculated by taking the expectation of $D^*(Y)$ with respect to the probability variate Y:

$$DF^* = E[D^*(Y)]. \tag{7.19}$$

For a discrete probability variate Y with the sample space $\mathcal{Y}$ defined by (7.13),

$$DF^* = \sum_{i=0}^{I} D^*(y_i)\kappa(y_i). \tag{7.20}$$

The expectation above may be interpreted in two ways: (i) as an evaluation over infinitely many repetitions of the forecast–decision process, wherein the forecast probability y varies from one decision time to the next and the relative frequency of each y is specified by the expected probability function κ, or (ii) as an evaluation of one-time forecast and decision, but before the forecast probability y is acquired, so that it must be treated as a variate Y whose "prediction" is specified by the probability function κ.

Finally, this optimal decision procedure with forecasts has an essential property, proven in Section 7.6.1: in terms of its disutility, DF^*, it is preferred at least as the procedure with a fixed prior probability, D^*, and preferred no more than the procedure with a perfect forecaster, D^{**}.

Theorem *(Order of disutilities)* *For any disutility vector* $(d_{00}, d_{01}, d_{10}, d_{11})$, *any prior probability g, and any conditional probability functions* (f_0, f_1) *of the BPF,*

$$0 \leq D^{**} \leq DF^* \leq D^*.$$

7.4.5 Value of Forecaster

When the disutilities are in monetary units, the *value of the forecaster* is defined as the difference between the evaluation without the forecaster and the evaluation with the forecaster:

$$VF = D^* - DF^*; \tag{7.21}$$

specifically, D^* is the minimum disutility of decision based on the prior probability, calculated according to equations (7.1)–(7.2), and DF^* is the integrated minimum disutility of decisions based on the posterior probabilities, calculated according to equations (7.16)–(7.20).

The VF is the maximum price the decider should be willing to pay for one forecast. The VF is never negative and is never greater than the value of the perfect forecast:

$$0 \leq VF \leq VPF. \tag{7.22}$$

7.4.6 Efficiency of Forecaster

For comparative analyses and discussions of forecast–decision systems, it is convenient and meaningful to summarize the contribution of the forecaster in terms of efficiency rather than value. The *efficiency of the forecaster* is defined as the quotient

$$EF = \frac{VF}{VPF} = \frac{D^* - DF^{**}}{D^* - D^{**}}. \tag{7.23}$$

The EF is bounded, $0 \leq EF \leq 1$. It represents the fraction of the clairvoyant's value that a given forecaster delivers (for a particular decision problem). More broadly, the EF measures how well the forecaster meets the need of the decider (for information about the predictand).

Two properties make the EF a comparative measure (across different forecasters for different deciders):

1. The EF is dimensionless (i.e., independent of the unit in which the disutilities of outcomes are measured).
2. The EF is invariant under a positive linear transformation of the disutilities (i.e., independent of the assigned disutility scale — Step 3 in Section 7.2.7: it can be [0, 100], or any other closed interval $[a, b]$ of real numbers, with $a < b$).

7.5 Informativeness of Forecaster

7.5.1 The Comparison Problem

Suppose there are two or more alternative forecasters (human experts or statistical models) of predictand W, and multiple deciders, each needing a forecast of W for a different detection-decision problem. In concept, every decider could determine the economic value of each forecaster, and then order the forecasters from the most valuable to the least valuable. If every decider arrived at the same order of forecasters, then this would be the preferred order from the viewpoint of a utilitarian society.

The binary relation of informativeness that formalizes such an order was defined verbally in Sections 5.5.1 and 6.3.1. Having derived the mathematical expression (7.21) for the value of the forecaster, we can revisit the informativeness relation and define it mathematically.

7.5.2 Mathematical Definition

Let us trace the calculation of VF via equation (7.21) to the input elements: D^* is calculated from $\mathbf{d} = (d_{00}, d_{01}, d_{10}, d_{11})$ and g via equations (7.1)–(7.2), whereas DF^* is calculated from $\mathbf{d}$, η, and κ via equations (7.16)–(7.20). In turn, η and κ are calculated from g, f_0, f_1 via equations (6.7)–(6.8). Thus, there are three original input elements: the vector of disutilities $\mathbf{d}$ which quantifies the decider's strength of preference for outcomes, the prior probability g which quantifies the prior uncertainty about the occurrence of an event, and two conditional probability functions (f_0, f_1) which characterize the forecaster. It follows that the value of the forecaster VF is a function of these original input elements; this can be indicated explicitly by rewriting equation (7.21) in the form

$$VF(\mathbf{d}, g; f_0, f_1) = D^*(\mathbf{d}, g) - DF^*(\mathbf{d}, g; f_0, f_1). \tag{7.24}$$

Now suppose there are two forecasters, A and B, characterized by (f_0^A, f_1^A) and (f_0^B, f_1^B), respectively. Furthermore, suppose there are multiple deciders, each having different **d** and g. Thus to say "for every rational decider", as per the verbal definitions of Sections 5.5.1 and 6.3.1, is equivalent to saying "for every **d** and g". Hence the following mathematical definition.

Definition   Forecaster A is said to be *more informative than* forecaster B if, for every disutility vector **d** and every prior probability g,

$$VF(\mathbf{d}, g; f_0^A, f_1^A) \geq VF(\mathbf{d}, g; f_0^B, f_1^B).$$

7.5.3   Informativeness Relation

The order of forecasters in terms of the informativeness relation does not always exist. Given two forecasters, it is possible that neither A is more informative than B, nor B is more informative than A. But if an order does exist, then it can be established in two ways: (i) by calculating the economic value of each forecaster for every **d** and g and testing directly the above definition; or (ii) by performing the verification analysis as detailed in Section 6.3. The essence of the verification analysis is this: the ROC of forecaster A is constructed from (f_0^A, f_1^A); the ROC of forecaster B is constructed from (f_0^B, f_1^B); the comparison theorem is applied to infer whether or not one forecaster is more informative than the other. This theorem is restated and proven in Section 7.6.3.

In summary, the Bayesian decision theory, presented in this chapter, defines the informativeness relation for comparing forecasters from the viewpoint of a utilitarian society; the Bayesian verification theory, presented in Chapter 6, provides a statistical method for establishing the existence of the informativeness relation. Thus the two theories cohere.

7.6   Concepts and Proofs

7.6.1   Order of Disutilities

Theorem *(Order of disutilities)*   *For any disutility vector* $(d_{00}, d_{01}, d_{10}, d_{11})$, *any prior probability g, and any conditional probability functions* (f_0, f_1) *of the BPF,*

$$0 \leq D^{**} \leq DF^* \leq D^*.$$

Proof: That $0 \leq D^{**} \leq D^*$ is demonstrated via geometric construction in Figure 7.2, which is explained in Sections 7.2.5 and 7.3.2. The two inequalities involving DF^* are demonstrated herein (with the understanding that Σ is over $y \in \mathcal{Y}$). First,

$$
\begin{aligned}
DF^* &= \sum D^*(y)\kappa(y) \\
&\geq \sum D^{**}(y)\kappa(y) \\
&= \sum \left[d_{00} + (d_{11} - d_{00})\eta(y)\right]\kappa(y) \\
&= d_{00} \sum \kappa(y) + (d_{11} - d_{00})g \sum f_1(y) \\
&= d_{00} + (d_{11} - d_{00})g = D^{**},
\end{aligned}
$$

where the inequality arises because $D^*(y) \geq D^{**}(y)$ for every probability $y \in \mathcal{Y}$; the fourth line results from the substitution $\eta(y)\kappa(y) = gf_1(y)$ prescribed by Bayes' theorem, equation (6.8); and each of the two sums equals 1. Second,

$$
\begin{aligned}
DF^* &= \sum D^*(y)\kappa(y) \\
&= \sum \left[\min\{D_0(y), D_1(y)\}\right]\kappa(y) \\
&\leq \min\left\{\sum D_0(y)\kappa(y), \sum D_1(y)\kappa(y)\right\} \\
&= \min\left\{\sum[d_{00} + (d_{01} - d_{00})\eta(y)]\kappa(y), \sum[d_{10} + (d_{11} - d_{10})\eta(y)]\kappa(y)\right\} \\
&= \min\left\{d_{00}\sum\kappa(y) + (d_{01} - d_{00})g\sum f_1(y), d_{10}\sum\kappa(y) + (d_{11} - d_{10})g\sum f_1(y)\right\} \\
&= \min\left\{d_{00} + (d_{01} - d_{00})g, d_{10} + (d_{11} - d_{10})g\right\} \\
&= \min\left\{D_0, D_1\right\} = D^*,
\end{aligned}
$$

where the inequality arises because the expectation of the minimum of two functions is not greater than the minimum of the expectations of these two functions; the fifth line results from the substitution $\eta(y)\kappa(y) = gf_1(y)$ prescribed by Bayes' theorem, equation (6.8); and each of the four sums equals 1. QED.

7.6.2 Integrated Minimum Disutility

An alternative calculation of the expectation (7.19) utilizes the detection set $\mathcal{Y}_1$, defined by equation (7.15), and its complement relative to $\mathcal{Y}$, which is $\mathcal{Y}_1^c = \mathcal{Y} - \mathcal{Y}_1$. Thus, expression (7.20) can be rewritten in the form

$$
\begin{aligned}
DF^* &= \sum_{y \in \mathcal{Y}} D^*(y)\kappa(y) \\
&= \sum_{y \in \mathcal{Y}_1^c} D_0(y)\kappa(y) + \sum_{y \in \mathcal{Y}_1} D_1(y)\kappa(y),
\end{aligned}
\tag{7.25}
$$

where $D_0(y)$ and $D_1(y)$ come from equation (7.16). The above expression shows that the integrated minimum disutility DF^* can be decomposed: the first summand represents the expected contribution of the occasion on which it is optimal not to take action, and the second summand represents the expected contribution of the occasion on which it is optimal to take action.

7.6.3 Comparison of Forecasters

The theorem of Section 6.3.4 on comparison of forecasters is restated and then proven using the mathematical definition of the relation "more informative than" stated in Section 7.5.2.

Theorem *(Comparison of forecasters)*   *Forecaster* A *is more informative than forecaster* B *if and only if the ROC outline of* A *is superior to the ROC outline of* B.

Proof: It proceeds in five steps.

1. *Value of forecaster.* As is to be proven in Exercise 3, the inequality in the definition of "more informative than" is equivalent to

$$
DF^*(\mathbf{d}, g; f_0^A, f_1^A) \leq DF^*(\mathbf{d}, g; f_0^B, f_1^B).
\tag{7.26}
$$

In the expression for DF^* given in Exercise 4.3, the first two summands are independent of the forecaster; therefore, they can be omitted, and the above inequality is equivalent to each of the following inequalities:

$$(1 - g)(d_{10} - d_{00})P^A(F) - g(d_{01} - d_{11})P^A(D) \leq (1 - g)(d_{10} - d_{00})P^B(F) - g(d_{01} - d_{11})P^B(D),$$

$$\frac{1-g}{g}\frac{d_{10} - d_{00}}{d_{01} - d_{11}}P^A(F) - P^A(D) \leq \frac{1-g}{g}\frac{d_{10} - d_{00}}{d_{01} - d_{11}}P^B(F) - P^B(D),$$

$$\frac{1-g}{g}\frac{1}{r}P^A(F) - P^A(D) \leq \frac{1-g}{g}\frac{1}{r}P^B(F) - P^B(D),$$

where r comes from equation (7.5). Introducing t defined in Exercise 2.1 yields

$$tP^A(F|t) - P^A(D|t) \leq tP^B(F|t) - P^B(D|t). \tag{7.27}$$

2. *Detection set.* As is to be proven in Exercise 2, expressions (7.15) and (6.11), which define set $\mathcal{Y}_1$, are equivalent. Hence, the following statements are equivalent:
   - for every rational decider;
   - for every **d** and g;
   - for every r and g, $0 < r < \infty$, $0 < g < 1$;
   - for every t, $0 < t < \infty$.

   Therefore, inequality (7.26) holds for every $(\mathbf{d}, g)$ if and only if inequality (7.27) holds for every t.

3. *ROC graph.* This is the set of points $\{(P(F|t), P(D|t)) : t \in (0, \infty)\}$. Let $\varphi : [0, 1] \to [0, 1]$ be a function such that $P(D|t) = \varphi(P(F|t))$ for every $t \in [0, \infty)$. Hence, the statement "the ROC outline of A is superior to the ROC outline of B" is equivalent to the statement: for every $p \in [0, 1]$,

$$\varphi^A(p) \geq \varphi^B(p). \tag{7.28}$$

4. *Proof that inequality (7.27) implies inequality (7.28).* For $t \in (0, \infty)$, let $\hat{t} \in (0, \infty)$ be a point such that $P^B(F|\hat{t}) = P^A(F|t)$. Then inequality (7.27) can evolve thusly:

$$t\, P^A(F|t) - \varphi^A(P^A(F|t)) \leq t\, P^B(F|t) - \varphi^B(P^B(F|t)),$$

$$t\, P^A(F|t) - \varphi^A(P^A(F|t)) \leq t\, P^B(F|\hat{t}) - \varphi^B(P^B(F|\hat{t})),$$

$$t\, P^A(F|t) - \varphi^A(P^A(F|t)) \leq t\, P^A(F|t) - \varphi^B(P^A(F|t)),$$

$$\varphi^A(p) \geq \varphi^B(p),$$

where the second inequality is possibly strengthened because $\hat{t}$ is the nonoptimal threshold for using the forecast from B, and in the last inequality $p = P^A(F|t)$.

5. *Proof that inequality (7.28) implies inequality (7.27).* For $t \in (0, \infty)$, let $\check{t} \in (0, \infty)$ be a point such that $P^A(F|\check{t}) = P^B(F|t)$, and let $p = P^B(F|t)$. Then inequality (7.28) can evolve thusly:

$$\varphi^A(P^B(F|t)) \geq \varphi^B(P^B(F|t)),$$

$$t\, P^B(F|t) - \varphi^A(P^B(F|t)) \leq t\, P^B(F|t) - \varphi^B(P^B(F|t)),$$

$$t\, P^A(F|\check{t}) - \varphi^A(P^A(F|\check{t})) \leq t\, P^B(F|t) - P^B(D|t).$$

Because $\check{t}$ is the nonoptimal threshold for using the forecast from A, the left side of the last inequality can be lowered by replacing $\check{t}$ with t:

$$t\, P^A(F|t) - \varphi^A(P^A(F|t)) \leq t\, P^A(F|\check{t}) - \varphi^A(P^A(F|\check{t})).$$

With the substitution $P^A(D|t) = \varphi^A(P^A(F|t))$, and via transitivity:

$$t\, P^A(F|t) - P^A(D|t) \le t\, P^B(F|t) - P^B(D|t).$$

QED.

7.6.4 Comparison of Predictors

The theorem of Section 5.5.4 on comparison of predictors for the Bayesian forecaster has an analogous proof. The only difference lies in the equation for DF^*: the discrete probability variate Y is replaced by the continuous predictor X, discrete functions κ and η on the sample space $\mathcal{Y}$ are replaced by continuous functions κ and π on the sample space $\mathcal{X}$, and the expectation with respect to X is defined in terms of the integral on $\mathcal{X}$. The proof in Section 7.6.3 remains intact.

7.6.5 Total Probability of Decision Error

The two types of errors defined in Section 7.2.2, false alarm F and missed event M, each has probability (equations (6.12)–(6.15)) that is conditional on an event, $W = 0$ or $W = 1$. Therefore, they may be combined into an unconditional decision error ER. For a fixed threshold $t \in (0, \infty)$, the total probability law yields

$$P(ER|t) = P(Y \in \mathcal{Y}_1(t)|W = 0)P(W = 0) + P(Y \in \mathcal{Y}_1^c(t)|W = 1)P(W = 1);$$

in the abbreviated notation,

$$P(ER|t) = P(F|t)(1 - g) + P(M|t)g.$$

This *total probability of decision error* has an interesting property, which is associated with a special case of the optimal decision rule — the subject of Exercise 12.

Theorem *(Total probability of decision error)*   *For any prior probability g, the total probability of decision error $P(ER|t)$ is minimal when the ratio of the opportunity losses $r = 1$; that is, when $t = (1 - g)/g$.*

Proof: It proceeds in four steps.

1. From expression for DF^* given in Exercise 4.4,

$$DF^* = C + (1 - g)(d_{10} - d_{00})P(F) + g(d_{01} - d_{11})P(M),$$

where

$$C = d_{00} + g(d_{11} - d_{00}).$$

2. Dropping out constant C, whose value is unaffected by decision, and reintroducing the threshold t into the probabilities, yields a characteristic

$$CDF^*(t) = (d_{10} - d_{00})P(F|t)(1 - g) + (d_{01} - d_{11})P(M|t)g.$$

The optimal decision rule (7.14) minimizes $CDF^*(t)$ by setting $t = (1/r)[(1 - g)/g]$, as given in Exercise 2.1.

3. If $r = 1$, then by definition (7.5), $d_{10} - d_{00} = d_{01} - d_{11} = \Delta$. Using Δ as a divisor on both sides of the above equation, and denoting $CDF^*(t)/\Delta = P(ER|t)$, yields

$$P(ER|t) = P(F|t)(1 - g) + P(M|t)g,$$

which is the definition of the total probability of decision error.

4. Because $CDF^*(t)$ attains the minimum at $t = (1/r)[(1 - g)/g]$, setting $r = 1$ implies that $P(ER|t)$ attains the minimum at $t = (1 - g)/g$. QED.

Bibliographical Notes

Publications by Blackwell (1951, 1953) and Blackwell and Girshick (1954) are the original sources of the informativeness relation and of the theorem on comparison of forecasters (comparison of experiments in their terminology). The proof presented herein derives from their ideas, but is cast at an elementary level.

Exercises

1 **Maximum VPF**. Consider expression (7.12). Suppose the disutilities $d_{00}, d_{01}, d_{10}, d_{11}$ are fixed, and one is interested in the sensitivity of *VPF* to the prior probability g. For this purpose, *VPF* is viewed as a function of g, and this can be shown explicitly by writing *VPF*(g). From the sensitivity analysis illustrated in Figure 7.4, it is known that

$$VPF(p^*) = \max_{0<g<1} VPF(g).$$

   1.1 Prove that each branch of expression (7.12) yields

$$VPF(p^*) = \frac{(d_{01} - d_{11})(d_{10} - d_{00})}{(d_{01} - d_{11}) + (d_{10} - d_{00})}.$$

   1.2 Identify the opportunity losses on the right side of the above equation; denote them by l_M and l_F. Then write an expression for *VPF*(p^*) in terms of two opportunity losses instead of four disutilities. Interpret the expression and draw a conclusion.

2 **Detection set**. In Chapter 7, the detection set is defined by expression (7.15). In Chapter 6, the detection set is defined by expression (6.11).
   2.1 Prove that the two definitions are equivalent. *Hint*: Start from (7.15), replacing the posterior probability $\eta(y)$ and the threshold probability for action p^* with appropriate expressions. Next manipulate the inequality to show that, when the prior probability g and the ratio of the opportunity losses r are fixed, the threshold t, which appears in (6.11), is given by

$$t = \frac{1}{r}\frac{1-g}{g}.$$

   2.2 To construct the ROC, all values of $t \in (0, \infty)$ must be considered. In light of the above relationship between r, g, and t, interpret the meaning of varying the threshold t from 0 to ∞.

3 **Informativeness relation**. Consider two inequalities, the elements of which are defined in Section 7.5:

$$VF\left(\mathbf{d}, g; f_0^A, f_1^A\right) \geq VF\left(\mathbf{d}, g; f_0^B, f_1^B\right),$$

$$DF^*\left(\mathbf{d}, g; f_0^A, f_1^A\right) \leq DF^*\left(\mathbf{d}, g; f_0^B, f_1^B\right).$$

   3.1 Prove that the two inequalities are equivalent (i.e., that one inequality holds if, and only if, the other inequality holds).
   3.2 The definition of "more informative than" is stated in terms of the first inequality. Restate it in terms of the second inequality.

4 **Integrated disutility**. Consider the integrated minimum disutility DF^* given by equation (7.25).
   4.1 Prove that it can be expressed in terms of the input elements $\left(\mathbf{d}, g; f_0, f_1\right)$ as follows:

$$DF^* = \sum_{y\in\mathcal{Y}_1^c} \left[d_{00}f_0\left(y\right)\left(1-g\right) + d_{01}f_1\left(y\right)g\right] + \sum_{y\in\mathcal{Y}_1} \left[d_{10}f_0\left(y\right)\left(1-g\right) + d_{11}f_1\left(y\right)g\right].$$

4.2 Prove that it can be expressed in terms of the performance probabilities (as they are defined in Section 6.3.2) as follows:

$$DF^* = (1 - g)\, d_{00} P(Q) + g d_{01} P(M) + (1 - g)\, d_{10} P(F) + g d_{11} P(D).$$

Hint. Assume that the threshold t has a fixed value, which is calculated from the given **d** and g via the relationship specified in Exercise 2 above, with r being calculated from **d** via relationship (7.5). With t fixed, the notation is abbreviated; for example, $P(D)$ stands for $P(D|t)$.

4.3 Prove that it can be expressed in terms of $P(F)$ and $P(D)$ only as follows:

$$DF^* = d_{00} + g(d_{01} - d_{00}) + (1 - g)\left(d_{10} - d_{00}\right) P(F) - g\left(d_{01} - d_{11}\right) P(D).$$

4.4 Prove that it can be expressed in terms of $P(F)$ and $P(M)$ only as follows:

$$DF^* = d_{00} + g(d_{11} - d_{00}) + (1 - g)(d_{10} - d_{00})P(F) + g(d_{01} - d_{11})P(M).$$

5 **Missing probability**. The owner of a radio tower must decide whether or not to purchase earthquake insurance. According to a risk analysis by civil engineers, the probability is 0.8 that the tower will be damaged by a critical earthquake — an earthquake of magnitude 8.5 or higher on the Richter scale with the epicenter within 50 km of the tower. The probability of damage by a weaker or more distant earthquake is practically nil. The replacement cost is \$900 000. The annual premium is \$10 000 for an insurance policy that covers the replacement cost. The owner has difficulty finding an estimate of the probability g of a critical earthquake occurring within a 1-year period.

5.1 Which decision would be preferred if $g = 0.004$?

5.2 Which decision would be preferred if $g = 0.012$?

5.3 Which decision would be preferred if $g = 0.016$?

5.4 Does the owner need to find the exact value of g in order to determine the preferred decision? What does he need to find?

6 **Missing disutility**. Upon returning home, a fisherman discovers he has lost his custom-made pocket knife. It may have fallen into the water ($W = 0$) or been left somewhere onshore ($W = 1$). In the former event, the knife is lost forever; in the latter event, it can be recovered if searched for. To decide whether to search ($a = 1$) or not to search ($a = 0$), the fisherman assesses the probability $P(W = 1) = 0.45$ and the disutilities d_{aw} of three outcomes,

$$d_{00} = 1,\ d_{01} = 1.6,\ d_{10} = 1.8,$$

but has difficulty assessing the disutility d_{11} of the fourth outcome.

6.1 Which decision would be preferred if $d_{11} = 0.3$?

6.2 Which decision would be preferred if $d_{11} = 0.5$?

6.3 Which decision would be preferred if $d_{11} = 0.7$?

6.4 Does the fisherman need to assess the exact value of d_{11} in order to determine the preferred decision? What does he need to assess?

Note. To this fisherman, the disutilities quantify the perceived value of the lost knife (d_{00}), the knife's value augmented by the regret he imagines experiencing should he learn that someone else found the knife and took it (d_{01}), the knife's value plus the search effort (d_{10}), the search effort offset partly by the joy of recovering the knife (d_{11}).

7 **Inferring the ratio of opportunity losses**. Dressed in academic regalia (gown, hood, and cap), and already on her way to the graduation ceremony, your friend heard a weather forecast calling for a 40% chance of rain during the day. She promptly entered a store and purchased an umbrella for $30. (She knew that the weather forecaster was well calibrated.)

7.1 Define the elements of this decision problem.

7.2 Define and interpret the ratio of the opportunity losses.

7.3 Suppose your friend made the optimal decision according to the detection model. Knowing this decision, infer a bound on the ratio of the opportunity losses. Is it a lower bound or an upper bound?

7.4 Calculate a bound on the amount your friend should have been willing to pay for a perfect weather forecast that day before purchasing the umbrella. Is it a lower bound, an upper bound, the greatest lower bound, or the least upper bound? Explain.

8 **Weather icons**. Forecasters of weather for the general public make decisions that map the forecast into an icon to be displayed on a website, on television, in a newspaper. For example, a sun stands for no rain, and a cloud stands for rain. The threshold probability for choosing a cloud differs among the forecasters. The National Weather Service (NWS) set it to 0.2, whereas the Weather Channel (WC) set it to 0.3 (*The Wall Street Journal*, 26–27 January 2013).

8.1 Infer the values of the ratio of the opportunity losses r that the NWS and WC presume to reflect the "preference" of the general public.

8.2 Choose a personal decision problem in which you would use the probability of rain occurrence to make a decision. (i) Define the elements of this decision problem. (ii) In the context of this problem, interpret each of the r values inferred in Exercise 8.1. (iii) Does any of these r values reflect your preference for outcomes in your decision problem? (iv) What would you suggest to the NWS and WC?

8.3 The threshold probabilities set by the NWS and WC are often said to reflect "wet bias". An official explained that the icons are intended to show "the expected conditions". (i) Do you agree with these statements? Justify your answer. (ii) If you disagree, then suggest a scientifically more appropriate explanation.

9 **Rescue search**. To scale a 14 300 foot-high mountain peak, a hiker leaves the lodge before dawn, and is expected to return by midnight. Sudden cold winds with snow showers often envelope the peak, sometimes for minutes, but sometimes for hours. When a hiker does not return, the rescue squad initiates the search in daytime, as soon as the weather permits, if the probability of finding the hiker alive is assessed to be a least 0.01. On average, the search costs $30 000. If the missing hiker is found alive, a benefactor donates $10 000 toward covering the cost.

9.1 Define elements of this decision problem, and state the assumptions needed to apply the detection-decision model.

9.2 Find the value of hiker's life imputed to the decision rule used by the rescue squad.

9.3 Find the value of a perfect forecast of the missing hiker's status at the time the decision (to search or not to search) is made, when the assessed probability of finding the hiker alive is (i) 0.05, (ii) 0.005. (iii) Interpret this *VPF* for the rescue squad.

10 **Automobile insurance**. A comprehensive insurance policy pays the owner the market value e (in dollars) of the auto lost due to a disastrous event such as theft, vandalism, fire, tornado, etc. The annual insurance premium m [$] depends on the auto's value (which declines with age and mileage), the area in which the owner resides (which influences the probability g of a disastrous event occurring within a year), and other factors (e.g., auto make, deductible amount, discount for insuring multiple autos). Each December, a taxicab operator must decide, for every auto she owns, to purchase or not to purchase a comprehensive insurance for the next year; the operator wants to minimize the expected cost per year. The autos she owns are specified as the options.

Assumptions. An auto may suffer only one disastrous event per year. The disutility equals the cost.

Options for autos:

    (A) $e = 65\,000$, $m = 95$, $g = 0.005$.
    (B) $e = 47\,000$, $m = 380$, $g = 0.009$.
    (C) $e = 15\,900$, $m = 142$, $g = 0.009$.
    (D) $e = 9400$, $m = 112$, $g = 0.01$.
    (E) $e = 5100$, $m = 93$, $g = 0.01$.
    (F) $e = 21\,800$, $m = 136$, $g = 0.005$.

10.1 Derive the expression for p^* in terms of e and m; interpret its meaning.

10.2 For every auto, calculate p^*, determine the optimal decision, and calculate the minimum expected cost per year.

10.3 Derive the expression for *VPF* in terms of e, m, g. For every auto, calculate *VPF*; interpret its meaning. Compare *VPF* with m; draw a conclusion.

10.4 The auto's value e declines with age, and so does the insurance premium m, but slower. Hence, there exists a threshold e^*, such that if $e \le e^*$, then it is not optimal to purchase comprehensive insurance. (Consistently with this rule, many owners do not carry the comprehensive insurance on older autos; likewise for collision insurance.) Derive the expression for e^* as a function of m and g. Calculate e^* for every auto.

10.5 To cover the administrative costs and to make a profit, the insurance company sets the premium m to 1.2 times the expected payment for a lost auto. For every auto, infer the value of probability g' of the disastrous event that the insurance company uses in its calculation of m. Compare g' with g — the given probability, which was assessed from the viewpoint of the insurance purchaser (the taxicab operator).

11 **Automobile insurance** — Exercise 10 *continued*. Let *VPF* denote the value of a perfect forecast of a disastrous event, which may occur only once per year and against which an auto can be insured by a comprehensive policy at the annual premium m. Prove that $VPF < m$.

Hint. The expression for *VPF* has two pieces; find an upper bound for each piece.

12 **Maximum *a posteriori* probability rule**. In some applications (e.g., in a radar system — see Exercise 6 in Chapter 5), the disutility vector **d** is unknown because the user is unknown, or because there will be many users, each having a different **d**. In such cases, the oft employed decision rule is the *maximum a posteriori probability* (MAP) rule:

$$a^*(y) = \begin{cases} 0 & \text{if } P(W = 0|Y = y) \ge P(W = 1|Y = y), \\ 1 & \text{if } P(W = 1|Y = y) > P(W = 0|Y = y). \end{cases}$$

12.1 Prove that this rule is equivalent to the optimal decision rule (7.14) whenever the threshold probability for action $p^* = 1/2$.

12.2 Prove that this rule is equivalent to presuming the ratio of the opportunity losses $r = 1$. In other words, this is a special case: the opportunity losses from a missed event and from a false alarm are identical.

Note. As a default decision rule, the MAP rule has an attractive property: it minimizes the total probability of decision error — as defined and proven in Section 7.6.5.

13 **Maximum likelihood rule** — Exercise 12 *continued.* When both the disutility vector **d** and the prior probability of an event occurring g are unknown, the oft employed decision rule is the *maximum likelihood* (ML) rule:

$$\alpha^*(y) = \begin{cases} 0 & \text{if } f_0(y) \geq f_1(y), \\ 1 & \text{if } f_1(y) > f_0(y). \end{cases}$$

Here, f_w is the probability function of the probability variate Y, conditional on the event $W = w$, for $w = 0, 1$, as defined and interpreted in Section 6.1.6. Recall that given $Y = y$, the object $f_w(y)$ is called the likelihood of event $W = w$; hence the name of this rule.

13.1 Prove that this rule is equivalent to the optimal decision rule (7.14) whenever $r = 1$ and $g = 1/2$.

13.2 Rewrite the ML rule in terms of the likelihood ratios. Next, state this decision rule in two sentences, employing the term *likelihood ratio on*. (Recall similar usage of this term in Sections 5.1.4 and 6.3.2.)

14 **Orchardist.** The Yakima Valley of central Washington is a land of orchards producing apples, pears, peaches, and other fruits. In the spring, the buds develop into blossoms, but can be damaged by frost — an event which reduces the fruit yield, and thereby the revenue. Every evening, from mid-March through May, the orchardist must decide whether or not to protect the buds from frost. Protection (total or partial) is achieved by deploying heaters, or wind machines (which circulate air in the orchard), or sprinklers (which cover tree branches with a layer of water), or a combination thereof; this costs, of course. You have been hired as a decision consultant by an orchardist who wants to minimize the expected cost per night in the following situation. The full yield of peaches has market value of $6200 per acre, and is lost when not protected during overnight frost. Deploying heaters during one night costs $420 per acre, and protects 90% of the yield. The climatic probability of overnight frost ($W = 1$) is 0.3. The decision is to be based on the forecast probability y of the frost, issued in the afternoon whenever overnight frost is possible. The conditional probability functions of the probability variate Y, estimated from a joint sample of (Y, W) collected during several frost seasons are:

y	0.2	0.4	0.6	0.8
$f_1(y)$	0.08	0.18	0.31	0.43
$f_0(y)$	0.50	0.29	0.16	0.05

14.1 *Preliminaries.* (i) Define the disutilities and assign their values. (ii) Calculate the ratio of the opportunity losses r, and the threshold probability for action p^*.

14.2 *Decision with prior probability.* (i) Find the optimal decision. (ii) Calculate the minimum disutility of decision. (iii) Calculate the value of the perfect forecaster; interpret it for the orchardist.

14.3 *Decisions with forecasts.* (i) Calculate the expected probability function κ and the probability calibration function η; graph functions κ and η; characterize in words the calibration skill (and any biases) of the forecaster. (ii) Determine the optimal decision function and the detection set. (iii) Calculate and graph the minimum disutilities of decisions. (iv) Calculate the integrated minimum disutility. (v) Calculate the value and the efficiency of the forecaster; interpret each measure for the orchardist.

14.4 *Decision errors.* (i) Calculate the threshold t defined in Exercise 2.1 and used in expression (6.11) for the detection set. (ii) Calculate the probability of detection, the probability of a false alarm, and the probability of a missed event. (iii) Calculate the total probability of decision error, according to the definition in Section 7.6.5.

Bibliographical note. This exercise conveys the gist of the orchardist's dilemma, but is a gross simplification of the actual decision problem, which was modeled by Katz et al. (1982). Firstly, the relationship between the freezing temperature and the yield reduction is complex. Secondly, many decisions may be necessary during the $2\frac{1}{2}$ months of the frost season. To make these decisions optimally requires a sequential decision model so that each decision takes into account the current state of the orchard (whose yield may have already been reduced by preceding frosts). A theory underlying such a sequential decision process with probabilistic forecasts was published by Krzysztofowicz and Long (1990b).

15 Orchardist — *continued*: MAP **and** ML **rules.** Read Exercises 12 and 13 to familiarize yourself with these rules. Next apply these rules to the data from Exercise 14 by completing Exercises 15.1 and 15.2 for each rule.

15.1 *Decision rule evaluation.* (i) Determine the decision function and the detection set. (ii) Calculate the disutilities of decisions. (iii) Calculate the integrated disutility. *Hint*: use functions η and κ from Exercise 14.3.

15.2 *Decision errors.* Repeat Exercise 14.4. *Hint*: use r and g values presumed by the given rule.

15.3 Compare the three decision rules (the optimal rule, the MAP rule, the ML rule) in terms of the following elements: (i) the integrated disutility; (ii) the threshold t; (iii) the probabilities $P(D|t)$, $P(F|t)$, $P(ER|t)$. (iv) Draw conclusions.

15.4 Use the three pairs of probabilities $(P(D|t), P(F|t))$ to graph the ROC of the forecaster. For each of the five points of this ROC, indicate the interval of threshold t values that generate this point. Draw conclusions.

    Hints. (i) What is happening as t increases from 0 to ∞?

      (ii) Which one is the optimal operating point for the orchardist?

      (iii) Why does this point offer the preferred trade-off? Explain it to the orchardist.

      (iv) Before completing this exercise, review Section 6.3.4.

16 Newsboy. Each day before dawn, the newsboy, who delivers newspapers by tossing them through his car window onto the front lawns of suburban houses, must decide whether or not to put the papers in plastic bags to protect them from rain before noon (by which time all papers are fetched by the subscribers). When he does not bag the papers and it does not rain, the delivery is successful; but if it does rain, he must purchase the second batch of papers, bag them, and redeliver, at the total cost of c cents per paper. When he bags the papers, at the cost of b cents per paper, the delivery is successful, regardless of the weather. Help the newsboy to minimize his expected cost due to the uncertainty about the occurrence of rain (and thereby to maximize his expected profit, given the costs specified in the option) in two cases: (i) given the prior probability $g = 0.45$ of occurrence of rain in the period 0600–1200 on any day in December (estimated from a long climatic record); (ii) given the forecast probability y assessed by a meteorologist before dawn on each day, for which the verification output is:

y	0.05	0.1	0.2	0.3	0.5	0.7	0.9
$\kappa(y)$	0.02	0.07	0.14	0.27	0.21	0.16	0.13
$\eta(y)$	0.08	0.09	0.17	0.26	0.58	0.74	0.83

Options: (A) $c = 36$, $b = 9$;

             (B) $c = 28$, $b = 8$;

             (C) $c = 21$, $b = 7$.

16.1 *Preliminaries.* (i) Define the disutilities and assign their values. (ii) Calculate the ratio of the opportunity losses r, and the threshold probability for action p^*. (iii) Graph p^* versus r for $r \in [0, 9]$; characterize in words the shape of this function.

16.2 *Decision with prior probability.* (i) Find the optimal decision. (ii) Calculate the minimum disutility of decision. (iii) Calculate the value of the perfect forecaster; interpret it for the newsboy. (iv) Graph *VPF* versus g for $g \in [0, 1]$.

16.3 *Decisions with forecasts.* (i) Graph functions κ and η; characterize in words the calibration skill (and any biases) of the forecaster. (ii) Determine the optimal decision function and the detection set. (iii) Calculate and graph the minimum disutilities of decisions. (iv) Calculate the integrated minimum disutility. (v) Calculate the value and the efficiency of the forecaster; interpret each measure for the newsboy.

  Bibliographical note. In the nineteenth and early twentieth centuries, as America grew, so did newspaper publishing, and the newsboy became a fixture of the city streets. He was idealized as a "young capitalist-in-training" (*The Wall Street Journal*, 7 October 2019). Every day he had to forecast the demand for news and stories of that day to decide the number of copies to purchase from the publisher for delivery to subscribers and for sale on street corners. Earning pennies, regardless of weather, he learned the virtues of work and resilience. Many renowned Americans got their first education in entrepreneurship as newsboys, among them Thomas Edison, Herbert Hoover, John Wayne, Warren Buffett (DiGirolamo, 2019).

17 **Newsboy** — *continued*: **negative forecast value.** A second newsboy from an adjacent suburb asks you to perform for him the decision analysis parallel to that in Exercise 16. Specifically, he asks you to use the same values of c and b, the same functions κ and η, but $g = 0.2$ because he has heard that in the coming winter the frequency of rain will be lower than the climatic mean due to the Southern Pacific Oscillation (the periodic warming, El Niño event, and cooling, La Niña event, of water in the tropical eastern Pacific; they produce wetter and drier weather, respectively).

17.1 *Decision with prior probability.* (i) Find the optimal decision a^*. (ii) Calculate D^*, D^{**}, *VPF*.

17.2 *Decisions with forecasts.* Given the newsboy's specification, what are $\mathcal{Y}_1$ and DF^* in comparison with those found in Exercise 16.3?

17.3 *Value of forecaster.* (i) Calculate *VF*. Is it negative? If so, then this result violates the property you learned in Section 7.4.5. (ii) Identify the reason for this violation, and describe the necessary correction to the decision analysis.

17.4 *Demonstration of incoherence.* Do not read this paragraph until you complete Exercise 17.3. (i) Recall the definition of a well-calibrated forecaster (Section 6.2.4). (ii) Recall the theorem on calibration-in-the margin of the Bayesian forecaster (Section 5.1.5); in parallel to its proof, prove that

$$E[\eta(Y)] = g.$$

(iii) Use this fact to calculate g from functions κ and η specified in Exercise 16. (iv) Compare the two values of g: the one just calculated with the one specified by the second newsboy. (v) Draw a conclusion.

Principle (Coherence of Bayesian forecast–decision theory). The prior probability g employed in every part of the decision analysis (Chapter 7) must be the same as that employed in the verification of forecasts (Chapter 6). This *coherence* guarantees that the economic value of the forecaster, *VF*, is never negative and is never greater than the value of the perfect forecaster, as the inequalities (7.22) assert.

18 **Newsboy** — *continued*: **calibration against new prior.** The second newsboy insists on using his prior probability $g = 0.2$, but has no access to the joint sample from which functions κ and η specified in Exercise 16 were obtained. Still, he wants to use optimally the same forecasts that the first newsboy uses. Help him by calculating new functions κ and η, given his g.

Hints. (i) You know g for which the specified κ and η were obtained.
(ii) Rearrange equation (6.8) to recover function f_1.
(iii) Rearrange equation (6.7) to recover function f_0.
(iv) Employ f_0, f_1 and the new g to calculate the new κ and η.

Mini-Projects

19 **Fault diagnosis** — Chapter 5, Exercise 7 *continued*. The manufacturer is considering two alternative tests of each assembled drone before shipping it to a customer. (i) A **laboratory test** — actually a series of tests which are *definitive* for all electronic and software components: they can diagnose every fault with certainty. (ii) A **flight test** — described in Exercise 7 of Chapter 5, which is *partial*: it outputs the posterior probability of the drone being faulty, given a predictor realization; hence, the uncertainty remains. Three decision problems must be solved. The disutilities of outcomes should be defined in terms of monetary costs (in dollars), which are specified in the option and are explained in each part.
Options for costs:

Option	c_n	c_s	c_r	c_l	c_f
(A)	195	145	210	140	75
(B)	180	130	200	120	80
(C)	185	160	140	125	60
(D)	190	170	205	110	55
(E)	205	190	175	115	70
(F)	210	180	190	105	65

Part A. *Decision to perform the laboratory test*. When the drone is not tested, it is shipped from the assembly line to a customer; if it is faultless, then this is the normal (the preferred) operation; but if it is faulty, then the customer returns it to the manufacturer, who pays the shipping cost c_n, tests the drone in the laboratory to diagnose the faults at the cost c_l, repairs it at the cost c_r, and ships it back to the customer at the cost c_s. When the drone is tested in the laboratory at the cost c_l, and is found faultless, then it is shipped to a customer, who finds it faultless as well; but if it is found faulty, then it is repaired at the cost c_r and shipped to a customer, who finds it faultless, of course.

19.1 *Preliminaries*. (i) Define the elements of this decision problem. (ii) Define the disutilities of outcomes in terms of the costs. (iii) Calculate the ratio of the opportunity losses r, and the threshold probability for action p^*. (iv) Interpret r and p^* for the factory manager.

19.2 *Decision with prior probability*. Should every drone be tested in the laboratory before being shipped to a customer? (i) Calculate the disutility of each decision. (ii) Find the minimum disutility of decision, and the optimal decision. (iii) Calculate the value of the perfect forecaster; interpret it for the factory manager. Repeat these tasks for each value of the prior probability $g : g_0, 2g_0, 3g_0$. (iv) Draw a conclusion.

Part B. *Decision to classify the drone after flight test*. The drone is tested in flight at the cost c_f, the distance x is measured, the posterior probability $P(W = 1 | X = x) = \pi(x)$ is calculated, and based on it, the drone is classified as faultless ($a = 0$) or faulty ($a = 1$). When it is classified as faultless, it is shipped to a customer; from then on, the remaining outcomes are the same as in Part A, sans laboratory test. When it is classified as faulty, it is tested in the laboratory to diagnose the faults at the cost c_l; if it is found faultless, then it is shipped to a customer who finds it faultless as well; if it is found faulty, then it is repaired at the cost c_r and shipped to a customer, who finds it faultless, of course.

Hints. (i) In the decision model with forecasts (Section 7.4), variate Y is to be replaced by variate X. This is allowable because the Bayesian forecaster is well calibrated against the prior probability it employs (recall Section 6.6.1). Thus, the posterior probability $\pi(x)$ from Chapter 5 can be used directly in the decision model, replacing the posterior probability $\eta(y)$ from Chapter 6.

(ii) Because X is a continuous predictor, its expected density function κ, defined by expression (5.1), is continuous. Therefore, the calculation of the integrated minimum disutility DF^* via equation (7.19), now in the form $DF^* = E[D^*(X)]$, would require numerical integration. To simplify the calculations, we restrict the sample space of X to the realizations used in Exercise 7.6 from Chapter 5, $\mathcal{X} = \{2, 4, 6\}$, and estimate the expected probability function of X:

$$\kappa(2) = 0.68, \quad \kappa(4) = 0.23, \quad \kappa(6) = 0.09, \quad \text{for } g = g_0;$$
$$\kappa(2) = 0.59, \quad \kappa(4) = 0.28, \quad \kappa(6) = 0.13, \quad \text{for } g = 2g_0;$$
$$\kappa(2) = 0.51, \quad \kappa(4) = 0.32, \quad \kappa(6) = 0.17, \quad \text{for } g = 3g_0.$$

19.3 *Preliminaries.* (i) Define the elements of this decision problem. (ii) Define the disutilities of outcomes in terms of the costs. (iii) Calculate the ratio of the opportunity losses r, and the threshold probability for action p^*.

19.4 *Decisions with forecasts.* (i) Determine the optimal decision function and the detection set for classifying the drone as faultless or faulty. (ii) Calculate the integrated minimum disutility. (iii) Calculate the value and the efficiency of the forecaster; interpret each measure for the factory manager. Repeat these tasks for each value of the prior probability g: $g_0, 2g_0, 3g_0$.

Part C. *To test or not to test.* The factory manager contemplates three alternatives: not to test the drone before shipping it to a customer (NT); to test every drone in the laboratory (LT); to test every drone in flight (FT). Rank the three alternatives in terms of their disutilities (which represent the expected costs).

19.5 Report the disutilities of the alternatives and the rankings for each value of the prior probability g: $g_0, 2g_0, 3g_0$.

19.6 Compare the results and draw conclusions.

20 **Quality control** — Chapter 5, Exercises 8, 9 *continued.* Having developed the Bayesian forecaster for the rolling mill (specified in the option), you are asked to develop a decision model for stopping the mill during an 8-hour shift. Sans stopping, the mill can produce 22 steel sheets, indexed by $j \in \{1, \ldots, 22\}$. But the mill has two states: in tune ($W_j = 0$), out of tune ($W_j = 1$). At the beginning of a shift, it is always in tune. After the jth steel sheet is outputted and its departure x is measured, the Bayesian forecaster calculates the posterior probability $\pi_j(x) = P(W_j = 1 | X = x)$ of the rolling mill being out of tune. Based on this probability, the factory manager wants to decide to stop ($a = 1$) or not to stop ($a = 0$) the rolling mill. Here is a summary of the outcomes.

When the decision is to stop, the mill produces one more sheet and then stops; the state of the mill is diagnosed, which takes about 44 minutes. If the mill is found in tune, then the production resumes, and the monetary outcome is the loss of profit from the steel sheets that could have been produced during the stoppage. If the mill is found out of tune, then it takes 65 minutes to service it and to resume the production; the monetary outcome is the cost of milling the last steel sheet (which is scrap), plus the cost of reprocessing the scrap into production steel, plus the loss of profit from the steel sheets that could have been produced if the mill operated normally.

When the decision is not to stop and the mill is in tune, the production continues normally. But if the mill is out of tune, then three more sheets are produced before the mill stops automatically because the departures grow rapidly: the penultimate one always exceeds 99 µm, which triggers stopping; these last three steel sheets are scrap; it takes about 109 minutes to diagnose, service, and restart the mill.

Based on the posterior probability of the out-of-tune state, the factory manager wants to make the decision that minimizes the expected reduction of the profit. The economic data are as follows.

Five sheets are produced from one metric ton of steel.

Cost of steel: $1740 per metric ton.

Cost of reprocessing scrap: $430 per metric ton.

Milling cost: $169 per steel sheet.

Sale price: $595 per steel sheet.

Assumptions. (i) The special cases of outcomes that may arise within the last 175 minutes of the shift are not considered in the decision model.

(ii) State any other assumptions in the formulation of the decision model, which you consider significant enough to be communicated to the factory manager.

Options for the Bayesian forecaster:

(E) Forecaster developed in Exercise 8 of Chapter 5.

(P) Forecaster developed in Exercise 9 of Chapter 5.

20.1 Preliminaries. (i) Define and calculate the disutilities of the decision-event pairs according to the incremental analysis (Section 7.2.6). (ii) Check the nontriviality conditions. (iii) Calculate the ratio of the opportunity losses and the threshold probability for action.

20.2 Consider two events: the rolling mill is out of tune while outputting the jth steel sheet in a shift, where $j = 2, 12$. For each of these two decision times: (i) identify the detection set $\mathcal{X}_1$ and its complement $\mathcal{X}_1^c$; and (ii) specify the decision function α_j^* on $\mathcal{X}$ using the above two subsets.

20.3 Suppose each of the three departures, $x = 29, 34, 41$ μm, was observed in each of the two steel sheets, $j = 2, 12$. For each of these six decision situations: (i) find the optimal decision $\alpha_j^*(x)$; (ii) calculate the minimum disutility of decision $D_j^*(x)$, and interpret it for the factory manager.

20.4 For each of the six decision situations identified by (j, x) in Exercise 20.3: (i) calculate the value of the perfect forecaster $VPF_j(x)$; (ii) interpret it for the factory manager.

Hint. In Section 7.3, replace the probability g by the forecast probability $\pi_j(x)$; interpret $VPF_j(x)$ accordingly.

20.5 Having operated the rolling mill according to your decision model, the factory manager came up with an improvement of the automatic stopping of the mill when it enters the out-of-tune state and the optimal decision is not to stop it: now, the mill produces only two more sheets (which are scrap). (i) Repeat Exercises 20.1–20.4. (ii) Compare the two solutions. (iii) Draw conclusions.

21 Snow removal — Chapter 6, Exercises 11, 12, 14 *continued*. Every day during the official "winter season", the manager of a city's Highway Department must decide whether or not the night shift of the snow removal crew should remain on duty. The major cost of this decision is overtime wages, $7000/night. There is also the cost of fuel, $1000/night, when plowing. Plowing of the roads at night is needed if the snowfall between the last evening plowing at 1900 and the first morning plowing at 0500 is 5 cm or more. If the night snowfall is 5 cm or more and there is no night plowing, roads cannot be cleared in time before the morning rush traffic. On such a day the city's business life comes to a halt. The tax revenue lost by the city because of one day of stoppage of business activities is $20 000.

The manager wants to make optimal decisions based on local snowfall forecasts. The forecast is made at 1800 Eastern Time on each day during the 100-day official winter season, and states the probability of an event. Use your results of the verification analysis specified in the option to perform the decision analysis for the manager.

Options for the forecaster:

(A) Forecaster verified in Exercise 11 of Chapter 6.

(B) Forecaster verified in Exercise 12 of Chapter 6.

(C) Forecaster verified on estimation sample in Exercise 14 of Chapter 6.

(D) Forecaster verified on validation sample in Exercise 14 of Chapter 6.

Forecasted event. In options (A) and (B), it is {snowfall during the night is at least 5 cm}, which is relevant to the manager. In options (C) and (D), assume that the relevant event is the forecasted event, as defined in Exercise 12 of Chapter 5.

21.1 Find the optimal decision on a day on which, due to the failure of the electric power system, the weather forecast is unavailable.

21.2 Find the optimal decision function that should be used on a day on which the weather forecast is available. Determine the detection set; identify it on a graph of the posterior probability versus the forecast probability (i.e., the graph of the probability calibration function).

21.3 Suppose the forecast probabilities on seven consecutive days are 0.6, 0.2, 0.8, 0.9, 0.4, 0.2, 0.1. Determine the optimal decision on each day.

21.4 Determine the upper bound on the economic value of a weather forecast to the city.

21.5 Determine how much (at most) the city should be willing to pay the meteorologist for one forecast.

21.6 Determine the efficiency of this weather forecaster and interpret it for the manager of the Highway Department.

8

Various Discrete Models

Two models are presented. They illustrate various extensions of the basic forecast–decision model. A Bayesian search model guides optimally the sequence of searches for a lost person (or object); the predictand is multicategory, the forecaster assesses judgmentally both inputs (prior and conditional probability functions) to the Bayesian revision model, and the decider's strength of preference for outcomes (finding, not finding) is quantified by the finite impulse function.

A flash-flood warning model offers analytic means for designing, operating, and evaluating a warning system. The three components of the model, monitor–forecaster–decider, mathematicize the inherent features of any rapid-onset natural hazard: it occurs intermittently; its probability is near zero for most of the time, then rises suddenly and rapidly; the lead time from a precursor to the event is short; and the risk of severe consequences (death, injury, property damage) is high without a timely warning and effective action.

8.1 Search Planning Model

This model is dedicated to searchers and rescuers. It is a forecast–decision model that undergirds the search planning process promulgated by the National Association for Search and Rescue (NASAR). Searchers and rescuers are mostly volunteers who devote their time, energy, knowledge, and skills to finding the lost person, to save life — often in rugged terrain and foul weather.

8.1.1 Search and Rescue Situation

Here is one search and rescue situation in the incident commander's own words (President of NASAR, Bill Pierce, 1988).

> a 66-year-old woman had wandered away from her home the afternoon of 19 April 1985, in Warren County, Virginia … Continuous searching was conducted for the next six days with two shifts a day, with resources like helicopters, air-scent search dog teams, hasty teams, grid searchers and numerous logistics personnel. By the time I was put in charge on April 28, there had been eight different incident commanders … With hundreds of searchers from dozens of federal, state and local groups participating, one could expect that the record keeping and tracking of resources and assignments would be a disaster. Such was not the case, however … Part of this planning process was to review the past shifts and complete the shifting PoA (probability of area) for the 10 sectors of the search area … The areas of initial low probability were now the highest … because they had neither been searched as much nor with the highest detecting resources … The shift was

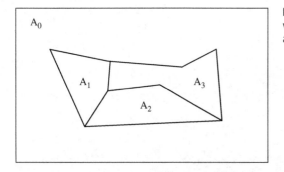

Figure 8.1 The search area A is the union of areas A_1, A_2, A_3, which are bounded and disjoint; the complement of A, denoted A_0 and dubbed "the rest of the world", is not searched.

implemented as planned and at 1245 hours, search teams found the subject, semiconscious and dehydrated, beneath a tree in sector 10 (the one with the highest shifted PoA). ... I became a firm believer in well-kept shift maps, good documentation of each shift, and the use of shifting PoA ...

Our objectives are (i) to extend the Bayesian revision model (Chapter 4) to quantify the evolving uncertainty about the location of the lost subject, and (ii) to adapt the basic decision model (Chapter 7) to provide the optimal answer to the decision dilemma: Where to search during the upcoming shift?

8.1.2 Events, Outcomes, Information

On a topographical map (Figure 8.1), the area to be searched, A, is bounded and then partitioned into $J \geq 2$ areas, $A_1, \ldots, A_J$, which searchers often call "sectors" regardless of their geometric shapes. By definition of the *partition*, areas $A_1, \ldots, A_J$ are disjoint and their union equals A. The complement of A, denoted $A_0 = A^c$, is the unbounded area exterior of A ("the rest of the world", so to speak).

The set of areas $\mathcal{A} = \{A_j : j = 0, 1, \ldots, J\}$ is given a second meaning: it is the sample space of the predictand, and j is the index of events, outcomes, sets, and probabilities. Their definitions follow.

A_j — event (predictand realization): the subject is in that area.

D_j — outcome of search (also predictor realization): the subject is detected in area A_j.

$D_j|A_j$ — the subject, who is hypothesized to be in area A_j, is detected during the search of that area. This outcome terminates the search.

$D_j^c|A_j$ — the subject, who is hypothesized to be in area A_j, is not detected during the search of that area. This is the outcome of an unsuccessful search.

I_0 — information about the incident gathered prior to the search (e.g., circumstances of subject's disappearance, time elapsed, subject's physical and mental conditions, surrounding terrain, prevailing weather, sundry clues).

I_j — information from the unsuccessful search of area A_j: knowledge and experience of searchers, search method, resources utilized (e.g., helicopters, air-scent search dogs), information gathered (e.g., regarding terrain features, potential obstacles and hazards to a walker, visibility range).

8.1.3 Judgmental Probabilities

To forecast probabilistically the location of the lost subject, two probability functions are assessed judgmentally. The first is assessed prior to the first search. The second is assessed after each unsuccessful search. The probabilities are defined for every $j \in \{0, 1, \ldots, J\}$:

$p_j = P(A_j|I_0)$ — the *prior probability* of the subject being in area A_j, assessed based on information I_0; in short, the prior *probability of area* (PoA), $0 < p_j < 1$.

$d_j = P(D_j|A_j, I_j)$ — the *probability* of detecting the subject during the search of area A_j, **conditional on the hypothesis** that the subject is in A_j, assessed based on information I_j; in short, the *probability of detection* (PoD), $0 \le d_j < 1$.

Because the search was unsuccessful, the probability d_j should quantify the collective judgment of the search team regarding the answer to the question (under the hypothesis that the subject is in area A_j): Given information I_j from the search, how certain are you that this search could have detected the subject?

For the Bayesian revision of the prior probability p_j, two other interpretations are needful:

$1 - d_j = P(D_j^c|A_j, I_j)$ — the *probability* of not detecting the subject during the search of area A_j, **conditional on the hypothesis** that the subject is in A_j, and given I_j.

$1 - d_j = P(D_j^c|A_j, I_j)$ — the *likelihood* of the subject being in area A_j, **conditional on** not detecting her during the search, D_j^c, and given I_j. This interpretation reverses the conditioning to reflect the reality of forecasting: D_j^c is known (the outcome of the search), while A_j is uncertain (must be forecasted).

Properties of PoD. These are logical implications of the definition that simplify the equations.

1. The search of area A_i cannot detect the subject in area A_j; thus for all $i, j \in \{0, 1, \dots, J\}$, if $i \neq j$ then

$$P(D_i|A_j, I_i) = 0 \iff P(D_i^c|A_j, I_i) = 1.$$

2. By design, area A_0 is not searched; thus

$$d_0 = P(D_0|A_0, I_0) = P(D_0|A_0) = 0 \iff P(D_0^c|A_0) = 1.$$

3. Likewise, if area A_j is not searched during a shift, then

$$d_j = P(D_j|A_j, I_j) = P(D_j|A_j) = 0 \iff P(D_j^c|A_j) = 1.$$

8.1.4 Assumptions

To obtain a forecasting model which is computationally simple, and therefore feasible for field operations, three assumptions are made.

1. The subject is stationary; that is, she remains in one of the areas $A_0, A_1, \dots, A_J$ during the entire search operation.
2. The probability of D_j^c, conditional on the hypothesis A_j, is determined based solely on information I_j, $j \in \{1, \dots, J\}$; in other words, prior information and information from the search of some other area do not affect $P(D_j^c|A_j, I_j)$.
3. Let $\mathbf{D}^c = (D_0^c, D_1^c, \dots, D_J^c)$ be the vector of outcomes from unsuccessful searches, and let $\mathbf{I} = (I_1, \dots, I_J)$ be the vector of information from these searches. Given $(I_0, \mathbf{I})$, the outcomes $D_0^c, D_1^c, \dots, D_J^c$ are *conditionally stochastically independent* (CSI) **relative to** each event A_j, for $j = 0, 1, \dots, J$. (For an introduction to conditional stochastic independence, review Section 4.4.3.)

Assumptions 2 and 3 allow us to factorize the joint conditional probability as follows:

$$\begin{aligned} P(\mathbf{D}^c|A_j, I_0, \mathbf{I}) &= P(D_0^c, D_1^c, \dots, D_J^c|A_j; I_0, I_1, \dots, I_J) \\ &= P(D_0^c|A_j, I_0)P(D_1^c|A_j, I_1) \cdots P(D_j^c|A_j, I_j) \cdots P(D_J^c|A_j, I_J) \\ &= P(D_j^c|A_j, I_j); \end{aligned} \tag{8.1}$$

the third equality results from the first property of the PoD.

8.1.5 Bayesian Revision Model

The Bayesian revision model formulated in Section 4.4.4 for a binary predictand is extended to a multicategory predictand whose predictor is the search outcome vector $\mathbf{D}^c$. Stage 1 revision is performed at the end of the first shift, if it is unsuccessful.

Input and Output

The input consists of the prior PoAs and the PoDs from the searches conducted during the first shift, if it is unsuccessful:

$$\{p_j : j = 0, 1, \dots, J\},$$
$$\{d_j : j = 1, \dots, J\}.$$

If an area was not searched, then its PoD is set to zero, according to Property 3.

The output consists of one probability (which plays the role of a scaling constant) and one probability function on the sample space $\mathcal{A}$:

$\kappa = P(\mathbf{D}^c|I_0, \mathbf{I})$ — the *expected probability* of the unsuccessful outcome vector $\mathbf{D}^c$, **given** the prior information I_0, and the search information vector $\mathbf{I}$.

$\pi_j = P(A_j|\mathbf{D}^c, I_0, \mathbf{I})$ — the *posterior probability* of the subject being in area A_j, **given** the unsuccessful outcome vector $\mathbf{D}^c$, the prior information I_0, and the search information vector $\mathbf{I}$; in short, the posterior PoA. (Searchers often refer to π_j as the "shifted" PoA.)

Stage 1 revision

From the total probability law,

$$\kappa = P(\mathbf{D}^c|I_0, \mathbf{I})$$
$$= \sum_{j=0}^{J} P(\mathbf{D}^c|A_j, I_0, \mathbf{I})P(A_j|I_0)$$
$$= \sum_{j=1}^{J} P(D_j^c|A_j, I_j)P(A_j|I_0) + P(D_0^c|A_0)P(A_0|I_0), \tag{8.2}$$

where the third equality results from factorization (8.1) induced by Assumptions 2 and 3, and the last summand is simplified by Property 2 of the PoD.

From Bayes' theorem, for $j = 0, 1, \dots, J$,

$$\pi_j = P(A_j|\mathbf{D}^c, I_0, \mathbf{I})$$
$$= \frac{P(\mathbf{D}^c|A_j, I_0, \mathbf{I})P(A_j|I_0)}{P(\mathbf{D}^c|I_0, \mathbf{I})}$$
$$= \frac{P(D_j^c|A_j, I_j)P(A_j|I_0)}{P(\mathbf{D}^c|I_0, \mathbf{I})}, \tag{8.3}$$

where the third equality results from factorization (8.1).

When the abbreviated notation defined earlier is employed, the operational formulae are obtained — for the expected probability of the unsuccessful outcome vector:

$$\kappa = \sum_{j=1}^{J} (1 - d_j)p_j + p_0; \tag{8.4}$$

and for the posterior PoA:

$$\pi_0 = \frac{p_0}{\kappa},$$

$$\pi_j = \frac{(1 - d_j)p_j}{\kappa}, \qquad j = 1, \dots, J. \tag{8.5}$$

Stage *n* revision

The derivation of equations for Stage 2 revision (after the second shift, if unsuccessful) is deferred to Exercise 2. Here, we exploit the *sequential revision principle* stated at the conclusion of Section 4.4.2. It justifies the conceptual algorithm for revising PoAs.

Step 1. New PoDs are assessed based on information from the searches conducted during the second shift.
Step 2. The posterior PoAs calculated in Stage 1 revision become the prior PoAs for Stage 2 revision.
Step 3. Formulae (8.4)–(8.5) are employed to calculate the revised PoAs, which are the posterior PoAs after the second shift.

This algorithm can be replicated at the end of every subsequent unsuccessful shift ($n = 2, 3, \dots$). It is formalized in the next section. But first, let us illustrate the working of formulae (8.4)–(8.5) and of the algorithm.

Example 8.1 *(Simultaneous searches)*

The search was planned for three areas in two shifts (Table 8.1). Area A_1 had the highest prior PoA, $p_1 = 0.4$, but after all three areas were searched simultaneously (and unsuccessfully) during the first shift, the posterior PoA for A_1 became the lowest, $\pi_1 = 0.108$ (because its search was judged to have a high PoD, $d_1 = 0.9$). The second shift had limited resources; therefore, only two areas were selected for searching, A_2 and A_3, because their posterior PoAs were the highest. (This decision rule is rationalized in Section 8.1.7.) After the unsuccessful searches, the posterior PoAs decreased for both areas (because their PoDs were relatively high, $d_2 = 0.8$ and $d_3 = 0.7$). Regarding the expected probability of the unsuccessful outcome, it was $\kappa = 0.37$ for the first shift, and increased to $\kappa = 0.543$ for the second shift. In tandem, the PoA for A_0 ("the rest of the world") increased from the prior value $p_0 = 0.1$ to the posterior value $\pi_0 = 0.497$ after two unsuccessful shifts.

The simplicity of calculations, the tabular display of all probabilities, and the intuitive logic of making decisions are the practical reasons for this judgmental forecasting procedure, coupled with the Bayesian revision model, being used by the search teams in field operations.

Table 8.1 Revision of the prior probability function $\{p_j\}$ into the posterior probability function $\{\pi_j\}$ via formulae (8.4)–(8.5) based on detection probabilities $\{d_j\}$: Stage 1 revision after the first shift searched areas A_1, A_2, A_3; Stage 2 revision after the second shift searched areas A_2, A_3.

Stage	A_j	p_j	d_j	$(1 - d_j)p_j$	π_j
1	A_0	0.1	0.0	0.10	0.270
	A_1	0.4	0.9	0.04	0.108
	A_2	0.2	0.6	0.08	0.216
	A_3	0.3	0.5	0.15	0.406
				$\kappa = 0.37$	
2	A_0	0.270	0.0	0.270	0.497
	A_1	0.108	0.0	0.108	0.199
	A_2	0.216	0.8	0.043	0.079
	A_3	0.406	0.7	0.122	0.225
				$\kappa = 0.543$	

8.1.6 Sequential Revision Equations

Let n ($n = 1, 2, 3, \ldots$) be the index of shifts (revision stages); let $D_{jn}^c, I_{jn}, d_{jn}, \kappa_n, \pi_{jn}$ be the elements which were defined for one shift and are now associated with the nth shift; let $\mathbf{D}_n^c = (D_{0n}^c, D_{1n}^c, \ldots, D_{Jn}^c)$, with $D_{0n}^c = D_0^c$, and $\mathbf{I}_n = (I_{1n}, \ldots, I_{Jn})$ be the vectors of outcomes and information from the nth shift; and let the three properties of PoD and the three assumptions about PoDs hold for every n.

Revision equations

Exploiting the pattern of equations (8.2)–(8.5) and the *sequential revision principle* (Section 4.4.2), the equations for Stage n revision can be induced. The expected probability of the unsuccessful outcomes from all searches conducted during the nth shift is

$$\kappa_n = P(\mathbf{D}_n^c | \mathbf{D}_1^c, \ldots, \mathbf{D}_{n-1}^c; I_0, \mathbf{I}_1, \ldots, \mathbf{I}_n). \tag{8.6}$$

The posterior PoA for area A_j ($j = 0, 1, \ldots, J$) after the nth shift is

$$\pi_{jn} = P(A_j | \mathbf{D}_1^c, \ldots, \mathbf{D}_n^c; I_0, \mathbf{I}_1, \ldots, \mathbf{I}_n). \tag{8.7}$$

When the total probability law and Bayes' theorem are employed as in equations (8.2)–(8.3), with $\pi_{j0} = p_j$ ($j = 0, 1, \ldots, J$), the operational equations for Stage n revision ($n = 1, 2, 3, \ldots$) are obtained:

$$\kappa_n = \sum_{j=1}^{J} (1 - d_{jn})\pi_{j(n-1)} + \pi_{0(n-1)}; \tag{8.8}$$

and

$$\pi_{0n} = \frac{\pi_{0(n-1)}}{\kappa_n},$$

$$\pi_{jn} = \frac{(1 - d_{jn})\pi_{j(n-1)}}{\kappa_n}, \qquad j = 1, \ldots, J. \tag{8.9}$$

Conditioning of probabilities

Expression (8.7) reveals an important property of the Bayesian revision model. Despite Assumptions 2 and 3, the posterior PoA for each area is conditional on (i) the unsuccessful outcomes from all searches conducted during n shifts, (ii) the prior information, and (iii) the information gathered from all areas searched during n shifts. In effect, the posterior probability function $(\pi_{0n}, \pi_{1n}, \ldots, \pi_{Jn})$ on the set $\mathcal{A}$ of areas quantifies the remaining uncertainty about the subject's location, given the entire search effort up till the nth revision.

Sequential search

Another property of the Bayesian revision model important for field operations is the flexibility of revising the PoAs in different search scenarios.

Example 8.2 *(Sequential search — Example 8.1 continued)*
Due to the limited resources at the beginning of the operation, the three areas could not be searched simultaneously but only sequentially — one area per shift. Therefore, the Stage 1 revision in Table 8.1 was replaced by a three-stage revision in Table 8.2. For the first shift, A_1 was selected based on the highest prior PoA: $\pi_{1,0} = 0.4$. For the second shift, A_3 was selected based on the highest posterior PoA from the first shift: $\pi_{3,1} = 0.469$. For the third shift, A_2 was selected based on $\pi_{2,2} = 0.408$. The posterior PoAs $\{\pi_{j3} : j = 0, 1, 2, 3\}$ after sequential searches by three shifts (Stage 3, Table 8.2) are identical with the posterior PoAs after simultaneous searches by one shift (Stage 1, Table 8.1). Figure 8.2 illustrates the sequential evolution of PoAs: at each Stage n, the assessed PoD after searching just one area affects the posterior PoAs for all areas.

Table 8.2 Three-stage revision of the prior probability function $\{\pi_{j0}\} = \{p_j\}$ into the posterior probability function $\{\pi_{j3}\}$ based on detection probabilities $\{d_{jn}\}$ given in Table 8.1 for Stage 1, but assessed under the *sequential search scenario*: each of the three shifts searched one area and afterward revised the PoAs; the areas were searched in the order A_1, A_3, A_2.

n	A_j	$\pi_{j(n-1)}$	d_{jn}	$(1 - d_{jn})\pi_{j(n-1)}$	π_{jn}
1	A_0	0.1	0.0	0.10	0.156
	A_1	0.4	0.9	0.04	0.063
	A_2	0.2	0.0	0.20	0.312
	A_3	0.3	0.0	0.30	0.469
			$\kappa_1 = 0.64$		
2	A_0	0.156	0.0	0.156	0.204
	A_1	0.063	0.0	0.063	0.082
	A_2	0.312	0.0	0.312	0.408
	A_3	0.469	0.5	0.234	0.306
			$\kappa_2 = 0.765$		
3	A_0	0.204	0.0	0.204	0.270
	A_1	0.082	0.0	0.082	0.109
	A_2	0.408	0.6	0.163	0.216
	A_3	0.306	0.0	0.306	0.405
			$\kappa_3 = 0.755$		

$\pi_{1,3}, \pi_{3,3}$ are inconsistent with π_1, π_3 from Stage 1 revision in Table 8.1 due to the rounding of numbers.
κ from Stage 1 revision in Table 8.1 equals the product $\kappa_1 \kappa_2 \kappa_3 : \kappa = 0.64 \times 0.765 \times 0.755 = 0.37$.

Exercise 4 asks you to redo Example 8.2 but with a different order of the searched areas. What conclusion do you anticipate? To generalize, consider any number of areas $J \geq 2$. The sequential search could be conducted for any of the $J!$ permutations of the areas. What conclusion would you anticipate?

Multiple searches

When area A_j is searched unsuccessfully multiple times during one shift, Assumption 3 is extended as follows. Given all information, the outcomes from multiple searches are CSI relative to event A_j. The implication is propitious. After N unsuccessful searches of A_j, with each search n ($n = 1, 2, \ldots, N$) contributing d_{jn} ($0 < d_{jn} < 1$) as the PoD, the *compounded* PoD is

$$d_j = 1 - (1 - d_{j1})(1 - d_{j2}) \cdots (1 - d_{jN}). \tag{8.10}$$

This d_j is interpretable as the probability of detecting the subject in area A_j at least once in N searches. Moreover, this d_j increases strictly with N — the property known as the *compounding effect* of repeated searches (or trials). Exercise 5 asks you to justify expression (8.10) and to prove the compounding effect.

Example 8.3   *(Multiple searches — Example 8.1 continued)*
Sufficient resources were available at the beginning of the operation to search A_1 once, and A_2 and A_3 twice during the first shift. The revision of the prior PoAs was performed at the end of the shift, in one stage, based on the PoDs from all five searches (Table 8.3). The posterior PoAs are identical with those obtained from the same five searches but conducted in two shifts (Stage 2, Table 8.1).

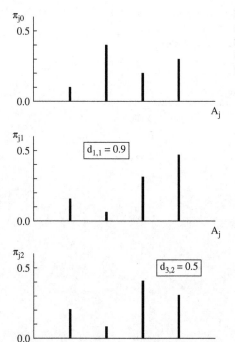

Figure 8.2 Evolution of the probability function $\{\pi_{jn} : j = 0, 1, 2, 3\}$ on the set of areas $\mathcal{A} = \{A_0, A_1, A_2, A_3\}$ from the prior function ($n = 0$) through three posterior functions ($n = 1, 2, 3$) resulting from the sequential search of areas $\{A_1, A_3, A_2\}$ with detection probabilities $(d_{1,1}, d_{3,2}, d_{2,3})$; all probabilities are reported in Table 8.2.

Termination of the search

The search is a sequential process of *random duration*. Eventually it terminates either for some practical reason or for one of the two probabilistic reasons.

1. The subject is detected, so that $P(D_{jn}|A_j, I_{jn}) = 1$ for some $j \in \{1, \ldots, J\}$, at some $n \in \{1, 2, 3, \ldots\}$.
2. The posterior probability π_{0n} for "the rest of the world" converges to 1 as the number of unsuccessful search shifts n increases; that is, it eventually becomes near certain that the subject is not within search area A. Figure 8.2 illustrates the initial increases of π_{0n}. Exercise 6 asks you to prove its convergence. (This is a version of Newman's principle of compounding evidence discussed in Section 4.4.6.)

8.1.7 Decision Model

Elements

Whereas the Incident Command Team must make many operational decisions to organize and conduct a search for the lost subject, the pivotal decision dilemma is: Where to search during the upcoming shift? To rationalize the answer is the purpose of the following decision model. It has six elements:

a_k — decision to search area $A_k, k \in \{1, \ldots, J\}$;
A_j — event: the subject is in that area, $j \in \{0, 1, \ldots, J\}$;
$\mathcal{A}$ — sample space of predictand, $\mathcal{A} = \{A_j : j = 0, 1, \ldots, J\}$;

Table 8.3 One-stage revision of the prior probability function $\{p_j\}$ into the posterior probability function $\{\pi_j\}$ based on detection probabilities $\{d_j\}$ given in Table 8.1 for Stages 1 and 2, but assessed under the *multiple searches scenario*: all 5 searches were conducted during one shift.

Stage	A_j	p_j	d_j	$(1 - d_j)p_j$	π_j
1	A_0	0.1	0.0	0.100	0.497
	A_1	0.4	0.9	0.040	0.199
	A_2	0.2	0.92	0.016	0.080
	A_3	0.3	0.85	0.045	0.224
				$\kappa = 0.201$	

π_2, π_3 are inconsistent with those from Stage 2 revision in Table 8.1 due to the rounding of numbers.

d_2, d_3 are compounded PoDs calculated from the PoDs assessed after each search and reported in Table 8.1:

$$d_2 = 1 - (1 - 0.6)(1 - 0.8) = 0.92,$$

$$d_3 = 1 - (1 - 0.5)(1 - 0.7) = 0.85.$$

κ equals the product of κ values calculated in Table 8.1:

$$\kappa = 0.37 \times 0.543 = 0.201.$$

p_j — prior probability of event A_j;

π_{jn} — posterior probability of event A_j after the nth shift of searches, $n \in \{1, 2, 3, \ldots\}$;

u — utility function that quantifies the strength of preference for the outcomes of the search.

The mission of the searchers is to save a life. Therefore, let us postulate that the utility of searching the right area (containing the subject) is indefinitely great relative to the utility of searching a wrong area. This strength of preference is captured mathematically by a *utility function* in the form:

$$u(a_k, A_j) = \begin{cases} 1 & \text{if } k = j, \\ 0 & \text{if } k \neq j. \end{cases} \tag{8.11}$$

This type of function on a countable domain is called a *finite impulse function*.

Optimal decision procedure

Any feasible decision a_k is evaluated in terms of its utility. At Stage n, the *utility of decision*, $U_n(a_k)$, is defined as the *expected utility of outcome* resulting from that decision.

At Stage 0, before the first shift, with only the prior probabilities being assessed,

$$U_0(a_k) = \sum_{j=0}^{J} u(a_k, A_j)P(A_j|I_0)$$

$$= \sum_{j=0}^{J} u(a_k, A_j)p_j = p_k. \tag{8.12}$$

The *optimal decision* a^* is the one that maximizes the expected utility of outcome:

$$U_0(a^*) = \max_{1 \leq k \leq J} U_0(a_k)$$

$$= \max_{1 \leq k \leq J} p_k. \tag{8.13}$$

The last equality means that the optimal decision a^* equals the *mode* under the prior probability function $\{p_j : j = 0, 1, \ldots, J\}$ on $\mathcal{A}$. In other words, it is optimal to search first the area to which the largest prior probability has been assigned. When resources are available to search several areas simultaneously, the *optimal decision rule* is: search the areas with the largest prior probabilities.

At Stage n, with the posterior probabilities after the nth unsuccessful shift determined, the procedure parallel to equations (8.12)–(8.13) leads to the parallel solution: the optimal decision a^* equals the *mode* under the posterior probability function $\{\pi_{jn} : j = 0, 1, \ldots, J\}$ on $\mathcal{A}$. This is a version of the *maximum a posteriori probability* (MAP) decision rule introduced in Exercise 12 of Chapter 7. When resources are available to search several areas simultaneously, the *optimal decision rule* is: search the areas with the largest posterior probabilities.

When π_{0n} is the largest probability, which may happen in a long search, as pointed out at the end of Section 8.1.6, the decision to terminate this search process and to initiate a new process with modified search area A should be considered.

8.1.8 Allocation of Resources

When several areas can be searched simultaneously, it is necessary to allocate the available resources. This operational problem is not modeled herein, but we should be aware of it because it is a concomitant decision problem under uncertainty: the criterion for allocation of resources should cohere to the criterion (8.12)–(8.13) of the expected utility maximization.

To show this concomitance, suppose the resources available to the first shift suffice to search two areas. By the optimal decision rule, these areas have the largest prior PoAs; suppose they are A_1 and A_2, having p_1 and p_2. The *ideal searches* would be exhaustive and would resolve the uncertainty. To resolve the uncertainty means either detecting the subject in A_1 or in A_2 and terminating the search, or concluding the unsuccessful searches with PoDs $d_1 = 1$ and $d_2 = 1$; that is, with being certain the subject is neither in A_1 nor in A_2.

The mathematical implication is threefold. First, $(1 - d_1)p_1 + (1 - d_2)p_2 = 0$; and with p_1 and p_2 being the largest, the expected probability of the unsuccessful outcome, κ given by expression (8.4), is minimized (over all combinations of two areas chosen from the set of J areas); equivalently, the expected PoD, which is $1 - \kappa$, is **maximized**. Second, because κ is minimized, the posterior PoA for A_0, which is π_0 given by expression (8.5), is **maximized**. Third, the posterior PoAs for A_1 and A_2, given by expression (8.5), are $\pi_1 = 0$ and $\pi_2 = 0$.

The vanished uncertainty about events A_1 and A_2 causes a concentration of the posterior probability mass on $J - 1$ events that remain uncertain: $\{A_0, A_3, \ldots, A_J\}$. If such ideal searches continued in the subsequent shifts, $n = 2, 3, \ldots$, then the posterior probability functions $\{\pi_{jn}\}$ on $\mathcal{A}$ would continue to concentrate and thus to exhibit increasing peaks — behavior that is coherent with the maximization of the expected utility $U_n(a_k) = \pi_{kn}$. (To observe these implications in sequences of numbers, work through Exercise 8.)

Whereas simultaneous ideal searches are rarely feasible operationally, they do reveal a conceptually simple, mathematically well-defined, and *asymptotically coherent allocation rule*: allocate the resources to the optimally selected areas in a manner that *maximizes the expected* PoD from the planned searches.

Bibliographical Note

This model owes its being to Greg Shea who, as the chairman of the Appalachian Search and Rescue Conference and a graduate student at the University of Virginia in 1990, shared with the author his experience and his article (Shea, 1988) in which he presented a new method for revising the PoAs. Formulae (8.4)–(8.5) are like his, but are derived via a somewhat different logic (that of the Bayesian revision theory of Section 4.4.2). His method, in turn, was based on the search theory presented in 1983 by John Bownds, Michael Ebersole, David Lovelock, and Daniel O'Connor. Dr. Bownds, who coined the term "the rest of the world", was a mathematician at the University of Arizona and later at the Oak Ridge National Laboratory in Tennessee, and an awardee of the National Association for Search and Rescue, in which he was involved in the years 1972–1993.

Adaptations of the Bayesian theory of sequential revision of probability (covered for a binary predictand in Chapter 4) to various search problems form collectively the *Bayesian search theory*. It has been applied in searches for lost persons (as in this chapter), sunken ships, missing airplanes, hidden treasures, undetonated bombs, and other vanished objects on land or in water.

8.2 Flash-Flood Warning Model

Flash floods, tornadoes, tsunamis, earthquakes, landslides, and avalanches are collectively called *rapid-onset natural hazards*. They occur intermittently and randomly. The probability of encountering one of them at a particular point and time is, *a priori*, near zero. It rises rapidly when precursors are observed, but the lead time is short and the uncertainty is large regarding the location, the timing and the magnitude of the event. To accomplish action that saves lives, prevents injuries, and reduces property damage, an automated warning system is needed.

The purpose of a mathematical model is to provide the scientific basis and analytic means for designing, operating, and evaluating the performance of a warning system. Our model, based on the Bayesian theories of Chapters 5 and 7, is tailored to warning systems for small communities exposed to flash floods — events which occur predominantly on mountain streams and for which the forecast lead time is less than 6 hours. Hence, if the floodplain dwellers are to accomplish action, a warning must be issued quickly based upon a probabilistic forecast that quantifies the uncertainty (the risk) at the onset of a heavy rainstorm. Here is one such a situation.

8.2.1 Flash-Flood Situation

The Big Thompson River originates in the Rocky Mountains in northern Colorado. It flows eastward through Estes Park (elevation 2287 m above sea level), then drops through a 40 km canyon into a valley (elevation 610 m) west of Loveland. On Saturday, 31 July 1976, a super cumulonimbus, reaching a height of 18 000 m, stalled over the Big Thompson watershed. Extremely heavy rainfall began at 1830; it lasted about 4 hours and its depth reached 30 cm in the center of the storm. The rate of river flow at the mouth of the canyon, normally 6 m^3/s, increased rapidly and crested at 870 m^3/s before midnight, then receded to about 60 m^3/s by morning. The velocity of flow peaked at 24 km/h. Being a popular tourist destination, the Big Thompson Canyon was filled with over 2500 people. No official warning was issued. By morning, 145 people perished and the damage to property (houses, cabins, motels, campgrounds, automobiles, bridges, etc.) was estimated at $40 million. Here are three situations and one conclusion reported by Gruntfest (1977) based on interviews with the survivors.

> At the lower end of the canyon, where it was not raining, it seemed hard to believe that a flood was possible. (p. 5)
>
> The people in the restaurant did not receive an official warning, but the water was rising in front.... Everyone moved to the proprietor's house adjacent to the restaurant ... All who were in the house were swept to their deaths. (p. 6)
>
> One state patrolman ... advised several people ... One couple he warned ... thought he was exaggerating the seriousness of the situation when he told them to get out of their car and climb. They died in the flood. (p. 7)
>
> The results indicate that climbing the canyon wall was the best action .... (p. iii)

8.2.2 Warning System Structure

The warning system is conceptualized as a cascade coupling of three components (Figure 8.3): the *monitor*, the *forecaster*, and the *decider*. The operation of this system is idealized as follows.

A flood is the portion of a hydrograph (a graph of the river stage versus time) above the flood stage, h_0, officially specified for a given river gauge (Figure 8.4). Flash floods occur intermittently. For this reason, the observing

Figure 8.3 Structure of the flash-flood warning system.

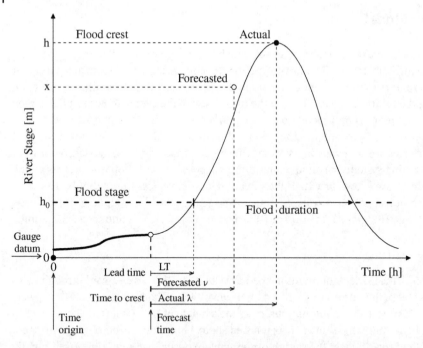

Figure 8.4 Elements of the flash-flood hydrograph: observed (heavy line up to the forecast time) and future (thin line after the forecast time); forecasted crest (v, x) and actual crest (λ, h).

network (rain gauges, river gauges), the forecaster, and the decider (along with the emergency management personnel) do not operate continuously. Rather, their operation is triggered when potential flood conditions are detected. To enable such detections, a system monitoring the weather operates on a timed basis. When a set of predefined conditions is observed, the monitor triggers operation of the forecaster. The observing network is activated, and a forecast of the flood hydrograph is prepared. This forecast is supplied to the decider (a floodplain manager or an emergency manager) who must then decide whether or not to issue a warning to the public.

Implementations are varied. For example, in one system, rain and river gauges are monitored on a timed basis, every 6 hours; when potential flood conditions are detected, gauges are interrogated remotely every hour; in another system, gauges are interrogated every 10 minutes. Still, a flash-flood forecast based solely on the observed rainfall could have the lead time as short as 30 minutes. To extend the lead time, it is necessary to forecast rainfall — one of the most difficult meteorological predictands: the errors of a deterministic forecast may be huge; especially the flood crest h may be misestimated. Hence, the resultant uncertainty should be quantified by the forecaster and accounted for by the decider.

8.2.3 Model of the Monitor

Forecast trigger
A forecast trigger is an event that (i) is likely to precede every flood, (ii) is routinely monitored or forecasted, and (iii) triggers preparation of the flood crest forecast. Here are two examples.

- Rainfall intensity during 2 hours, recorded by a monitoring gauge at a mountaintop within the watershed, exceeds a specified threshold (e.g., 3 cm/h).
- Forecasted average rainfall over the watershed in the next 6 hours exceeds 5 cm.

Notwithstanding such technical definitions, the actual trigger may simply be an alert person who observes heavy rainfall and reports it. This happened to the operator of water projects affected by the Big Thompson flood.

"This seemingly simple 'Alert' or sensing of the situation was the most critical and necessary step in the entire emergency operational response" (Berling, 1978, p. 39).

Hardware and software

Hardware and software used in monitoring and forecasting must have their reliability quantified because a failure is consequential. For instance, a flash flood developing rapidly during nighttime may occur undetected because of a failure of the monitoring gauge; preparation of the flood crest forecast may be impossible because of a failure of the computer code. In the model, the reliability of all hardware and software is represented in aggregate as a contributor to the reliability of the monitor. (This simplification has no bearing on the evaluation of system performance because it is immaterial which component has caused failure.)

Diagnosticity and reliability

In order to characterize the performance of the monitor statistically, two binary variates are defined.

T — trigger indicator, whose realization is $t \in \{0, 1\}$:

$$T = 0 \iff \text{trigger is not observed},$$
$$T = 1 \iff \text{trigger is observed}.$$

W — flash-flood indicator, whose realization is $w \in \{0, 1\}$:

$$W = 0 \iff \text{flood does not occur}, h \leq h_0;$$
$$W = 1 \iff \text{flood occurs}, h > h_0.$$

The random vector (T, W) defines three observable events, as depicted in Table 8.4: a flood not preceded by a trigger $(0, 1)$, a trigger not followed by a flood $(1, 0)$, and a trigger followed by a flood $(1, 1)$. These events are observable in the sense that their occurrences over a period of time can be counted. Let N_{tw} denote the count of the event $(T = t, W = w)$. From these counts, two conditional probabilities can be estimated:

$$\gamma = P(W = 1 | T = 1) = \frac{N_{11}}{N_{10} + N_{11}}, \tag{8.14}$$

$$\rho = P(T = 1 | W = 1) = \frac{N_{11}}{N_{01} + N_{11}}. \tag{8.15}$$

The probability γ $(0 < \gamma \leq 1)$ characterizes the *diagnosticity* of the monitor. For example, $\gamma = 1$ means that every trigger is followed by a flood; in other words, the monitor provides a perfect diagnosis of every flood situation.

The probability ρ $(0 < \rho \leq 1)$ characterizes the *reliability* of the monitor. For example, $\rho = 1$ means that every flood is preceded by a trigger; in other words, the monitor never fails to signal the oncoming flood.

The probabilities γ and ρ are independent of each other. A monitor having perfect diagnosticity may be unreliable simply because it is frequently down and does not observe the trigger when needed. A perfectly

Table 8.4 Observable events $\{(t, w)\}$ and their counts N_{tw} that characterize the performance of a monitor.

		w	
		0	1
t	0		N_{01}
	1	N_{10}	N_{11}

reliable monitor will have a low diagnosticity when the trigger is observed not only before every flood but also on numerous other occasions. In general, γ and ρ depend on the definition of the trigger event, whereas ρ depends also on the reliability of hardware and software.

8.2.4 Model of the Forecaster

Stage-time coordinates

The flash flood is a continuous and finite-duration process that occurs intermittently. Therefore, to analyze consistently every flood, the stage-time coordinates must be uniquely defined. Figure 8.4 illustrates the definitions and assumptions that follow. The *flood stage* is set at an elevation h_0 above the gauge datum. A flood begins at the instant the hydrograph up-crosses h_0 and ends at the instant the hydrograph down-crosses h_0. The *time origin* is reset to zero for every flood at the instant the trigger is observed. The *forecast time* is the instant up to which the meteorologic and hydrologic observations for preparing the forecast have been collected. For the purpose of our model, a flash flood is described in terms of the *actual flood crest*, h, measured from the gauge datum, and the *actual time to crest*, λ, measured from the forecast time.

Flash-flood forecast

A hydrologic forecasting model is typically deterministic and operates on a discrete time scale. Its main input is a time series of rainfall observed within the watershed up to the forecast time and estimated for the future several hours. Its output is a time series of river stage estimates for several hours beyond the forecast time. For the purpose of our model, a simplified forecast suffices. It is a deterministic forecast that supplies a point estimate x of the crest height h, and a point estimate v of the time to crest λ, as shown in Figure 8.4.

Bayesian forecaster

The Bayesian forecaster of Chapter 5 is adapted to produce a probabilistic forecast of the flash-flood occurrence. The premise is that the higher the estimated crest x, the larger the probability of the hydrograph crossing the flood stage h_0, as one may visualize with the aid of Figure 8.4.

Let W be the *predictand* — the flash-flood indicator defined in Section 8.2.3.

Let X be the *predictor* — the estimator of the flood crest, whose realization x is supplied by a hydrologic model; X is a continuous variate having the sample space $\mathcal{X}$.

Let T be the trigger indicator defined in Section 8.2.3. Because a trigger must be observed before the forecaster can begin to operate, the event $T = 1$ conditions every element of the Bayesian forecaster.

The inputs into the Bayesian forecaster are conditional on the trigger as follows.

$g = P(W = 1 | T = 1)$ is the *prior probability* of flood occurrence, $W = 1$, given trigger $T = 1$; it is equal to the diagnosticity of the monitor, $g = \gamma$, defined by equation (8.14).

$f_w(x) = p(x | W = w, T = 1)$ for all $x \in \mathcal{X}$ and $w = 0, 1$. For a fixed $w \in \{0, 1\}$, the object f_w is the *density function* of the flood crest estimator X, **conditional on the hypothesis** $W = w$, and given $T = 1$.

The outputs from the Bayesian forecaster are likewise conditional on the trigger as follows.

$\kappa(x) = p(x | T = 1)$ for all $x \in \mathcal{X}$. The function κ is the *expected density function* of X, given $T = 1$.

$\pi(x) = P(W = 1 | X = x, T = 1)$ is the *posterior probability* of the flood occurrence, $W = 1$, **conditional on** the flood crest estimate $X = x$, where $x \in \mathcal{X}$, and given $T = 1$.

With the above adaptations, equation (5.1) for the expected density function may be rewritten as

$$\kappa(x) = f_0(x)(1 - \gamma) + f_1(x)\gamma; \tag{8.16}$$

and equation (5.3) for the *Bayesian forecaster* may be rewritten as

$$\pi(x) = \left[1 + \frac{1 - \gamma}{\gamma} \frac{f_0(x)}{f_1(x)} \right]^{-1}. \tag{8.17}$$

Three special cases are enlightening. (i) If the monitor has perfect diagnosticity, so that $\gamma = 1$, then $\pi(x) = 1$ regardless of the estimate x. (ii) If the estimator X is uninformative, so that $f_0 = f_1$, then $\pi(x) = \gamma$ for every $x \in \mathcal{X}$; in other words, the trigger observed by the monitor becomes the sole basis for deciding to issue or not to issue a warning. (iii) If the estimator X is perfect, so that the sample spaces $\mathcal{X}_0$ and $\mathcal{X}_1$ under f_0 and f_1 do not intersect, then either $\pi(x) = 0$ or $\pi(x) = 1$. (Recall Section 5.5.)

The *time to crest* is not incorporated into our model. Nonetheless, the reader may try to sketch a concept for a model that would quantify the joint uncertainty about (λ, h), given (v, x). Inasmuch as all four variables are continuous, the forecasting theory presented in Chapter 10 would be applicable (after some extension).

8.2.5 Model of the Decider

The floodplain manager, or the emergency manager, faces a decision problem under uncertainty like that modeled in Chapter 7. Having received the probability $\pi(x)$ of the flood occurrence, with the estimated time to crest v, he must make a binary decision:

$$a = 0 \iff \text{do not issue a warning,}$$
$$a = 1 \iff \text{issue a warning.}$$

Thereafter the event takes place: the flood does not occur, $W = 0$, or the flood occurs, $W = 1$.

Outcomes and disutilities

Each decision–event pair (a, w) leads to an outcome whose degree of undesirability is measured in terms of the disutility d_{aw}. The disutilities $(d_{00}, d_{01}, d_{10}, d_{11})$ are assessed subjectively by following the five-step procedure of Section 7.2.7. The first step is to identify and describe the consequences which comprise each outcome. They are multidimensional: economic, social, behavioral; examples are listed in Table 8.5.

Economic consequences (costs, losses, damage) are measurable in monetary units, and there are models for estimating them. Social and behavioral consequences have been studied extensively, but the results are mostly

Table 8.5 Consequences comprising the outcome of each decision–event pair (a, w) for floodplain dwellers.

(a, w):	System state • consequence
$(0, 0)$:	Quiet state (Q) • none
$(1, 1)$:	Detection (D) • cost of protective actions (e.g., floodproofing, evacuation) • unavoidable damage to properties • satisfaction from having taken action that was needed • confirmation of the credibility of the warning system
$(1, 0)$:	False warning (F) • cost of protective actions • loss from the interruption of productive activities • regret from having taken action that was not needed • lowered credibility of the warning system
$(0, 1)$:	Missed flood (M) • damage to properties • deaths and injuries • spectrum of feelings as some were more lucky than others • lowered credibility of the warning system

qualitative. The primary result to remember is that the consequences (even the economic ones) depend upon the response of the floodplain dwellers to a warning (correct or false) and their capacities for coping with a flood that arrives without a warning (a miss). Some consequences are long-term; for example, a lowered credibility of the warning (after a false one) may reduce the response to a future warning — the "cry wolf" effect, which may increase deaths, injuries, and property damage from a future flood. In short, human response is a major determinant of consequences; it varies across communities, it depends on people's experience with floods; it should be considered in the assessment of disutilities.

Optimal warning rule

The optimal decision procedure of Section 7.4.3 is now adapted. First, given the disutility vector $(d_{00}, d_{01}, d_{10}, d_{11})$, equations (7.5)–(7.6) specify the threshold probability for action p^*. Second, based on the theorem of Section 6.6.1, the Bayesian forecaster (8.17) is assumed to be well calibrated against the prior probability $g = \gamma$. Third, interpreting $\alpha^*(x) \in \{0, 1\}$ as the optimal decision, given the flood crest estimate $x \in \mathcal{X}$, the optimal decision rule (7.14) can be restated as the *optimal warning rule*:

$$\alpha^*(x) = \begin{cases} 0 & \text{if } \pi(x) \le p^*, \\ 1 & \text{if } \pi(x) > p^*. \end{cases} \tag{8.18}$$

This rule can be tailored in two ways. First, purpose-specific warnings may be desirable for various types of recipients, such as the general public, managers of commercial establishments, and operators of water projects (e.g., dams, canals). For each type of recipients, there may be a distinct disutility vector, and consequently a distinct threshold probability for action, p^*.

Second, if a sequence of forecasts can be prepared before the occurrence of a flash flood, each forecast revising the previous one, then a two-stage alarm may be desirable. For example, a policy of the US National Weather Service calls for issuing first a watch (having longer lead time and higher uncertainty) and then a warning (having shorter lead time and lower uncertainty). Such a policy would require two disutility vectors, each evaluating outcomes resulting from a different decision: (i) to issue or not to issue a watch, and (ii) to issue or not to issue a warning. Then two thresholds p^* would be determined; and rule (8.18) would be expanded to decide the stage of the alarm.

8.2.6 Performance of Forecaster

Performance states

The vector (T, a, W) of binary indicators of the states of the trigger T, warning a, and flood W can take on five values which define four performance states of the flash-flood warning system (Table 8.6). The decision tree in Figure 8.5 depicts the paths through which each state can be reached.

Table 8.6 Performance states of a flash-flood warning system: detection (*D*), false warning (*F*), missed flood (*M*), quiet (*Q*).

		w	
t	*a*	0	1
0	0		*M*
1	0	*Q*	*M*
1	1	*F*	*D*

Conditional ROC

After the branch $T = 1$, the decision tree matches the tree in Figure 7.1 for the detection problem. Therefore, the informativeness of the forecaster, which operates only when $T = 1$, can be characterized by its ROC. For the Bayesian forecaster outputting a strictly increasing function π, which is the case here, the probabilities of detection and of false warning are specified by equations (5.13) and (5.14). They are adapted by introducing the conditioning on $T = 1$. For any $x \in \mathcal{X}$,

$$
\begin{aligned}
P(D|x) &= P(a = 1 | W = 1, T = 1) \\
&= P(X > x | W = 1, T = 1) \\
&= 1 - F_1(x);
\end{aligned}
\tag{8.19}
$$

$$
\begin{aligned}
P(F|x) &= P(a = 1 | W = 0, T = 1) \\
&= P(X > x | W = 0, T = 1) \\
&= 1 - F_0(x).
\end{aligned}
\tag{8.20}
$$

Above, F_w is the conditional distribution function corresponding to the conditional density function f_w defined in Section 8.2.4, for $w = 0, 1$. The ROC of the forecaster using predictor X can now be constructed numerically via the algorithm of Section 5.5.4. This ROC is **conditional on** $T = 1$; aside from this interpretation, it has the same four properties as those stated in Section 5.5.4.

8.2.7 Performance of Monitor–Forecaster

A monitor is characterized by the diagnosticity γ and the reliability ρ. A forecaster coupled to a monitor is characterized by the conditional ROC. The objective now is to characterize statistically the joint performance of the monitor and the forecaster, assuming that the decider operates optimally. For this problem, the vantage is the first chance node from which branches $T = 0$ and $T = 1$ emanate (Figure 8.5).

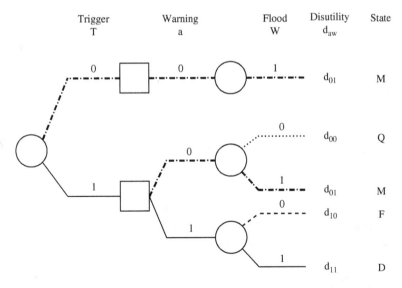

Figure 8.5 Decision tree of the flash-flood warning problem, and four performance states of the warning system: detection (*D*), false warning (*F*), missed flood (*M*), quiet (*Q*).

Expected numbers

Inasmuch as the marginal probabilities of branches $T = 0$ and $T = 1$ are undefinable, we resort to counting the observable events and states over a time interval of N years. It may be $N = 1$ year, or several decades, say $N \in \{30, 50, 100\}$ years, because flash floods are rare. Let

N_W — number of flash floods,
N_T — number of triggers,
N_D — number of detections,
N_F — number of false warnings.

The expected number of floods $E(N_W)$ in N years may be estimated from a historical record. Then all floods occurring in N years may be stratified into those not preceded by a trigger, $T = 0$, and those preceded by a trigger, $T = 1$. Their expected numbers are

$$E(N_W|T = 0) = (1 - \rho)E(N_W),$$
$$E(N_W|T = 1) = \rho \, E(N_W). \tag{8.21}$$

Also,

$$E(N_W|T = 1) = \gamma \, E(N_T). \tag{8.22}$$

It follows that

$$E(N_T) = \frac{\rho}{\gamma}E(N_W). \tag{8.23}$$

In other words, the expected number of triggers equals the product of the expected number of floods and the ratio of the reliability to the diagnosticity of the monitor.

Performance trade-off characteristic

For the performance states, the mappings between the probabilities and the expectations, conditional on $X = x$, are as follows. The expected number of detections is

$$\begin{aligned} E(N_D|x) &= E(N_W)P(a = 1|W = 1, \, T = 1)P(T = 1|W = 1) \\ &= E(N_W)P(D|x)\rho \\ &= \rho P(D|x)E(N_W). \end{aligned} \tag{8.24}$$

The expected number of false warnings is

$$\begin{aligned} E(N_F|x) &= E(N_T)P(a = 1|W = 0, \, T = 1)P(W = 0|T = 1) \\ &= \frac{\rho}{\gamma}E(N_W)P(F|x)(1 - \gamma) \\ &= \frac{\rho}{\gamma}(1 - \gamma)P(F|x)E(N_W). \end{aligned} \tag{8.25}$$

Definition   For any positive number $E(N_W)$, a graph of the expected number of detections $E(N_D|x)$ versus the expected number of false warnings $E(N_F|x)$ for all predictor values $x \in \mathcal{X}$ is called the *performance trade-off characteristic* (PTC) of the monitor–forecaster component of the warning system.

In essence, equations (8.24)–(8.25) modify the conditional ROC of the forecaster to account for the performance of the monitor — its diagnosticity γ and reliability ρ. Figures 8.6 and 8.7 show an example. The resultant PTC has five main properties that should guide its interpretation.

Figure 8.6 Conditional receiver operating characteristic (ROC) of the forecaster for a warning system; R^* is the optimal operating point $(P(F|x^*), P(D|x^*))$.

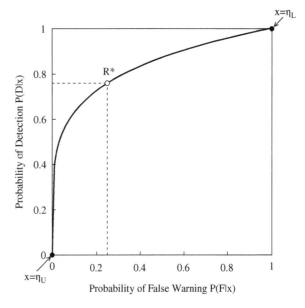

Figure 8.7 Performance trade-off characteristic (PTC) of the monitor–forecaster: the mapping of the conditional ROC from Figure 8.6 for a warning system in which $\gamma = 0.6$, $\rho = 0.9$, and $E(N_W) = 50$ in $N = 100$ years; R^* is the optimal operating point $(E(N_F|x^*), E(N_D|x^*))$.

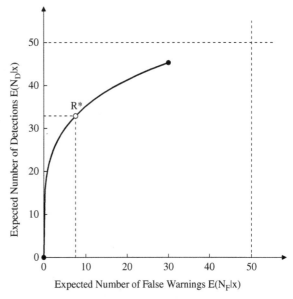

1. The PTC displays all *trade-offs* between the expected number of detections $E(N_D|x)$ and the expected number of false warnings $E(N_F|x)$ that a given monitor–forecaster offers to all deciders.
2. The PTC of a *perfect monitor–forecaster* consists of one operating point, $(0, E(N_W))$.
3. The PTC of a monitor coupled with an *uninformative forecaster* consists of two operating points, $(0, 0)$ and $((\rho/\gamma)(1 - \gamma)E(N_W), \rho E(N_W))$.
4. The PTC of a monitor coupled with an informative forecaster is a strictly concave function between the two points specified in Property 3.
5. The shape of the PTC is invariant to $E(N_W)$, which is merely a scaling constant for both axes. It may be set conveniently to any number; for example, $E(N_W) \in \{1, 10, 100\}$ floods, with the implication that N changes accordingly.

Design and analysis

The conditional ROC and the PTC can serve as graphical aids to engineering design and analysis of a flash-flood warning system. For example, an engineer may be considering several alternative designs for the monitor and for the forecaster, each having different $(\gamma, \rho; f_0, f_1)$. The alternatives can be compared succinctly in terms of their conditional ROCs and their PTCs.

Once a design for monitor–forecaster is chosen, the two characteristics provide aids to analyzing the implications of the optimal warning rule (8.18) for a particular community. With the disutilities assessed, and the threshold probability for action p^* determined, the unique threshold $x^* = \pi^{-1}(p^*)$ may be found in $\mathcal{X}$. This threshold fixes the *optimal operating point* on the conditional ROC, $(P(F|x^*), P(D|x^*))$, and on the PTC, $(E(N_F|x^*), E(N_D|x^*))$, which constitutes the optimal trade-off for that community. This trade-off is optimal in the sense of being determined by the strength of preference for outcomes encoded in the disutilities for that community.

8.2.8 System Evaluation

The objective is to evaluate the long-term performance of a warning system. Per the decision tree in Figure 8.5, this evaluation is decomposed into three steps: two conditional evaluations, given $T = 1$ and given $T = 0$, and their synthesis into the total evaluation.

Conditional evaluations

With the trigger being observed, $T = 1$, the succedent decision tree matches the tree in Figure 7.1 for the detection problem. Therefore, the evaluation procedure of Section 7.4.4 is adapted by replacing the sample space $\mathcal{Y}$ with $\mathcal{X}$, the predictor Y with X, realization $y \in \mathcal{Y}$ with $x \in \mathcal{X}$, function η with π from equation (8.17), and function κ with that from equation (8.16). Accordingly, given $X = x$, equation (7.16) yields the disutility of decision, $D_a(x)$, for $a = 0, 1$; and equation (7.17) yields the *minimum disutility of decision*,

$$D^*(x) = \min \{D_0(x), D_1(x)\}. \tag{8.26}$$

For a continuous predictor X with the sample space $\mathcal{X} = (\eta_L, \eta_U)$, and with the unique threshold $x^* = \pi^{-1}(p^*) \in \mathcal{X}$, the *integrated minimum disutility*, being defined as the expectation of $D^*(X)$ with respect to the predictor X, is given by the integral

$$DF^* = \int_{\mathcal{X}} D^*(x)\kappa(x)\,dx$$

$$= \int_{\eta_L}^{x^*} D_0(x)\kappa(x)\,dx + \int_{x^*}^{\eta_U} D_1(x)\kappa(x)\,dx. \tag{8.27}$$

The analytic solution takes the form

$$DF^* = d_{00} + \gamma(d_{01} - d_{00}) + (1 - \gamma)(d_{10} - d_{00})[1 - F_0(x^*)] - \gamma(d_{01} - d_{11})[1 - F_1(x^*)]. \tag{8.28}$$

Its derivation is deferred to Exercise 16. It is worth noting that equation (8.27) parallels equation (7.25) for a discrete predictand. The solutions are parallel too: one may compare expression (8.28) with the expression given in Exercise 4.3 of Chapter 7.

Without the trigger being observed, $T = 0$, neither the forecaster nor the decider are activated. Consequently, no timely warning is issued, $a = 0$, and a flood that occurs, $W = 1$, causes disutility d_{01}.

Total evaluation

The *total disutility* of the warning system over the long term, say N years, should account for the expected number of floods not preceded by the trigger, $E(N_W|T = 0)$ given by expression (8.21), and the expected number of triggers, $E(N_T)$ given by expression (8.23). The measure of performance of the warning system over N years of operation is the *expected integrated minimum disutility*, TDF^* (the *total disutility* for short):

$$
\begin{aligned}
TDF^* &= d_{01} E(N_W|T = 0) + DF^* E(N_T) \\
&= d_{01}(1 - \rho)E(N_W) + \frac{\rho}{\gamma}DF^* E(N_W) \\
&= \left[(1 - \rho)d_{01} + \frac{\rho}{\gamma}DF^*\right] E(N_W).
\end{aligned}
\tag{8.29}
$$

It turns out that $E(N_W)$ is merely a scaling constant; it may be set arbitrarily, for convenience of the interpretation; for example, $E(N_W) \in \{1, 10, 100\}$ floods. $E(N_W) = 10$ imparts the interpretation: TDF^* is the total disutility after 10 floods.

The weights that combine the evaluations from the two initial branches of the decision tree are specified entirely by the performance probabilities of the monitor, γ and ρ. Naturally, $\gamma > 0$ and $\rho > 0$ are required of any monitor. Exercise 17 asks you to determine and interpret the limiting expressions for TDF^*, when the monitor becomes partially perfect ($\gamma = 1$ or $\rho = 1$) or perfect ($\gamma = 1$ and $\rho = 1$).

System efficiency

Inasmuch as a warning system cannot eliminate the consequences of a flash flood (deaths, injuries, property damage) but can only reduce them, a relative measure of performance is convenient and meaningful. Such a measure, defined in Section 7.4.6 as the efficiency, is adapted herein.

Floods occurring without a warning system over N years result in the expected disutility of outcome

$$
D^* = d_{01}E(N_W).
\tag{8.30}
$$

A perfect warning system would always warn floodplain dwellers about the impending flood with the lead time sufficient to accomplish the optimal response. Over N years of system operation, the expected disutility of outcome would be

$$
D^{**} = d_{11}E(N_W).
\tag{8.31}
$$

Hence, per equation (7.23), the *efficiency of a warning system* is defined as the quotient

$$
EF = \frac{D^* - TDF^*}{D^* - D^{**}}.
\tag{8.32}
$$

After expressions (8.29), (8.30) and (8.31) are inserted into this definition,

$$
EF = \rho \frac{d_{01} - DF^*/\gamma}{d_{01} - d_{11}},
\tag{8.33}
$$

where DF^* comes from equation (8.28). The EF is bounded, $0 \leq EF \leq 1$. It represents the fraction of the disutility of a flood that a given warning system prevents relative to the maximum preventable disutility.

As such, EF is a comparative measure. It may be used to rank the alternative designs of a warning system for a community. It may be used to rank the warning systems in different communities to establish the priority for improving them.

Bibliographical Note

The recountal of the Big Thompson flash flood is based on the report by Gruntfest (1977) and the article by Berling (1978). The flash-flood warning model is a simplified version of the theory formulated by Krzysztofowicz (1993); detailed models of some components were developed by Kelly and Krzysztofowicz (1994) and by Krzysztofowicz et al. (1994). A systematic analysis of the psychological effects of false warnings was published by Breznitz (1984).

Exercises

1 **Assumptions of the forecaster**. The model for forecasting the location of the lost subject is based on three assumptions (Section 8.1.4). Select two of them that, in your opinion, are likely to be violated in search operations. Offer arguments for (i) possible causes of each violation, (ii) the possible extent of each violation, and (iii) the possible impact of each violation on the success of a search. *Hint*: Does a violation affect the calibration of the PoA? Does the forecaster become overconfident or underconfident?

2 **Stage 2 revision**. Write the total probability law and Bayes' theorem for Stage 2 revision of PoAs, in parallel to equations (8.2)–(8.3), but with subscripts indicating the revision stage, as in equations (8.6)–(8.7).

3 **Stage 3 revision**. Equations (8.8)–(8.9) are written in a recursive form. Rewrite them for $\kappa_1, \kappa_2, \kappa_3$ and for $\pi_{j1}, \pi_{j2}, \pi_{j3}$ in a form that includes only the prior PoAs, $\{p_j\}$, and the PoDs, $\{d_{jn} : n = 1, 2, 3\}$. *Note*: The symbol κ_1 may appear in equations for Stage 2 revision; the symbols κ_1, κ_2 may appear in equations for State 3 revision.

4 **Sequential search**. (i) Redo Example 8.2 from Section 8.1.6 for the scenario in which the areas are searched in the order specified in the option. (ii) Compare your results with those reported in Table 8.2. (iii) Draw a conclusion. (iv) Generalize it to any number of areas $J \geq 2$. *Hint*: Inasmuch as each of the J areas is to be searched once, there are $J!$ permutations of areas. Should the incident commander be concerned about this? *Options for the order of areas*:

(123) A_1, A_2, A_3;   (213) A_2, A_1, A_3;   (312) A_3, A_1, A_2;
(231) A_2, A_3, A_1;   (321) A_3, A_2, A_1.

5 **Multiple searches**. (i) Write the probabilistic reasoning that justifies expression (8.10) for the compounded PoD. (ii) Prove that the compounded PoD increases strictly with the number N of unsuccessful searches — the property known as the *compounding effect* of repeated trials.

6 **Convergence of PoA for "the rest of the world"**. Prove that as the number of unsuccessful search shifts n increases without bound, the posterior PoA for the "rest of the world", π_{0n} given by expression (8.9), converges to 1.

7 **Optimal decision procedure**. Write the optimal decision procedure for Stage n of the search, in parallel to the procedure (8.12)–(8.13) for Stage 0.

8 **Nearly ideal searches**. Reread Section 8.1.8. (i) Redo Example 8.2 from Section 8.1.6 for the sequence of nearly ideal searches with PoDs: $d_{1,1} = 0.98$, $d_{3,2} = 0.98$, $d_{2,3} = 0.98$. (ii) Graph the posterior probability functions $\{\pi_{jn} : j = 0, 1, 2, 3\}$ for $n = 1, 2, 3$. (iii) Write your observations about the sequence $\{1 - \kappa_n : n = 1, 2, 3\}$ of the expected PoDs, and about the sequences $\{\pi_{jn} : n = 1, 2, 3\}$ of the posterior PoAs for $j = 0, 1, 2, 3$. (iv) Do these sequences exhibit, approximately, the implications of the ideal searches?

9 **Search paradox**. Two unsuccessful searches of area A_3 (Figure 8.1) were conducted during one shift. The first team searched every square yard of a contiguous subarea, which constitutes 30% of A_3. Therefore, it concluded that its PoD was $d_{3,1} = 0.3$. Aware of this search method, the second team searched every square yard of the adjacent and contiguous 30% of A_3. Likewise, it concluded that its PoD was $d_{3,2} = 0.3$. At the end

of the shift, one search manager inferred that, because the two teams searched exhaustively 60% of area A_3, the PoD is $d_3 = 0.6$. Another search manager was aware of the search method, but applied formula (8.10) and found a different value of d_3 (calculate it).

9.1 Which PoD value is incorrect? Are they both incorrect? Provide a reasoned answer.

9.2 Do the incoherent values of d_3 imply that something is wrong with the mathematical model for revising the PoAs. If yes, then point out the error. If no, then trace out the source of this incoherence.

9.3 Propose a solution that would preclude this paradox from occurring in future searches.
Hint. Suppose only one team searched A_3 during the shift, and reported PoD of 0.3 sans revealing to the manager its method of searching and assessing PoD. Thus, no incoherence, no problem. Is this correct?

10 **Long search: one more shift**. The search in Warren County, Virginia, recounted in Section 8.1.1, terminated successfully on the ninth day. In the morning, areas A_5, A_8, A_9, A_{10} were selected for simultaneous searches based on the posterior PoAs after the preceding 17 shifts:

j	0	1	2	3	4	5	6	7	8	9	10
$\pi_{j,17}$	0.219	0.023	0.010	0.028	0.031	0.131	0.069	0.030	0.109	0.153	0.197

The 18th shift found the subject in A_{10}. To build intuition about the working of the Bayesian revision model, let us modify this scenario.

10.1 Write your observations about the posterior PoAs $\{\pi_{j,17}\}$. What is peculiar about them? What inferences do they suggest?

10.2 Based on these PoAs and on all information $\{I_0, I_1, \dots, I_{17}\}$, the search manager decided to concentrate the limited resources on searches of A_5, A_8, A_9 in order to get high PoDs. They were assessed after the unsuccessful searches: $d_{j,18} = 0.7$ for $j = 5, 8, 9$. Calculate the posterior PoAs.

10.3 Write your observations about the posterior PoAs $\{\pi_{j,18}\}$. What is peculiar about them? What inferences do they suggest?

10.4 Imagine yourself as the search manager. (i) What would be your search plan for the 19th shift? Justify it. (ii) What would be your contingency plan for beyond the 19th shift? *Hint*: State your decision conditional on each hypothesized outcome and the PoD value from the next shift search.

11 **Long search: termination**. Read the preface to Exercise 10. This exercise has the same purpose but a different scenario.

11.1 Write your observations about the posterior PoAs $\{\pi_{j,17}\}$. What is peculiar about them? What inferences do they suggest?

11.2 Based on these PoAs, the search manager selected $A_5, A_6, A_8, A_9, A_{10}$ for searching, and the resources were great enough to conduct almost exhaustive searches. They were unsuccessful, but the PoDs were high: $d_{6,18} = 0.7$; $d_{j,18} = 0.8$ for $j = 5, 8$; $d_{j,18} = 0.9$ for $j = 9, 10$. Calculate the posterior PoAs.

11.3 Write your observations about the posterior PoAs $\{\pi_{j,18}\}$. What is peculiar about them? What inferences do they suggest?

11.4 Imagine yourself as the search manager. What would be your decision based on these posterior PoAs. Justify it.

12 **Lost siblings**. At sleep time (2130) after the 4 July celebrations at a summer camp in the Appalachian mountains, two sibling campers, a 13-year-old boy and an 11-year-old girl, were not found in their cabins. The camp director notified the sheriff, who in turn requested help from a search and rescue organization. By early morning, the command post was set in a cabin. Two search areas, A_1 and A_2, were delineated. The prior PoAs were

assessed, $p_1 = 0.4$ and $p_2 = 0.5$, and the searches began. Imagine receiving a call from a friend, who happened to serve as the incident commander, asking for your help with the "analytic" part of the search.

12.1 At the end of the first shift: a team with air-scent search dogs returned from A_2 and assessed the PoD at 0.8; a hasty-search team returned from A_1 and assessed the PoD at 0.3. Revise the PoAs, and recommend the order in which the areas should be searched again.

12.2 During the second shift, two searches of A_1 were conducted: the first by a large team with PoD equal to 0.55, and the second by a small team with PoD equal to 0.35. Revise the PoAs and recommend the order in which the areas should be searched again.

12.3 On the next day, the third shift employed the grid search method twice over A_2; the assessed PoDs were 0.6 and 0.9. Simultaneously, a hasty-search team spent half of the shift searching A_1, and assessed the PoD at 0.25. Revise the PoAs, and recommend a search plan for the next shift.

12.4 On the fourth shift, all available resources were deployed in A_1. At 1830, nearly 48 hours after the siblings had left the camp, they were found alive at the bottom of a ravine, exhausted and hungry, but well hydrated by the brook nearby. (i) At the beginning of this shift, what was the probability of the siblings not being in A_1? (ii) In retrospect: Which shift had the largest expected probability of success? Which shift had the smallest expected probability of success?

12.5 After the search terminated, you were asked to explain the forecasting procedure and the decision rule to the novice searchers. (i) Graph the trajectory of the PoAs, from the prior to the terminal posterior (after the fourth shift) for each area: A_0, A_1, A_2. (ii) Explain any trends and fluctuations of each trajectory by identifying the causes: the decisions implemented by the search managers, and the PoDs assessed by the search teams. (iii) Explain the rationale behind the recommendations of search areas.

13 **Calibration assumption**. The optimal warning rule (8.18) assumes that the Bayesian forecaster (8.17) is well calibrated against the prior probability $g = \gamma$. (i) State this assumption mathematically. (ii) Write the optimal warning rule without this assumption. (iii) Explain the advantage of this assumption.

14 **Expected numbers of events**. Write a mathematical justification of each conditional expectation given by equations (8.21)–(8.22).

15 **Expected number of false warnings**. (i) Prove that the expected number of false warnings may be larger than the expected number of flash floods (in N years). (ii) Is this strict inequality possible for every $x \in \mathcal{X}$? (iii) Create two numerical examples, one illustrating the inequality $>$, and one illustrating the inequality $<$.

16 **Integrated minimum disutility**. (i) Beginning with equation (8.27), which defines DF^*, derive equation, (8.28) for calculating DF^*. (ii) Prove that equation (8.28) can be expressed in terms of the decision error probabilities $P(F|x^*)$ and $P(M|x^*)$.

17 **Limiting cases of total disutility**. Given expression (8.29) for the total disutility of the warning system, TDF^*, find the expression for TDF^* in each of the three limiting cases of the monitor's performance: (i) perfect diagnosticity, $\gamma = 1$; (ii) perfect reliability, $\rho = 1$; (iii) perfect monitor, $\gamma = 1$ and $\rho = 1$. *Hint*: When $\gamma = 1$, determine first its effect on $\pi(x)$ and on DF^*.

18 **Relationship between** PTC **and** TDF^*. Prove that the total disutility of the warning system, TDF^* given by equation (8.29), can be expressed in terms of the optimal trade-off point on the PTC, $(E(N_F|x^*), E(N_D|x^*))$, as follows:

$$TDF^* = \left[d_{01} + \frac{\rho}{\gamma}(1 - \gamma)d_{00} \right] E(N_W) + (d_{10} - d_{00})E(N_F|x^*) - (d_{01} - d_{11})E(N_D|x^*).$$

19 Warning system efficiency. (i) Beginning with equation (8.32), derive equation (8.33). (ii) Define mathematically a perfect warning system, and show that its efficiency is $EF = 1$.

20 Analysis of a warning system. After experiencing 9 flash floods in 45 years, the town hired a hydrological–meteorological consultancy which designed a warning system. The estimates of its parameters are specified in the options for the monitor, the forecaster, and the decider. The planned lifetime of the system (for its operation and evaluation) is 30 years. Imagine you have been hired by the emergency manager for the town to evaluate the system design.

Options for the monitor:

(M1) $\gamma = 0.7$, $\rho = 0.9$;   (M3) $\gamma = 0.4$, $\rho = 0.8$;
(M2) $\gamma = 0.6$, $\rho = 0.7$;   (M4) $\gamma = 0.3$, $\rho = 0.6$.

Options for the forecaster:

(EX) F_0 is EX(0.2, 2), F_1 is EX(0.8, 2);
(WB) F_0 is WB(1.1, 2.9, 5), F_1 is WB(1.8, 2.9, 5);
(IW) F_0 is IW(1.3, 2.8, 4), F_1 is IW(2.2, 2.8, 4);
(LW) F_0 is LW(1.5, 2.7, 3), F_1 is LW(2.6, 2.7, 3);

Note. Parameters α_0, α_1, η are in meters. Model EX is the subject of Exercise 3 in Chapter 5. Models WB, IW, LW are the subjects of Exercise 4 in Chapter 5. If one of these exercises was performed earlier and the model matches the option specified herein, then some results may be reused.

Options for the decider:

(D1) $d_{10} = 15$, $d_{11} = 80$;   (D3) $d_{10} = 25$, $D_{11} = 80$;
(D2) $d_{10} = 20$, $d_{11} = 75$;   (D4) $d_{10} = 40$, $D_{11} = 70$.

Note. In every option, $d_{00} = 0$ and $d_{01} = 100$.

20.1 Write the expressions for the conditional density functions f_w and the conditional distribution functions F_w of the flood crest estimator X ($w = 0, 1$).

20.2 Write the expression for the likelihood ratio function f_0/f_1 and simplify it. Write the expression for function π, and validate the monotonicity of π; recall Section 5.3.4. Derive the inverse function π^{-1}.

20.3 Construct and graph the conditional ROC of the forecaster and the PTC of the monitor–forecaster.

20.4 Determine (i) the ratio of the opportunity losses r, (ii) the threshold probability for action p^*, (iii) the corresponding threshold crest estimate x^*, and (iv) the optimal operating point R^* on the conditional ROC and on the PTC; show R^* on each graph. (v) Interpret the optimal trade-off that R^* represents on each graph. *Reminder*: The system has a planned lifetime.

20.5 Evaluate system performance by calculating (i) the integrated minimum disutility DF^*, (ii) the total disutility TDF^*, and (iii) the efficiency of the warning system EF. (iv) Interpret EF for the emergency manager.

Mini-Projects

21 Warning system for Milton. The town of Milton is located on the West Branch of the Susquehanna River in central Pennsylvania. The watershed above the river gauge covers 17 223 km². Suppose a flash-flood warning system for Milton is being designed. The trigger will be observed when the river stage exceeds threshold h_T ($h_T < h_0$). The flood crest estimator will have its two conditional distributions drawn from the Weibull family $WB(\alpha_w, \beta_w, \eta)$, for $w = 0, 1$. Two alternative systems under consideration have the following parameter estimates:

System	h_T	γ	ρ	α_0	β_0	α_1	β_1	η
S1	3.4 m	0.80	1.00	0.86	2.35	2.30	2.11	4.57 m
S2	4.6 m	0.90	0.89	0.34	1.85	1.88	2.16	5.49 m

The flood stage h_0, the expected number of floods $E(N_W)$ per year (estimated based on the flood record from 1889 to 1975), the expected warning lead time $E(LT)$, and the disutility vector are specified in the option. The two alternative thresholds h_T affect the parameter estimates in a way that creates trade-offs. In particular, a higher threshold h_T reduces the expected lead time $E(LT)$, but increases the diagnosticity γ of the monitor (because a higher h_T is more diagnostic of the incoming flood), and decreases the reliability ρ (because the monitor observes the river stage every 6 h, and it is possible for a rapidly rising river to exceed both h_T and h_0 within a 6-h interval, in which case the flood occurs without the trigger).

Options for the flood stage:
(O) The official flood stage.

$h_0 = 5.8$ m, $E(N_W) = 0.53$/year, $E(LT) = 9$ h for S1, $E(LT) = 5$ h for S2;
$d_{00} = 0$, $d_{01} = 100$, $d_{10} = 10$, $d_{11} = 75$, for S1;
$d_{00} = 0$, $d_{01} = 100$, $d_{10} = 5$, $d_{11} = 80$, for S2.

(P) The practical flood stage: almost all the town is located above it.

$h_0 = 6.7$ m, $E(N_W) = 0.36$/year, $E(LT) = 15$ h for S1, $E(LT) = 11$ h for S2;
$d_{00} = 0$, $d_{01} = 100$, $d_{10} = 30$, $d_{11} = 60$, for S1;
$d_{00} = 0$, $d_{01} = 100$, $d_{10} = 20$, $d_{11} = 70$, for S2.

For each alternative system, do Exercises 21.1–21.5; then complete Exercise 21.6.

21.1 Write the expressions for the conditional density functions f_w and the conditional distribution functions F_w of the flood crest estimator X ($w = 0, 1$).

21.2 Write the expression for the likelihood ratio function f_0/f_1 and simplify it. Write the expression for function π, and validate the monotonicity of π; recall Section 5.3.4. Derive the inverse function π^{-1}.

21.3 Construct and graph the conditional ROC of the forecaster and the PTC of the monitor–forecaster.

21.4 Determine (i) the ratio of the opportunity losses r, (ii) the threshold probability for action p^*, (iii) the corresponding threshold crest estimate x^*, and (iv) the optimal operating point R^* on the conditional ROC and on the PTC; show R^* on each graph.

21.5 Evaluate system performance by calculating (i) the integrated minimum disutility DF^*, (ii) the total disutility TDF^* per year, and (iii) the efficiency of the warning system EF.

21.6 Perform a comparative analysis of the two alternative system designs; frame your explanations as if you were presenting them to a client — an emergency manager for Milton. In particular, do the following. (i) Explain the trade-offs on the conditional ROC and on the PTC that each alternative offers. (ii) For each alternative, explain the optimal trade-off implied by the operating point R^* on the PTC; include the expected lead time of the flood warning as the third dimension of the trade-off. (iii) Interpret and compare the values of the performance measures for each alternative: DT^*, TDF^*, EF. (iv) Recommend the alternative to be implemented; rationalize your recommendation.

Bibliographical note. The source of this exercise is the article by Krzysztofowicz et al. (1994).

22 **Synergistic effect of dam and forecast**. Connellsville, a town in southwestern Pennsylvania, embraces the banks of the Youghiogheny River — a tributary of the Monongahela River. The watershed above the river gauge covers 3434 km². The flood stage is $h_0 = 3.7$ m. The river flows are partially controlled by the Youghiogheny reservoir, located 47.3 km upstream and completed in October 1943. The reservoir is operated by the US Army Corps of Engineers and serves multiple purposes. It has a capacity of 313 304 387 m³, which equals 42% of the average annual runoff at the dam, and closes a watershed of 1124 km², which constitutes 33% of the watershed above Connellsville. Two flash-flood warning systems for Connellsville are to be analyzed and compared. System S0 is hypothetical, without any influence of the Youghiogheny dam on flood flows. Its parameters were estimated based on data from the years 1910–1942. System S1 is the present one, with the dam. Its parameters were estimated based on the prior sample of the flood crest H from the years 1943–1986 and the joint sample of the crest estimator X and the flood crest H from the years 1984–1986. The two conditional distributions of the crest estimator belong to the Weibull family: $WB(\alpha_w, \beta_w, \eta)$, for $w = 0, 1$. The statistics of the floods are:

System	Sample years	N	N_W	Per year $E(N_W)$
S0	1910–1942	33	22	0.67
S1	1943–1986	44	22	0.5

The estimates of parameters for the two systems are:

System	γ	ρ	α_0	β_0	α_1	β_1	η
S0	0.79	0.95	0.08	0.51	1.24	4.15	3.35 m
S1	0.79	0.95	0.08	0.51	0.92	4.21	3.35 m

Options for the disutility vector:

(D1)	$d_{01} = 15$,	$d_{11} = 80$;	(D3)	$d_{01} = 35$,	$d_{11} = 70$;
(D2)	$d_{01} = 10$,	$d_{11} = 85$;	(D4)	$d_{01} = 25$,	$d_{11} = 75$.

Note. In every option, $d_{00} = 0$ and $d_{01} = 100$.

The objective of this exercise is to determine the synergistic gains from a flash-flood warning system coupled with a flood-control dam upstream. For each system, pre-dam and post-dam, carry out Exercises 22.1–22.5; then complete Exercise 22.6.

22.1 Write the expressions for the conditional density functions f_w and the conditional distribution functions F_w of the flood crest estimator X ($w = 0, 1$).

22.2 Write the expression for the likelihood ratio function f_0/f_1 and simplify it. Write the expression for function π, and validate the monotonicity of π; recall Section 5.3.4. Derive the inverse function π^{-1}.

22.3 Construct and graph the conditional ROC of the forecaster and the PTC of the monitor–forecaster.

22.4 Determine (i) the ratio of the opportunity losses r, (ii) the threshold probability for action p^*, (iii) the corresponding threshold crest estimate x^*, and (iv) the optimal operating point R^* on the conditional ROC and on the PTC; show R^* on each graph.

22.5 Evaluate system performance by calculating (i) the integrated minimum disutility DF^*, (ii) the total disutility TDF^* per year, and (iii) the efficiency of the warning system EF.

22.6 Perform a comparative analysis of system S1 with the dam against system S0 without the dam. In particular, do the following. (i) Analyze the influence of the dam on the conditional ROC of the forecaster; compare the feasible trade-offs between $P(D|x)$ and $P(F|x)$. (ii) Analyze the influence of the dam on the PTC of the monitor–forecaster; compare the feasible trade-offs between $E(N_D|x)$ and $E(N_F|x)$. (iii) Analyze the shift, induced by the dam, of the optimal operating point R^* on the conditional ROC and on the PTC. (iv) Analyze the influence of the dam on the performance measures: DT^*, TDF^*, EF. (v) Summarize the overall effect of the dam on the performance of the flash-flood warning system. In a nutshell: Are there any synergistic gains from coupling a structural solution (a dam) with an information–technology solution (a warning system)?

Bibliographical note. The source of this exercise is the article by Krzysztofowicz et al. (1994).

Part III

Continuous Models

9

Judgmental Forecasting

In many situations, the only feasible way or the most expeditious way of obtaining a forecast of a continuous predictand is to call upon the expert's judgment. An expert can gather predictive information, identify sources of uncertainty, and quantify the uncertainty. This chapter formalizes the judgmental task of quantification of uncertainty about a continuous predictand. It introduces the concept of a *judgmental distribution function*, prescribes the procedure for assessing a distribution function judgmentally, details the task of modeling parametrically the judgmental distribution function, and briefly treats two problems: forecasting by a group of experts, and adjusting judgmentally any specified distribution.

9.1 A Perspective on Forecasting

9.1.1 Prototypical Forecasting Problems

Below is a partial list of prototypical forecasting problems in which an expert may be called upon to assess judgmentally a distribution function of a continuous predictand. Each problem is sketched by identifying the predictand and some salient characteristics.

- The *time* needed to complete a creative or complex activity such as designing a new product, building a prototype, testing a prototype, starting a production process, setting up a distribution network.
- The *cost* of a unique project or of a new technology; the capital needed to start a new venture. It is necessary to contemplate all the possible technical and organizational obstacles that may affect the cost and that, therefore, cause the uncertainty.
- The *demand* for a new product. It may be necessary to consider not only the current preferences of consumers but also the future preferences as they may be shaped by an advertising campaign.
- The *prime interest rate* in a year, in two years, or in several years. This forecasting problem requires the expert to anticipate, *inter alia*, the monetary policy decisions of the Federal Reserve Board.
- The *exchange rate* between two currencies in a year, in two years, or in several years. This forecasting problem requires the expert to anticipate, *inter alia*, the international trade balance, significant economic events, significant political events.
- *Meteorological variates* such as the rainfall or snowfall over an area within a period of time. Forecasting the amount of rain or snow, as opposed to forecasting just the occurrence, remains one of the most challenging problems for atmospheric scientists.
- *Geological variates* such as the amount of ore, coal, oil, or gas in a geological formation (referred to as "proved reserves"), the concentration of a contaminant in an aquifer, or the distance from the source to the boundary of an underground plume.

9.1.2 Characteristics of Forecasting Problems

In general, the forecasting problem in which the judgment of an expert is potentially valuable has one or more of the following characteristics. (i) The predictand is new or unique, so that there are no data that could be used to construct and estimate a statistical forecasting model. (ii) The process that generates the realization of a predictand is understood only qualitatively. Thus, it is infeasible to construct a mathematical model, but it may be feasible to construct a conceptual model that an expert may employ, as in a mental simulation process, to arrive at a forecast. (iii) The model outputs only a deterministic forecast — a point estimate of the predictand — but the expert recognizes the uncertainty and is capable of quantifying it judgmentally. (iv) The mathematical model of the phenomenon being forecasted is simplistic — it does not include all predictors and processes. Consequently, an expert may be able to improve upon the output from the model. For instance, a meteorologist may understand influences of lakes and mountains on the local weather — influences which are not captured in a large-scale (global) numerical weather prediction model. Based on this knowledge, he may adjust the model output to produce, on average, a superior local weather forecast.

9.1.3 Elements of Methodology

Suppose an expert is called upon to make a forecast of some continuous predictand. A single number, "the best estimate", is often offered as the forecast. Such a forecast does not convey the degree of certainty the expert may feel about this estimate; and the degree of an expert's certainty may vary from occasion to occasion (from week to week in financial forecasting, or from field to field in geological forecasting). Therefore, the quantification of this degree of certainty may provide additional information valuable for decision making. This chapter presents a methodology through which such a quantification could be accomplished. The result is the probabilistic forecast specified in terms of a distribution function of the predictand.

The subsequent sections describe procedures that answer four questions:

- How to prepare a probabilistic forecast judgmentally.
- How to estimate a parametric distribution function based on judgmental assessments.
- How to reconcile or combine probabilistic forecasts of the same predictand, prepared by several experts, into a single distribution function.
- How to adjust judgmentally a specified distribution function.

9.2 Judgmental Distribution Function

9.2.1 Definition of Judgmental Distribution Function

Let W denote the predictand — a variate whose realization must be forecasted. It is a continuous variate having the sample space $\mathcal{W}$; its realization is denoted w, where $w \in \mathcal{W}$. Ideally, the uncertainty about W is characterized in terms of a continuous distribution function G of W, as sketched in Figure 9.1.

Definition A *judgmental distribution function G* of a continuous predictand W is the numerical measure of the *degree of certainty* about the occurrence of **all** events $\{W \leq w : w \in \mathcal{W}\}$, given information ($I$), knowledge ($K$), and experience ($E$) possessed by the expert at the time of forecast preparation. At every point $w \in \mathcal{W}$, function G specifies the probability assigned by the expert to event $\{W \leq w\}$:

$$G(w) = P(W \leq w | I, K, E).$$

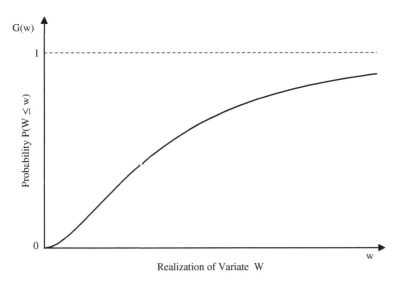

Figure 9.1 Idealized distribution function G of a continuous variate W.

What distinguishes then a judgmental distribution function from any other distribution function of W is the conditioning on I, K, E, which is unique to the expert but may vary with time. For simplicity, this conditioning is not included in the notation $G(w)$, but it should be remembered.

Assessing G according to its definition is infeasible because there are infinitely many points $w \in \mathcal{W}$ and thus infinitely many events of the form $\{W \leq w\}$ to which the probabilities $G(w)$ must be assigned. To formulate a feasible, albeit approximate, assessment technique, a pointwise representation of the continuous function G is sought in terms of a few quantiles of W.

Definition The *judgmental quantile* of W corresponding to probability $p \in (0, 1)$ is a realization $y_p \in \mathcal{W}$ such that event $\{W \leq y_p\}$ is judged to have the probability

$$G(y_p) = P(W \leq y_p | I, K, E) = p.$$

For short, y_p is called the *p-probability quantile* of W (or just the *p-quantile*); synonyms are the *p-probability fractile* of W (or just the *p-fractile*), and the *100p-percentile* of W.

Definition The *pointwise representation* of the judgmental distribution function G is a set of I distinct points

$$\{(y_p, p) : p \in \mathcal{P}\},$$
$$\mathcal{P} = \{p : p = p_1, \dots, p_I; 0 < p_1 < \cdots < p_I < 1\},$$

such that for each probability $p \in \mathcal{P}$, the quantile $y_p \in \mathcal{W}$ has been assessed judgmentally by the forecaster.

Figure 9.2 provides an illustration with $I = 3$ and $\mathcal{P} = \{0.5, 0.25, 0.75\}$; the probabilities are listed in the **order** in which the quantiles are assessed.

9.2.2 Assessment Procedure

To obtain a pointwise representation of the judgmental distribution function G via the *quantile technique*, the forecaster should follow a seven-step procedure.

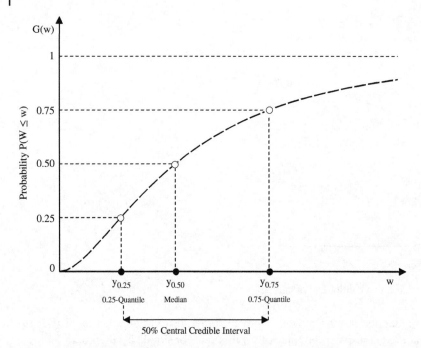

Figure 9.2 Pointwise representation of the distribution function *G* of a continuous variate *W* in terms of three quantiles.

Step 1. Define precisely the predictand *W* and its sample space $\mathcal{W}$.

Step 2. Gather, organize, and analyze all predictive information.

Step 3. Judge the **veracity** of the information; judge the **predictive strength** of the information; form an overall judgment of the degree of certainty about the predictand *W*, given the information. (Information is veracious if it can be taken at its face value; for instance, a thermometer reading −5°C versus a skier reporting that it felt like −5°C at the mountain top. Predictive strength refers to the degree of stochastic dependence between information and events $\{W \leq y_p\}$; for instance, a high unemployment rate is a predictor of the number of applicants to a graduate school, but a high inflation rate is not.)

Step 4. Fix the set of probabilities $\mathcal{P}$: at least three values of *p*, for example,

$$p = 0.5,\ 0.25,\ 0.75,\ \text{or}$$

$$p = 0.5,\ 0.1,\ 0.9;$$

in a more thorough approach, five values of *p*, for example,

$$p = 0.5, 0.25, 0.125, 0.75, 0.875,\ \text{or}$$

$$p = 0.5, 0.25, 0.1, 0.75, 0.9.$$

(Further guidelines for fixing the probabilities are stated below, in the examples, and in the exercises.)

Step 5. For each probability *p*, assess the quantile y_p.

Step 6. Validate the internal coherence of the assessed quantiles.

Step 7. Use the assessed quantiles to estimate a parametric model for *G*.

Figure 9.2 illustrates $(y_{0.25}, y_{0.5}, y_{0.75})$. The quantile $y_{0.5}$ is the *median*:

$$P(W < y_{0.5}) = P(W > y_{0.5}) = 0.5.$$

The quantiles $y_{0.25}$ and $y_{0.75}$ are the *first quartile* and the *third quartile*, respectively:

$$P(W \le y_{0.25}) = 0.25, \qquad P(W \le y_{0.75}) = 0.75.$$

The two quartiles define the 50% *central credible interval* $(y_{0.25}, y_{0.75})$ about the median, with the property

$$P(y_{0.25} < W < y_{0.75}) = 0.5.$$

It is this credible interval that quantifies, and conveys to the decider, the uncertainty about W on a particular occasion. The larger the uncertainty in the forecaster's judgment, the wider the credible interval should be.

The interval $(y_{0.25}, y_{0.75})$ need not be symmetric about the median. In fact, one would expect that its width $[y_{0.75} - y_{0.25}]$ and skew $[y_{0.5} - y_{0.25} \ne y_{0.75} - y_{0.5}]$ vary from occasion to occasion, thereby reflecting the forecaster's judgment on each occasion.

The two quartiles define also the *tail credible intervals*. When the sample space of W is a bounded open interval $W = (\eta_L, \eta_U)$, the 25% *lower tail credible interval* is $(\eta_L, y_{0.25})$, and the 25% *upper tail credible interval* is $(y_{0.75}, \eta_U)$.

In Step 4, the values of probability p may be fixed to suit the application, although consideration must be given to the difficulty of the judgmental tasks. Beginning with $p = 0.5$ is preferable because assessing the median requires bisection (of the sample space into two equally likely events) — the easiest judgmental task. Similarly, $p = 0.25$, 0.75, 0.125, 0.875 lead to conditional bisection tasks. Both types of tasks, *bisection* and *conditional bisection*, are detailed in Section 9.2.4.

9.2.3 Predictive Information

An example of the forecasting problem is introduced first in order (i) to provide context for the assessment of quantiles, (ii) to illustrate structuring of the predictive information, and (iii) to explain "the scientific use of the imagination" (recall Section 4.1.3) in judgmental forecasting of a continuous predictand.

Business venture
Toward the end of the internet technology bubble, *circa* 2000, two ventures (Web-Grocer and Web-Van) made a splashy start in large cities (San Francisco and Atlanta among them). They were envisioned as high-tech operations, which would revolutionize shopping for groceries by American urbanites; they were featured on the cover of *Time* magazine. Their business model was as follows. Grocery supplies would be stored within the city, in a huge refrigerated warehouse, with conveyor belts and bagging machines. A customer would order groceries via a website from home or office. Items ordered would be automatically retrieved and bagged in the warehouse. A van would deliver the bags to the specified address at the appointed time. The customer would pay for groceries (marked up) and delivery. The euphoria ended two years later in bankruptcies: each venture lost nearly $1 billion of investors' money. The reason was typical for start-ups: overestimated demand and underestimated cost. (*Post factum.* Some 15 years later, companies that deliver groceries have learnt to operate frugally, with only a website, cars, and drivers who double as shoppers on foot, purchasing the ordered items in local grocery stores and making deliveries (*The Wall Street Journal*, 12–13 March 2016).)

Forecasting problem
There were two major predictands: the cost of building and operating a high-tech warehouse, and the demand for a web grocer. If the uncertainty about each predictand were quantified in terms of a distribution function, then one could derive a distribution function of the profit and use it to decide whether or not to start the venture; and if so, then at what capacity — the decision problem studied in Chapter 13. Imagine being tasked with predicting the weekly demand, W, for the web grocer in a large city (choose one). You have no past data to construct a quantitative model because the venture is new. Thus, you proceed by constructing a conceptual model to guide your reasoning.

Conceptual model

It may be just a mental model. Or it may be a verbalized model. Here is a sketch of one. It consists of a list of predictors that you would estimate sequentially and then combine algebraically to come up with an estimate of W.

- Population of the city.
- Number of households (HH) in the city.
- Fraction of HH having access to the internet.
- Fraction of those HH preferring to shop for groceries online. (Here you need to predict a change of **social behavior**: from shopping on foot to shopping on screen.)
- Fraction of those HH willing to pay extra for groceries: the markups of items, fees for deliveries, tips for drivers. (Here you need to predict a change of **economic behavior**: the customer's preferred trade-off between extra cost and extra free time plus convenience.)
- The weekly average amount spent by those HH on groceries.

Suppose the estimates of the six predictors made by an expert are:

$$1\,500\,000, \ 500\,000, \ 0.7, \ 0.2, \ 0.1, \ 450 \ \$.$$

Then the estimate of W — the weekly demand for the web grocer in terms of dollars spent (excluding the markups, delivery fees, tips) — is

$$500\,000 \times 0.7 \times 0.2 \times 0.1 \times 450 = \$3\,150\,000.$$

Note that this is a sequential application of the *anchoring and adjustment* heuristic introduced in Section 4.4.7.

Influence diagram

In general, the list of predictors may be long, and the influence of one on the others may be more complicated than sequential. In such a case, the conceptual model may be structured graphically as an influence diagram (or a block diagram, or a network diagram, or a flowchart) to depict the predictors and their influences. Figure 9.3 shows an example. (The influence diagram does have a formal definition in decision theory; but here it may be constructed imaginatively.)

Mental simulation

This is the reasoning process that follows the conceptual model from the input (the first predictor) to the output (the predictand), while (i) making inevitable assumptions, (ii) estimating each predictor, and (iii) performing calculations (exactly or approximately, or mentally). In the example above, the expert began with the known population count, assumed 3 persons per household, and then estimated each of the remaining four predictors.

Envisaging uncertainty

Different assumptions and different estimates of predictors obviously yield different estimates of W. Their range should reflect the uncertainty about W. How to ensure that it does? Focus your thinking on the potential **sources of uncertainty**, and account for them by expanding the mental simulations — their range and number. Here is a partial guide.

- Think of all the model assumptions that may be wrong.
- Think of all the predictor estimates that may be wrong.
- Think of all the predictors that may be informative but are omitted.
- Think of all the influences that may be too simplistic.
- Think of all the rare, extreme, exogenous events (e.g., recession, inflation, unemployment, hurricane, epidemic) that may affect the demand — the surprises, so to speak.

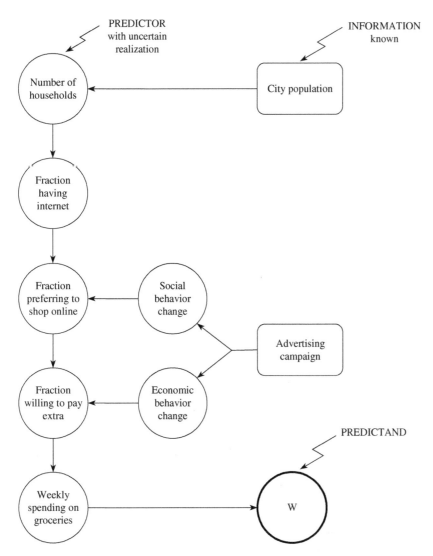

Figure 9.3 Influence diagram of a conceptual model for judgmental forecasting of the weekly demand, *W*, for a web grocer.

Scenarios

These are imagined (hypothesized); they help to bring the future to mind; they are invoked by forecasters to simplify (to discretize) the reasoning about the future. For us, a scenario is what generates a subset of mental simulations that yield estimates of *W* within the same interval of the sample space $\mathcal{W}$. This interval is defined in terms of quantiles of *W*. Ergo, the scenario and the interval (as an event) should be assigned the same probability. Cognitive psychologists suggest a heuristic: the "easier" an event can be reached (imagined) in mental simulation, the higher its judgmental probability.

9.2.4 Assessment of Quantiles

Detailed below is the reasoning process that leads to the assessment of quantiles $(y_{0.25}, y_{0.5}, y_{0.75})$. Figure 9.4 schematically depicts the events that are considered in this process. In application, this process should be adapted to the problem at hand. The adaptation may consist of (i) suitable interpretations of the variate and the events,

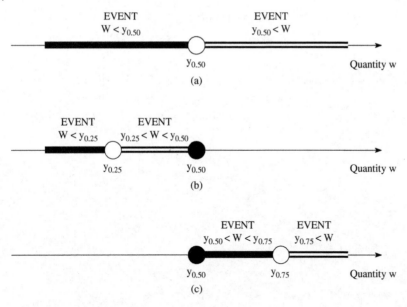

Figure 9.4 Schema for assessing (a) the median $y_{0.50}$, (b) the 0.25-quantile $y_{0.25}$, and (c) the 0.75-quantile $y_{0.75}$ of a continuous variate W.

(ii) assessment of quantiles other than $y_{0.25}$ and $y_{0.75}$, and (iii) assessment of more quantiles, for example $y_{0.125}$ and $y_{0.875}$, which specify the tails of the distribution function.

Assessment of the 0.5-quantile (median)

1. Consider **all** possible scenarios which may generate realizations of the variate W. Suppose one of these scenarios will come true, but, of course, you are not certain which one.

2. Assess an estimate of W; denote it $y_{0.5}$. The estimate $y_{0.5}$ should be such that in your judgment the following two events are *equally likely* to occur (Figure 9.4a):

Realization of W	Realization of W
will be *below*	will be *above*
your estimate:	your estimate:
$W < y_{0.5}$	$y_{0.5} < W$

3. Verify judgmentally that the estimate $y_{0.5}$ indeed has the property of the median. To accomplish this, (i) identify a *reference event* whose probability can be deduced logically and is equal to $\frac{1}{2}$ (e.g., an outcome from a toss of a coin), and (ii) compare your degree of certainty about each of the events above with your degree of certainty about the reference event. Is event $W < y_{0.5}$ as likely as obtaining a tail, and event $y_{0.5} < W$ as likely as obtaining a head, in one toss of a coin? If the answer is yes, then your estimate $y_{0.5}$ is indeed the median of W. If event $y_{0.5} < W$ appears to you more (less) likely than a head, then your estimate $y_{0.5}$ should be adjusted upward (downward).

Assessment of the 0.25-quantile (first quartile)

1. Hypothesize that $W < y_{0.5}$. Consider **only those** scenarios which may generate realizations smaller than the median $y_{0.5}$. Suppose one of these scenarios will come true, but, of course, you are not certain which one.

2. **Conditional on the hypothesis** that the realization of W will be below $y_{0.5}$, assess an estimate $y_{0.25}$ such that in your judgment the following two events are *equally likely* to occur (Figure 9.4b):

Realization of W	Realization of W
will be *below*	will be *above*
your estimate:	your estimate:
$W < y_{0.25}$	$y_{0.25} < W$

3. Verify judgmentally that the estimate $y_{0.25}$ indeed has the property of the 0.25-quantile. As in the verification of the median, you may help yourself by comparing the two conditional events to the outcomes from a toss of a coin. **Conditional on the hypothesis** that event $W < y_{0.5}$ will occur, is event $W < y_{0.25}$ as likely as obtaining a tail, and event $y_{0.25} < W$ as likely as obtaining a head, in one toss of a coin? If the answer is yes, then your estimate $y_{0.25}$ is indeed the first quartile of W. If, **conditional on the hypothesis** that $W < y_{0.5}$, event $y_{0.25} < W$ appears to you more (less) likely than a head, then your estimate $y_{0.25}$ should be adjusted upward (downward).

Assessment of the 0.75-quantile (third quartile)

1. Hypothesize that $y_{0.5} < W$. Consider **only those** scenarios which may generate realizations larger than the median $y_{0.5}$. Suppose one of these scenarios will come true, but, of course, you are not certain which one.
2. **Conditional on the hypothesis** that the realization of W will be above $y_{0.5}$, assess an estimate $y_{0.75}$ such that in your judgment the following two events are *equally likely* to occur (Figure 9.4c):

Realization of W	Realization of W
will be *below*	will be *above*
your estimate:	your estimate:
$W < y_{0.75}$	$y_{0.75} < W$

3. Verify judgmentally that the estimate $y_{0.75}$ indeed has the property of the 0.75-quantile. As in the verification of the median, you may help yourself by comparing the two conditional events to the outcomes from a toss of a coin. **Conditional on the hypothesis** that event $y_{0.5} < W$ will occur, is event $W < y_{0.75}$ as likely as obtaining a tail, and event $y_{0.75} < W$ as likely as obtaining a head, in one toss of a coin? If the answer is yes, then your estimate $y_{0.75}$ is indeed the third quartile of W. If, **conditional on the hypothesis** that $y_{0.5} < W$, event $y_{0.75} < W$ appears to you more (less) likely than a head, then your estimate $y_{0.75}$ should be adjusted upward (downward).

9.2.5 Validation of Coherence

Having assessed the quantiles, you should validate that, in your judgment, the quantiles jointly obey the rules of probability. Toward this end check two conditions, depicted schematically in Figure 9.5.

Validation of Central Credible Interval

Consider two events.

Realization of W will be *inside* the 50% central credible interval:

$$y_{0.25} < W < y_{0.75}$$

Realization of W will be *outside* the 50% central credible interval:

$$W < y_{0.25} \quad \text{or} \quad y_{0.75} < W$$

Are the two events equally likely? Is each event as likely as obtaining a head in one toss of a coin? If not, then adjust the quantiles.

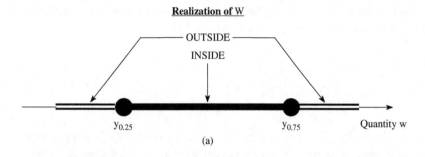

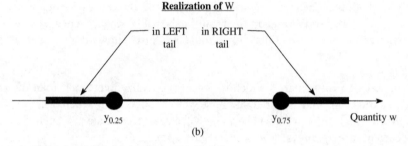

Figure 9.5 Schema for validating the coherence of (a) the 50% central credible interval, and (b) the 25% tail credible intervals, left and right.

Validation of Tail Credible Intervals

Consider the two extreme events:

Realization of W	Realization of W
will be in the	will be in the
left tail:	*right* tail:
$W < y_{0.25}$	$y_{0.75} < W$

Are the two events equally likely? Is each event as likely as obtaining two heads in two tosses of a coin? If not, then adjust the quantiles.

9.2.6 Judgmental Task

This is a "warm-up" task, to get the first feel for the assessment of quantiles. It is to be performed intuitively, sans any preparation and aid, and sans gathering any predictive information. Rely on your general knowledge and follow the reasoning process detailed in the preceding section.

Note 1. You may view this task as a friendly mini-test from geography and history: you are not required to answer with a single number if you do not know it. Instead, you are asked to assess five numbers that reflect your uncertainty about the true answer.

Note 2. If you know the realization w of the variate W with certainty, that is, sans violating Cromwell's rule (recall Section 4.2.4), then your judgmental distribution function is a unit step function: $P(W < w) = 0$ and $P(W \leq w) = 1$. Skip that variate.

Note 3. The extreme probabilities, 0.01 and 0.99, are not those recommended in Step 4 of Section 9.2.2. The reason for that recommendation is revealed in Section 11.5.3.

An analysis of the task and the usage of the verification table are covered in Chapter 11. It may be read after completing the task, or after studying this chapter, or after studying Chapter 10. The modeling and estimation of a parametric distribution function based on the assessed quantiles are covered in the next section and are the subject of Exercise 3.

Task 9.1 (*Judgmental distribution function*)   Consider three variates:

1. The total number of eggs produced in the USA per year (unit: millions).
2. The mean annual precipitation in Charlottesville, Virginia (unit: inches).
3. The year of birth of John Adams, second president of the USA.

Assessment Task: Assess judgmentally and write down your estimates of five quantiles for each variate.

	Quantile				
Variate	$y_{0.01}$	$y_{0.25}$	$y_{0.50}$	$y_{0.75}$	$y_{0.99}$
Eggs [millions]					
Precip. [inches]					
Year					

Verification Task: Read Exercise 3 to find out the realization w for each variate.
For each variate (row), write "1" under the event that occurred.
For each event (column), sum the number of occurrences.

	Event					
	Ω_1	Ω_2	Ω_3	Ω_4	Ω_5	Ω_6
Variate	$w \le y_{0.01}$	$y_{0.01} < w \le y_{0.25}$	$y_{0.25} < w \le y_{0.50}$	$y_{0.50} < w \le y_{0.75}$	$y_{0.75} < w \le y_{0.99}$	$y_{0.99} < w$
Eggs						
Precip.						
Year						
Sum						

9.3   Parametric Distribution Function

The few quantiles assessed by an expert provide only a crude representation of the distribution function G of a continuous variate W. However, they suffice to estimate a parametric model for G, which characterizes uncertainty about W completely and conveniently. In particular, the model allows one to calculate the judgmental nonexceedance probability $P(W \le w)$ at any point $w \in \mathcal{W}$ by the interpolation between, or the extrapolation beyond, the assessed points.

9.3.1   Modeling Procedure

To obtain a parametric model for G, we follow, in essence, the distribution modeling methodology detailed in Chapter 3. The methodology consists of six steps.

Step 1. Construct the *pointwise distribution function* of W thusly. Suppose the expert assessed I quantiles. Arrange the assessed quantiles in ascending order of numerical values, so that the corresponding probabilities form a strictly increasing sequence:

$$p_1 < p_2 < \cdots < p_I.$$

Pair each quantile with its probability to form a set of I points:

$$\{(y_p, p) : p = p_1, p_2, \ldots, p_I\}.$$

This is the pointwise distribution function of W.

Step 2. Specify the sample space $\mathcal{W}$ of W (Section 3.3).

Step 3. Hypothesize a parametric model for G (Section 3.4).

Step 4. Estimate the parameters of the hypothesized model: (i) use the set of I points as if it were an empirical distribution function of W; (ii) use the method of least squares, or the method of uniform distance (Section 3.5 and Appendix B), or the method of quantiles–moments (Section 9.3.2).

Step 5. Evaluate the goodness of fit of the hypothesized model by calculating the MAD (Section 3.6).

Step 6. Accept the hypothesized model, or reject it and return to Step 3.

Note. The quantiles assessed judgmentally do not constitute a sample of the variate W. Thus, estimation methods such as the method of moments or the method of maximum likelihood are unapplicable, and no statistical goodness-of-fit test can be performed.

9.3.2 Gaussian Models

Motivation

In some applications, it is desirable or convenient to hypothesize one of the four Gaussian models for G (Section 2.6). Their advantage is a simple and transparent method for estimating parameters: the method of *quantiles–moments* (QM). It is formulated first for the case wherein an expert assessed three quantiles $(y_a, y_{0.5}, y_b)$, with

$$0 < a < 0.5 < b < 1,$$

and is next extended to the case wherein an expert assessed five quantiles $(y_c, y_a, y_{0.5}, y_b, y_d)$, with

$$0 < c < a < 0.5 < b < d < 1.$$

Basic Relations

We shall write $W \sim \mathrm{NM}(\mu, \sigma)$ to say that the variate W has a normal distribution with mean μ and variance σ^2 (Section 2.6.2). The standard transform (Section 2.6.3)

$$Z = \frac{W - \mu}{\sigma} \tag{9.1}$$

yields the standard normal variate $Z \sim \mathrm{NM}(0, 1)$. Applying this transform to quantiles yields a relationship between quantile y_p of W and the corresponding quantile z_p of Z:

$$z_p = \frac{y_p - \mu}{\sigma}. \tag{9.2}$$

In the estimation stage, μ and σ are the parameters to be estimated, y_p has been assessed judgmentally, and z_p can be calculated as (Section 2.6.1)

$$z_p = Q^{-1}(p), \qquad 0 < p < 1, \tag{9.3}$$

where Q^{-1} denotes the inverse of the standard normal distribution function Q. Inasmuch as Q and Q^{-1} do not have closed forms, Tables 2.2 and 2.3 give approximate formulae for calculating $Q(z)$ and z_p. In particular, $z_{0.5} = 0$ and

$$z_{0.25} = -0.675, \qquad z_{0.75} = 0.675,$$
$$z_{0.125} = -1.150, \qquad z_{0.875} = 1.150.$$

The Normal Model

The normal model, introduced in Section 2.6.2, is summarized in Table 9.1. This model may be hypothesized when the sample space of W is unbounded (or can be reasonably approximated by an unbounded space), and the density function g is symmetric (at least approximately).

   Three quantiles. To estimate the parameters from three quantiles $(y_a, y_{0.5}, y_b)$, proceed as follows.

1. Validate the necessary condition for the symmetry of the density function: the median $y_{0.5}$ must be approximately equal to the arithmetic mean of quantiles y_a and y_b, whenever $b = 1 - a$.
2. Estimate parameters μ and σ according to the expressions given in Table 9.1.

   The expression for the mean μ follows from the symmetry of the normal distribution, which implies that the mean of W equals the median of W. The expression for the standard deviation σ follows from equation (9.2). By evaluating this equation first at $p = a$ and next at $p = b$, and then by forming the difference, we obtain

$$z_b - z_a = (y_b - y_a)/\sigma.$$

Therefrom an expression for σ can be found, as given in Table 9.1.

Table 9.1   Estimation of the normal distribution from quantiles.

Abbreviation: $\mathrm{NM}(\mu, \sigma)$

Parameters: location $-\infty < \mu < \infty$, $\qquad$ scale $\sigma > 0$

Sample space: $-\infty < w < \infty$

Density function

$$g(w) = \frac{1}{\sigma\sqrt{2\pi}} \exp\left(-\frac{1}{2}\left(\frac{w - \mu}{\sigma}\right)^2\right)$$

Distribution function

$$G(w) = Q\left(\frac{w - \mu}{\sigma}\right)$$

Necessary condition

$$y_{0.5} = (y_a + y_b)/2 \qquad \text{if } b = 1 - a$$

Estimators of parameters

$$\mu = y_{0.5}$$
$$\sigma = \frac{y_b - y_a}{z_b - z_a}$$

Linear equation

$$y_p = \sigma z_p + \mu, \qquad 0 < p < 1$$

Moments

$$E(W) = \mu$$
$$Var(W) = \sigma^2$$

Five quantiles. To estimate the parameters from five quantiles $(y_c, y_a, y_{0.5}, y_b, y_d)$ proceed as follows.

1. Validate the necessary conditions for the symmetry of the density function:

$$y_{0.5} = (y_a + y_b)/2 \qquad \text{if } b = 1 - a,$$
$$y_{0.5} = (y_c + y_d)/2 \qquad \text{if } d = 1 - c. \tag{9.4}$$

2. Estimate parameters μ and σ. There are two alternative procedures.

    (i) ***The QM method***. Set $\mu = y_{0.5}$ and calculate

    $$\sigma_1 = (y_b - y_a)/(z_b - z_a),$$
    $$\sigma_2 = (y_d - y_c)/(z_d - z_c). \tag{9.5}$$

    The normal model is plausible if σ_1 is close to σ_2; then their geometric mean, $\sigma = (\sigma_1 \sigma_2)^{1/2}$, is a reasonable estimate.

    (ii) ***The LS method***. Rewrite equation (9.2) in the form

    $$y_p = \sigma z_p + \mu, \tag{9.6}$$

    which shows that σ is the slope and μ is the intercept of a linear relationship between y_p and z_p. Estimate σ and μ via the LS method (Appendix B) using the set of points

    $$\{(y_p, z_p) : p = c, a, 0.5, b, d\}.$$

The QM method justifies our statement in Section 9.2.2 that a central credible interval quantifies the uncertainty about W. To wit, suppose $a = 0.25$, $b = 0.75$, so that $z_b - z_a = 0.675 - (-0.675) = 1.35$, and $\sigma = (y_{0.75} - y_{0.25})/1.35$. This shows that when $W \sim \text{NM}(\mu, \sigma)$, its standard deviation is proportional to the width $(y_b - y_a)$ of a central credible interval. Hence, σ and $(y_b - y_a)$ are equivalent measures of uncertainty.

The Log-Normal Model

The log-normal model, introduced in Section 2.6.4, is summarized in Table 9.2. This model may be hypothesized when the sample space of W is a bounded-below interval (η, ∞), and the density function g has a positive skew. The model is obtained by applying a *logarithmic transform*

$$X = \ln(W - \eta), \qquad \eta < W, \tag{9.7}$$

and then assuming that $X \sim \text{NM}(\mu, \sigma)$.

To estimate the parameters μ and σ from quantiles, each assessed quantile y_p is transformed into the quantile

$$x_p = \ln(y_p - \eta), \qquad 0 < p < 1, \tag{9.8}$$

and then one of the methods (QM or LS) for estimating the parameters of the normal distribution is applied to the transformed quantiles. The necessary condition for the log-normal density function can be stated in terms of the assessed quantiles $(y_a, y_{0.5}, y_b)$ as follows: the shifted median $y_{0.5}$ must be approximately equal to the geometric mean of the shifted quantiles y_a and y_b, whenever $b = 1 - a$.

The Log-Ratio Normal Model

The log-ratio normal model, introduced in Section 2.6.5, is summarized in Table 9.3. This model may be hypothesized when the sample space of W is a bounded open interval (η_L, η_U). The density function g may have a negative skew or a positive skew; it may also be symmetric unimodal or symmetric bimodal. The model is obtained by applying a *log-ratio transform*

$$X = \ln \frac{W - \eta_L}{\eta_U - W}, \qquad \eta_L < W < \eta_U, \tag{9.9}$$

and then assuming that $X \sim \text{NM}(\mu, \sigma)$.

Table 9.2 Estimation of the log-normal distribution from quantiles.

Abbreviation: $LN(\mu, \sigma, \eta)$

Parameters: location $-\infty < \mu < \infty$,　　scale $\sigma > 0$,　　shift $-\infty < \eta < \infty$

Sample space: $\eta < w < \infty$

Density function

$$g(w) = \frac{1}{(w - \eta)\sigma\sqrt{2\pi}} \exp\left(-\frac{1}{2}\left(\frac{\ln(w - \eta) - \mu}{\sigma}\right)^2\right)$$

Distribution function

$$G(w) = Q\left(\frac{\ln(w - \eta) - \mu}{\sigma}\right)$$

Necessary condition

$$y_{0.5} - \eta = [(y_a - \eta)(y_b - \eta)]^{1/2} \qquad \text{if } b = 1 - a$$

Transformation of quantiles

$$x_p = \ln(y_p - \eta), \qquad p = a, 0.5, b$$

Estimators of parameters

$$\mu = \quad x_{0.5}$$
$$\sigma = \frac{x_b - x_a}{z_b - z_a}$$

Linear equation

$$x_p = \sigma z_p + \mu, \qquad 0 < p < 1$$

Moments

$$E(W) = \exp(\sigma^2/2 + \mu) + \eta$$
$$Var(W) = [\exp(\sigma^2) - 1]\exp(\sigma^2 + 2\mu)$$

To estimate the parameters μ and σ from quantiles, each assessed quantile y_p is transformed into the quantile

$$x_p = \ln\frac{y_p - \eta_L}{\eta_U - y_p}, \qquad 0 < p < 1, \tag{9.10}$$

and then one of the methods (QM or LS) for estimating the parameters of the normal distribution is applied to the transformed quantiles. The necessary condition for the log-ratio normal density function can be stated in terms of the assessed quantiles $(y_a, y_{0.5}, y_b)$ as follows: the ratio transform of the median $y_{0.5}$ must be approximately equal to the geometric mean of the ratio transforms of quantiles y_a and y_b, whenever $b = 1 - a$.

The Reflected Log-Normal Model

The reflected log-normal model, introduced in Section 2.6.6, is summarized in Table 9.4. This model may be hypothesized when the sample space of W is a bounded-above interval $(-\infty, -\eta)$, and the density function g has a

Table 9.3 Estimation of the log-ratio normal distribution from quantiles.

Abbreviation: $\mathrm{LR1}(\eta_L, \eta_U)\text{-}\mathrm{NM}(\mu, \sigma)$

Parameters: location $-\infty < \mu < \infty,$ scale $\sigma > 0,$ shift $-\infty < \eta_L < \eta_U < \infty$

Sample space: $\eta_L < w < \eta_U$

Density function

$$g(w) = \frac{\eta_U - \eta_L}{(w - \eta_L)(\eta_U - w)\sigma\sqrt{2\pi}} \exp\left(-\frac{1}{2\sigma^2}\left(\ln\frac{w - \eta_L}{\eta_U - w} - \mu\right)^2\right)$$

Distribution function

$$G(w) = Q\left(\frac{1}{\sigma}\left(\ln\frac{w - \eta_L}{\eta_U - w} - \mu\right)\right)$$

Necessary condition

$$\frac{y_{0.5} - \eta_L}{\eta_U - y_{0.5}} = \left[\left(\frac{y_a - \eta_L}{\eta_U - y_a}\right)\left(\frac{y_b - \eta_L}{\eta_U - y_b}\right)\right]^{1/2} \qquad \text{if } b = 1 - a$$

Transformation of quantiles

$$x_p = \ln\frac{y_p - \eta_L}{\eta_U - y_p}, \qquad p = a, 0.5, b$$

Estimators of parameters

$$\mu = \quad x_{0.5}$$
$$\sigma = \frac{x_b - x_a}{z_b - z_a}$$

Linear equation

$$x_p = \sigma z_p + \mu, \qquad 0 < p < 1$$

Moments

$$E(W) = \eta_U - (\eta_U - \eta_L)E\left(\frac{1}{e^X + 1}\right)$$
$$Var(W) = (\eta_U - \eta_L)^2 Var\left(\frac{1}{e^X + 1}\right)$$

negative skew. The model is obtained by applying a *logarithmic transform*

$$X = \ln(-W - \eta), \qquad W < -\eta, \tag{9.11}$$

and then assuming that $X \sim \mathrm{NM}(\mu, \sigma)$.

To estimate the parameters μ and σ from quantiles, each assessed quantile y_p is transformed into the quantile

$$x_p = \ln(-y_p - \eta), \qquad 0 < p < 1, \tag{9.12}$$

and then one of the methods (QM or LS) for estimating the parameters of the normal distribution is applied to the transformed quantiles. The necessary condition for the reflected log-normal density function can be stated in

Table 9.4 Estimation of the reflected log-normal distribution from quantiles.

Abbreviation: $LN(\mu, \sigma, \eta, -1)$

Parameters: location $-\infty < \mu < \infty$, scale $\sigma > 0$, shift $-\infty < \eta < \infty$

Sample space: $-\infty < w < -\eta$

Density function

$$g(w) = \frac{1}{(-w - \eta)\sigma\sqrt{2\pi}} \exp\left(-\frac{1}{2}\left(\frac{\ln(-w - \eta) - \mu}{\sigma}\right)^2\right)$$

Distribution function

$$G(w) = 1 - Q\left(\frac{\ln(-w - \eta) - \mu}{\sigma}\right)$$

Necessary condition

$$-y_{0.5} - \eta = \left[(-y_a - \eta)(-y_b - \eta)\right]^{1/2} \qquad \text{if } b = 1 - a$$

Transformation of quantiles

$$x_p = \ln(-y_p - \eta), \qquad p = a, 0.5, b$$

Estimators of parameters

$$\mu = x_{0.5}$$
$$\sigma = \frac{x_b - x_a}{z_{1-b} - z_{1-a}}$$

Linear equation

$$x_p = \sigma z_{1-p} + \mu, \qquad 0 < p < 1$$

Moments

$$E(W) = -\exp(\sigma^2/2 + \mu) - \eta$$
$$Var(W) = \lfloor\exp(\sigma^2) - 1\rfloor \exp(\sigma^2 + 2\mu)$$

---

terms of the assessed quantiles $(y_a, y_{0.5}, y_b)$ as follows: the shifted median $y_{0.5}$ must be approximately equal to the geometric mean of the shifted quantiles y_a and y_b, whenever $b = 1 - a$.

Example 9.1 (*Reflected log-normal model*)

Recall the forecasting problem of a web grocer described in Section 9.2.3. Suppose the forecaster assessed five quantiles of W — the weekly demand for a web grocer in terms of dollars spent by the customers. The fixed set of probabilities was

$$\mathcal{P} = \{p : p = 0.125, 0.25, 0.5, 0.75, 0.875\};$$

the assessed quantiles y_p, in 10^6 \$/week, are reported in Table 9.5, and the pointwise distribution function is graphed in Figure 9.6. Having compared this graph with graphs of the four Gaussian distribution functions in Figures 2.1–2.4, the forecaster hypothesized the reflected log-normal model $LN(\mu, \sigma, \eta, -1)$ with the sample space $\mathcal{W} = (-\infty, -\eta)$, assessed its upper bound $-\eta = 7$ [10^6 \$/week], and accepted the unbounded left tail as an approximation. The calculations that follow use the expressions from Table 9.4.

Table 9.5 Modeling the judgmental distribution function G of predictand W; the hypothesized G is LN(μ, σ, η, -1); the parameter estimation method is quantiles–moments (QM).

Probability	p	0.125	0.25	0.5	0.75	0.875
Assessed quantile	y_p	0.9	2.2	3.15	3.9	4.4
Transformed quantile	x_p	1.808	1.569	1.348	1.131	0.956
Quantile $Q^{-1}(1-p)$	z_{1-p}	1.150	0.675	0	−0.675	−1.150
Hypothesized DF	$G(y_p)$	0.09	0.26	0.5	0.74	0.87
MAD	$|p - G(y_p)|$	**0.035**	0.01	0	0.01	0.005

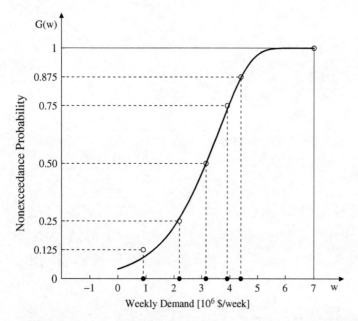

Figure 9.6 The assessed distribution function $\{(y_p, p) : p \in \mathcal{P}\}$ (pointwise), and the parametric distribution function G (continuous); G is reflected log-normal LN(μ, σ, η, -1) on the sample space $(-\infty, 7)$, with parameter estimates $\mu = 1.348$, $\sigma = 0.346$, $\eta = -7$; the estimation method is quantiles–moments (QM); MAD = 0.035 at point (0.9, 0.125).

The two necessary conditions for the negative skew of the density function have identical left side:

$$-y_{0.5} - \eta = -3.15 + 7 = 3.85;$$

the right sides are:

$$\text{for } a = 0.25, \quad b = 0.75, \quad [(-2.2 + 7)(-3.9 + 7)]^{1/2} = 3.86;$$
$$\text{for } c = 0.125, \quad d = 0.875, \quad [(-0.9 + 7)(-4.4 + 7)]^{1/2} = 3.98.$$

The closeness of the three numbers to each other allows us to accept the model as plausible (i.e., as having the correct negative skew).

The parameters (μ, σ) are estimated via the QM method (Table 9.5). First, each assessed quantile y_p of predictand W is transformed into quantile x_p of variate X defined by transform (9.12). Second, the estimate of the mean μ of X is

$$\mu = x_{0.5} = 1.348.$$

Third, the estimate of the standard deviation σ of X is calculated from two central credible intervals:

$$\sigma_1 = \frac{x_{0.75} - x_{0.25}}{z_{0.25} - z_{0.75}} = \frac{1.131 - 1.569}{-0.675 - 0.675} = 0.324,$$

$$\sigma_2 = \frac{x_{0.875} - x_{0.125}}{z_{0.125} - z_{0.875}} = \frac{0.956 - 1.808}{-1.150 - 1.150} = 0.370,$$

$$\sigma = [0.324 \times 0.370]^{1/2} = 0.346.$$

Because 0.324 is close enough to 0.370, their geometric mean is a reasonable estimate of σ. The model of the predictand is now specified as $W \sim \text{LN}(1.348, 0.346, -7, -1)$.

The distribution function of W takes the form

$$G(w) = 1 - Q\left(\frac{\ln(-w + 7) - 1.348}{0.346}\right), \qquad w \in \mathcal{W} = (-\infty, 7).$$

The graph of this parametric continuous distribution function is superposed on the graph of the assessed pointwise distribution function in Figure 9.6. Visually, the fit of G to the points appears very good. To obtain the MAD, $G(y_p)$ is calculated for every $p \in \mathcal{P}$ in Table 9.5. The result, MAD = 0.035, reaffirms the visual impression of a very good fit.

9.4 Group Forecasting

When several experts individually have assessed quantiles $(y_c, y_a, y_{0.5}, y_b, y_d)$ of the same predictand W and the assessments are not identical, the question arises what could be done next. Two approaches are available: reconciling and combining.

9.4.1 Reconciling Assessments

Reconciliation should be attempted first whenever discrepancies among individual assessments are large. For in such instances one would like to bring to light the reasons behind the diverse judgments so that they can be weighted accordingly in the final assessment. Reconciliation of individual assessments is best accomplished at a workshop designed and conducted by a facilitator. The workshop should provide a process of consensus building. Individual experts should be invited to elaborate on the reasons for their assessments, encouraged to share information, and then asked to revise their quantiles.

This process is dubbed an *estimate–talk–estimate* process. It is a version of the *nominal group process* — one of several methods of problem-solving by small groups. Its premise is that (i) the post-discussion judgments tend to be more polarized ("sharpened"), or less dispersed, than the pre-discussion judgments, and (ii) the post-discussion estimates tend to converge closer to the correct values than do estimates made by individuals alone or by other group processes. Still, the consensus may not emerge, in which case the facilitator should be prepared to close the debate and arrive at the final assessment by algorithmic means. This brings us to the second approach.

9.4.2 Combining Assessments

Combining individual assessments is an algorithmic approach. Individual experts' quantiles are not revised, but are inputted into a mathematical procedure that outputs a single vector of quantiles. A variety of procedures may be found in the literature. The procedure described below derives from the theory of voting (justified in Section 9.8).

A democratic procedure for collective choice between two alternatives is the *majority rule* voting. Among its properties is one stating that the majority rule minimizes the expected number of disappointed voters.

The decisional disappointment is defined as the case of one voting for alternative A but the group choosing alternative B, or vice versa.

An analogous democratic procedure for collective choice of a real number from a set $\mathcal{Y}$ (countable or uncountable) is the *median rule* voting. This rule shares several properties with the majority rule, including minimization of the expected number of disappointed experts. In the definition of decisional disappointment, the alternatives now represent subsets of numbers:

$$A = \{y : y \leq \bar{y}, y \in \mathcal{Y}\},$$
$$B = \{y : y \geq \bar{y}, y \in \mathcal{Y}\},$$

where $\bar{y} \in \mathcal{Y}$ is the number chosen by the group. An expert choosing a number, in effect, casts a vote for either A or B.

Suppose there are M experts indexed by m ($m = 1, \ldots, M$), and each expert m has prepared judgmentally a probabilistic forecast of a continuous predictand W by assessing several quantiles, say $(y_c(m), y_a(m), y_{0.5}(m), y_b(m), y_d(m))$. The *median rule* for combining the assessed quantiles operates as follows.

Step 1. For every probability value $p \in \{c, a, 0.5, b, d\}$, arrange the M quantiles in ascending order of numerical values:

$$y_{p(1)} \leq y_{p(2)} \leq \cdots \leq y_{p(M)},$$

where $y_{p(m)}$ denotes the p-probability quantile having rank m ($m = 1, \ldots, M$).

Step 2. For every probability value $p \in \{c, a, 0.5, b, d\}$, find the median of the M quantiles:

$$\bar{y}_p = \begin{cases} y_{p((M+1)/2)} & \text{if } M \text{ is odd,} \\ \frac{1}{2}\left[y_{p(M/2)} + y_{p((M+2)/2)}\right] & \text{if } M \text{ is even.} \end{cases}$$

Step 3. Form a vector of group quantiles,

$$(\bar{y}_c, \bar{y}_a, \bar{y}_{0.5}, \bar{y}_b, \bar{y}_d),$$

which constitutes a probabilistic forecast of W by the group of M experts.

The group quantiles may be treated in the same way as the individual experts' quantiles. In particular, they may be used to estimate parameters of any hypothesized model for the distribution function G of W, and may be subjected to verification as described in Chapter 11.

9.5 Adjusting Distribution Function

Suppose a model outputted distribution function G of predictand W based on current realizations of predictors; the model itself (a physically based model, a statistical model, a learning model) had been developed from past data. But if some new information I pertinent to W has just been acquired, then an expert may intuit (via "the scientific use of imagination" explained in Section 9.2.3) an improvement to the forecast by adjusting G. (Recall an example in Section 9.1.2, characteristic (iv).)

The need for adjusting G arises also when the process that generates the realization of W becomes nonstationary and, consequently, the data used for model development may be unrepresentative of future realizations of W. If a forecaster with experience (E) and knowledge (K) of the process (e.g., inflation, pandemic, weather) detects the nonstationarity (I), then he may be capable of adjusting G.

Here is a procedure for judgmentally adjusting any specified distribution function G of predictand W.

Step 1. Fix the set of probabilities $\mathcal{P}$, as in Step 4 of Section 9.2.2.

Step 2. For each $p \in \mathcal{P}$, calculate the quantile $w_p = G^{-1}(p)$.

Step 3. For each $p \in \mathcal{P}$, adjust w_p judgmentally upward or downward by following the reasoning process detailed in Section 9.2.4. Let y_p denote the adjusted quantile of W.

Step 4. Treat the set $\{(y_p, p) : p \in \mathcal{P}\}$ as the pointwise distribution function of W. Input it to the modeling procedure of Section 9.3.1 to obtain a revised parametric model for G.

This procedure is yet another application of the *anchoring and adjustment* heuristic introduced in Section 4.4.7; the "anchor" is the specified quantile w_p, and the output is the adjusted quantile y_p.

The resultant G may be viewed as a judgmental "posterior" distribution function of W, which is conditional on (i) the model, (ii) the data from which the model was developed, (iii) the current realizations of predictors, and (iv) the input (I, K, E) from the forecaster.

9.6 Applications

There have been many applications of the judgmental forecasting procedure described in this chapter. Three of them are summarized below.

9.6.1 Auditing Financial Statements

In public accounting, the task of an external auditor would ideally be to determine the accuracy of dollar amounts presented in a financial statement of an organization. For large organizations, this would be an arduous task. Therefore, a process of auditing had been developed that aids the auditor to arrive at an opinion regarding the "fairness" of the reported amount. An amount is "fair" if it is judged to be not significantly misleading based on the generally accepted accounting principles. To quantify the uncertainty associated with this judgment, Solomon (1982) turned to the judgmental distribution functions.

The experts were 103 practicing auditors representing large public accounting firms in the USA and having experience in auditing of at least 2 years, with the mean of 4 years. Prior to the experiment, they underwent training in the quantile technique.

The predictand W was the "true" dollar amount of an account balance. There were six predictands, one for each of six cases, which had been developed based on actual audits of business firms.

The predictive information provided to the expert was like that in a real audit: (i) description of the firm and the specific account, (ii) data regarding transactions and balances, (iii) description of the internal control systems in the firm, and (iv) results of compliance tests — the number and type of errors detected.

The judgmental distribution functions. Seven quantiles of W were assessed for probabilities

$$p = 0.50, \ 0.25, \ 0.75, \ 0.1, \ 0.9, \ 0.01, \ 0.99.$$

They were assessed for the six cases by individual auditors and by three-person audit teams. The assessment was conducted at an accounting firm on a business day and took about 2 hours. In total, 449 distribution functions were assessed and then verified against the actual realizations of W. A subset of the verification results is reported in Chapter 11.

9.6.2 Forecasting Net Income

The activities of an oil company — searching for, producing, refining, and marketing of oil and gas — require huge investments while the returns are highly uncertain (e.g., on average in the United States, out of 10 wells drilled in exploration, only one becomes commercially viable). Hence, the corporate after-tax annual income (net income) is uncertain. Davidson and Cooper (1980) describe the development of a system at the Getty Oil Company for forecasting net income in terms of a distribution function. The Getty management asked for uncertainty quantification after the economic downturn and a fire in one of its refineries caused their estimate of the net income in 1973 to be erroneous.

The forecast system. The forecast is made as part of the annual planning process in November, is revised in June, and is verified in December. After planning premises are negotiated by corporate managers and division managers, each division formulates its plan and forecast. The *predictand* is the before-tax annual income of the division, W. Next, a corporate management group aggregates these forecasts by taking into account the inter-division dependencies and by employing a corporate tax model; the output is a distribution function of the net income.

The assessment procedure. To gain support of the management, the procedure needed to quantify the essence of uncertainty about W while being simple, intuitive, and easy to learn. Therefore, it requires only three estimates: $(y_p, y_{0.5}, y_{1-p})$ for $0 < p < 0.5$. They correspond roughly to three quantiles of W, but Getty analysts dubbed them low view, plan view, and high view estimates, respectively. Estimate $y_{0.5}$ comes from their normal planning process, while the other two estimates are assessed largely judgmentally so that (y_p, y_{1-p}) is interpretable as the $(1 - 2p)$-probability central credible interval (CCI) of W. Initially, $p = 0.05$, giving the 90% CCI; but with experience, the analysts found that $p = 0.1$, giving the 80% CCI, made the assessments easier. They also learned to assess an asymmetric CCI, whenever appropriate.

The parameter estimation. From the three estimates of W, the mean μ and the standard deviation σ of W are calculated via empirically developed formulae based on simulations. For the net income, the normal model is assumed after it had been validated through simulation. (We would hypothesize the four Gaussian models and use theoretic formulae from Section 9.3.2 for each division's W and for the net income.)

9.6.3 Forecasting Precipitation Amount

Precipitation amount, as opposed to precipitation occurrence, remains one of the most difficult meteorological predictands. This motivated the development of a human–computer system to aid a forecaster in the judgmental quantification of uncertainty that involves meteorologic inference and probabilistic reasoning. The system was tested operationally in the Weather Service Forecast Office in Pittsburgh, Pennsylvania, during 5 years, from 1 August 1990 to 31 July 1995 (Krzysztofowicz et al., 1993).

The experts were some 15 professional meteorologists with experience in forecasting weather, who rotated duties at the office. Prior to the experiment, they participated in one-day training in the quantile technique.

The predictand W was the average precipitation amount (in inches) over an area (a river basin) accumulated in 24 hours. The forecast was made twice a day for 24-h periods beginning at 0700 and 1900 local time, for two river basins: (i) the Lower Monongahela River basin above Connellsville, which covers 3429 km^2 in Pennsylvania and Maryland, and (ii) the Upper Allegheny River basin above the Kinzua dam, which covers 5853 km^2 in Pennsylvania and New York.

The predictive information comprised (i) observations from ground stations, radiosondes, satellites, and radar, (ii) guidance forecasts in the form of graphic and alphanumeric outputs from numerical weather prediction models, (iii) conclusions from subjective analyses of local weather data, and (iv) conclusions from comparisons of different model outputs. If there is a consensus among models on the evolution of the weather system, then the degree of the forecaster's certainty about W usually increases. If not, then the larger the disagreements, the smaller the degree of certainty.

The judgmental distribution functions. Three quantiles of W were assessed: $(y_{0.5}, y_{0.25}, y_{0.75})$. They were used in the LS method to estimate the parameters of the Weibull distribution (Section C.3.2). This model had been shown to fit well to the empirical distribution functions of W constructed from climatic samples for river basins in the northeastern United States. The Weibull distribution of W served as an input to probabilistic forecasting of river stages with lead times up to 48 hours. During the 5 years, Pittsburgh meteorologists prepared 6159 forecasts of W. Their verification against the actual realizations of W, and the conclusions, are reported in Chapter 11.

9.7 Judgment, Data, Analytics

We conclude this chapter by addressing the question: Why learn judgmental forecasting in the era of data and analytics? The answer comprises four arguments, which derive from the characteristics of forecasting problems and the needs of human deciders.

1. ***The necessity of judgmental forecasts***. This arises in forecasting problems with little or no data, and with simplistic or no mathematical models — the characteristics identified in Section 9.1.2.
2. ***The assessment of prior distributions***. In some forecasting problems there exist data as well as qualitative information, and it is potentially advantageous to use both. This calls for combining, or fusing, two kinds of predictive information. For this purpose, an expert needs to process mentally the qualitative information I and to assess judgmentally a "prior" distribution function $G(\cdot|I, K, E)$ of predictand W. Next, this function is revised based on data by a Bayesian forecaster — the subject of Chapter 10.
3. ***The adjustment of model forecasts***. This arises in forecasting problems identified in Section 9.5, wherein an expert can either improve a forecast from the model or recognize an erroneous forecast outputted by a faulty model. (For, as experienced analysts caution, "the data are not infallible", no matter how "big" they are.)
4. ***The support for human deciders***. "As data rises, instinct is still the key." Under this heading, *The Wall Street Journal* (19–20 October 2019) reported results of a global survey of business executives. Two-thirds of them admitted to ignoring the output from data analysis or computer model whenever "it contradicted their intuition". The article commented:

> An overreliance on data can also numb the intellect and dull decision-making skills. Multiple studies suggest that managers who bury themselves in data lose their ability to see how their decisions play out in the real world.

The article also quoted Jeff Bezos, the founder of Amazon.com, who opined:

> If you can make a decision with analysis, you should do so. But it turns out in life that your most important decisions are always made with instinct and intuition, taste, heart.

These quotations reaffirm the adage of systems engineers and decision scientists: the analysis should be done not for but with the client. In particular, it should allow for the client's, or his expert's, intuitive input. This is exactly what the judgmental forecasting is for in the domain of uncertainty quantification.

9.8 Concepts and Proofs

9.8.1 Group Decision Making

Suppose a group of M experts indexed by m ($m = 1, \ldots, M$) must collectively make a decision. Beforehand, it must choose a *decision rule* for combining individual decisions into a group decision. This choice, however, constitutes a decision problem in itself and so a philosophical issue of an infinite regress emerges (Plott, 1972, p. 83). A pragmatic solution is to choose a decision rule and then to justify it by demonstrating that it satisfies certain plausible criteria. Two mutually consistent decision rules are justified in this way.

9.8.2 Majority Rule

For a binary choice, that is, the choice between two alternatives A and B, the *majority rule* determines the group decision thusly: A is chosen if {number of votes for A} $\geq t$, where

$$t = \begin{cases} \dfrac{M+1}{2} & \text{if } M \text{ is odd,} \\[2mm] \dfrac{M+2}{2} & \text{if } M \text{ is even.} \end{cases}$$

The majority rule satisfies two plausible criteria of collective decision making. One is that of *decisional equality* (Schofield, 1972, p. 62) which requires that individuals be counted equally in formal voting. The second is that of *decisional disappointment* (Rae, 1969; Curtis, 1972; Schofield, 1972).

Decisional disappointment

In a binary choice problem, each group member m faces four possible outcomes:

    I. m votes for A, but B is chosen,
   II. m votes for B, but A is chosen,
  III. m votes for A, and A is chosen.
  IV. m votes for B, and B is chosen.

Assuming that outcomes I and II are equally disappointing, the probability of disappointment to m is $p_D = P(\mathrm{I}) + P(\mathrm{II})$. Minimization of p_D for an individual, or minimization of the expected number of disappointed individuals, establishes the second plausible criterion for choosing the decision rule. Curtis (1972) proved two facts.

Theorem (*Majority rule*)   *Majority rule minimizes the expected number of disappointed voters.*

Theorem (*Majority rule with exchangeable voters*)   *If all voters are independent of each other and identical in the sense that every voter has, a priori, the same probability p of voting for A (and probability $1 - p$ of voting for B), then the majority rule minimizes the probability of disappointment p_D for every voter.*

9.8.3 Median Rule

For a numerical decision, that is, the decision concerning the value of real variable y from a set $\mathcal{Y}$ (countable or uncountable), the *median rule* determines the group value $\bar{y}$ as a median of the individual values $\{y(m) : m = 1, \dots, M\}$. Specifically, let

$$y_{(1)} \leq y_{(2)} \leq \cdots \leq y_{(M)}$$

be the sequence of M individual values arranged in ascending order, with $y_{(m)}$ denoting the value having rank m ($m = 1, \dots, M$). The median of these values is defined as

$$\bar{y} = \begin{cases} y_{((M+1)/2)} & \text{if } M \text{ is odd,} \\[2mm] \dfrac{1}{2}\left[y_{(M/2)} + y_{((M+2)/2)}\right] & \text{if } M \text{ is even.} \end{cases}$$

Now it will be demonstrated that, in a certain well-defined sense, this median rule is consistent with the majority rule. Let the two alternatives subject to the majority rule voting be defined as subsets of real numbers such that, for any fixed number $\bar{y} \in \mathcal{Y}$,

$$A = \{y : y \leq \bar{y}, y \in \mathcal{Y}\},$$
$$B = \{y : y \geq \bar{y}, y \in \mathcal{Y}\}.$$

The majority rule for choosing between alternatives A and B implies the median rule for choosing the value of y. Formally, for M odd,

$$t = \frac{M+1}{2} \implies y_{(t)} = y_{((M+1)/2)} = \bar{y}.$$

For M even, if the group chooses A, then

$$t = \frac{M+2}{2} \implies y_{(t)} = y_{((M+2)/2)};$$

if the group chooses B, then $t = (M+2)/2$ applied to the reversed order (descending) of the sequence of M individual values implies the rank

$$M - t + 1 = M - \frac{M+2}{2} + 1 = \frac{M}{2},$$

which retrieves value $y_{(M/2)}$. Hence, any number y such that

$$y_{(M/2)} < y < y_{((M+2)/2)}$$

satisfies the majority rule; in particular, number $\bar{y}$.

Bibliographical Notes

The quantile technique has been described and tested by many researchers, notably by Murphy and Winkler (1974), and Winkler and Murphy (1979), who used it effectively with professional forecasters in real forecasting settings. A general overview of various assessment techniques and practical issues experienced by decision analysts working with clients was published by Spetzler and Staël von Holstein (1975). The "warm-up" Task 9.1 parallels the experiment of Alpert and Raiffa (1982).

Mental simulation, the associated heuristics, and the potential biases in probabilities assessed via poorly constructed scenarios were investigated by Kahneman and Tversky (1982). The phenomenon of polarization of post-discussion judgments was described by Myers and Lamm (1975). That the estimate–talk–estimate group process is superior in converging to correct probability was demonstrated experimentally by Gustafson et al. (1973). The formal definition of exchangeability may be found in the books of de Finetti (1974), Berger (1985), and Bernardo and Smith (1994).

Exercises

1 **Combining standard deviation estimates**. Under the hypothesis that $W \sim \mathrm{NM}(\mu, \sigma)$, the quantiles–moments (QM) method of estimating σ recommends taking the geometric mean of estimates σ_1 and σ_2, each calculated via equation (9.5) from a different pair of quantiles.

   1.1 Modify equation (9.5) so that it yields two estimates of the variance of W. Write an equation for their geometric mean.

   1.2 Create a numerical example wherein, for some predictand W, an expert assessed five quantiles $\{y_p : p = c, a, 0.5, b, d\}$ which satisfy approximately the symmetry conditions (9.4). (i) Use these quantiles to calculate $(\sigma_1, \sigma_2, \sigma)$ and $(\sigma_1^2, \sigma_2^2, \sigma^2)$. Are σ and σ^2 coherent? (ii) Recalculate σ and σ^2 using the arithmetic mean. Are σ and σ^2 coherent?

   1.3 Write a logical–mathematical justification for combining two (or more) different estimates of σ or σ^2 via the geometric mean rather than the arithmetic mean.

2 **Probability–quantile assessment**. Let W be an outcome variate from an investment — a monetary amount being a loss ($W < 0$) or a gain ($W > 0$) relative to the status quo ($W = 0$). An investor assessed p_o — the probability of a loss, and y_p — the p-probability quantile of W ($0 < p_o < p < 1, p_o < 0.5$).

2.1 Assuming that $W \sim NM(\mu, \sigma)$, derive the equations for μ and σ in terms of p_o and y_p. *Note*: This is a version of the QM method for parameter estimation.

2.2 Calculate the values of (μ, σ) from the values of (p_o, y_p) specified in the option.
Options:

(A) $p_o = 0.10$, $y_{0.4} = 3000$;   (D) $p_o = 0.11$, $y_{0.9} = 5000$;

(B) $p_o = 0.05$, $y_{0.3} = 4000$;   (E) $p_o = 0.09$, $y_{0.7} = 6000$;

(C) $p_o = 0.3$, $y_{0.8} = 9000$;   (F) $p_o = 0.07$, $y_{0.6} = 8000$.

3 **General knowledge test** — *continued*. This exercise is a continuation of Task 9.1 in which you quantified your uncertainty about three continuous variates. Their realizations, w, are as follows. Eggs: 97 000 millions (in 2018); Precipitation: 45 inches; Year: 1735. Your task now is to model the judgmental distribution function of the variate designated in the option.

3.1 Graph the pointwise representation of your judgmental distribution function (which is specified by the five quantiles you have assessed). Do the quantiles imply a symmetric or asymmetric density function? If asymmetric, is the implied density function skew to the right or skew to the left? Explain your inference.

3.2 Hypothesize a parametric model for your judgmental distribution function; choose it from the family of four Gaussian models (Section 9.3); justify your choice. *Hints*: Check the necessary conditions for the validity of each model; judge the degree to which the conditions are met; for the models on bounded sample spaces, the bounds must be assessed beforehand.

3.3 Write the expressions for the hypothesized distribution function and density function.

3.4 Estimate the location parameter μ and the scale parameter σ of the hypothesized model by the quantiles–moments (QM) method.

3.5 Evaluate the goodness of fit of the hypothesized model; report the MAD.

3.6 Graph the parametric distribution function of the variate, along with the assessed quantiles; indicate the MAD; conclude whether or not this parametric model provides a suitable representation of your judgmental distribution function. Graph the corresponding parametric density function of the variate; conclude whether or not this density function exhibits the symmetry or the asymmetry you postulated in Exercise 3.1.

Options for the variate:

(E) The total number of eggs produced in the USA per year.

(P) The mean annual precipitation in Charlottesville, Virginia.

(Y) The year of birth of John Adams, second president of the USA.

4 **Forecasting demand for new product**. Motor scooters were popular in Europe after World War II, providing inexpensive and convenient personal transportation in cities; movie stars and opera singers rode them in Rome and Paris; Italy and France produced several models. Some seventy years later, their popularity is reviving, especially among young adults. But the phenomenon is recent, and sales data are too sparse for statistical forecasting of future demand.

A motorcycle manufacturing company, having a new motor scooter on the drawing board, hired a marketing expert who combined data with sociological analyses to come up with a judgmental forecast of W — the demand for the new motor scooter in the USA during the next calendar year. The forecast is in terms of five judgmentally assessed quantiles $(y_{0.125}, y_{0.25}, y_{0.5}, y_{0.75}, y_{0.875})$ whose values are listed in the option.

4.1 Graph the pointwise representation of this judgmental distribution function. Do the quantiles imply a symmetric or asymmetric density function? If asymmetric, is the implied density function skew to the right or skew to the left? Explain your inference.

4.2 Hypothesize a parametric model for the judgmental distribution function; choose it from the family of four Gaussian models (Section 9.3); justify your choice. *Hints*: Check the necessary conditions for the validity of each model; judge the degree to which the conditions are met; for the models on bounded sample spaces, the bounds must be assessed beforehand.

4.3 Write the expressions for the hypothesized distribution function and density function.

4.4 Estimate the location parameter μ and the scale parameter σ of the hypothesized model by the quantiles–moments (QM) method.

4.5 Evaluate the goodness of fit of the hypothesized model; report the MAD.

4.6 Graph the parametric distribution function of the predictand superposed on the pointwise distribution function; indicate the MAD; conclude whether or not this parametric model provides a suitable representation of the expert's judgmental distribution function. Graph the corresponding parametric density function of the predictand; conclude whether or not this density function exhibits the symmetry or the asymmetry you postulated in Exercise 4.1.

Options for the assessed quantiles :

    (A) (1200, 2100, 3100, 4200, 5400);   (C) (1800, 2300, 3000, 4100, 5000);

    (B) (2100, 2400, 2900, 3700, 5600);   (D) (2000, 2700, 3400, 3800, 4200).

5 **Quantiles–moments method on sample space** $(-\infty, \infty)$. An expert assessed three quantiles $(y_a, y_{0.5}, y_b)$ of predictand W for probabilities such that $0 < a < 0.5 < b < 1$. The hypothesized parametric model of W is specified in the option.

5.1 Derive equations for the location parameter β and the scale parameter α of the model in terms of the three quantiles and their probabilities. *Hint*: Review the derivation of the QM method for estimating the parameters of the NM(μ, σ) model.

5.2 Show that the standard deviation of W is proportional to the width $(y_b - y_a)$ of the credible interval:

$$\sigma = Var^{1/2}(W) = \frac{y_b - y_a}{c_{ab}};$$

write the formula for constant c_{ab}, which depends on the probabilities (a, b). (Note the parallelism between this expression and the one for σ in the NM(μ, σ) model.)

5.3 Calculate the values of constant c_{ab} for (a, b) equal to $(0.25, 0.75)$ and $(0.125, 0.875)$. Compare them with the values of constant $z_b - z_a$ in the NM(μ, σ) model.

Options for the distribution:

    (LG) $W \sim LG(\beta, \alpha)$;   (GB) $W \sim GB(\beta, \alpha)$;

    (LP) $W \sim LP(\beta, \alpha)$;   (RG) $W \sim RG(\beta, \alpha)$.

6 **Forecasting demand** — Exercise 5 *continued*. The forecasting problem is as described in Exercise 4. Your task is to apply the parametric model for which you derived the QM equations in Exercise 5. The values of the five assessed quantiles $(y_{0.125}, y_{0.25}, y_{0.5}, y_{0.75}, y_{0.875})$ are listed in the option. *Note*: This option must match the option that was designated for Exercise 5.

6.1 Graph the pointwise representation of the judgmental distribution function G of W.

6.2 Hypothesize the parametric model for G — the one you worked with in Exercise 5. Write the expressions for its distribution function and density function.

6.3 Estimate the location and the scale parameters by the quantiles–moments (QM) method whose equations you derived in Exercise 5. *Hint*: Consult the QM method in Section 9.3.2 regarding the estimation from five quantiles.

6.4 Evaluate the goodness of fit of the hypothesized model; report the MAD.

6.5 Graph the parametric distribution function of W superposed on the pointwise distribution function; indicate the MAD; conclude whether or not this parametric model provides a suitable representation of the expert's judgmental distribution function. Graph the corresponding parametric density function of W.

Options for the assessed quantiles:

(LG)	(1000, 1800, 2600, 3500, 4400);		(GB)	(1700, 1900, 2300, 3000, 4500);
(LP)	(1200, 1700, 2500, 3400, 4100);		(RG)	(1600, 2200, 2700, 3000, 3400).

7 **Quantiles–moments method on sample space** (η, ∞). An expert assessed three quantiles $(y_a, y_{0.5}, y_b)$ of predictand W for probabilities such that $0 < a < 0.5 < b < 1$. The hypothesized parametric model of W is specified in the option. Derive equations for the scale parameter α and the shape parameter β of the model in terms of the three quantiles and their probabilities.

Hints. (i) Inasmuch as the sample space $\mathcal{W} = (\eta, \infty)$ must be specified beforehand, the shift parameter η should be assumed to have a known value. (ii) Consult the notes for the parametric model in Section C.3 regarding the transform of the variate through which the model is derived. (iii) Review the equations of the QM method for the Gaussian models in Section 9.3.2 and note the use of the transforms.

Options for the distribution:

(WB)	$W \sim \text{WB}(\alpha, \beta, \eta)$;		(LW)	$W \sim \text{LW}(\alpha, \beta, \eta)$;
(IW)	$W \sim \text{IW}(\alpha, \beta, \eta)$;		(LL)	$W \sim \text{LL}(\alpha, \beta, \eta)$.

8 **Forecasting demand** — Exercise 7 *continued*. The forecasting problem is as described in Exercise 4. Your task is to apply the parametric model for which you derived the QM equations in Exercise 7. The values of the five assessed quantiles $(y_{0.125}, y_{0.25}, y_{0.5}, y_{0.75}, y_{0.875})$ are listed in the option. *Note*: This option must match the option that was designated for Exercise 7.

8.1 Graph the pointwise representation of the judgmental distribution function G of W.

8.2 Hypothesize the parametric model for G — the one you worked with in Exercise 7. Write the expressions for its distribution function and density function. Specify the sample space $\mathcal{W}$ of W; assess its lower bound η (see Section 3.3.2).

8.3 Estimate the scale and the shape parameters by the quantiles–moments (QM) method whose equations you derived in Exercise 7. *Hint*: Consult the QM method in Section 9.3.2 regarding the estimation from five quantiles.

8.4 Evaluate the goodness of fit of the hypothesized model; report the MAD.

8.5 Graph the parametric distribution function of W superposed on the pointwise distribution function; indicate the MAD; conclude whether or not this parametric model provides a suitable representation of the expert's judgmental distribution function. Graph the corresponding parametric density function of W.

Options for the assessed quantiles:

(WB)	(1000, 1200, 1600, 2000, 3000);		(LW)	(1400, 1900, 2400, 3000, 3600);
(IW)	(1300, 1800, 2300, 3200, 6400);		(LL)	(1200, 1500, 1900, 2500, 3200).

9 **Adjusting forecast of lead time**. A retailer having stores throughout the continental United States orders merchandise from manufacturers in Southeast Asia. The *lead time* of an order, W, is the number of days elapsed between the date of placing the order and the date of receiving the merchandise at the warehouse in Los Angeles. Based on a sample of realizations of predictand W, collected during 3 years of relatively stationary global trade activities prior to 2020, the distribution function G of W was modeled and estimated (as prescribed in Chapter 3); it is specified in the option.

But the pandemic, which began in 2020, abruptly changed the dynamics of manufacturing and shipping activities worldwide — resulting in delays. The chief merchandising officer (CMO), responsible for managing the retailer's inventory, suddenly faced novel uncertainty and no past data to quantify it. After consulting with the manufactures, shippers, and analysts, the CMO judgmentally adjusted three quantiles $\{w_p : p = 0.5, 0.2, 0.8\}$ of W to account for new information thusly: 65 days were added to $w_{0.5}$, 45 days were added to $w_{0.2}$, and 95 days were added to $w_{0.8}$.

  9.1 Apply the distribution adjustment procedure of Section 9.5 to obtain the revised parametric model for G. For estimation of the parameters, use the method specified in the option. Report the values of (i) the original quantiles w_p of W calculated from the original distribution function G; (ii) the adjusted quantiles y_p; (iii) the parameters of the revised distribution function G; and (iv) the revised quantiles w_p of W calculated from the revised G.

  9.2 Graph in one figure (i) the two parametric distribution functions of W (the original and the revised); and (ii) the three adjusted points $\{(y_p, p): p = 0.2, 0.5, 0.8\}$ from which the revised parameters were estimated. (iii) Characterize in words the impact of the three judgmental adjustments on the shape of G.

  9.3 The CMO's preference is to have probability 0.9 of receiving the merchandise for the Christmas season no later than 1 November. What was the latest date to place the orders that satisfy this preference? Calculate it for orders placed (i) before the pandemic, and (ii) during the pandemic.

Options for the distribution:

  (WB) WB(22, 0.81, 61);   (LW) LW(2.96, 2.57, 63);

  (IW) IW(9.7, 1.23, 62);   (LL) LL(11, 1.32, 64).

Options for the estimation method:

  (LS) Least squares method.

  (UD) Uniform distance method (using DFit or other software).

Mini–Projects

10 **Score from the final exam**. Suppose the final examination at the end of this course will be written, closed-book, and comprehensive — asking for definitions, assumptions, theorems, proofs, derivations, expressions, algorithms, properties, interpretations. The score will measure the percentage of correct and complete answers. Plan your study in preparation for this examination; then make a probabilistic forecast of your score. Toward this end, perform and report the following.

  10.1 Identify all the variables that may influence your performance on this examination and thereby your score; construct an influence diagram.

  10.2 Make a probabilistic forecast of your score: assess five quantiles of the score corresponding to the probabilities 0.125, 0.25, 0.5, 0.75, 0.875.

  10.3 Graph the pointwise representation of your judgmental distribution function (which is specified by the five quantiles you have assessed). Do the quantiles imply a symmetric or asymmetric density function of the score? If asymmetric, is the implied density function skew to the right or skew to the left? Explain your inference.

10.4 Hypothesize two parametric models for your judgmental distribution function; these should be models whose sample space is a bounded open interval (Sections 9.3.2, C.4); justify your hypotheses. *Hint*: Narrow the hypotheses by following the guidelines from Section 3.4.2.

10.5 Write the expressions for the hypothesized distribution functions and density functions. Assess the bounds of the sample space. Estimate the parameters of each hypothesized model by the method specified in the option.

10.6 Evaluate the goodness of fit of each hypothesized model; report the MAD. Choose the better of the two models, and use it in the subsequent exercises.

10.7 Graph the parametric distribution function of the score, along with the assessed quantiles; indicate the MAD; conclude whether or not this parametric model provides a suitable representation of your judgmental distribution function. Graph the corresponding parametric density function of the score; conclude whether or not this density function exhibits the symmetry or the asymmetry you postulated in Exercise 10.3.

10.8 Suppose the mapping of the score W to the letter grade is as follows: $W \in [90, 100] \Rightarrow$ A, $W \in [80, 90) \Rightarrow$ B, $W \in [70, 80) \Rightarrow$ C, $W \in [60, 70) \Rightarrow$ D, $W \in [0, 60) \Rightarrow$ F. Calculate the probability of receiving each letter grade.

10.9 Calculate the score value that has the probability $1 - p$ of being exceeded; perform the calculations for $p = 0.1, 0.3, 0.5, 0.7, 0.9$. Identify the median of the score, the 40% central credible interval, and the 80% central credible interval. *Hint*: Use the expression for the quantile function corresponding to the distribution function.

Options for the estimation method:

(LS) Least squares method.

(UD) Uniform distance method (using DFit or other software).

11 **Forecast of DJIA**. For five consecutive business days, gather news from the US stock market and record daily movements of the Dow Jones Industrial Average (DJIA): open, close, low, high values. (*The Wall Street Journal*, Section B: Business & Finance, is one source of such information.) On the last day of this information-gathering period, consider the predictand W — the closing DJIA value after the next three trading days.

11.1 Organize and analyze all predictive information you have gathered. Summarize it in a written, tabular, or graphical format.

11.2 Make your judgmental forecast of W in terms of five quantiles.

11.3 Graph the pointwise representation of your judgmental distribution function. Do the quantiles imply a symmetric or asymmetric density function of the predictand? If asymmetric, is the implied density function skew to the right or skew to the left? Explain your inference.

11.4 Hypothesize two parametric models for your judgmental distribution function; consider all models from Section 9.3 and Appendix C; justify your hypotheses. *Hint*: Narrow the hypotheses by following the guidelines from Section 3.4.2.

11.5 Write the expressions for the hypothesized distribution functions and density functions. If the model has a bounded sample space, then assess the bounds. Estimate the parameters of each hypothesized model by the method specified in the option.

11.6 Evaluate the goodness of fit of each hypothesized model; report the MAD. Choose the better of the two models, and use it in the next exercise.

11.7 Graph the parametric distribution function of the predictand, along with the assessed quantiles; indicate the MAD; conclude whether or not this parametric model provides a suitable representation of your judgmental distribution function. Graph the corresponding parametric density function of the predictand; conclude whether or not this density function exhibits the symmetry or the asymmetry you postulated in Exercise 11.3.

11.8 Observe the realization of W, report its value, and indicate it on the graphs of the parametric distribution and density functions.

Options for the estimation method:

(LS) Least squares method.

(UD) Uniform distance method (using DFit or other software).

12 Group forecast of market value. A manufacturer of supercomputers wants to acquire a credit company through which it could offer direct and customized financing of its products to prospective buyers. Four accountants have been poring over the financial statements and accounting books of the credit company to determine its market value, which is not apparent because of the many complex transactions involving financial derivatives. At the end of a 5-day marathon, each of the four accountants presented his estimate of the company's value. The estimates are vastly different. You are called to serve as a facilitator. You study Section 9.4 of this chapter and proceed as follows.

12.1 You explain to the accountants the concept of a judgmental distribution function, and ask each of them to assess quantiles $(y_{0.05}, y_{0.25}, y_{0.5}, y_{0.75}, y_{0.95})$ of the company's value. Write instructions for assessing $y_{0.5}, y_{0.25}, y_{0.05}$.

12.2 The quantiles (in millions of dollars) assessed by the accountants are specified in the option. (Accountants C and D had to depart before completing their assessments; hence the dots.) Combine the individual quantiles according to the median rule.

12.3 Graph the pointwise representations of the five judgmental distribution functions, each specified by four or five quantiles: four functions from the individuals and one function from the group. (You may connect the points of each function with line segments.)

12.4 Hypothesize two parametric models for the group distribution function; choose them from the family of four Gaussian models (Section 9.3) in two ways. (i) First, narrow the hypotheses by following the guidelines from Section 3.4.2. (ii) Second, apply the necessary conditions for the validity of each model; judge the degree to which the conditions are met to narrow the hypotheses. *Hint*: For the models on bounded sample spaces, the bounds must be assessed beforehand. (iii) Third, compare the hypotheses arrived at in two ways. Are they the same? If not, then narrow them further.

12.5 Write the expressions for the hypothesized distribution functions and density functions. Estimate the parameters of each hypothesized model by the method specified in the option.

12.6 Evaluate the goodness of fit of each hypothesized model; report the MAD. Choose the better of the two models, and use it in the next exercise.

12.7 Graph the parametric distribution function superposed on the pointwise distribution function; indicate the MAD; conclude whether or not this parametric model provides a suitable representation of the judgmental group distribution function. Graph the corresponding parametric density function.

12.8 Just as you finish the analysis, you receive a telephone call. The manufacturer's board of directors is about to offer $32 million for the credit company, and wants to know the probability of the company's value being below this offer. Provide the answer.

12.9 A few minutes later, you receive a second question: What would be the largest offer guaranteeing the probability 0.6 of the company's value being above the offer. Provide the answer.

Options for the team of accountants:

(T1)	Accountant A	$(7, 15, 19, 22, 26)$	(T3)	Accountant A	$(10, 15, 18, 28, 42)$
	Accountant B	$(-19, -3, 13, 36, 59)$		Accountant B	$(12, 19, 22, 24, 32)$
	Accountant C	$(\ldots, 12, 23, 28, 43)$		Accountant C	$(\ldots, 17, 28, 31, 38)$
	Accountant D	$(-8, 4, 11, 20, \ldots)$		Accountant D	$(9, 12, 16, 26, \ldots)$
(T2)	Accountant A	$(-6, 6, 12, 19, 38)$	(T4)	Accountant A	$(14, 30, 32, 34, 39)$
	Accountant B	$(-2, 3, 7, 12, 33)$		Accountant B	$(11, 24, 41, 43, 47)$
	Accountant C	$(\ldots, -4, 1, 8, 22)$		Accountant C	$(\ldots, 33, 36, 41, 44)$
	Accountant D	$(-9, 1, 3, 7, \ldots)$		Accountant D	$(8, 18, 28, 35, \ldots)$

Options for the estimation method:

(QM) Quantiles–moments method.

(LS) Least squares method.

(UD) Uniform distance method (using DFit or other software).

13 **Forecast of time needed to launch a company**. Aspiring entrepreneurs exhibit a tendency to underestimate the amount of venture capital and the length of time needed to launch a company. Driven by passion for their ideas about new products or services, they let their judgments constrict: they envision only optimistic scenarios, unrepresentative of all possible sources of uncertainty (the roadblocks, so to speak). Our objective is to formulate a method for forecasting probabilistically W — the time needed to launch a company (in short, the execution time).

Noam Wasserman, professor at Harvard Business School, surveyed 100 Business School graduates who founded companies during the past 10 years (How an entrepreneur's passion can destroy a startup, *The Wall Street Journal*, 25 August 2014, pp. R1–R2). One of his questions was: How the actual time needed to start a company compared with the planned time. In 17% of cases the actual time was shorter, in 4% on target, in 79% longer, and in 41% at least twice as long. Could this information about experience of 100 founders be used in forecasting W by an aspiring entrepreneur? Exercises 13.1–13.6 explain a method and ask you to implement it. Then Exercise 13.7 asks you to modify this method and to implement it.

13.1 *Assumptions.* Let W be a continuous variate on a bounded sample space $\mathcal{W} = (\eta_L, \eta_U)$, and let y be an estimate of W. The percentages from the survey are interpreted as statistical probabilities of events; however, $P(W = y) = 0.04$ is invalid for the continuous W; also, the answer "on target" was most likely approximate. Hence, the probability 0.04 is assigned to an interval of W centered on y; let it be $(0.9y, 1.1y]$. With these assumptions, the survey and the estimate y for a particular company yield three points of the distribution function G of W:

$P(W \le 0.9\,y) = 0.17,$

$P(W \le 1.1\,y) = 0.17 + 0.04 = 0.21,$

$P(W \le 2\,y) = 1 - 0.41 = 0.59.$

The estimate y of the execution time forecasted turns out to be, approximately, the 0.19-probability quantile of W — the evidence of a huge underestimation bias.

13.2 *Assessments.* Suppose that when preparing the business plan an aspiring entrepreneur assesses judgmentally the estimate y of W and the bounds of its sample space: η_L below the most optimistic execution time, and η_U above the most pessimistic execution time. The values assessed by an entrepreneur are listed in the option. Together with the above three probabilities, they specify the pointwise representation of the *judgmental–statistical* distribution function G of W. Graph it.

13.3 *Modeling.* After perusing the graphs of the distribution functions in Appendix C, the class of log-ratio distributions (Section C.4) was selected as the source of hypothesized models because it offers a variety of shapes on the bounded open interval $\mathcal{W} = (\eta_L, \eta_U)$. The hypothesized model is specified in the option. Write the expressions for the hypothesized distribution function and density function.

13.4 *Estimation.* Estimate the parameters of the hypothesized model by the method specified in the option.

13.5 *Evaluation.* Evaluate the goodness of fit of the hypothesized model: report the MAD.

13.6 *Results.* Graph the parametric distribution function of W superposed on the pointwise representation of the judgmental–statistical distribution function. Graph the corresponding parametric density function of W. Use the model to calculate the quantiles of the time needed to launch the company: $w_{0.1}$, $w_{0.25}$, $w_{0.5}$, $w_{0.75}$, $w_{0.9}$. Draw conclusions.

13.7 *Modification.* Modify the above method in any way you think is reasonable. The modification may pertain to the assumptions, the assessed estimates, the hypothesized parametric model, or to any other element. Keep in mind the objective: to formulate a method for forecasting W that combines the statistical probabilities from the survey with the judgmentally assessed estimates of W. (i) Describe and rationalize your modification. (ii) Implement it for a hypothetical entrepreneur. (iii) Compare the results with those from Exercise 13.6. (iv) Draw conclusions.

Note. Exercises 13.1–13.2 omit the probabilistic interpretation of y. This is a deficiency — the result of a common business practice to produce estimates which are undefined mathematically, and are called casually (and meaninglessly) the "best" estimates or the "most likely" estimates. Consequently, there is a tacit assumption in Exercise 13.2 that the aspiring entrepreneur assesses y in a way compatible with the ways the survey respondents assessed theirs, so that the statistical probabilities from the survey apply to a future start-up. (To replace y with a mathematically defined and properly assessed quantile y_p of W would be our next objective. This is one modification you may consider.)

Options for assessed estimates (in months):

(A) $y = 23$, $\eta_L = 12$, $\eta_U = 66$; (C) $y = 35$, $\eta_L = 18$, $\eta_U = 105$;

(B) $y = 29$, $\eta_L = 15$, $\eta_U = 84$; (D) $y = 41$, $\eta_L = 21$, $\eta_U = 120$.

Options for the distributions:

(LR-LG) $W \sim LR1(\eta_L, \eta_U)\text{-}LG(\beta, \alpha)$.

(LR-LP) $W \sim LR1(\eta_L, \eta_U)\text{-}LP(\beta, \alpha)$.

(LR-GB) $W \sim LR1(\eta_L, \eta_U)\text{-}GB(\beta, \alpha)$.

(LR-RG) $W \sim LR1(\eta_L, \eta_U)\text{-}RG(\beta, \alpha)$.

Options for the estimation method:

(LS) Least squares method.

(UD) Uniform distance method (using DFit or other software).

10

Statistical Forecasting

A *continuous predictand* (such as the demand volume for a product, price of a commodity, volume of runoff) is to be forecasted *repeatedly* with a specified *lead time*. A continuous *predictor* is available — a variate containing predictive information or being itself a point estimator of the predictand (such as the output from a deterministic predictive model). Records of past realizations of the predictand and the predictor exist. How can these records be used to build a statistical model for forecasting (i.e., for quantifying the uncertainty about) future realizations of the predictand?

This chapter lays down a Bayesian theory for statistical forecasting of a continuous predictand. First, the theoretic structure of the *Bayesian forecaster* is formulated. Next, a parametric model, called the Bayesian Gaussian forecaster (BGF), is derived. Subsequently, procedures are presented for parameter estimation, model validation, predictor selection, communication of forecasts, and forecasting the sum.

10.1 Bayesian Forecaster

10.1.1 Variates

Let W be the *predictand* — a variate whose realization must be forecasted. It is a continuous variate having the sample space $\mathcal{W}$; its realization is denoted w, where $w \in \mathcal{W}$.

Let X be the *predictor* — a variate whose realization provides information that reduces uncertainty about the predictand. Suppose X is a continuous variate having the sample space $\mathcal{X}$; its realization is denoted x, where $x \in \mathcal{X}$.

10.1.2 Input Elements

With p denoting a generic density function, the inputs into the Bayesian forecaster are defined as follows.

$g(w) = p(w)$ for all $w \in \mathcal{W}$. The function g is the *prior density function* of the predictand W. This function quantifies the uncertainty about the predictand W that exists before a realization of the predictor X is available. Equivalently, it characterizes the natural variability of the predictand.

$f(x|w) = p(x|W = w)$ for all $x \in \mathcal{X}$ and $w \in \mathcal{W}$. For a fixed $w \in \mathcal{W}$, the object $f(\cdot|w)$ is the *density function* of the predictor X, **conditional on the hypothesis** that the actual realization of the predictand is $W = w$. For a fixed $x \in \mathcal{X}$, the object $f(x|\cdot)$ is the *likelihood function* of the predictand W, **conditional on a realization** of the predictor $X = x$. All possible realizations x of the predictor (and there are infinitely many of them), generate a *family of likelihood functions*, denoted by f. The family of likelihood functions quantifies the stochastic dependence between predictor X and predictand W. Equivalently, it characterizes the informativeness of the predictor with respect to the predictand. The term "informativeness" is defined precisely in Section 10.4.

Probabilistic Forecasts and Optimal Decisions, First Edition. Roman Krzysztofowicz.
© 2025 John Wiley & Sons Ltd. Published 2025 by John Wiley & Sons Ltd.
Companion website: www.wiley.com/go/ProbabilisticForecastsandOptimalDecisions1e

10.1.3 Output Elements

The outputs from the Bayesian forecaster are defined as follows.

$\kappa(x) = p(x)$ for all $x \in \mathcal{X}$. The function κ is the *expected density function* of the predictor X; it may also be called the marginal density function of X.

$\phi(w|x) = p(w|X = x)$ for all $w \in \mathcal{W}$ and $x \in \mathcal{X}$. For a fixed $x \in \mathcal{X}$, the object $\phi(\cdot|x)$ is the *posterior density function* of the predictand W, **conditional on a realization** of the predictor $X = x$.

10.1.4 Theoretic Structure

The prior density function g and the family of likelihood functions f carry information about the prior uncertainty and the informativeness of the predictor into the Bayesian forecaster. The output from the Bayesian forecaster is derived in two steps.

First, the *expected density function* of predictor X is obtained from the *total probability law*; at any point $x \in \mathcal{X}$,

$$\kappa(x) = \int_{\mathcal{W}} f(x|w)g(w)\, dw. \tag{10.1}$$

Second, the *posterior density function* of the predictand W, conditional on a realization of the predictor $X = x$, is obtained from *Bayes' theorem*; at any point $w \in \mathcal{W}$,

$$\phi(w|x) = \frac{f(x|w)g(w)}{\kappa(x)}. \tag{10.2}$$

Equations (10.1)–(10.2) define the theoretic structure of the *Bayesian forecaster* of a continuous predictand. In essence, this Bayesian forecaster revises the prior density function g of the predictand W based on a given realization of the predictor $X = x$. The output of this revision is the posterior density function $\phi(\cdot|x)$, which constitutes a probabilistic forecast of the predictand W.

The primary challenge en route to an operational technique is modeling and estimation of functions g and f, and then derivation of expressions for functions κ and ϕ. Figure 10.1 shows a schema of this process. We shall follow it to obtain a particular model.

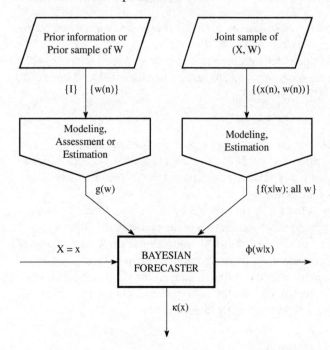

Figure 10.1 A schema of the Bayesian forecaster of continuous predictand W using continuous predictor X.

10.2 Bayesian Gaussian Forecaster

One convenient and useful implementation of the Bayesian forecaster occurs when all density functions are Gaussian. Thus each sample space is the unbounded interval: $\mathcal{W} = (-\infty, \infty)$ and $\mathcal{X} = (-\infty, \infty)$.

10.2.1 Prior Density Function

Let the prior mean and the prior variance of the predictand W be denoted by

$$E(W) = M,$$
$$Var(W) = S^2.$$
(10.3)

The prior density function g is assumed to be normal, so that at any point w $(-\infty < w < \infty)$,

$$g(w) = \frac{1}{S\sqrt{2\pi}} \exp\left(-\frac{1}{2}\left(\frac{w - M}{S}\right)^2\right).$$
(10.4)

10.2.2 Family of Conditional Density Functions

The stochastic dependence between predictor X and predictand W is assumed to be described by a *normal-linear model*: conditional on the hypothesis that the actual realization of the predictand is $W = w$, for any w $(-\infty < w < \infty)$,

$$X = aw + b + \Theta,$$
(10.5)

where a, b are parameters, and Θ is a variate (called the *noise*, the *disturbance*, or the *residual*) being stochastically independent of W, and having a normal density function with zero mean and positive variance ($\sigma^2 > 0$),

$$E(\Theta) = 0,$$
$$Var(\Theta) = \sigma^2.$$
(10.6)

It follows that given any hypothesis $W = w$ $(-\infty < w < \infty)$, the conditional mean and the conditional variance of predictor X are

$$E(X|W = w) = aw + b,$$
$$Var(X|W = w) = \sigma^2;$$
(10.7)

and the conditional density function $f(\cdot|w)$ of predictor X is normal, so that at any point x $(-\infty < x < \infty)$,

$$f(x|w) = \frac{1}{\sigma\sqrt{2\pi}} \exp\left(-\frac{1}{2}\left(\frac{x - aw - b}{\sigma}\right)^2\right).$$
(10.8)

The derivation of this equation is the subject of Exercise 1.

10.2.3 Expected Density Function

When expressions (10.4) and (10.8) are inserted into the total probability law (10.1) and the integral is solved, the result is as follows. The mean and the variance of the predictor X are

$$E(X) = aM + b,$$
$$Var(X) = a^2S^2 + \sigma^2;$$
(10.9)

and the expected density function κ of the predictor X is normal, so that at any point x $(-\infty < x < \infty)$,

$$\kappa(x) = \frac{1}{\sqrt{a^2S^2 + \sigma^2}\sqrt{2\pi}} \exp\left(-\frac{1}{2}\left(\frac{x - aM - b}{\sqrt{a^2S^2 + \sigma^2}}\right)^2\right). \tag{10.10}$$

An alternative derivation of this equation is the subject of Exercise 2.

10.2.4 Posterior Density Function

When expressions (10.4), (10.8), and (10.10) are inserted into Bayes' theorem (10.2) and the terms are combined appropriately, the result is as follows. The prior parameters M, S and the likelihood parameters a, b, σ yield three posterior parameters A, B, T as follows:

$$A = \frac{aS^2}{a^2S^2 + \sigma^2}, \qquad B = \frac{M\sigma^2 - abS^2}{a^2S^2 + \sigma^2}, \tag{10.11}$$

$$T^2 = \frac{\sigma^2 S^2}{a^2S^2 + \sigma^2}. \tag{10.12}$$

These parameters, in turn, specify the posterior mean and the posterior variance of the predictand W:

$$E(W|X = x) = Ax + B,$$
$$Var(W|X = x) = T^2. \tag{10.13}$$

Finally, the posterior density function $\phi(\cdot|x)$ of the predictand W is normal, so that at any point w $(-\infty < w < \infty)$,

$$\phi(w|x) = \frac{1}{T\sqrt{2\pi}} \exp\left(-\frac{1}{2}\left(\frac{w - Ax - B}{T}\right)^2\right). \tag{10.14}$$

10.2.5 Distribution Functions

With P denoting a probability, the distribution functions corresponding to the density functions g and $\phi(\cdot|x)$ are defined as follows.

$G(w) = P(W \leq w)$ for all $w \in \mathcal{W}$. The function G is the *prior distribution function* of the predictand W.

$\Phi(w|x) = P(W \leq w|X = x)$ for all $w \in \mathcal{W}$ and $x \in \mathcal{X}$. For a fixed $x \in \mathcal{X}$, the object $\Phi(\cdot|x)$ is the *posterior distribution function* of the predictand W, **conditional on a realization** of the predictor $X = x$.

With Q denoting the standard normal distribution function (which is defined in Section 2.6.1, and whose approximation is given in Table 2.2), the prior distribution function is specified by the expression

$$G(w) = Q\left(\frac{w - M}{S}\right), \tag{10.15}$$

and the posterior distribution function is specified by the expression

$$\Phi(w|x) = Q\left(\frac{w - Ax - B}{T}\right). \tag{10.16}$$

10.2.6 Quantile Functions

Let Q^{-1} denote the standard normal quantile function, the inverse of the function Q (which is defined in Section 2.6.1, and whose approximation is given in Table 2.3).

Definition   For any number p $(0 < p < 1)$, the *p-probability prior quantile* of W is the point $w_p \in \mathcal{W}$ such that $G(w_p) = p$. Equivalently, the point w_p is such that $w_p = G^{-1}(p)$, and G^{-1} is called the *prior quantile function* of W.

Replacing $G(w_p)$ with expression (10.15) yields

$$Q\left(\frac{w_p - M}{S}\right) = p,$$

$$\frac{w_p - M}{S} - Q^{-1}(p).$$

Therefrom the prior quantile function of W takes the form

$$w_p = G^{-1}(p)$$

$$= M + SQ^{-1}(p)$$

$$= M + Sz_p, \tag{10.17}$$

where $z_p = Q^{-1}(p)$ is the p-probability standard normal quantile. When $p = 0.5$, we have $z_{0.5} = 0$ and consequently $w_{0.5} = M$; that is, the prior median of W equals the prior mean of W.

Definition   For any number p $(0 < p < 1)$, the *p-probability posterior quantile* of W, conditional on $X = x$, is the point $w_p \in \mathcal{W}$ such that $\Phi(w_p|x) = p$. Equivalently, the point w_p is such that $w_p = \Phi^{-1}(p|x)$, and $\Phi^{-1}(\cdot|x)$ is called the *posterior quantile function* of W, conditional on $X = x$.

Replacing $\Phi(w_p|x)$ with expression (10.16) yields

$$Q\left(\frac{w_p - Ax - B}{T}\right) = p,$$

$$\frac{w_p - Ax - B}{T} = Q^{-1}(p).$$

Therefrom the posterior quantile function of W takes the form

$$w_p = \Phi^{-1}(p|x)$$

$$= Ax + B + TQ^{-1}(p),$$

$$= Ax + B + Tz_p, \tag{10.18}$$

where $z_p = Q^{-1}(p)$ is the p-probability standard normal quantile. This shows that for a fixed p, the posterior quantile w_p is a linear function of the predictor realization x; this function is obtained by shifting the posterior mean $Ax + B$ vertically by Tz_p units. When $p = 0.5$, we have $z_{0.5} = 0$ and consequently $w_{0.5} = Ax + B$; that is, the posterior median of W equals the posterior mean of W.

10.2.7 Central Credible Intervals

Definition   For any number p $(0 < p < 0.5)$, the $(1 - 2p)$-*probability central credible interval* (CCI) of W is the bounded open interval (w_p, w_{1-p}).

This interval is central because the probabilities p and $1 - p$ are symmetrical about probability 0.5. The CCI may be used to communicate uncertainty, as detailed in Section 10.5. The *width* of the CCI, $w_{1-p} - w_p$, is a measure of uncertainty. Using equation (10.17), one can find the width of the $(1 - 2p)$-*probability prior* CCI:

$$w_{1-p} - w_p = S(z_{1-p} - z_p)$$
$$= 2Sz_{1-p}, \tag{10.19}$$

because $z_p = -z_{1-p}$ (Section 2.6.1). Using equation (10.18), one can find the width of the $(1 - 2p)$-*probability posterior* CCI:

$$w_{1-p} - w_p = T(z_{1-p} - z_p)$$
$$= 2Tz_{1-p}. \tag{10.20}$$

This posterior CCI has two properties. (i) Its width is independent of the predictor realization x, an implication of the homoscedasticity of the posterior distribution function. (ii) It is never wider than the prior CCI because $T \leq S$, as demonstrated in Section 10.4.

10.3 Estimation and Validation

The Bayesian Gaussian forecaster requires values of five parameters: two prior parameters (M, S) and three likelihood parameters (a, b, σ), which need to be estimated. It is derived from four assumptions: two distributional assumptions and two structural assumptions, which should be validated.

10.3.1 Estimation of Prior Parameters

When the prior distribution function G is assessed judgmentally via the quantile technique, the values of M and S are calculated from the assessed quantiles as described in Section 9.3.2.

When a record of past realizations of the predictand W is available, it should be organized into a *prior sample*,

$$\{w(n) : n = 1, \ldots, N'\},$$

from which M and S can be estimated. We employ the *maximum likelihood estimates* (which are equal to the sample mean and the sample variance):

$$\hat{M} = \frac{1}{N'} \sum_{n=1}^{N'} w(n), \tag{10.21a}$$

$$\hat{S}^2 = \frac{1}{N'} \sum_{n=1}^{N'} [w(n) - \hat{M}]^2. \tag{10.21b}$$

10.3.2 Estimation of Likelihood Parameters

The available record of past realizations of predictor X and predictand W should be organized into a *joint sample*,

$$\{(x(n), w(n)) : n = 1, \ldots, N\},$$

from which a, b, σ can be estimated. We employ the *maximum likelihood estimates* (which for a, b are equal to the least squares estimates):

$$\bar{x} = \frac{1}{N} \sum_{n=1}^{N} x(n), \qquad \bar{w} = \frac{1}{N} \sum_{n=1}^{N} w(n), \tag{10.22a}$$

$$\hat{a} = \frac{\sum_{n=1}^{N} x(n)w(n) - N\bar{x}\,\bar{w}}{\sum_{n=1}^{N} w^2(n) - N\bar{w}^2}, \qquad \hat{b} = \bar{x} - \hat{a}\bar{w}, \tag{10.22b}$$

$$\hat{\sigma}^2 = \frac{1}{N} \sum_{n=1}^{N} \theta^2(n). \tag{10.22c}$$

Before executing the last equation, one must obtain a sample of residuals:

$$\{\theta(n) : n = 1, \dots, N\},$$

where for every n ($n = 1, \dots, N$) the residual $\theta(n)$ is calculated from the relationship

$$\theta(n) = x(n) - \hat{a}w(n) - \hat{b}. \tag{10.23}$$

This relationship follows from equation (10.5) by solving it for the residual variate Θ, substituting parameters a, b by their estimates $\hat{a}, \hat{b}$, and then applying it to every realization $(x(n), w(n))$ in the joint sample to get every realization $\theta(n)$.

10.3.3 Validation of Assumptions

The BGF derives from four assumptions: two assumptions about marginal distribution functions, and two assumptions about the structure of the stochastic dependence. Once the parameters have been estimated, these four assumptions should be validated.

1. *Normality*: $W \sim NM(M, S)$. The prior distribution function G of the predictand W must be normal (Gaussian). This can be validated in three ways:
   (a) Graph the parametric distribution function G together with either the judgmental distribution function or the empirical distribution function, as appropriate. Then judge visually the goodness of fit.
   (b) Calculate the MAD (Section 3.6).
   (c) Perform the Kolmogorov-Smirnov goodness-of-fit test (Section 3.6), if G is estimated from a prior sample.
2. *Normality*: $\Theta \sim NM(0, \sigma)$. The distribution function of the residual variate Θ must be normal (Gaussian). This can be validated in three ways, as described above.
3. *Linearity*: $E(X|W = w) = aw + b$. The regression of X on W must be linear. This can be validated by superposing the estimated regression line on the scatterplot of the joint sample, $x(n)$ versus $w(n)$ for all n ($n = 1, \dots, N$), and judging visually whether or not the contours of the scatterplot form ellipses that hug the regression line.
4. *Homoscedasticity*: $Var(X|W = w) = Var(\Theta) = \sigma^2$. The variance of the residual variate Θ must be constant for all values w. This can be validated by making a scatterplot of $\theta(n)$ versus $w(n)$ for all n ($n = 1, \dots, N$), and judging visually whether or not the scatter of points about the horizontal axis (recall that the mean of the residual variate is zero) is (i) random on both sides, positive and negative, and (ii) approximately the same for all values of w.

It is also prudent to check the correctness of the numerical calculations by validating the assumptions that $E(\Theta) = 0$, and that $E(W\Theta) = 0$ because W and Θ must be stochastically independent of each other. These two assumptions are satisfied empirically if

$$\sum_{n=1}^{N} \theta(n) = 0, \qquad \sum_{n=1}^{N} w(n)\theta(n) = 0.$$

10.3.4 Fusion of Information

Reflecting on the above procedures, it is apparent that the prior distribution function and the family of likelihood functions are estimated and validated independently of each other. In a broad sense, the BGF allows us to *fuse information* from different sources, or from samples of different sizes — a phenomenal advantage because there are forecasting problems in which the joint sample is smaller than the prior sample: $N < N'$. This is called *sample asymmetry*, and is illustrated in Section 10.6.2, where $N = 7$ and $N' = 11$. Of course, it is possible that $N = N'$, so that the prior sample is just the marginal sample of the joint sample. It is also possible that the two samples are disjoint — a situation which arises, for example, when the forecaster originally developed for one predictand must be adapted to another (similar) predictand, for which only a prior sample is available.

Classical statistical techniques, such as the linear regression method, cannot handle *asymmetric samples* — so they force us to ignore the large prior sample. In effect, these techniques throw away vast amounts of information about the predictand. In contrast, the Bayesian forecaster can use effectively both samples. It extracts information from each sample into the parameters (M, S) and (a, b, σ) of functions g and f, and then optimally fuses information according to the total probability law (10.1) and Bayes' theorem (10.2).

10.4 Informativeness of Predictor

10.4.1 The Concept

Suppose there are two candidate predictors, X_1 and X_2, each of which could be employed in the Bayesian Gaussian forecaster, and we wish to select one. This selection problem may be framed from the viewpoint of rational deciders — those who will use forecasts in decision models developed based on the principles of Bayesian theory. This theory, detailed in Section 12.10, prescribes the criterion: select a predictor that maximizes the economic value of the forecaster for every rational decider who will use the forecasts. Hence the following definition.

Definition   Predictor X_1 is said to be *more informative than* predictor X_2 with respect to predictand W if, for every rational decider, the economic value of the forecaster using predictor X_1 is **at least as high as** the economic value of the forecaster using predictor X_2.

10.4.2 Posterior Variance

Under the forecast outputted by the BGF, the measure of uncertainty about the predictand is the posterior variance $Var(W|X = x) = T^2$ given by equation (10.12):

$$T^2 = \frac{\sigma^2 S^2}{a^2 S^2 + \sigma^2}.$$

When each, the numerator and the denominator, is divided by $\sigma^2 S^2$, we obtain

$$T^2 = \left(\frac{a^2}{\sigma^2} + \frac{1}{S^2} \right)^{-1}.$$

Three properties of T^2 are apparent: (i) it is never greater than the prior variance S^2, a fact demonstrated explicitly in the next section; (ii) it does not depend on the predictor value x, but only on the parameters a, σ of the family of likelihood functions; and (iii) the influence of the likelihood parameters a, σ is separated from the influence of the prior variance S^2. Hence, one may anticipate that a^2/σ^2 provides a unique characterization of the predictor. This is indeed the case.

10.4.3 Sufficiency Characteristic

Parameter a is the slope of the regression of X on W, so $|a|$ is the magnitude of the slope, irrespective of its sign. Parameter σ is the standard deviation of the residual variate about the regression line.

Definition   Under the Bayesian Gaussian forecaster, $|a|$ is a measure of *signal* carried by the predictor, σ is a measure of *noise* in the predictor, and the "signal-to-noise" ratio,

$$SC = \frac{|a|}{\sigma}, \tag{10.24}$$

is called the *sufficiency characteristic* of predictor X with respect to predictand W.

The SC is dimensional: its unit equals the reciprocal of the unit of W. It is nonnegative and unbounded, $0 \le SC < \infty$. It is a measure of *informativeness* of predictor X with respect to predictand W — as the theorem in Section 10.4.5 asserts.

The posterior variance T^2 may now be rewritten in terms of the prior variance S^2 of the predictand and the sufficiency characteristic SC of the predictor:

$$T^2 = (SC^2 + S^{-2})^{-1}. \tag{10.25}$$

For a fixed S, the graph of T versus SC (to be constructed in Exercise 4) is a decreasing, convex function, with the domain $[0, \infty)$ and the range $[0, S]$.

The range of T arises from the two limits of SC, which characterize the three limiting cases of the likelihood regression (Figure 10.2).

1. If $|a| > 0$ and $\sigma \to 0$, then $SC \to \infty$. In the limit, $T = 0$, so predictor X eliminates the uncertainty about predictand W. Such a predictor is called *perfect*.
2. If $|a| \to 0$ or $\sigma \to \infty$, then $SC \to 0$. In the limit, $T = S$, so predictor X does not reduce the uncertainty about predictand W. Such a predictor is called *uninformative*.

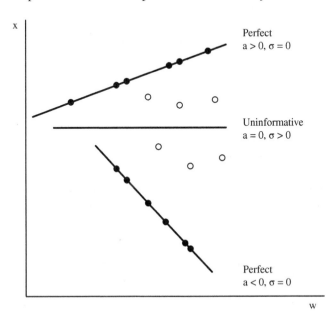

Figure 10.2   Examples of the three limiting cases of the likelihood regression of predictor X on predictand W: two perfect predictors and one uninformative predictor (when the joint sample size is $N = 6$).

10.4.4 Informativeness Score

In some situations, it may be desirable to compare the informativeness of forecasts of different predictands. Here are three such situations. (i) The Yellowstone River near Billings, Montana, receives in January a forecast of the runoff volume during April–September, and in May a forecast of the runoff volume during May–September. Not only is the forecast lead time different, but also the predictand. (ii) A retailer wants to compare the informativeness of forecasts of the same quantity, say seasonal sales of a merchandise, for different stores. (iii) A weather service wants to compare the informativeness of forecasts of different quantities, such as the mean seasonal temperature and the mean seasonal precipitation, in order to prioritize research on improving the forecasts.

To admit meaningful cross-predictand comparisons, a measure must be dimensionless and must account for the *a priori* "degree of difficulty" of forecasting each predictand. For the predictand $W \sim NM(M, S)$, the standard deviation S is a measure of *dispersion* of the prior distribution around the mean M, while the reciprocal $1/S$ is a measure of *concentration* of the prior distribution around the mean M. Intuitively, small S (large $1/S$) suggests an "easy-to-forecast" predictand, whereas large S (small $1/S$) suggests a "difficult-to-forecast" predictand. This is a rationale behind the following score.

Definition   Under the Bayesian Gaussian forecaster, which employs predictor X with the sufficiency characteristic SC to forecast predictand W with the prior standard deviation S,

$$IS = \left[\left(\frac{SC}{1/S} \right)^{-2} + 1 \right]^{-\frac{1}{2}} = \left[\left(\frac{aS}{\sigma} \right)^{-2} + 1 \right]^{-\frac{1}{2}} \tag{10.26}$$

is called the *informativeness score* of predictor X with respect to predictand W.

The *IS* is a strictly increasing transformation of *SC* scaled by $1/S$. To illustrate the effect, let $SC = 1/10$; then $S = 10$ yields $IS = 0.707$, whereas $S = 100$ yields $IS = 0.995$. Hence, this predictor with a fixed signal-to-noise ratio is "more informative" for forecasting the second predictand (the "difficult" one) than for forecasting the first predictand (the "easy" one). The principal properties of the informativeness score are as follows:

1. The IS is dimensionless.
2. The IS is bounded: $0 \leq IS \leq 1$, with $IS = 1$ for a perfect predictor, and $IS = 0$ for an uninformative predictor.
3. The IS is a measure of *informativeness* of predictor X with respect to predictand W (as the next theorem asserts).
4. The IS is interpretable as a *Bayesian estimator* of the absolute value of the correlation between predictor X and predictand W; as proven in Section 10.9.1,

$$Cor(X, W) = (\text{sign of } a) \, IS.$$

This estimator of the correlation is Bayesian because information from two different sources may be used to estimate the component parameters: (a, σ) come from the joint sample of (X, W), whereas S comes from a prior sample of W or from a judgmental assessment of the prior distribution of W. Hence, this correlation need not equal the empirical correlation estimated from the joint sample alone. (When there is no prior information, and (a, σ, S) are estimated from the joint sample of (X, W), then IS^2 equals the popular R^2 of the linear regression of X on W.)

10.4.5 Comparison Theorem

We can now return to the primary purpose of the above analysis, which is the comparison of predictors. The theory underlying this comparison is fully presented in Section 12.10, after the forecaster and the decider are coupled into a F–D system. The practical result is this.

Theorem (*Comparison of predictors*)   *Let W be the predictand, and let X_i ($i = 1, 2$) be two of its predictors, each characterized by $SC_i = |a_i|/\sigma_i$.*

1. *If $SC_1 > SC_2$, then predictor X_1 is more informative than predictor X_2.*

   Let S be the prior standard deviation of W, and let (S, SC_i) be used in equations (10.25) and (10.26) to calculate T_i and IS_i, respectively $(i = 1, 2)$.

2. *The statistics SC_i, IS_i, T_i, are equivalent measures of the informativeness of predictor X_i with respect to predictand W, in that*

$$SC_1 > SC_2 \iff IS_1 > IS_2 \iff T_1 < T_2.$$

In conclusion, the BGF offers its developer two useful theoretic properties. First, each of the three measures (SC, IS, T) is sufficient for ordering two or more predictors in terms of the *informativeness relation*. Second, the induced order of predictors is *complete*: all candidate predictors can be ranked, and the predictor ranked first is the most informative among them; that is, it is the most valuable economically for every rational decider who might use the forecasts.

When it comes to the comparison of predictors for different predictands, the only meaningful measure is *IS*, as explained in Section 10.4.4. However, the precise interpretation of the induced order in terms of the economic values requires a more nuanced conceptual framework. (It may be found in Krzysztofowicz (1992).)

10.5 Communication of Probabilistic Forecast

10.5.1 Sophisticated Deciders

Under the Bayesian Gaussian forecaster, the probabilistic forecast of a continuous predictand is completely specified by the three posterior parameters (A, B, T) and the expression for either

the posterior density function (10.14), or
the posterior distribution function (10.16), or
the posterior quantile function (10.18).

To communicate the probabilistic forecast routinely to sophisticated deciders, we can furnish each decider with the three expressions and the three parameter values (A, B, T), so that every time a new forecast is prepared, only the predictor value x needs to be disseminated. The decider can then employ any of the three expressions, as needed, in a decision model.

10.5.2 General Users

To communicate forecast uncertainty to the general public, a format simpler than the entire distribution function may be sufficient, or even preferable. One simple format consists of three prior or posterior quantiles of the predictand W:

$$(w_p, w_{0.5}, w_{1-p}), \qquad 0 < p < 0.5;$$

they specify the median $w_{0.5}$ of W, and the $(1 - 2p)$-probability central credible interval (w_p, w_{1-p}) of W. The probability p should be chosen judiciously, considering the nature of the predictand and the purpose of the forecast; for instance, $p \in \{0.05, 0.1, 0.125, 0.2, 0.25\}$.

Each forecast being disseminated should be accompanied by a two-part interpretation. To illustrate, let $p = 0.1$ so that the forecast is $(w_{0.1}, w_{0.5}, w_{0.9})$. The first part interprets the median $w_{0.5}$:

> There is an even chance that the realization of W
> will either *exceed* or *not exceed* $w_{0.5}$.

The second part pertains to the outer quantiles $(w_{0.1}, w_{0.9})$ and may interpret either the quantiles or the central credible interval. The interpretation of the quantile $w_{0.1}$ may focus on the nonexceedance event:

> There is a 10% chance that the realization of W
> will *not exceed* $w_{0.1}$.

Or it may focus on the exceedance event, if that interpretation is directly relevant to making intuitive decisions:

> There is a 90% chance that the realization of W
> will *exceed* $w_{0.1}$.

The interpretation of the central credible interval $(w_{0.1}, w_{0.9})$ is this:

> There is an 80% chance that the realization of W
> will fall *inside* the interval $(w_{0.1}, w_{0.9})$.

10.6 Application

10.6.1 Predictand and Predictor

Each year the US Department of Agriculture, in cooperation with other agencies, prepares a sequence of five forecasts of runoff volumes during the snowmelt season. The snowmelt process extends over several months: from January to May in Arizona, from April to September in Montana. Forecasts are issued at the beginning of each month from January through May for over 500 river gauging stations in 11 western states. The first forecast is thus prepared 5–9 months before the actual runoff can be observed. The primary users of these forecasts are farmers, ranchers, orchardists, hydropower companies, and water supply authorities.

The main predictors are measurements of snowpack (depth and density) in the mountainous portion of the drainage basin, the soil moisture in late fall (when the ground freezes and thus becomes impervious), and the antecedent river flow. The forecasting system employs a hydrologic model of the drainage basin or a multivariate technique (e.g., a multiple linear regression). It outputs a deterministic forecast in the form of an estimate x (a real number) of the unknown runoff volume W.

Our objective is to process this **deterministic forecast** into a **probabilistic forecast**. For this purpose, we shall treat the estimate x as a realization of a continuous predictor X, and shall employ the Bayesian Gaussian forecaster to obtain a posterior distribution function $\Phi(\cdot|x)$ of the continuous predictand W, conditional on the estimate $X = x$.

In a broad sense, hydrologists use their knowledge of the physical processes to make the "best" estimate x of W, whereas we use our knowledge of the Bayesian theory to quantify the uncertainty about W, given their estimate x. Therefore, if the probabilistic forecast, specified by the posterior distribution function $\Phi(\cdot|x)$, has an economic value to deciders, then this is a synergistic value of the physical–statistical modeling.

10.6.2 Samples

Exercise 18 includes samples of W and X extracted from the database of the US Department of Agriculture for two gauging stations in the Snake River basin, Idaho: (i) the Weiser station on the Weiser River, which closes a drainage area of 1460 mi^2, and (ii) the Twin Springs station on the Boise River, which closes a drainage area of 830 mi^2. The usual snowmelt season at each station extends from April to July. The runoff volume during that season, both estimated and actual, is reported in thousands of acre-feet (10^3 ac-ft). (An acre-foot is the volume

Table 10.1 Prior sample of the predictand W and joint sample of the predictor–predictand vector (X, W) for snowmelt runoff forecasting: January forecast; Weiser, Idaho; runoff volumes $x(n)$ and $w(n)$ are in 10^3 ac-ft.

Year	n	$x(n)$	$w(n)$
1973	11		234.7
1974	10		625.8
1975	9		549.2
1976	8		459.6
1977	7	225.1	38.8
1978	6	499.0	472.9
1979	5	269.1	275.8
1980	4	289.9	407.6
1981	3	318.9	283.8
1982	2	405.7	589.1
1983	1	492.7	608.4

that would cover 1 acre to a depth of 1 foot. It is a practical unit for farmers, as it admits direct mapping of the volume of water that is available to the number of acres that can be irrigated to a desired depth.)

The example presented below is for Weiser and uses a portion of the samples (Table 10.1): the prior sample of the predictand W from $N' = 11$ years, and the joint sample of the predictor–predictand vector (X, W) from $N = 7$ years; the predictor is the estimator from the beginning of January. Thus the *lead time* of the forecast is 7 months (from the time of forecast issuance in January to the end of the snowmelt season in July, when the realization of W becomes known).

10.6.3 Estimation and Validation

Table 10.2 reports the estimates of the prior parameters (M, S) and the likelihood parameters (a, b, σ) as well as the calculated values of the posterior parameters (A, B, T). Figures 10.3–10.6 illustrate the estimation and the validation steps.

The normal model NM(M, S) for the prior distribution function G of predictand W may appear to fit the empirical distribution function poorly (Figure 10.3), especially in the right tail, with MAD = 0.103. However, the Kolmogorov-Smirnov statistic $D = 0.149$ (Table 10.3) is much smaller than the critical value $C = 0.307$ at the significance level $\alpha = 0.20$; thus the model is not rejected. This can be explained by the sample size: $N' = 11$; it is small and thus it prods the statistical test "to tolerate" a larger D (and us "to tolerate" a larger MAD) so as to avoid the Type I error — rejecting the model when it is true.

Table 10.2 Estimates of parameters of the Bayesian Gaussian forecaster obtained from samples listed in Table 10.1; snowmelt runoff forecasting; January forecast; Weiser, Idaho.

M	S	a	b	σ	A	B	T
413.25	177.39	0.456	182.70	55.64	1.488	−139.34	100.47

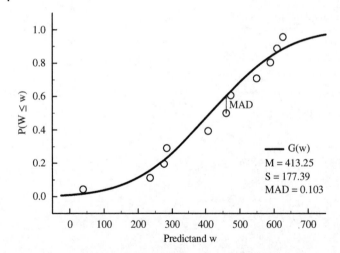

Figure 10.3 Prior distribution function G of predictand W: the estimated parametric distribution function, which is NM(M, S), overlaid on the empirical distribution function constructed from the prior sample of size $N' = 11$; snowmelt runoff forecasting; Weiser, Idaho.

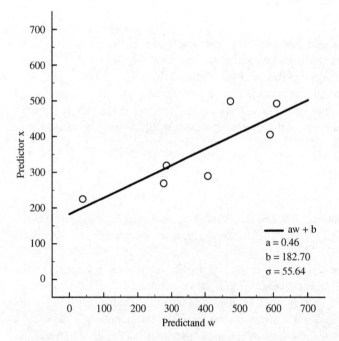

Figure 10.4 Regression of predictor X on predictand W: the estimated linear regression superposed on the scatterplot of the joint sample of size $N = 7$; snowmelt runoff forecasting; January forecast; Weiser, Idaho.

The normal model NM$(0, \sigma)$ for the distribution function of the residual variate Θ offers a similar level of goodness of fit (Figure 10.5) and an identical conclusion from the Kolmogorov-Smirnov test (Table 10.3).

The assumptions of linearity (Figure 10.4) and homoscedasticity (Figure 10.6) appear to hold as well.

Whereas the four assumptions of the BGF cannot be rejected *vis-à-vis* the data, it must be recognized, and indeed stressed, that this is an approximate model: under the normal distribution function, the predictand W has the unbounded sample space $(-\infty, \infty)$, whereas in reality the runoff volume has a bounded-below sample space (η, ∞),

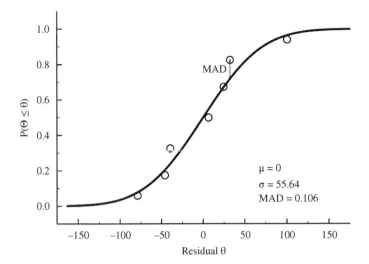

Figure 10.5 Distribution function of residual variate Θ from the regression of predictor *X* on predictand *W* (Figure 10.4): the estimated parametric distribution function, which is NM(0, σ), overlaid on the empirical distribution function constructed from the sample of size *N* = 7; snowmelt runoff forecasting; January forecast; Weiser, Idaho.

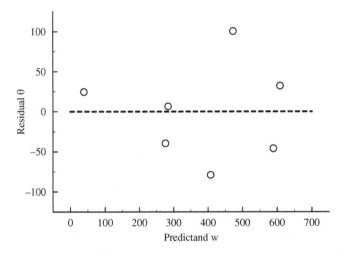

Figure 10.6 Validation of homoscedasticity of residual variate Θ from the regression of predictor *X* on predictand *W* (Figure 10.4); sample size *N* = 7; snowmelt runoff forecasting; January forecast; Weiser, Idaho.

Table 10.3 Validation of the estimated distribution functions for the Bayesian Gaussian forecaster; snowmelt runoff forecasting; January forecast; Weiser, Idaho.

		Sample		Kolmogorov-Smirnov test	
Variate	**Distribution**	**size**	**MAD**	**D**	**C at $\alpha = 0.20$**
W	NM(*M*, *S*)	11	0.103	0.149	0.307
Θ	NM(0, σ)	7	0.106	0.190	0.381

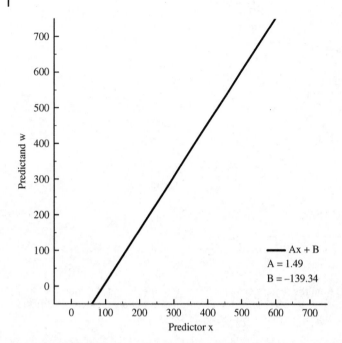

Figure 10.7 Posterior mean of predictand, $E(W|X = x)$, as a function of predictor realization x; snowmelt runoff forecasting; January forecast; Weiser, Idaho.

with $\eta > 0$. One should check, therefore, that G(0) is close to zero; here, the normal model NM(413.25, 177.39) yields G(0) = 0.0099, which seems acceptable for this type of application.

10.6.4 Distribution and Density Functions

The posterior mean $E(W|X = x)$ is a linear function of the predictor realization x (Figure 10.7), whereas the posterior variance $Var(W|X = x)$ is constant. Hence, given a predictor realization x, the posterior distribution function $\Phi(\cdot|x)$ is obtained by translating a fixed normal distribution function with mean 0 and variance T^2 to the location $Ax + B$ (Figure 10.8); likewise, the posterior density function $\phi(\cdot|x)$ is obtained by translating a fixed normal density function with mean 0 and variance T^2 to the location $Ax + B$ (Figure 10.9).

An insight offered by the above relationships is this: the BGF extracts information from the predictor X, and uses it to revise the prior density function g of W, in two ways. (i) It makes the mean of W adaptive: in contrast to the fixed prior mean M, the posterior mean $Ax + B$ varies with predictor realization x, which in turn varies from year to year because it reflects the state of the drainage basin in January of a particular year (recall Section 10.6.1). Visually, the location of the posterior density function varies from year to year, as illustrated in Figure 10.9 with $x = 250$ and $x = 450$. (ii) It reduces the uncertainty about W: the posterior variance T^2 is smaller than the prior variance S^2. Visually, each posterior density function in Figure 10.9 is thinner and taller than the prior density function. (Equivalently, the probability mass under the density function $\phi(\cdot|x)$ is concentrated more than it is under the density function g.)

10.6.5 Quantiles and Credible Intervals

Given the three distribution functions from Figure 10.8, with the parameter values in Table 10.2, let us create three probabilistic forecasts for the general public in the simple format recommended in Section 10.5.2. Suppose

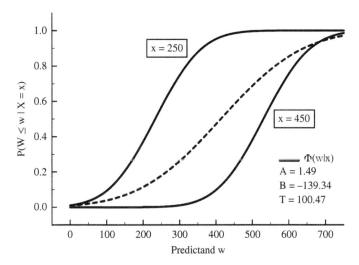

Figure 10.8 Two posterior distribution functions $\Phi(\cdot|x)$ of predictand W (solid lines), which are NM($Ax + B, T$), each conditional on a different realization of predictor, $x = 250$ and $x = 450$, contrasted with the prior distribution function G of W (broken line), which comes from Figure 10.3; snowmelt runoff forecasting; January forecast; Weiser, Idaho.

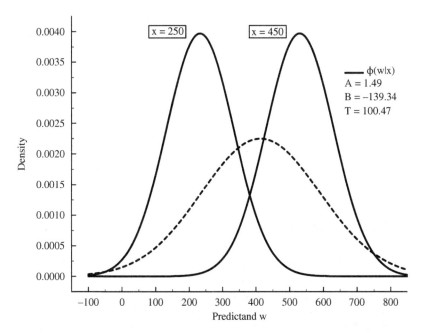

Figure 10.9 Two posterior density functions $\phi(\cdot|x)$ of predictand W (solid lines), contrasted with the prior density function g of W (broken line), each corresponding to a distribution function shown in Figure 10.8; snowmelt runoff forecasting; January forecast; Weiser, Idaho.

we choose $p = 0.2$, so the simple format is $(w_{0.2}, w_{0.5}, w_{0.8})$ — three quantiles of the runoff volume in 10^3 ac-ft, which convey the median $w_{0.5}$ and the 0.6-probability central credible interval $(w_{0.2}, w_{0.8})$.

Calculations. Proceed according to Sections 10.2.6–10.2.7. First, find the values of the corresponding standard normal quantiles: $z_{0.2} = -0.842$, $z_{0.5} = 0.0$, $z_{0.8} = 0.842$. Second, calculate the prior quantiles (equation (10.17)).

Table 10.4 Three probabilistic forecasts for the general public in the simple format $(w_{0.2}, w_{0.5}, w_{0.8})$ — three quantiles of the runoff volume in 10^3 ac-ft, which convey the median $w_{0.5}$ and the 0.6-probability central credible interval $(w_{0.2}, w_{0.8})$; snowmelt runoff forecasting; January forecast; Weiser, Idaho.

Quantiles	$w_{0.2}$	$w_{0.5}$	$w_{0.8}$	$w_{0.8} - w_{0.2}$
prior quantiles	263.9	413.2	562.6	298.7
posterior quantiles:				
given $x = 250$	148.1	232.7	317.3	169.2
given $x = 450$	445.7	530.3	614.9	169.2

Third, calculate the posterior quantiles (equation (10.18)), given $x = 250$ and given $x = 450$. Finally, calculate the width of each CCI (equations (10.19)–(10.20)). The results are compiled in Table 10.4.

   Interpretation of prior quantiles. In any year, there is an even chance that the runoff volume will either *exceed* or *not exceed* 413.2×10^3 ac-ft; there is a 20% chance that the runoff volume will *not exceed* 263.9×10^3 ac-ft; there is a 20% chance that the runoff volume will *exceed* 562.6×10^3 ac-ft; and there is a 60% chance that the runoff volume will fall *between* 263.9×10^3 and 562.6×10^3 ac-ft.

   Interpretation of posterior quantiles in a year in which the estimate made by hydrologists in January of the April–July runoff volume is 250×10^3 ac-ft. There is an even chance that the runoff volume will either *exceed* or *not exceed* 232.7×10^3 ac-ft; there is a 20% chance that the runoff volume will *not exceed* 148.1×10^3 ac-ft; there is a 20% chance that the runoff volume will *exceed* 317.3×10^3 ac-ft; and there is a 60% chance that the runoff volume will fall *between* 148.1×10^3 and 317.3×10^3 ac-ft.

   A parallel interpretation can be attached to the posterior quantiles in a year in which $x = 450 \times 10^3$ ac-ft. Then the two impacts of x can be noted and characterized (recall Section 10.6.4). (i) The posterior median is adaptive — it varies with x. But x itself does not admit a consistent interpretation: for example, $x = 250$ is greater than the posterior median $w_{0.5} = 232.7$, whereas $x = 450$ is smaller than the posterior median $w_{0.5} = 530.3$. (ii) The 0.6-probability posterior CCI is narrower than the 0.6-probability prior CCI: 169.2 constitutes 57% of 298.7. To a user, this may be a substantial reduction of uncertainty about the runoff volume, especially because it is achieved at the lead time of 7 months.

10.7 Forecaster of the Sum of Two Variates

A distributor of automotive parts owns two warehouses, one serving the eastern states and another serving the western states. To decide how many parts to order from a manufacturer, the distributor needs a forecast of the total demand; however, it may be logical, and easier, to forecast the demand in each region and then to aggregate the forecasts. For instance, each regional demand could be forecasted by a different expert, who has experience with the peculiarities of that market. But when each forecast is in the form of a distribution function, we need more than just the addition operation to **aggregate** them into a single distribution function of the total demand. Hence the following theory.

10.7.1 Problem Formulation

Let W_1 and W_2 be two continuous variates. Each variate W_j ($j = 1, 2$) is characterized in terms of its first two moments, the mean and the variance:

$$\mu_j = E(W_j), \qquad \sigma_j^2 = Var(W_j).$$

Suppose one is interested in the variate W, which is the sum

$$W = W_1 + W_2. \tag{10.27}$$

The first task is to derive the moments of the sum W in terms of the known moments of the summands; the general solution is presented. The second task is to derive the distribution function of the sum W in terms of the known distribution functions of the summands; only the solution for normally distributed summands is presented. In both tasks, the first step is to characterize the association and the dependence between the two summands.

10.7.2 Correlation Between the Summands

Pearson's product-moment *correlation* between the two variates, W_1 and W_2, is defined by

$$\gamma = Cor(W_1, W_2) = \frac{Cov(W_1, W_2)}{\sqrt{Var(W_1)}\sqrt{Var(W_2)}}, \tag{10.28}$$

where $Cov(W_1, W_2)$ stands for the covariance of W_1 and W_2. The correlation is bounded as $-1 \leq \gamma \leq 1$. The variates W_1 and W_2 are said to be *positively correlated* if $0 < \gamma \leq 1$, *negatively correlated* if $-1 \leq \gamma < 0$, and *uncorrelated* if $\gamma = 0$. The correlation may be qualified as *strong* when $|\gamma|$ is close to 1, or as *weak* when $|\gamma|$ is close to 0.

The correlation γ is properly interpreted as a *measure of association* between two variates. More specifically, γ measures the strength of *linear association* between W_1 and W_2 — their tendency to vary together in the same direction (if $\gamma > 0$) or in the opposite directions (if $\gamma < 0$) of their joint sample space (a plane or its subset). Association must not be confused with dependence.

Definition The variates W_1 and W_2 are *stochastically independent* (SI) of each other (*independent*, for short) if their joint, conditional, and marginal density functions satisfy any of the three conditions, for all possible realizations w_1 and w_2:

$$g(w_1, w_2) = g_1(w_1)g_2(w_2),$$
$$g_{12}(w_1|w_2) = g_1(w_1),$$
$$g_{21}(w_2|w_1) = g_2(w_2).$$

The three conditions are equivalent: any one holds if and only if the other two hold. (i) The first condition — the factorization of the joint density function in terms of the marginal density functions — informs us that W_1 and W_2 are *mutually independent*, as multiplication is commutative. (ii) The second condition reads: W_1 is independent of W_2. (iii) The third condition reads: W_2 is independent of W_1. When none of these conditions holds, the variates W_1 and W_2 are said to be *stochastically dependent* (SD) on each other.

Theorem *(Independence implies zero correlation)* If variates W_1 and W_2 are independent, then they are uncorrelated.

Proof: The definition of covariance is

$$Cov(W_1, W_2) = E(W_1 W_2) - E(W_1)E(W_2).$$

If W_1 and W_2 are independent, then (as Exercise 5 directs you to show)

$$E(W_1 W_2) = E(W_1)E(W_2),$$

which implies $Cov(W_1, W_2) = 0$. This, in turn, implies $\gamma = 0$. QED.

The reverse is not always true. That is, if W_1 and W_2 are uncorrelated, then it is not always true that they are independent. Whether or not $\gamma = 0$ implies independence is contingent upon the form of the joint distribution of W_1 and W_2.

10.7.3 Moments of the Sum

The first two moments of the sum W can be derived as follows. The mean of W is

$$E(W) = E(W_1 + W_2)$$
$$= E(W_1) + E(W_2).$$

Thus, using the abbreviated notation,

$$\mu = \mu_1 + \mu_2. \qquad (10.29)$$

The variance of W is

$$Var(W) = E(W^2) - E^2(W)$$
$$= Var(W_1) + Var(W_2) + 2Cov(W_1, W_2),$$

(where the derivation of the second line is deferred to Exercise 6). Next, expressing the covariance in terms of the correlation (10.28) yields

$$Var(W) = Var(W_1) + Var(W_2) + 2Cor(W_1, W_2)\sqrt{Var(W_1)}\sqrt{Var(W_2)}.$$

Finally, using the abbreviated notation,

$$\sigma^2 = \sigma_1^2 + \sigma_2^2 + 2\gamma\sigma_1\sigma_2. \qquad (10.30)$$

10.7.4 Distribution of the Sum

Two theorems are stated for normally distributed variates: the first theorem for independent variates; the second theorem for dependent variates.

Theorem *(Sum of independent normal variates)* *If the variates W_1 and W_2 are stochastically independent, and each W_j has a normal distribution,*

$$W_j \sim NM(\mu_j, \sigma_j), \qquad j = 1, 2,$$

then the sum W has a normal distribution:

$$W \sim NM(\mu, \sigma),$$
$$\mu = \mu_1 + \mu_2,$$
$$\sigma^2 = \sigma_1^2 + \sigma_2^2.$$

Theorem *(Sum of dependent normal variates)* *If the variates W_1 and W_2 are stochastically dependent, and their joint distribution is bivariate normal, then each W_j has a normal distribution,*

$$W_j \sim NM(\mu_j, \sigma_j), \qquad j = 1, 2,$$

and the sum W has a normal distribution:

$$W \sim NM(\mu, \sigma),$$
$$\mu = \mu_1 + \mu_2,$$
$$\sigma^2 = \sigma_1^2 + \sigma_2^2 + 2\gamma\sigma_1\sigma_2.$$

It is apparent that $\gamma = 0$ makes the expressions for σ^2 in the two theorems identical. One may thus infer that under the bivariate normal distribution, the correlation is not only a measure of association but also a measure of dependence:

$$\gamma > 0 \iff W_1, W_2 \text{ are } positively \text{ } dependent,$$
$$\gamma < 0 \iff W_1, W_2 \text{ are } negatively \text{ } dependent,$$
$$\gamma = 0 \iff W_1, W_2 \text{ } are \text{ } independent.$$

Inasmuch as $\sigma_1 > 0$ and $\sigma_2 > 0$, it is apparent also that

$$\gamma > 0 \implies \sigma^2 > \sigma_1^2 + \sigma_2^2,$$
$$\gamma < 0 \implies \sigma^2 < \sigma_1^2 + \sigma_2^2.$$

These inequalities may have profound practical implications.

Example 10.1   *(Two warehouses)*

Continuing the example from the introductory paragraph of Section 10.7, consider weekly demands for two automotive parts, A and B. Suppose for each part, $\sigma_1 = 200$, $\sigma_2 = 300$, but $\gamma = 0.4$ for part A and $\gamma = -0.4$ for part B. Then $\sigma = 422$ for part A, $\sigma = 286$ for part B, and $\sigma = 361$ for each part when $\gamma = 0$. Clearly, the magnitude of uncertainty about the total demand W for a part is affected by the correlation γ between the demand W_1 from the eastern states and the demand W_2 from the western states: positive correlation increases the uncertainty about W, whereas negative correlation decreases the uncertainty about W, relative to W from the uncorrelated demands. Hence the following principle.

> **Diversification principle**. If you wish to minimize the uncertainty about the sum W (equivalently, minimize the variability of the sum W), then diversify your activity in a way that makes the summands W_1 and W_2 negatively correlated.

This effect of the correlations, in turn, propagates to optimal decisions in the management of inventory, as we shall learn in Chapter 13 (Exercises 12, 13).

The task remaining in this chapter is to show how the above two theorems can be integrated into the Bayesian Gaussian forecaster.

10.8   Prior and Posterior Sums

10.8.1   Prior Distribution of the Sum

Suppose that the forecast of each predictand W_j ($j = 1, 2$) is made in terms of its prior distribution, as described in Section 10.2, so that $W_j \sim NM(M_j, S_j)$. Assigning

$$\mu_j = M_j, \qquad \sigma_j = S_j, \qquad j = 1, 2, \tag{10.31}$$

gives four parameters needed for the first theorem (sum of independent normal variates). The fifth parameter, needed for the second theorem (sum of dependent normal variates), is the *prior correlation* γ between W_1 and W_2. Its definition and the given prior moments yield the equation

$$\gamma = \frac{Cov(W_1, W_2)}{S_1 S_2} = \frac{M_{12} - M_1 M_2}{S_1 S_2}, \tag{10.32}$$

for which the expectation of the product, $M_{12} = E(W_1 W_2)$, remains to be estimated.

Estimation. The available record of past simultaneous realizations of W_1 and W_2 should be organized into a *prior joint sample*,

$$\{(w_1(n),\ w_2(n)) : n = 1, \ldots, N'\}, \tag{10.33}$$

from which M_{12} can be estimated. We employ the *maximum likelihood estimate*:

$$\hat{M}_{12} = \frac{1}{N'} \sum_{n=1}^{N'} w_1(n)w_2(n). \tag{10.34}$$

When inserted into equation (10.32), it yields the maximum likelihood estimate $\hat{\gamma}$ of the prior correlation. (Recall the invariance property of the maximum likelihood estimator, Section 2.4.3.)

Fusion of information. The size N' of the prior joint sample may be smaller than either N'_1 or N'_2 — the sizes of the prior marginal samples of W_1 and W_2, respectively, which need not be identical. In short, this may be possible: $N' < \min \{N'_1, N'_2\}, N'_1 \neq N'_2$. When such a *sample asymmetry* exists, equation (10.32) allows us to fuse information from three samples of different sizes; no realization of W_1 or W_2 is wasted.

10.8.2 Validation of Bivariate Normality

Validation of bivariate normality should be performed whenever $\gamma \neq 0$ and the second theorem is to be applied. It exploits two facts. (i) The factorization of the prior joint density function into the conditional and the marginal density functions:

$$g(w_1, w_2) = g_{21}(w_2|w_1)g_1(w_1). \tag{10.35}$$

(ii) A property of the normal family of density functions: g is bivariate normal if and only if each g_1 and $g_{21} (\cdot|w_1)$ for every $w_1 (-\infty < w_1 < \infty)$ is univariate normal. Inasmuch as g_1 is known to be normal, it suffices to validate the normality of the family of the conditional density functions g_{21}. The procedure parallels modeling and validation of the family f in Sections 10.2.2 and 10.3.3. Therefore, it is compressed here into three steps.

1. Hypothesize a *normal-linear model* of the stochastic dependence between W_2 and W_1 (as in Section 10.2.2):

$$W_2 = c_2 w_1 + d_2 + \Xi_2, \tag{10.36}$$

   with $E(\Xi_2) = 0$, $Var(\Xi_2) = \tau_2^2$, and $\Xi_2 \sim NM(0, \tau_2)$.
2. Calculate the maximum likelihood estimates of the parameters from the already known parameter values and the theoretic relationships:

$$c_2 = \gamma \frac{S_2}{S_1}, \qquad d_2 = M_2 - \gamma M_1 \frac{S_2}{S_1}, \tag{10.37}$$

$$\tau_2^2 = S_2^2(1 - \gamma^2). \tag{10.38}$$

3. Validate the three assumptions of the normal-linear model (as in Section 10.3.3) on the prior joint sample (10.33):

   Normality: $\Xi_2 \sim NM(0, \tau_2)$.
   Linearity: $E(W_2|W_1 = w_1) = c_2 w_1 + d_2$.
   Homoscedasticity: $Var(W_2|W_1 = w_1) = Var(\Xi_2) = \tau_2^2$.

If these assumptions are judged valid, at least approximately, then it is reasonable to accept the bivariate normal density function for (W_1, W_2) and to apply the second theorem for the prior sum $W = W_1 + W_2$.

Note on factorization. There exists, of course, an equivalent factorization of $g(w_1, w_2)$ in the form $g_{12}(w_1|w_2) g_2(w_2)$, for which the above validation procedure can be rewritten. (Exercise 7 directs you to do this.) Being equivalent, the two procedures should lead to the same conclusion.

10.8.3 Posterior Distribution of the Sum

Suppose that the forecast of each predictand W_j ($j = 1, 2$) comes from a different BGF, who supplies the posterior parameters (A_j, B_j, T_j) and the predictor realization $X_j = x_j$ for the posterior density function $\phi_j(\cdot|x_j)$. Our task is to formulate an *aggregator* — a procedure that applies one of the theorems for the sum $W = W_1 + W_2$. Toward this end, the theory underlying an aggregator is sketched, the simplifying assumptions are stated, and two workable aggregators are formulated.

Theoretic structure of an aggregator

The most general aggregator is the *bivariate Bayesian forecaster*. Letting w and x be vectors, $w = (w_1, w_2)$ and $x = (x_1, x_2)$, with the sample spaces $\mathcal{W} = \mathcal{W}_1 \times \mathcal{W}_2$ and $\mathcal{X} = \mathcal{X}_1 \times \mathcal{X}_2$, respectively, the Bayesian forecaster (10.1)–(10.2) becomes

$$\kappa(x_1, x_2) = \int_{\mathcal{W}_1} \int_{\mathcal{W}_2} f(x_1, x_2|w_1, w_2)g(w_1, w_2)dw_2\,dw_1, \tag{10.39}$$

$$\phi(w_1, w_2|x_1, x_2) = \frac{f(x_1, x_2|w_1, w_2)g(w_1, w_2)}{\kappa(x_1, x_2)}. \tag{10.40}$$

Figure 10.10(a) visualizes the structure of equation (10.40) using arrows to depict the *stochastic dependence* (SD) between the four variates. A general solution for ϕ would require substantial additional modeling, while we wish to employ only the models already developed. To identify the possibilities, each input density function is factorized. The function g has factorization (10.35). The function f has the factorization

$$f(x_1, x_2|w_1, w_2) = f_{21}(x_2|w_1, w_2, x_1)f_1(x_1|w_1, w_2),$$

which suggests possible simplifications of the *likelihood dependence structure*. One such a simplification is this.

Definition   (i) **Conditional on** variate W_1, variate X_1 is *stochastically independent* of variate W_2 if for all possible realizations w_1, w_2, x_1,

$$f_1(x_1|w_1, w_2) = f_1(x_1|w_1).$$

(ii) **Conditional on** variate W_2, variate X_2 is *stochastically independent* of variates W_1, X_1 if for all possible realizations w_1, w_2, x_1, x_2,

$$f_{21}(x_2|w_1, w_2, x_1) = f_2(x_2|w_2).$$

When both definitions hold, the conditional joint density function f takes the product form:

$$f(x_1, x_2|w_1, w_2) = f_2(x_2|w_2)f_1(x_1|w_1). \tag{10.41}$$

In short, we say: predictors X_1, X_2 are *conditionally stochastically independent* (CSI) **relative to** predictands W_1, W_2. This is all the theory needed for two aggregators.

Aggregator with total conditional stochastic independence

Two assumptions are made about the stochastic dependence between the variates under density functions g and f, visualized in Figure 10.10(b).

- Prior structure: W_1, W_2 are stochastically independent.
- Likelihood structure: X_1, X_2 are conditionally stochastically independent.

Under these assumptions, it can be shown (see Exercise 9) that the posterior joint density function of (W_1, W_2) takes the form

$$\phi(w_1, w_2|x_1, x_2) = \phi_2(w_2|x_2)\phi_1(w_1|x_1). \tag{10.42}$$

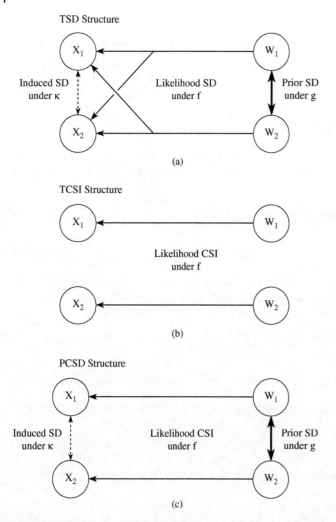

Figure 10.10 Stochastic dependence (SD) structures between predictands (W_1, W_2) and predictors (X_1, X_2): (a) total stochastic dependence (TSD) that requires equation (10.40) for the general aggregator; (b) total conditional stochastic independence (TCSI) that leads to aggregator (10.42); (c) partial conditional stochastic dependence (PCSD) that leads to aggregator (10.44). The SD between X_1 and X_2 is induced by the total probability law, equation (10.39).

In other words, W_1 and W_2 are conditionally stochastically independent *a posteriori*. This is the simplest aggregator: merely assigning

$$\mu_j = A_j x_j + B_j, \qquad \sigma_j = T_j, \qquad j = 1, 2, \tag{10.43}$$

gives four parameters needed to apply the first theorem for the posterior sum $W = W_1 + W_2$, which is now conditional on $(X_1 = x_1, X_2 = x_2)$.

Aggregator with partial conditional stochastic dependence

Several forms of partial conditional stochastic dependence are possible. The one adopted herein derives from two assumptions about the stochastic dependence between the variates under density functions g and f, visualized in Figure 10.10(c).

- Prior structure: W_1, W_2 are stochastically dependent.
- Likelihood structure: X_1, X_2 are conditionally stochastically independent.

Under these assumptions, it is proven in Section 10.9.3 that the posterior joint density function of (W_1, W_2) takes the form

$$\phi(w_1, w_2 | x_1, x_2) = \xi(w_1, w_2; x_1, x_2)\phi_2(w_2 | x_2)\phi_1(w_1 | x_1), \tag{10.44}$$

where ξ is the *density weighting function*, which quantifies the stochastic dependence between W_1 and W_2 a *posteriori*.

The concept behind the resultant aggregator is this. The density functions inputted to the bivariate BGF, function g (equation (10.35)) and function f (equation (10.41)), are bivariate normal. Therefore, the posterior joint density function ϕ is bivariate normal. The task is to find the *posterior (conditional) correlation*, for all possible realizations x_1 and x_2:

$$\gamma_c = Cor(W_1, W_2 | X_1 = x_1, X_2 = x_2).$$

It turns out, as is proven in Section 10.9.4, that the density weighting function ξ contains the formula for the square of the posterior correlation, which is remarkably simple:

$$\gamma_c^2 = 1 - \frac{1 - \gamma^2}{1 - \gamma^2 \cdot IS_1^2 \cdot IS_2^2}; \tag{10.45}$$

in it, IS_j is the informativeness score of predictor X_j with respect to predictand W_j, calculated via equation (10.26) for $j = 1, 2$; γ is the prior correlation between W_1 and W_2, calculated via equation (10.32); and the sign of γ_c is the same as the sign of γ.

Together, expressions (10.43) and (10.45) give five parameters needed to apply the second theorem for the posterior sum $W = W_1 + W_2$, which is now conditional on $(X_1 = x_1, X_2 = x_2)$.

Conditional correlation and predictor informativeness

Formula (10.45) admits an insightful interpretation. In essence, it revises the unconditional correlation γ between the predictands W_1 and W_2 into a conditional correlation γ_c, based on the informativeness of predictors which are employed by the two BGFs. Hence, several properties can be inferred.

1. The limiting cases of a predictor (recall Section 10.4.3) imply the following limiting cases of the conditional correlation: (i) when each predictor is perfect,

   $$\text{perfect } X_1, X_2 \quad \Longleftrightarrow \quad IS_1 \cdot IS_2 \to 1 \quad \Longrightarrow \quad \gamma_c \to 0;$$

   (ii) when one predictor is uninformative ($j = 1$ or $j = 2$),

   $$\text{uninformative } X_j \quad \Longleftrightarrow \quad IS_j \to 0 \quad \Longrightarrow \quad \gamma_c \to \gamma.$$

2. The conditional correlation γ_c has the same sign as the unconditional correlation γ, is bounded as

   $$0 \leq |\gamma_c| \leq |\gamma|,$$

   and $|\gamma_c|$ is a strictly decreasing function of the product of the informativeness scores: as $IS_1 \cdot IS_2$ increases from 0 to 1, $|\gamma_c|$ decreases from $|\gamma|$ to 0 (as Exercise 10 directs you to graph). Ergo, W_1 and W_2 become conditionally stochastically independent.

The most consequential is the strict monotonicity of function (10.45). To illustrate it, let us compare three forecasting systems having the same $\gamma = 0.7$, but different IS_j ($j = 1, 2$):

IS_1	IS_2	γ_c
0.45	0.45	0.692
0.45	0.99	0.660
0.99	0.99	0.191

We have effectively proven and illustrated the following theorem.

Theorem (*Informativeness – Correlation*)   *In the **bivariate Bayesian Gaussian forecaster** with conditionally independent predictors and dependent predictands W_1 and W_2, the absolute value of the posterior (conditional) correlation $|\gamma_c|$ between W_1 and W_2 decreases whenever the informativeness score IS_j of at least one predictor increases. In the limit, when both predictors become perfect, W_1 and W_2 become conditionally uncorrelated (and conditionally independent).*

It follows that aggregator (10.42) may be considered as a reasonable approximation whenever both predictors are highly informative, even though the predictands are dependent.

10.9   Concepts and Proofs

10.9.1   Informativeness Score as Correlation

Theorem (*Informativeness score as correlation*)   *Let IS be the informativeness score of the predictor X with respect to the predictand W, defined by expression (10.26), and let $Cor(X, W)$ be Pearson's product-moment correlation coefficient between X and W. Then*

$$Cor(X, W) = (\text{sign of } a) IS.$$

Proof: Expressing X in terms of the normal-linear model (10.5), the covariance of X and W is

$$
\begin{aligned}
Cov(X, W) &= E(XW) - E(X)E(W) \\
&= E[(aW + b + \Theta)W] - E(aW + b + \Theta)E(W) \\
&= aE(W^2) + bE(W) + E(\Theta)E(W) - aE^2(W) - bE(W) - E(\Theta)E(W) \\
&= a[E(W^2) - E^2(W)] \\
&= a\,Var(W),
\end{aligned}
$$

where in the third line, $E(\Theta W) = E(\Theta)E(W)$ because Θ is stochastically independent of W. With $Var(W)$ and $Var(X)$ given by expressions (10.4) and (10.9), respectively,

$$
\begin{aligned}
Cor(X, W) &= \frac{Cov(X, W)}{\sqrt{Var(X)}\,\sqrt{Var(W)}} \\
&= \frac{aS^2}{\left(a^2 S^2 + \sigma^2\right)^{1/2} S} \\
&= (\text{sign of } a)\left(\frac{\sigma^2}{a^2 S^2} + 1\right)^{-\frac{1}{2}} \\
&= (\text{sign of } a) IS.
\end{aligned}
$$

QED.

10.9.2 Bivariate Normal Distribution

The joint distribution of the variates W_1 and W_2 is *bivariate normal* with parameters $(\mu_1, \sigma_1; \mu_2, \sigma_2; \gamma)$, as they are defined for the second theorem of Section 10.7.4, whenever the joint density function h of W_1 and W_2 is of the form

$$
h(w_1, w_2) = \frac{1}{2\pi\sigma_1\sigma_2\sqrt{1-\gamma^2}} \exp\left\{ \frac{-1}{2(1-\gamma^2)} \left[\left(\frac{w_1 - \mu_1}{\sigma_1}\right)^2 \right. \right.
$$
$$
\left. \left. -2\gamma\left(\frac{w_1-\mu_1}{\sigma_1}\right)\left(\frac{w_2-\mu_2}{\sigma_2}\right) + \left(\frac{w_2-\mu_2}{\upsilon_2}\right)^2 \right] \right\} \tag{10.46}
$$

for all $(w_1, w_2) \in \mathcal{W}_1 \times \mathcal{W}_2$, with $\mathcal{W}_j = (-\infty, \infty)$ for $j = 1, 2$.

10.9.3 Density Weighting Function

Aggregator (10.44) of Section 10.8.3 is based on two assumptions, which are necessary for two theorems: the first one is general — it applies to any bivariate Bayesian forecaster; the second one (in Section 10.9.4) is specific — it applies to the bivariate BGF.

Theorem *(Density weighting function)*   *Consider the* **bivariate Bayesian forecaster** *(10.39)–(10.40). Suppose (i) the predictands (W_1, W_2) are dependent and have a prior joint density function (10.35), and (ii) the predictors (X_1, X_2) are conditionally independent and have a conditional joint density function (10.41). Then (W_1, W_2) have the posterior joint density function (10.44), in which the density weighting function takes the form*

$$
\xi(w_1, w_2; x_1, x_2) = \frac{\kappa_2(x_2)}{\kappa_{21}(x_2|x_1)} \frac{g_{21}(w_2|w_1)}{g_2(w_2)}. \tag{10.47}
$$

Proof: It proceeds in two steps.

1. Given functions $g = g_{21} \cdot g_1$ and $f = f_2 \cdot f_1$, and supposing $\kappa = \kappa_{21} \cdot \kappa_1$, to be demonstrated in Step 2, Bayes' theorem (10.40) yields

$$
\phi(w_1, w_2|x_1, x_2) = \frac{f_2(x_2|w_2)g_{21}(w_2|w_1)}{\kappa_{21}(x_2|x_1)} \frac{f_1(x_1|w_1)g_1(w_1)}{\kappa_1(x_1)}
$$
$$
= \frac{\kappa_2(x_2)}{\kappa_{21}(x_2|x_1)} \frac{g_{21}(w_2|w_1)}{g_2(w_2)} \frac{f_2(x_2|w_2)g_2(w_2)}{\kappa_2(x_2)} \phi_1(w_1|x_1)
$$
$$
= \xi(w_1, w_2; x_1, x_2)\phi_2(w_2|x_2)\phi_1(w_1|x_1),
$$

where ϕ_1 and ϕ_2 follow their definition (10.2), and ξ is the product of the first and second quotients in the second line.

2. To show that $\kappa = \kappa_{21} \cdot \kappa_1$ and that κ_{21} can be obtained from the given functions, the total probability law (10.39) is applied:

$$
\kappa(x_1, x_2) = \int_{\mathcal{W}_1} \int_{\mathcal{W}_2} f_2(x_2|w_2)g_{21}(w_2|w_1)f_1(x_1|w_1)g_1(w_1)\, dw_2\, dw_1
$$
$$
= \int_{\mathcal{W}_1} \left[\int_{\mathcal{W}_2} f_2(x_2|w_2)g_{21}(w_2|w_1)\, dw_2\right] \frac{f_1(x_1|w_1)g_1(w_1)}{\kappa_1(x_1)} \kappa_1(x_1)\, dw_1
$$
$$
= \int_{\mathcal{W}_1} \xi_{21}(x_2|w_1)\phi_1(w_1|x_1)\, dw_1 \cdot \kappa_1(x_1)
$$
$$
= \kappa_{21}(x_2|x_1)\kappa_1(x_1),
$$

where

$$\xi_{21}(x_2|w_1) = \int_{w_2} f_2(x_2|w_2) g_{21}(w_2|w_1) \, dw_2,$$

$$\kappa_{21}(x_2|x_1) = \int_{w_1} \xi_{21}(x_2|w_1) \phi_1(w_1|x_1) \, dw_1.$$

The first equation proves the factorization $\kappa = \kappa_{21} \cdot \kappa_1$, supposed in Step 1. The last two equations show the two-step derivation of κ_{21}, which is handy for proving the theorem of Section 10.9.4. QED.

10.9.4 Conditional Correlation

Theorem *(Conditional correlation)* *If the theorem of Section 10.9.3 applies to the **bivariate Bayesian Gaussian forecaster**, then the density weighting function ξ prescribes the posterior (conditional) correlation γ_c between W_1 and W_2 in the form (10.45).*

Proof: It proceeds in six steps, which explain the logic, while the algebraic manipulations in Step 6 are deferred to Exercise 11.

1. *Preliminaries.* By equation (10.47), the function ξ is specified by four univariate normal density functions. The BGF which supplies ϕ_2 can supply also g_2, equation (10.4), and κ_2, equation (10.10). The function g_{21} has its parameters specified by equations (10.37)–(10.38). With $\equiv$ meaning the identity of objects written in different symbols, the three known functions are:

$$g_2(w_2) \equiv \text{NM}(M_2, S_2), \tag{10.48}$$

$$g_{21}(w_2|w_1) \equiv \text{NM}(c_2 w_1 + d_2, \tau_2), \tag{10.49}$$

$$\kappa_2(x_2) \equiv \text{NM}\left(a_2 M_2 + b_2, \sqrt{a_2^2 S_2^2 + \sigma_2^2}\right). \tag{10.50}$$

It remains to derive the function κ_{21}.

2. *Derivation via pattern.* Let us first learn how to derive an expected density function by following the pattern of the known solution for normal density functions inputted to the total probability law:

$$\kappa(x) = \int_w f(x|w) g(w) \, dw.$$

Given g specified by equations (10.3)–(10.4), and f specified by equations (10.7)–(10.8), the solution is κ specified by equations (10.9)–(10.10). Symbolically,

$$\text{NM}\left(aM + b, \sqrt{a^2 S^2 + \sigma^2}\right) \equiv \int_w \text{NM}(aw + b, \sigma) \cdot \text{NM}(M, S).$$

Pattern: To get the mean $E(X)$, replace w by M; to get the variance $Var(X)$, multiply a^2 by S^2 and add σ^2.
Hint: In each of the following two steps, cover the solution and attempt to derive it yourself by following this pattern.

3. *Derivation of ξ_{21}.* Given $g_{21}(w_2|w_1)$ from Step 1, and

$$f_2(x_2|w_2) \equiv \text{NM}(a_2 w_2 + b_2, \sigma_2)$$

from the second BGF, insert them into the integral in Step 2 of the proof in Section 10.9.3, and derive

$$\xi_{21}(x_2|w_1) \equiv \text{NM}\left(a_2(c_2 w_1 + d_2) + b_2, \sqrt{a_2^2 \tau_2^2 + \sigma_2^2}\right).$$

4. *Derivation of κ_{21}.* Given $\xi_{21}(x_2|w_1)$ derived above, and

$$\phi_1(w_1|x_1) \equiv NM(A_1 x_1 + B_1, T_1)$$

from the first BGF, insert them into the integral in Step 2 of the proof in Section 10.9.3, and derive

$$\kappa_{21}(x_2|x_1) \equiv NM\left(a_2(c_2(A_1 x_1 + B_1) + d_2) + b_2, \sqrt{a_2^2 c_2^2 T_1^2 + a_2^2 \tau_2^2 + \sigma_2^2}\right). \tag{10.51}$$

5. *Bivariate normal ϕ.* All four functions that define ξ via equation (10.47) are now specified by expressions (10.48)–(10.51). When ξ with these expressions, along with ϕ_1 and ϕ_2 with their expressions (10.14), are inserted into equation (10.44), the function ϕ is obtained in the form of a product of six univariate normal density functions — two unconditional (g_2, κ_2), and four conditional ($g_{21}, \kappa_{21}, \phi_1, \phi_2$). When simplified to a single natural exponential function, ϕ derived from equation (10.44) must have the same *structure* as the function h specified by expression (10.46) has because both are bivariate normal. It is now straightforward to identify the correspondence for four parameters — the means and the variances under h and under ϕ; this was done in equations (10.43). The fifth parameter under ϕ is the conditional correlation γ_c, which must correspond to the unconditional correlation γ under h.

6. *Derivation of γ_c.* It is sufficient to concentrate on the scaling constant of the bivariate normal density function. By Step 5 above,

$$\phi = \frac{\kappa_2}{\kappa_{21}} \cdot \frac{g_{21}}{g_2} \cdot \phi_2 \cdot \phi_1.$$

In analogy to function h, function ϕ on the left side must have the scaling constant in the form

$$L = \frac{1}{2\pi T_1 T_2 \sqrt{1 - \gamma_c^2}}.$$

It must result from the scaling constants of the univariate normal density functions whose product constitutes the right side of the above equation for ϕ:

$$R = \frac{\frac{1}{\sqrt{a_2^2 S_2^2 + \sigma_2^2}\sqrt{2\pi}}}{\frac{1}{\sqrt{a_2^2 c_2^2 T_1^2 + a_2^2 \tau_2^2 + \sigma_2^2}\sqrt{2\pi}}} \cdot \frac{\frac{1}{\tau_2 \sqrt{2\pi}}}{\frac{1}{S_2 \sqrt{2\pi}}} \cdot \frac{1}{T_2 \sqrt{2\pi}} \cdot \frac{1}{T_1 \sqrt{2\pi}}.$$

Expression (10.45) can now be derived by (i) substituting T_j by expression (10.12) for $j = 1, 2$; (ii) substituting c_2 and τ_2 by expressions (10.37)–(10.38); (iii) solving the equation $L = R$ for $1 - \gamma_c^2$; (iv) rearranging the terms so that they match the right side of equation (10.26), and substituting it by the left side, IS_j, for $j = 1, 2$. QED.

Bibliographical Notes

The concept of informativeness was originated by Blackwell (1951, 1953) in the context of sufficient comparisons of statistical experiments. It was adapted to the comparison of Bayesian Gaussian forecasters, for which the sufficiency characteristic and the informativeness score were derived, by Krzysztofowicz (1987, 1992). The concept of conditional stochastic independence received a fundamental treatment in the books by de Finetti (1974) and by Bernardo and Smith (1994). The aggregators in Section 10.8 are based on Bayesian models developed by Krzysztofowicz (1985). That any bivariate (or multivariate) density function can always be written as the product of the univariate (marginal) density functions and a density weighting function is a known fact in the theory of multivariate distributions (Kotz and Seeger, 1991).

Exercises

1   **Conditional density function**. Consider the linear model specified by expressions (10.5)–(10.6). Let ξ denote the density function of the residual variate Θ; let θ denote its realization $(-\infty < \theta < \infty)$.
    1.1 Write an expression for the density function ξ of Θ.
    1.2 With w held fixed, equation (10.5) may be viewed as a transformation between variate Θ and variate X. Given the expression for the density function ξ of Θ, derive the expression for the density function $f(\cdot|w)$ of X.
    Hint. Apply the theory from Section 3.8; compare your answer with expression (10.8).

2   **Expected density function**. Consider variates X, W, Θ, and what is known about them from the assumptions that support the models for g and f; these assumptions imply a linear equation

$$X = aW + b + \Theta.$$

2.1 Write an expression for the density function of Θ.
2.2 Derive the expression for the density function of V, where $V = aW + b$.
2.3 Derive the expression for the density function κ of X.
    Hint. Make use of the theorem pertaining to the "sum of independent normal variates" (Section 10.7.4). Compare your answer with expression (10.10).

3   **Dimensional analysis**. Given the units of the predictor X and of the predictand W, derive (via dimensional analysis of the appropriate equations) the units of the sufficiency characteristic SC and of the posterior standard deviation T.
    Options:
    (A) X — precipitation depth [mm], W — runoff volume [m³].
    (B) X — distance traveled [km], W — fuel consumption [l/100 km].
    (C) X — timber log diameter [cm], W — log weight [kg].

4   **Posterior standard deviation**. Construct a graph of the posterior standard deviation T versus the sufficiency characteristic SC, for a fixed value of the prior standard deviation S.
    Options:

    (A) $S = 2$,   (C) $S = 4$,   (E) $S = 6$,
    (B) $S = 3$,   (D) $S = 5$,   (F) $S = 7$.

5   **Expectation of the product**. Complete the proof of the theorem in Section 10.7.2 by showing that if W_1 and W_2 are independent variates, then $E(W_1 W_2) = E(W_1)E(W_2)$.

6   **Variance of the sum**. Consider the expression for $Var(W)$ in Section 10.7.3. Starting from the first line (the definition), derive the second line.

7   **Validation of bivariate normality**. Rewrite the three steps of the validation procedure of Section 10.8.2 for the equivalent factorization mentioned in the note below the procedure. *Hint*: Start by defining a logical parallel notation.

8   **Bivariate prior density function**. Use the information provided in Section 10.8.2 to write the expressions for three prior density functions.

8.1 The marginal density function g_1 of W_1.

8.2 The conditional density function $g_{21}(\cdot|w_1)$ of W_2 for every w_1 ($-\infty < w_1 < \infty$). *Hint*: Consult Section 10.2.2 for a parallel task.

8.3 The joint density function g of (W_1, W_2). Simplify the expression to the form containing a single natural exponential function. *Note*: The result is an expression for the bivariate normal density function with the parameters $(M_1, S_1; M_2, S_2; \gamma)$; see Section 10.9.2.

9   Total conditional stochastic independence. Beginning with the bivariate Bayesian forecaster (10.39)–(10.40), incorporate in it the two assumptions of the aggregator, and derive equation (10.42).

10   Conditional correlation. Let X_j be a predictor having the informativeness score IS_j ($j = 1, 2$). Consider the product $t = IS_1 \cdot IS_2$ to be the sole argument of function (10.45).

10.1 Use expression (10.45) to construct a graph of the conditional correlation γ_c versus the product t, for a fixed value of the unconditional correlation γ, specified in the option.

10.2 Reflect on the above graph. Offer an intuitive explanation of its monotonicity: Why does the increasing informativeness of a predictor reduce the magnitude of the conditional correlation $|\gamma_c|$ between the predictands? *Note*: Under the BGF, reduced $|\gamma_c|$ is equivalent to reduced strength of conditional stochastic dependence (positive or negative).

Options:

(A) $\gamma = 0.4$,	(C) $\gamma = 0.6$,	(E) $\gamma = 0.8$,
(B) $\gamma = 0.5$,	(D) $\gamma = 0.7$,	(F) $\gamma = 0.9$.

11   Derivation of conditional correlation. Derive expression (10.45) for the conditional correlation γ_c by following the instructions stated in Step 6 of the proof in Section 10.9.4.

12   Density weighting function. In parallel to equation (10.44), the prior joint density function g of (W_1, W_2) can be constructed from the prior marginal density functions g_j of W_j, for $j = 1, 2$, and a density weighting function ξ, which quantifies the stochastic dependence between W_1 and W_2 *a priori*:

$$g(w_1, w_2) = \xi(w_1, w_2)g_2(w_2)g_1(w_1).$$

12.1 Derive the general equation for ξ.

12.2 Suppose $W_j \sim NM(M_j, S_j)$ for $j = 1, 2$. Derive the parametric expression for ξ in the form of a single natural exponential function.

12.3 Validate the expression for ξ by showing that $\xi \cdot g_2 \cdot g_1$ equals g, which is the bivariate normal density function.

Hints. For the general equation, consult Section 10.9.3. For parametric derivation and validation, read Exercise 8 above.

13   Auto parts. At the beginning of each month, the manager of a taxicab service in a large city forecasts the monthly demand for engine oil (in liters). The predictor is the total distance (in thousands of kilometers) traveled by the fleet of taxicabs since the last service of each vehicle. From the records, the manager estimated the parameters of the Bayesian Gaussian forecaster. (The parameter values are different for each month of the year because the demand for taxicabs is nonstationary.) For June: $M = 780$, $S = 120$; $a = 0.73$, $b = 94$, $\sigma = 65$.

13.1 Calculate the values of the posterior parameters A, B, T, the sufficiency characteristic SC, and the informativeness score IS.

13.2 Graph the posterior mean $E(W|X = x)$ as a function of x; in the same figure, graph the 50% posterior central credible interval as a function of x, and the 98% posterior central credible interval as a function of x; the domain of x in all graphs should be the interval $[E(X) - 3Var^{1/2}(X), E(X) + 3Var^{1/2}(X)]$.

13.3 Graph in the same figure the prior density function g and the posterior density functions $\phi(\cdot|600)$ and $\phi(\cdot|900)$; compare them.

13.4 Graph in the same figure the prior distribution function G and the posterior distribution functions $\Phi(\cdot|600)$ and $\Phi(\cdot|900)$; compare them.

14 **Auto parts** — *continued*: **correlated demands**. In an adjacent city, taxicab service B has the same forecasting problem as taxicab service A in Exercise 13. Its Bayesian Gaussian forecaster (BGF) for June has parameter values: $M = 910$, $S = 140$; $a = 0.52$, $b = 37$, $\sigma = 71$.

The two services merged. The new manager wants to retain the BGF of each service, and also wants to get a probabilistic forecast of the monthly total demand for engine oil, $W = W_1 + W_2$, the sum of the demands of services A and B, respectively. The prior joint sample of (W_1, W_2) compiled from the records of both services is specified in the option.

Options for the prior joint sample:

(P) $\{(762, 873), (745, 813), (752, 907), (790, 858), (803, 958), (857, 993), (913, 974), (840, 918), (706, 852)\}$.

(N) $\{(719, 958), (732, 973), (754, 929), (763, 902), (787, 871), (805, 852), (821, 869), (847, 841), (861, 813)\}$.

14.1 Estimate the prior correlation between W_1 and W_2. Calculate the posterior (conditional) correlation. Comment on the difference.

14.2 Write expressions for the prior mean and the prior variance of W. Write expressions for the prior density function g of W, and the prior distribution function G of W.

14.3 Calculate the values of all parameters needed for the posterior distribution of W, conditional on the realizations of predictors $(X_1 = x_1, X_2 = x_2)$. Write expressions for the posterior mean and the posterior variance of W.

14.4 Write expressions for the posterior density function $\phi(\cdot|x_1, x_2)$ of W, and the posterior distribution function $\Phi(\cdot|x_1, x_2)$ of W.

14.5 Graph in the same figure the prior density function g and the posterior density functions $\phi(\cdot|600, 400)$ and $\phi(\cdot|900, 700)$; compare the three functions.

14.6 Suppose the monthly demands for engine oil from the two taxicab services became uncorrelated. Repeat Exercise 14.5.

14.7 Compare the figures from Exercises 14.5 and 14.6. Characterize in words the effect of correlation between the two demands on the density function of the total demand, *a priori* and *a posteriori*.

15 **Treasury notes**. The US Department of Treasury issues bonds, such as the 10-year Treasury note. The yield on bond varies with time. When the bond yield falls, the bond price rises, and vice versa. Thus, an investor can profit by buying bonds at a low price and selling them later at a high price, provided he forecasted the yield's decline correctly. Economists and investment analysts routinely forecast the yield on the 10-year Treasury note — our predictand W. For example, *The Wall Street Journal* (10 June 2019) reported a survey of some 50 economists who at the beginning of January, when the yield was 2.68%, made deterministic forecasts (point estimates) of W at the end of June (a lead time of 6 months). The median estimate was 2.93%, the range of estimates was [2.50%, 3.77%]. By 10 June, the actual yield was 2.09%, well below the range of estimates; and only nine economists forecasted the yield's decline from its January value. Clearly, bond yield is not an easy predictand; hence, it is apt for probabilistic forecasting.

Let us design a hypothetical BGF for a slightly reframed paradigm. At the beginning of each month, an investor makes a probabilistic forecast of W, which explicitly quantifies the uncertainty in the form of a judgmental distribution function G, by following the procedures of Chapter 9. From a monthly survey of economists and other investors, *à la Wall Street Journal*, the median of individual estimates, x, is extracted; it is considered a realization of predictor X; its role is to revise the judgmental (now prior) distribution function G to the extent merited by the informativeness of X with respect to W. For this purpose, a joint sample of (X, W) has been collected over 12 months.

Options for the prior parameters [in %]:

(A) $M = 1.8, S = 0.9$; (C) $M = 3.5, S = 1.2$;

(B) $M = 2.4, S = 1.7$; (D) $M = 4.6, S = 0.8$.

Options for the joint sample [in %]:

(K) $\{(2.83, 4.05), (3.56, 4.34), (3.91, 4.82), (2.78, 5.13), (3.48, 3.79), (2.57, 3.16), (3.04, 2.77), (2.64, 2.08), (3.29, 2.33), (2.32, 1.06), (2.25, 1.64), (3.07, 1.31)\}$.

(L) $\{(4.17, 1.67), (3.04, 1.48), (3.61, 1.10), (4.76, 3.29), (3.45, 3.80), (3.38, 2.79), (4.39, 4.06), (3.67, 5.12), (2.93, 2.09), (3.95, 2.52), (4.92, 5.34), (4.75, 4.51)\}$.

15.1 Estimate the likelihood parameters a, b, σ.

15.2 Calculate the values of the posterior parameters A, B, T, the sufficiency characteristic SC, and the informativeness score IS.

15.3 Graph the posterior median of W as a function of x; in the same figure, graph the 50% and the 80% posterior central credible intervals of W, each as a function of x.

15.4 Graph in the same figure the prior distribution function G of W, and a posterior distribution function $\Phi(\cdot|x)$ of W; choose the realization x so that the distinction between the two functions is clear; characterize the distinction in words.

15.5 Calculate the prior and the posterior probabilities of W being below x. Calculate the prior and the posterior 0.8-probability central credible intervals of W.

15.6 What assumptions are necessary about X and W (beyond those listed in Section 10.3.3) to justify this application of the BGF? How could these assumptions be validated?

16 **Hotel guests: nonstationary predictand.** The owner of a grand hotel in a large city wants to forecast probabilistically demand W_j, measured in hundreds of guest-nights during month $j \in \{1, \ldots, 12\}$, in order to optimally plan the number of workers needed. Hotel occupancy records were used to estimate the prior mean M_j and the prior standard deviation S_j of W_j for $j = 1, \ldots, 12$. Independently, at the beginning of each month, a consultancy processes consumer survey data and economic model outputs to calculate an outlook index x. It measures consumers' short-term outlook for the US economy: labor market, business conditions, personal finances. Generally, $x > 0$ means an optimistic outlook, and $x < 0$ means a pessimistic outlook. As a variate, X is associated positively with consumer spending, including that for leisure travel, in the coming few months. Hence, the hotelier wants to use X as the predictor of W_j three months ahead. For this purpose, he collected joint realizations $(x(j), w(j))$ during the last calendar year ($j = 1, \ldots, 12$). Specifically, $x(1)$ is the index from October and $w(1)$ is the demand in January; and so on. You have been hired to develop the BGF for this problem and to make a forecast for the month specified in the option.

Note. The outlook index is hypothetical. However, several surveys and indices gauging the consumers' outlook for various elements of the US economy are reported routinely, for example, by the University of Michigan (consumer sentiment) and by the Conference Board (consumer confidence, consumer expectations). Their combination does serve as a predictor of consumer spending (*The Wall Street Journal*, 29 June 2022).

Hints. (i) The values of the outlook index from different months may be treated as realizations of the same predictor X, which is normally distributed.

(ii) There are 12 predictands $W_1, \ldots, W_{12}$, each being normally distributed, but not identically.

(iii) To model the **nonstationarity** of the demand, standardize each W_j to obtain the variate $Z \sim \mathrm{NM}(0, 1)$. This transformed predictand is identically distributed in each month. Likewise, standardize the realization $w(j)$ of W_j to obtain the realization $z(j)$ of Z.

(iv) Develop the BGF for the predictor–predictand pair (X, Z). Having obtained the posterior distribution function $\Phi(\cdot|x)$ of Z, derive the posterior distribution function $\Phi_j(\cdot|x)$ of W_j by applying the derived distribution theory (Section 3.8). Likewise for the posterior density function.

Prior moments (in hundreds of guest-nights) and joint realizations:

j	1	2	3	4	5	6	7	8	9	10	11	12
M_j	36	21	25	37	49	58	61	67	52	47	41	53
S_j	7	5	8	9	11	20	18	19	17	12	14	21
$x(j)$	1.2	−3.3	−1.8	−1.1	−2.8	1.7	3.9	3.1	3.6	6.1	5.5	6.9
$w(j)$	37	19	20	34	46	56	62	71	60	58	47	65

Options for the forecast:

(J) $j = 1, x = -2.4$; (Y) $j = 5, x = 2.9$; (S) $j = 9, x = 7.6$;

(F) $j = 2, x = -4.3$; (E) $j = 6, x = 4.7$; (O) $j = 10, x = 5.4$;

(M) $j = 3, x = -3.1$; (L) $j = 7, x = 5.8$; (N) $j = 11, x = 3.1$;

(A) $j = 4, x = -1.1$; (T) $j = 8, x = 6.6$; (D) $j = 12, x = -3.8$.

16.1 Standardize the realizations of the predictand. Estimate the likelihood parameters a, b, σ (in the regression of X on Z).

16.2 Calculate the values of the posterior parameters A, B, T, the sufficiency characteristic SC, and the informativeness score IS (for the forecast of Z, conditional on $X = x$).

16.3 Write expressions for the posterior distribution function $\Phi(\cdot|x)$ and the posterior density function $\phi(\cdot|x)$ of Z.

16.4 Derive expressions for the posterior distribution function $\Phi_j(\cdot|x)$ and the posterior density function $\phi_j(\cdot|x)$ of W_j, for any $j \in \{1, \ldots, 12\}$.

16.5 Graph in the same figure the prior distribution function G_j of W_j, and the posterior distribution function $\Phi_j(\cdot|x)$ of W_j, given j and x specified in the option.

16.6 Graph in the same figure the prior density function g_j of W_j, and the posterior density function $\phi_j(\cdot|x)$ of W_j, given j and x specified in the option.

Mini-Projects

17 **Small-car market share forecasting**. To decide the number of cars to be ordered from an automaker, the dealer needs to forecast the demand for cars. Then the total demand needs to be disaggregated into types of cars. This is a **hierarchical forecasting** problem. Here is a subproblem. The predictand is W — the market share of small cars [%], which are the compact cars (whose wheelbase is usually between 100 and 105 inches) plus the subcompact cars (still smaller). Data from Edmunds.com suggest a positive association between W and X — the national average price of regular grade gasoline (in dollars per gallon). Use all the data from Table 10.5 to develop the BGF for this problem.

Note. Use the joint realizations, x and w, from the same month. Consequently, the forecast lead time will be zero. To obtain a forecast of W with, say, 6-month lead time, X would have to be forecasted 6 months ahead. This would be a **two-stage forecasting** problem, which you could model thusly.

Stage 1. Using some predictor Y (e.g., an estimator of X outputted by a multivariate econometric model), whose realization y is available with the lead time of 6 months, obtain the posterior density function $\phi_X(\cdot|y)$ of X.

Stage 2. Using predictor X, obtain a family of posterior density functions $\phi(\cdot|x)$ of W for all $x \in \mathcal{X}$.

Integrator. Apply the total probability law to obtain the predictive density function $\pi(\cdot|y)$ of W:

$$\pi(w|y) = \int_{\mathcal{X}} \phi(w|x)\phi_X(x|y)\,dx.$$

This exercise is limited to Stage 2. But we do have a design for the complete Bayesian forecast system.

17.1 Formulate a BGF. Estimate the prior parameters and the likelihood parameters; then calculate the posterior parameters. Calculate the sufficiency characteristic and the informativeness score. Graph the following elements: (i) the estimated parametric prior distribution function of the predictand overlaid on the empirical distribution function; (ii) the estimated linear regression of the predictor on the predictand superposed on the scatterplot of the joint sample; (iii) the estimated parametric distribution function of the residual variate overlaid on the empirical distribution function.

17.2 Validate the four assumptions of the BGF, as described in Section 10.3.3. Document all steps of the validation procedure. Draw conclusions as to the degree of validity of each assumption.

17.3 Illustrate the output from the BGF as follows.

    17.3.1 Graph the posterior mean $E(W|X = x)$ as a function of x; in the same figure, graph the 50% posterior central credible interval as a function of x, and the 90% posterior central credible interval as a function of x; the domain of x in all graphs should be the interval [1.5, 4.5] \$/gal.

    17.3.2 Prepare a probabilistic forecast for general users (car dealers), as described in Section 10.5.2, in terms of three posterior quantiles with $p = 0.1$, given $x = \$3.20$/gal. Write a concise interpretation of this forecast for the dealers.

    17.3.3 Graph in the same figure the prior density function g and the posterior density functions $\phi(\cdot|x_1)$ and $\phi(\cdot|x_2)$, where $x_1 = \$2.40$/gal, $x_2 = \$3.90$/gal; compare the three functions.

    17.3.4 Graph in the same figure the prior distribution function G and the posterior distribution functions $\Phi(\cdot|x_1)$ and $\Phi(\cdot|x_2)$, where $x_1 = \$2.40$/gal, $x_2 = \$3.90$/gal; compare the three functions.

17.4 The graphs created in Exercise 17.3.1 illustrate the positive stochastic dependence of W on X. An article in *The Wall Street Journal* (25 May 2011) stated: "Car buyers' interest in compacts rises along with gas prices." (i) Does this statement for the general public agree with the results from your BGF? How would you summarize the gist of your results for the general public or a car dealer? (ii) Suggest a rationale for this behavior of car buyers. *Hint*: Calculate the extra cost per month to the average US driver due to the rise of the gasoline price from its minimum to its maximum in the period 2007–2008.

18 **Snowmelt runoff forecasting.** Consider the problem of forecasting seasonal snowmelt runoff volume in the western United States (Sections 10.6.1–10.6.2) and the available data (Tables 10.6–10.9). Your task is to develop a Bayesian Gaussian forecaster for the specific application, which is designated by the option.

Options for station and prior sample:

(W) Weiser on the Weiser River.

    (13) Prior sample from 13 years, 1971–1983.

(21) Prior sample from 21 years, 1963–1983.
(31) Prior sample from 31 years, 1953–1983.

(T) Twin Springs on the Boise River.

(13) Prior sample from 13 years, 1971–1983.
(21) Prior sample from 21 years, 1963–1983.
(33) Prior sample from 33 years, 1951–1983.
(72) Prior sample from 72 years, 1912–1983.

Options for forecast lead time:

(1) Forecast made in January; lead time of 7 months.
(2) Forecast made in February; lead time of 6 months.
(3) Forecast made in March; lead time of 5 months.
(4) Forecast made in April; lead time of 4 months.
(5) Forecast made in May; lead time of 3 months.

A complete option includes three specifications. For instance, (T)(21)(3, 4, 5) commands you to develop a forecaster for the Twin Springs station, using the prior sample from 21 years (1963–1983), to forecast the runoff volume based on estimates available at the beginning of March, April, and May; this last specification means that, in essence, you must develop three forecasters, one for each month (although some elements may be common to all three). In every option, collate and use the joint sample from 13 years (1971–1983).

18.1 Formulate a BGF. Estimate the prior parameters and the likelihood parameters; then calculate the posterior parameters. Calculate the sufficiency characteristic and the informativeness score. Graph the following elements: (i) the estimated parametric prior distribution function of the predictand overlaid on the empirical distribution function; (ii) the estimated linear regression of the predictor on the predictand superposed on the scatterplot of the joint sample; (iii) the estimated parametric distribution function of the residual variate overlaid on the empirical distribution function.

18.2 Validate the four assumptions of the BGF, as described in Section 10.3.3. Document all steps of the validation procedure. Draw conclusions as to the degree of validity of each assumption.

18.3 Illustrate the output from the BGF as follows.

    18.3.1 Graph the posterior mean $E(W|X = x)$ as a function of x; in the same figure, graph the 50% posterior central credible interval as a function of x, and the 90% posterior central credible interval as a function of x; the domain of x in all graphs should be the interval $[E(X) - 3Var^{1/2}(X), E(X) + 3Var^{1/2}(X)]$.

    18.3.2 Prepare a probabilistic forecast for general users, as described in Section 10.5.2, in terms of three posterior quantiles with $p = 0.05$, in a year in which $x = 543$. Write a concise interpretation of this forecast for the users.

    18.3.3 Graph in the same figure the prior density function g and the posterior density functions $\phi(\cdot|x_1)$ and $\phi(\cdot|x_2)$, where $x_1 = M - 100$, $x_2 = M + 100$, and M is the prior mean $E(W)$; compare the three functions.

    18.3.4 Graph in the same figure the prior distribution function G and the posterior distribution functions $\Phi(\cdot|x_1)$ and $\Phi(\cdot|x_2)$, where $x_1 = M - 100$, $x_2 = M + 100$, and M is the prior mean $E(W)$; compare the three functions.

19 **Snowmelt runoff forecasting** — *continued*: **comparison of predictors**. It seems plausible that as the forecast lead time decreases from 7 months (when the predictor is the estimate of the runoff volume made in January) to 3 months (when the predictor is the estimate of the runoff volume made in May), the informativeness of the predictor (and hence the informativeness of the forecaster) increases. To validate this hypothesis, complete the following exercises for the designated option (station and prior sample).

19.1 Estimate the likelihood parameters (a, b, σ) of the BGF from the joint sample for each of the five lead times.

19.2 Calculate the sufficiency characteristic, the posterior standard deviation, and the informativeness score for each lead time; indicate the unit of each measure.

19.3 Graph the informativeness score versus the lead time. Interpret the result, and draw a conclusion. (If the result is inconsistent with the hypothesis, then offer a plausible explanation.)

20 **Sum of runoff volumes** — Exercise 18 *continued*. A regional water authority bought rights to seasonal snowmelt runoffs at two gauging stations in Idaho: the Weiser station on the Weiser River and the Twin Springs station on the Boise River (see Sections 10.6.1–10.6.2 for the background and Tables 10.6–10.9 for the available data). The authority wants to develop plans for storing and then supplying water to farms and towns based on the total snowmelt runoff volume in a season. Hence your task: develop a Bayesian Gaussian forecaster of the sum of seasonal snowmelt runoff volumes at the two stations.

It is assumed that Exercise 18 has already been completed for one option, which designated the station, the prior sample, the lead time; for example: (W)(13)(3). For this exercise, a second option should designate the other station, any prior sample, the same lead time; for example, (T)(72)(3). Let the predictor–predictand pairs be (X_1, W_1) for Weiser and (X_2, W_2) for Twin Springs; let $W = W_1 + W_2$; other symbols should be indexed accordingly.

20.1 Complete Exercise 18.1 for the second predictand.

20.2 Estimate the prior correlation between W_1 and W_2. Calculate the posterior (conditional) correlation. Comment on the difference.

20.3 Write expressions for the prior mean and the prior variance of W. Write expressions for the prior density function g of W, and the prior distribution function G of W.

20.4 Compile the values of all parameters needed for the posterior distribution of W, conditional on the realizations of predictors $(X_1 = x_1, X_2 = x_2)$. Write expressions for the posterior mean and the posterior variance of W.

20.5 Write expressions for the posterior density function $\phi(\cdot | x_1, x_2)$ of W, and the posterior distribution function $\Phi(\cdot | x_1, x_2)$ of W.

20.6 Illustrate the output from the BGF as follows.

    20.6.1 Prepare a probabilistic forecast for general users, as described in Section 10.5.2, in terms of three posterior quantiles with $p = 0.1$, in a year in which $x_1 = M_1 + 110$ and $x_2 = M_2 + 130$, where M_j is the prior mean of W_j $(j = 1, 2)$. Write a concise interpretation of this forecast for the user.

    20.6.2 Graph in the same figure the prior density function g and two posterior density functions $\phi(\cdot | x_1, x_2)$, given $(x_1, x_2) = (M_1 - 100, M_2 - 100)$ and $(x_1, x_2) = (M_1 + 100, M_2 + 100)$; compare the three functions.

    20.6.3 Graph in the same figure the prior distribution function G and two posterior distribution functions $\Phi(\cdot | x_1, x_2)$, given the realizations (x_1, x_2) specified in Exercise 20.6.2; compare the three functions.

    20.6.4 Suppose W_1 and W_2 became uncorrelated. Repeat Exercise 20.6.2.

    20.6.5 Compare the figures from Exercises 20.6.2 and 20.6.4. Characterize in words the effect of the correlation between the seasonal snowmelt runoff volumes at the two stations on the density function of the total volume, *a priori* and *a posteriori*.

Table 10.5 National average price of regular grade gasoline [$/gallon] and market share of small cars [%].

2007			2008					
Month	Price	Share	Month	Price	Share	Year	Month	Share
1	2.25	16.25	1	3.13	17.30	2009	2	19.95
2	2.26	16.23	2	3.15	18.00		6	20.10
3	2.40	16.25	3	3.17	18.30		8	27.50
4	2.70	17.45	4	3.35	20.05	2010	2	21.00
5	3.00	18.00	5	3.56	24.50		9	19.00
6	3.17	20.05	6	3.82	26.25			
7	3.00	20.00	7	4.10	25.00			
8	2.95	17.90	8	3.83	22.75			
9	2.65	17.55	9	3.77	20.00			
10	2.70	17.50	10	3.25	19.80			
11	2.99	17.50	11	2.56	20.00			
12	3.15	17.40	12	1.75	18.00			

Source: Edmunds.com.

Table 10.6 Measurements of the seasonal (April–July) snowmelt runoff volume in the Weiser River at Weiser, Idaho; the volumes are in 10^3 ac-ft.

Year	Volume	Year	Volume	Year	Volume
1953	513.7	1963	336.1	1973	234.7
1954	361.8	1964	607.0	1974	625.8
1955	387.3	1965	523.7	1975	549.2
1956	436.0	1966	207.6	1976	459.6
1957	553.0	1967	389.4	1977	38.8
1958	583.3	1968	212.6	1978	472.9
1959	273.0	1969	520.5	1979	275.8
1960	357.1	1970	419.8	1980	407.6
1961	274.2	1971	656.1	1981	283.8
1962	350.7	1972	423.9	1982	589.1
				1983	608.4

Table 10.7 Estimates of the seasonal (April–July) snowmelt runoff volume in the Weiser River at Weiser, Idaho; the estimates are in 10^3 ac-ft and were made at the beginning of each month, from January through May.

Year	Month	Estimate	Year	Month	Estimate	Year	Month	Estimate
1971	1	565.9	1975	1	298.6	1979	1	269.1
	2	631.5		2	319.0		2	248.4
	3	578.2		3	417.2		3	364.4
	4	627.4		4	445.9		4	343.7
	5	594.6		5	470.4		5	352.0
1972	1	422.3	1976	1	413.1	1980	1	289.9
	2	471.5		2	437.6		2	298.2
	3	529.0		3	433.5		3	318.9
	4	483.8		4	462.2		4	397.6
	5	451.0		5	474.4		5	370.9
1973	1	344.4	1977	1	225.1	1981	1	318.9
	2	344.4		2	139.1		2	265.0
	3	307.5		3	90.0		3	294.0
	4	299.3		4	90.0		4	215.3
	5	303.4		5	69.6		5	153.9
1974	1	643.8	1978	1	499.0	1982	1	405.7
	2	721.7		2	535.8		2	418.2
	3	654.5		3	450.0		3	534.1
	4	711.7		4	417.2		4	496.8
	5	617.6		5	458.1		5	534.7
						1983	1	492.7
							2	451.3
							3	513.4
							4	538.2
							5	569.7

Table 10.8 Measurements of the seasonal (April–July) snowmelt runoff volume in the Boise River at Twin Springs, Idaho; the volumes are in 10^3 ac-ft.

Year	Volume	Year	Volume	Year	Volume
1912	798.8	1936	681.8	1960	530.8
1913	695.8	1937	368.6	1961	403.2
1914	652.4	1938	789.5	1962	641.8
1915	348.4	1939	382.6	1963	577.1
1916	877.0	1940	508.8	1964	597.1
1917	791.5	1941	414.3	1965	1032.7
1918	685.3	1942	522.2	1966	378.4
1919	545.6	1943	1092.9	1967	615.7
1920	558.4	1944	368.7	1968	408.8
1921	953.1	1945	597.9	1969	769.3
1922	751.5	1946	741.2	1970	709.5
1923	575.5	1947	640.5	1971	1029.3
1924	249.8	1948	645.3	1972	914.5
1925	760.5	1949	649.1	1973	421.2
1926	295.0	1950	778.8	1974	1080.9
1927	840.0	1951	810.4	1975	770.5
1928	726.4	1952	884.1	1976	669.3
1929	423.1	1953	754.2	1977	167.2
1930	417.0	1954	754.3	1978	730.5
1931	295.6	1955	555.8	1979	389.9
1932	662.6	1956	983.9	1980	693.6
1933	559.6	1957	800.9	1981	462.2
1934	309.0	1958	823.5	1982	930.5
1935	544.4	1959	539.6	1983	905.6

Table 10.9 Estimates of the seasonal (April–July) snowmelt runoff volume in the Boise River at Twin Springs, Idaho; the estimates are in 10^3 ac-ft and were made at the beginning of each month, from January through May.

Year	Month	Estimate	Year	Month	Estimate	Year	Month	Estimate
1971	1	819.3	1975	1	564.5	1979	1	459.8
	2	939.2		2	630.9		2	425.9
	3	899.2		3	750.4		3	466.4
	4	965.8		4	850.0		4	392.1
	5	932.5		5	869.9		5	371.8
1972	1	712.7	1976	1	723.8	1980	1	547.6
	2	845.9		2	697.3		2	642.3
	3	905.8		3	717.2		3	655.8
	4	885.9		4	750.4		4	689.6
	5	879.2		5	743.8		5	669.3
1973	1	599.4	1977	1	418.4	1981	1	588.1
	2	586.1		2	232.4		2	486.8
	3	546.2		3	119.5		3	425.9
	4	512.9		4	112.8		4	419.2
	5	512.9		5	99.6		5	425.9
1974	1	852.6	1978	1	783.6	1982	1	824.8
	2	905.8		2	803.5		2	804.6
	3	816.8		3	836.7		3	851.8
	4	956.3		4	750.4		4	905.9
	5	962.9		5	790.2		5	966.8
						1983	1	804.6
							2	791.0
							3	811.3
							4	899.2
							5	905.9

11

Verification of Forecasts

Periodically, a set of forecasts should be verified against actual realizations of the predictand in order to evaluate, and to track over time, the performance of the forecaster. Taking the viewpoint of a forecast user — the rational decider who bears the consequences of decisions based on forecasts — the Bayesian verification theory prescribes two necessary and sufficient attributes of probabilistic forecasts: *calibration* and *informativeness*.

For probabilistic forecasts (judgmental or statistical) in the form of distribution functions of a continuous predictand, the verification problem is theoretically involved. Therefore, this chapter presents only the necessary, or specialized (not general), measures of calibration and measures of informativeness. Still, they offer feedback to the forecaster and statistics to the decider regarding the "goodness" of probabilistic forecasts.

11.1 Data and Inputs

Several verification measures are defined using pointwise representation of a distribution function of the predictand. This is a natural representation of the judgmental forecast. A discretization of a continuous distribution function is prescribed to get a pointwise representation of the statistical forecast.

11.1.1 Variates and Samples

Let W be the *predictand* — a continuous variate whose sample space is $\mathcal{W}$, and whose realization w is forecasted; $w \in \mathcal{W}$.

Let Y_p be the *quantile variate* — a variate who realization y_p is the p-probability quantile of W assessed judgmentally by the forecaster; $y_p \in \mathcal{W}$ for $p \in (0, 1)$.

Let $\mathcal{P}$ be a set of probabilities fixed in Step 4 of the quantile technique (Section 9.2.2) for which the forecaster assesses quantiles:

$$\mathcal{P} = \{p : p = p_1, \ldots, p_I; \, 0 < p_1 < \cdots < p_I < 1\}. \tag{11.1}$$

Let $\mathbf{Y} = \{Y_p : p \in \mathcal{P}\}$ be the vector of quantile variates; its realization $\mathbf{y} = \{y_p : p \in \mathcal{P}\}$ is the vector of quantiles that constitute the *pointwise representation* of the judgmental distribution function G of W, as defined in Section 9.2.1.

Probabilistic forecasts can be verified only after the forecaster has assessed quantiles and recorded predictand realization on many occasions. Then one can form a *sample of joint realizations* of the quantile vector $\mathbf{Y}$ and the predictand W:

$$\{(\mathbf{y}(n), \, w(n)) : n = 1, \ldots, N\}, \tag{11.2}$$

where n is the index of realizations (forecasting occasions), N is the *sample size*, $\mathbf{y}(n) = \{y_p(n) : p \in \mathcal{P}\}$ is the vector of quantiles assessed on the nth occasion, and $w(n)$ is the realization of the predictand on the nth occasion.

For example, when $I = 3$ and $\mathcal{P} = \{0.25, 0.5, 0.75\}$, the vector of quantile variates is $\mathbf{Y} = (Y_{0.25}, Y_{0.5}, Y_{0.75})$ and its realization on the nth occasion is $\mathbf{y}(n) = (y_{0.25}(n), y_{0.5}(n), y_{0.75}(n))$. The joint realization of $(\mathbf{Y}, W)$ on the nth occasion is

$$(\mathbf{y}(n), w(n)) = (y_{0.25}(n), y_{0.5}(n), y_{0.75}(n); w(n)).$$

11.1.2 Discretization Algorithm

The algorithm is applicable whenever the probabilistic forecast of a continuous predictand W takes the form of a continuous distribution function of W. It may come from any forecaster, but for the sake of specificity it comes here from the Bayesian Gaussian forecaster (BGF) of Chapter 10. The purpose of the algorithm is to replace the posterior distribution function $\Phi(\cdot|x)$ of W by its *pointwise representation* that matches in form the representation of the judgmental distribution function in Section 11.1.1.

Step 0. Given is a joint sample of the predictor–predictand pair (X, W), on which the verification of forecasts produced by the BGF is to be performed:

$$\{(x(n), w(n)) : n = 1, \ldots, N\}.$$

Step 1. Fix the set $\mathcal{P}$ of probabilities, defined by expression (11.1), in a manner similar to that prescribed in Step 4 of Section 9.2.2; however, different or more values of p may be fixed if they are relevant to a particular decision problem. The sample size N needs to be considered as well because it must be larger whenever the extreme p values are closer to 0 and 1.

Step 2. From the BGF with specified values of the posterior parameters A, B, T, get expression (10.18) for $\Phi^{-1}(\cdot|x)$ — the posterior quantile function of W, conditional on $X = x$.

Step 3. For every $n \in \{1, \ldots, N\}$, take $x(n)$ and calculate I posterior quantiles of W:

$$y_p(n) = \Phi^{-1}(p|x(n)), \qquad p = p_1, \ldots, p_I.$$

Step 4. For every $n \in \{1, \ldots, N\}$, match the I quantiles with the realization $w(n)$ and form the vector

$$(\mathbf{y}(n), w(n)) = (y_{p_1}(n), \ldots, y_{p_I}(n); w(n)).$$

Collate all N vectors into a *sample of joint realizations* of the quantile vector $\mathbf{Y}$ and the predictand W:

$$\{(\mathbf{y}(n), w(n)) : n = 1, \ldots, N\}.$$

This joint sample is like the joint sample (11.2). Hence, each may be inputted into the common verification method.

11.1.3 Aggregate Verifications

Ideally, the joint sample (11.2) would be **large** and **homogeneous**, in the sense that one forecaster provided quantiles of one predictand on many occasions under stationary conditions. Whereas these properties are approximately satisfied in some forecasting problems (e.g., daily minimum temperatures forecasted with the lead time of 24 h by a model on 30 consecutive days), they are lacking in other problems. For instance, in Task 9.1, one student assessed quantiles of three predictands on one occasion. Thus the sample size for forecaster–predictand is $N = 1$. Yet, with 100 students in a class, 300 forecasts were made. Could not these forecasts be verified in aggregate to characterize performance of the group, to get a sense of human forecasting capabilities? In line with this idea, four types of verifications are defined based on the level of joint sample **aggregation**.

1. *Verification for an expert with respect to a predictand.* This is based on a sample of forecasts of the same predictand prepared by one expert on many occasions.
2. *Verification for an expert.* This is based on a sample of forecasts of different predictands prepared by one expert on one or many occasions. (The predictands are assumed to be exchangeable.)
3. *Verification for a group of experts with respect to a predictand.* This is based on a sample of forecasts of the same predictand prepared by different experts on one or many occasions. (The experts are assumed to be exchangeable.)
4. *Verification for a group of experts.* This is based on a sample of forecasts of different predictands prepared by different experts on one or many occasions. (The experts and the predictands are assumed to be jointly exchangeable.)

Exchangeability is defined formally in the Bayesian probability theory. Its intuitive meaning here is this. In type 2 verification, the exchangeability of predictands means that if type 1 verification were performed for each predictand, then all sets of the verification measures would have identical values, implying identical performance of the expert with respect to each predictand. The reader should state a similar interpretation for type 3 and type 4 verifications.

Inasmuch as the very reason for an aggregate verification is the small sample size for type 1 verification, the exchangeability is not validated empirically. Instead, the degree of its validity should be reasoned by an analyst, and then should mollify the interpretation of the verification results: to whom and to what they apply.

11.2 Calibration

11.2.1 The Concept

Suppose an expert had assessed quantiles $(y_{0.25}, y_{0.5}, y_{0.75})$ of some continuous predictand W whose realization w was next observed. It is not meaningful to evaluate the "goodness" of such a single forecast. In particular, it is neither meaningful to compare the realization w with any of the quantiles, nor meaningful to conclude that the forecaster was "right" or "wrong" in any sense. After all, as long as p was neither 0 nor 1, each quantile conveyed the possibility of observing $w < y_p$ or $y_p < w$.

The performance of a probabilistic forecaster may be verified only based on a joint sample (11.2); and the verification measures should answer two sets of questions — those posed later in Section 11.3, and these:

- Can the two numbers (y_p, p) given by a forecaster be interpreted: the nonexceedance event $W \le y_p$ has probability p?
- Can the set of $2I$ numbers $\{(y_p, p) : p \in P\}$ be interpreted as I points of a distribution function of W? Can it be used directly (with only some interpolation and extrapolation) in a decision model?

In order that users may be assured of the affirmative answers to these questions, and thus may take the $2I$ numbers at their face values, *normative interpretation* of the forecast must be maintained consistently over time or across predictands. It is the uniqueness of this interpretation that makes the distribution function a means of communicating uncertainty from a forecaster to a decider, from an analyst to the public, from one individual to another. Consistent interpretability requires *good calibration* — an attribute of probabilistic forecasts that should be verified.

11.2.2 Order Probabilities

A general definition of the well-calibrated forecaster of a continuous predictand W, which parallels the definition of Chapter 6 for a binary predictand, is stated in Section 11.7.1. A verification method based on that definition would be involved. Therefore, we present only an *empirical method* for verifying in-the-large the calibration attribute of forecast quantiles. This method rests on five premises.

1. For a fixed probability $p \in \mathcal{P}$, the quantile y_p specifies the event $W \leq y_p$. When y_p is different on each forecasting occasion, the forecasted event $W \leq y_p$ is different on each forecasting occasion as well.
2. To recognize this, the quantile y_p on a particular occasion is viewed as a realization of the quantile variate Y_p. Thus on occasion n, the variates (Y_p, W) have realization $(y_p(n), w(n))$, which may be different for every n $(n = 1, \ldots, N)$.
3. With two variates, one can define the *order event*: $W \leq Y_p$. Its probability, called the *order probability*, is denoted by

$$R(p) = P(W \leq Y_p), \qquad p \in \mathcal{P}. \tag{11.3}$$

This is the probability of observing the order $W \leq Y_p$ on any forecasting occasion. The function R is nondecreasing on $\mathcal{P}$.
4. Given the joint sample (11.2), the estimator of the order probability, for a fixed $p \in \mathcal{P}$, is the relative frequency:

$$R(p) = \frac{\text{number of joint realizations with } \{w(n) \leq y_p(n)\}}{\text{sample size } N}. \tag{11.4}$$

5. Even though the forecasted event $W \leq y_p$ is different on each occasion, the probability p attached to it is invariant. In effect, the forecaster states that

$$p = P(W \leq Y_p), \qquad p \in \mathcal{P}. \tag{11.5}$$

Should not $R(p) \approx p$ for sufficiently large N?

11.2.3 Distribution Calibration Function

To characterize the performance of the forecaster with respect to her ability to maintain consistently the normative interpretation of the quantiles, two terms are introduced.

Definition A graph of the order probability $R(p)$ versus the forecast probability p, for all $p \in \mathcal{P}$, is called the *distribution calibration function* (DCF) of the forecaster.

Definition The forecaster of a continuous predictand is said to be *well calibrated in-the-large* if for all $p \in \mathcal{P}$,

$$R(p) = p.$$

The qualifier "in-the-large" is meant to remind us that this calibration measure is not in the form of conditional probabilities, which would be outputted from the Bayesian processor of forecast. Because it is independent of the assessed quantile vector **y**, this DCF is only a necessary measure of calibration (and, therefore, a weaker measure than the PCF derived in Chapter 6 for a binary predictand).

For the minimal set $\mathcal{P} = \{0.25, 0.50, 0.75\}$, Table 11.1 summarizes the calibration analysis, and Figure 11.1 shows two examples of the DCF. Inasmuch as p can be any real number from the interval $(0, 1)$, and R is a nondecreasing function of p, it is permissible to draw an interpolating curve (or line segments) from the corner $(0, 0)$, through the points $(p, R(p))$, to the corner $(1, 1)$.

11.2.4 Generic Calibration Functions

When forecasts are prepared judgmentally, the shape of the DCF reveals biases in judgments. It is instructive, therefore, to learn how to interpret the distribution calibration functions for two generic cases (Figure 11.1).

Table 11.1 Summary of the calibration analysis for the minimal (just three quantiles) probabilistic forecasts of a continuous predictand.

Order event	$W \leq Y_{0.25}$	$W \leq Y_{0.5}$	$W \leq Y_{0.75}$
Forecast probability	0.25	0.50	0.75
Order probability	$R(0.25)$	$R(0.50)$	$R(0.75)$

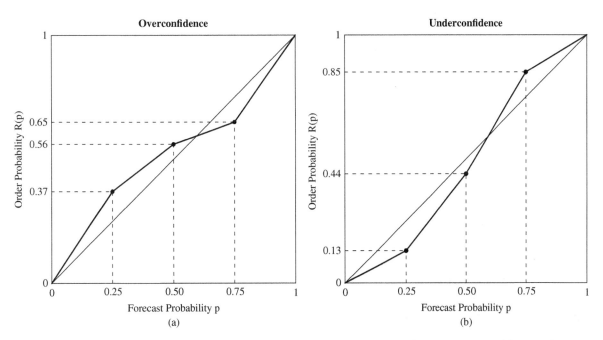

Figure 11.1 The distribution calibration function (DCF) R of the forecaster who provides three p-probability quantiles ($p = 0.25, 0.50, 0.75$) of a continuous predictand W. Neither of the two forecasters is well calibrated: (a) is overconfident, (b) is underconfident.

A. *Overconfidence.* The DCF is concave for $p < \frac{1}{2}$ and convex for $p > \frac{1}{2}$. It implies that the three forecast quantiles are, in-the-large, biased thusly.
  - The 0.25-quantile is too high, as $P(W \leq Y_{0.25}) = 0.37$, while it should equal 0.25.
  - The median is also a bit too high, as $P(W \leq Y_{0.5}) = 0.56$.
  - The 0.75-quantile is too low, as $P(W \leq Y_{0.75}) = 0.65$, while it should equal 0.75.

  Manifestly, the forecaster's estimates $y_{0.5}$ are not, on average, the medians, but instead the 0.56-quantiles; and the 50% central credible intervals $(y_{0.25}, y_{0.75})$ are, on average, too narrow: only $65 - 37 = 28\%$ of realizations w fall inside them instead of 50%. A too narrow central credible interval implies that the degree of uncertainty about W conveyed by the forecast is lower than the degree of uncertainty actually observed. Ergo, the forecaster's judgments have a tendency toward *overconfidence*. The advice for future assessments would be: spread the quantiles $y_{0.25}$ and $y_{0.75}$.

B. *Underconfidence.* The DCF is convex for $p < \frac{1}{2}$ and concave for $p > \frac{1}{2}$. It implies that the three forecast quantiles are, in-the-large, biased thusly. The forecaster's estimates $y_{0.5}$ are not, on average, the medians, but instead the

0.44-quantiles; and the 50% central credible intervals $(y_{0.25}, y_{0.75})$ are, on average, too wide: $85 - 13 = 72\%$ of realizations w fall inside them instead of 50%. A too wide central credible interval implies that the degree of uncertainty about W conveyed by the forecast is higher than the degree of uncertainty actually observed. Ergo, the forecaster's judgments have a tendency toward *underconfidence*. The advice for future assessments would be: contract the quantiles $y_{0.25}$ and $y_{0.75}$.

11.2.5 Purposes of Calibration Function

The DCF serves two purposes. First, it answers the basic question of a decider: Can the probability p attached to the quantile y_p be taken at its face value? If the forecaster is not well calibrated then the answer is negative. In such a case, the DCF offers also a solution: the decider can replace p with $R(p)$; in effect, the p-probability quantile, y_p, provided by the forecaster, becomes the $R(p)$-probability quantile, $y_{R(p)}$, for the decider. This process of mapping p into $R(p)$ is called *recalibration*.

Second, the DCF provides feedback to the forecaster. If the forecaster is not well calibrated, then: (i) an expert who prepared the judgmental forecasts may learn to adjust her judgments of uncertainty in the future (and thereby to improve her metacognitive skill — as discussed in Section 6.5); (ii) an analyst who developed the forecasting model may identify the sources of miscalibration to determine whether or not an alteration of the model formulation or of the parameter estimates is warranted.

11.2.6 Calibration Score

For lesser purposes (e.g., for monitoring the forecaster's performance over time, on a monthly, quarterly, or yearly time scale), it may be convenient to compress a function into a score and then to compare the scores.

Looking at Figure 11.1, the visual perception of the degree of calibration is equivalent to the visual perception of the goodness of fit of the distribution calibration function R to the main diagonal. This analogy makes the *uniform distance* — a measure of goodness of fit (recall Section 3.6.2) — an apt measure here as well.

The *calibration score* is defined as the maximum absolute difference between the distribution calibration function and the line of perfect calibration:

$$CS = \max_{p \in P} |R(p) - p|. \tag{11.6}$$

The CS is bounded, $0 \leq CS < 1$, with $CS = 0$ implying perfect calibration in-the-large and CS near 1 implying the worst possible calibration in-the-large.

From a practical point of view, considering the approximate nature of judgmental assessments and of relative frequencies (11.4), a forecaster achieving $CS < 0.05$ may be viewed as approximately well calibrated in-the-large. Each DCF in Figure 11.1 yields $CS = 0.12$. The lowest CS that some weather forecasters achieved consistently from month to month, while making forecasts daily, was around 0.03. (This and other examples are presented in Section 11.6.)

11.2.7 Uniform Calibration

There is an alternative *empirical method* for verifying in-the-large the calibration of any forecaster who outputs a continuous distribution function of predictand W. The method exploits the *probability integral transform* (PIT), defined already in Exercise 1 of Chapter 3. Given a prior distribution function G of W, the PIT outputs variate $U = G(W)$ whose distribution is uniform, $UN(0, 1)$, as defined in Section 2.5. This is a necessary property of the marginal distribution of W. It is logical, therefore, to require this property of every conditional distribution of W.

Definition   The forecaster of a continuous predictand W, who provides conditional distribution functions from the family $\{\Phi(\cdot|x) : x \in \mathcal{X}\}$, is said to be *uniformly well calibrated* if, for every $x \in \mathcal{X}$, the variate

$$U = \Phi(W|x)$$

has the uniform distribution $UN(0, 1)$.

This PIT between W and U has an attractive property: although the forecast distribution of W is conditional on $X = x$, the distribution of U is not. To show this, let us derive $H(\cdot|x)$ — the distribution function of U, conditional on $X = x$. For any $u \in (0, 1)$,

$$H(u|x) = P(U \le u|X = x)$$
$$= P(\Phi(W|x) \le u)$$
$$= P(W \le \Phi^{-1}(u|x))$$
$$= \Phi(\Phi^{-1}(u|x)|x) = u.$$

Hence, the variate U is uniform and stochastically independent of the variate X. Exercise 1 asks you to derive the corresponding conditional density function of U. Another attractive property of the above PIT is this theorem.

Theorem *(Equivalent calibrations)*   *The forecaster of a continuous predictand is well calibrated in-the-large if and only if he is uniformly well calibrated.*

The proof is deferred to Section 11.7.3. But an intuitive explanation of this equivalence rests of the observation that neither calibration type gauges the calibration of a forecaster against the standard — a prior distribution function of the predictand W.

11.2.8   Uniform Calibration Score

The stochastic independence between U and X implies that, in repetitive forecasting of W, every realization (x, w) of the predictor–predictand pair (X, W) should yield a realization $u = \Phi(w|x)$ of the identical uniform variate U. This fact undergirds the following algorithm for verification of uniform calibration (equivalently, of calibration in-the-large). The algorithm is detailed for the BGF, for the sake of specificity, but can be readily adapted to any forecaster.

Step 0.   Given is a joint sample of the predictor–predictand pair (X, W), on which the verification of forecasts is to be performed:

$$\{(x(n), w(n)) : n = 1, \dots, N\}.$$

Step 1.   From the BGF with specified values of the posterior parameters A, B, T, get expression (10.16) for $\Phi(w|x)$.

Step 2.   For every $n \in \{1, \dots, N\}$, take $(x(n), w(n))$ and calculate the nonexceedance probability:

$$u(n) = \Phi(w(n)|x(n)).$$

Collate the sample of the variate U:

$$\{u(n) : n = 1, \dots, N\}.$$

Step 3.   Use this sample and the meta-Gaussian plotting positions (Section 3.2) to construct the empirical distribution function of the variate U:

$$\{(u_{(n)}, p_n) : n = 1, \dots, N\}.$$

Step 4. Evaluate the goodness of fit of the UN(0, 1) distribution function to the empirical distribution function of U: the better the fit, the better the uniform calibration. Any of the three procedures described in Section 3.6 may be applied. (i) The graphical comparison. (ii) The K-S goodness-of-fit test. (iii) The uniform distance measure: relabel MAD as *UCS* — the *uniform calibration score*; calculate it:

$$UCS = \max_{1 \leq n \leq N} |p_n - u_{(n)}|. \tag{11.7}$$

The *UCS* is bounded, $0 \leq UCS < 1$, with $UCS = 0$ implying perfect uniform calibration and *UCS* near 1 implying the worst possible uniform calibration.

To adapt the algorithm to any forecaster, replace the symbol $\Phi(\cdot|x(n))$ by the symbol $\Phi(\cdot|n)$, which denotes the distribution function of W on the nth forecasting occasion, $n \in \{1, \ldots, N\}$. In particular, when an expert prepared N forecasts of W via the quantile technique (Section 9.2), $\Phi(\cdot|n)$ would be a parametric distribution function of W fitted to the pointwise representation of the judgmental distribution function (as described in Section 9.3) on the nth forecasting occasion.

Whereas *UCS* is a convenient summary measure for tracking the calibration of an expert, it does not offer feedback to the expert regarding specific judgmental biases. Hence, the *UCS* does not replace the DCF.

However, when the number of forecasts N is too small to reliably estimate the DCF (at three or more probabilities $p \in \mathcal{P}$), the *UCS* can still be calculated because it is robust: it possesses all the properties of the maximum absolute difference (MAD) and of the Kolmogorov-Smirnov (K-S) statistic (recall Section 3.6).

Example 11.1 *(BGF of snowmelt runoff)*

Section 10.6 describes the Bayesian Gaussian forecaster of the seasonal snowmelt runoff volume at Weiser, Idaho, and reports its parameter values (Table 10.2). The present task is to verify empirically the uniform calibration of this forecaster on the joint sample (Table 10.1) of the predictor–predictand pair (X, W) from $N = 7$ years (1977–1983); this is the same sample from which the likelihood parameters of the BGF were estimated. Table 11.2 reports the output from the verification algorithm, and Figure 11.2 shows the fit of the uniform distribution function, $P(U \leq u)$ $= u$ for $0 < u < 1$, to the empirical distribution function $\{(u_{(n)}, p_n) : n = 1, \ldots, 7\}$. With the sample size $N = 7$ being small and $UCS = 0.115$ implying an adequate fit, one may conclude that this forecaster is approximately well calibrated in-the-large. Exercise 3 asks you to execute the K-S test and to determine the significance level of this conclusion.

Note. The singularly large *UCS* in Table 11.2 is due to the near identicalness of realizations $u_{(4)}$ and $u_{(5)}$, which causes the empirical distribution function to jump steeply — an artifact of the small sample. As sample size increases, the height of each jump decreases, and *UCS* inevitably decreases if the forecaster is uniformly well calibrated.

Table 11.2 Output from the uniform calibration algorithm applied to the joint sample of (X, W) from Table 10.1; Bayesian Gaussian forecaster of the snowmelt runoff; January forecast; Weiser, Idaho; $UCS = 0.115$.

Year	n	(n)	$u_{(n)}$	p_n	$|p_n - u_{(n)}|$
1977	7	1	0.0592	0.0598	0.0006
1978	6	2	0.0975	0.1743	0.0768
1981	3	3	0.3043	0.3267	0.0224
1983	1	4	0.5576	0.5	0.0576
1979	5	5	0.5580	0.6733	**0.1153**
1980	4	6	0.8748	0.8257	0.0491
1982	2	7	0.8927	0.9402	0.0475

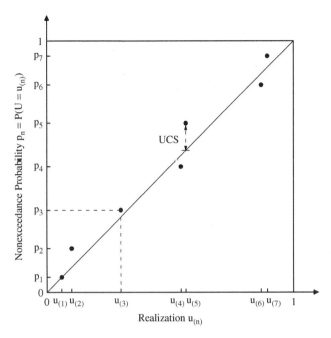

Figure 11.2 The *uniform calibration function* of the forecaster: a graph of the empirical distribution function $\{(u_{(n)}, p_n) : n = 1, \ldots, 7\}$ of the variate U, calculated in Table 11.2; it yields $UCS = 0.115$; Bayesian Gaussian forecaster of the snowmelt runoff; January forecast; Weiser, Idaho.

11.3 Informativeness

11.3.1 The Concept

What is the forecaster's ability *to predict* the realization of a continuous variate W? What is the forecaster's ability *to reduce uncertainty* about the predictand W? What is the forecaster's ability *to provide valuable information* for making decisions? These are equivalent questions that probe the second attribute of the forecaster, which is called *informativeness*.

The ultimate measure of informativeness is the economic value of the forecaster to rational deciders — those who make optimal decisions based on forecasts using decision models developed according to the principles of Bayesian decision theory, which we shall study in later chapters. When two or more forecasters provide forecasts of the same predictand, their economic values to a particular decider can be determined, and then used to order the forecasters. However, this preference order may be different for every decider.

Enter the concept of informativeness, already defined and operationalized in three other contexts (Sections 5.5, 6.3, 10.4). Its purpose is to establish the order of forecasters that corresponds to the order of their economic values for every rational decider. Hence the following definition.

Definition Forecaster A is said to be *more informative than* forecaster B with respect to predictand W if, for every rational decider, the economic value of forecasts provided by A is **at least as high as** the economic value of forecasts provided by B.

The order of forecasters in terms of the *informativeness relation* does not always exist. Given two forecasters, it is possible that neither A is more informative than B, nor B is more informative than A. But if an order does exist, then it can be established via a statistical measure provided by the Bayesian verification theory; it is not necessary to determine the economic values of forecasters A and B for every rational decider!

With the concept in place, our objective is to formulate an informativeness measure for verification of forecasts. Its purpose is to tell us whether the economic value has increased or decreased from one set of forecasts to another, and in relation to perfect forecasts (which establish an upper bound on the economic value) and uninformative forecasts (whose economic value is zero).

11.3.2 Informativeness Measures

Sufficient statistics

Often, though not always, all predictive information contained in a vector of quantiles **y** may be summarized in two sufficient statistics:

- a *predictor of central tendency* of W, for example, the median of W;
- a *predictor of uncertainty about* W, for example, a central credible interval (CCI) of W — its width, or its skew, or both.

Such an idea was already exploited in Chapter 9 (Section 9.3 and Exercises 5, 7) where a set of I points $\{(y_p, p): p \in \mathcal{P}\}$ representing the judgmental distribution function was "replaced" by a model having only two parameters, for example, the NM(μ, σ) model. Its two parameters, the mean μ and the standard deviation σ, constitute the sufficient statistics of the judgmental distribution function of a continuous predictand.

A two-dimensional measure of informativeness is theoretically involved. Therefore, we adopt a one-dimensional measure, already derived in Chapter 10: the informativeness score, *IS*. It is sufficient for some types of forecasts but not for others. Here is the explanation.

Single sufficient statistic

The type of probabilistic forecast of a continuous predictand W for which a single statistic constitutes a sufficient measure of informativeness is the posterior distribution function outputted from the Bayesian Gaussian forecaster (BGF): this statistic is *SC* or *IS* or *T*, as defined in Section 10.4. To explain this simplicity, observe that the posterior standard deviation T of the predictand W is constant across all forecasting occasions. Ergo, each forecast reduces the uncertainty about W from its prior magnitude S (the prior standard deviation of W) to its posterior magnitude T by the constant value $T - S$. Consequently, the two sufficient statistics postulated above collapse into one. And because S is independent of the forecaster, while T may differ from one forecaster to another, T is a sufficient measure for ordering the forecasters in terms of the informativeness relation; and so are the equivalent measures, *SC* and *IS*.

Qualified interpretation of *IS*

A concomitant property of forecasts outputted from the BGF is easy to judge visually: for any probability $p \in (0, 0.5)$, the $(1 - 2p)$-probability posterior CCI has constant width and is symmetrical about the posterior median of W. This is the *benchmark property* of forecasts that qualifies the interpretation of *IS*.

- If the forecasts exhibit the benchmark property, at least approximately, then one can say: the *IS* is a sufficient measure of informativeness of the set of forecasts.
- If the forecasts do not exhibit the benchmark property, then one can only say: the *IS* is a measure of informativeness of the set of medians.

With these qualifications in mind, we are ready to calculate *IS* for the purpose of verifying (sufficiently or partially) the informativeness of a set of probabilistic forecasts of a continuous predictand.

11.3.3 Informativeness Score: Gaussian Model

This algorithm is based on the *normal-linear likelihood model*; it is to be fully executed if its assumptions are validated in Step 2. Otherwise, it branches out to a generalized algorithm presented in Section 11.3.4.

Step 0. (Samples). From the joint sample (11.2) of the quantile vector **Y** and the predictand W, extract the joint sample of the median variate $Y_{0.5}$ and the predictand W:

$$\{(y_{0.5}(n), w(n)) : n = 1, \ldots, N\}.$$

Simplify the notation: replace $Y_{0.5}$ by Y, and replace $y_{0.5}(n)$ by $y(n)$. Henceforth, Y is treated as an *estimator* of W; and the joint sample of (Y, W) is

$$\{(y(n), w(n)) : n = 1, \ldots, N\}. \tag{11.8}$$

Next, collate a prior sample of W:

$$\{w(n) : n = 1, \ldots, N'\}. \tag{11.9}$$

When this sample is just the marginal sample of the joint sample, $N' = N$. When there exist realizations of W which have not been forecasted, they should be included, so that $N' > N$. (When verifying forecasts from the BGF, this may be the same prior sample from which the prior mean and variance of W were estimated in Section 10.3.1.)

Note. The symbol Y is not only a simplification, but also a generalization — Y may be any quantile variate or any other estimator of W.

Step 1. (Likelihood parameters). The *normal-linear model* (10.5)–(10.7) is adapted to quantify the stochastic dependence between Y and W. Retaining all other symbols for the sake of analogy,

$$E(Y|W = w) = aw + b,$$
$$Var(Y|W = w) = \sigma^2. \tag{11.10}$$

Estimate the likelihood parameters a, b, σ from the joint sample (11.8) by applying the procedures from Section 10.3.2.

Step 2. (Validation of assumptions). Apply the procedures from Section 10.3.3 to validate assumptions 2, 3, 4 of the normal-linear model: normality, linearity, homoscedasticity. If one or more of these assumptions are violated to the extent that would render the model invalid (not even approximately valid), then exit this algorithm, and proceed to Section 11.3.4.

Step 3. (Prior standard deviation). Estimate the prior standard deviation S of predictand W from the prior sample (11.9), as detailed in Section 10.3.1.

Step 4. (Informativeness score). Calculate the *informativeness score* of estimator Y with respect to predictand W:

$$IS = \left[\left(\frac{aS}{\sigma}\right)^{-2} + 1\right]^{-\frac{1}{2}}. \tag{11.11}$$

The properties of this *IS* are listed below its definition — equation (10.26) in Section 10.4.4.

11.3.4 Informativeness Score: Meta-Gaussian Model

This is a generalized algorithm to be executed whenever the validation procedure in Step 2 of Section 11.3.3 reveals one or more of the following properties of the joint sample of (Y, W): (i) nonnormality of the residual variate; (ii) nonlinearity of the regression; (iii) heteroscedasticity of the residual variate. The algorithm is based on the *meta-Gaussian likelihood model*, which allows for (i) any marginal distribution functions of Y and W, and (ii) a particular type of nonlinearity and heteroscedasticity. Its theory is beyond the scope of this book. Its application is included here because it offers a versatile tool as well as an exemplar of the capabilities of advanced models.

Step 0. (Samples). Repeat Step 0 from Section 11.3.3.

Step 1. (Empirical distribution functions). Construct two empirical distribution functions, $\check{H}$ of Y and $\check{G}$ of W, by applying the algorithm of Section 3.2 with the meta-Gaussian plotting positions. Use the marginal sample of Y from the joint sample (11.8) to get the function

$$\check{H}(y_{(n)}) = p_n, \qquad n = 1, \ldots, N.$$

Use the prior sample (11.9) of W to get the function

$$\check{G}(w_{(n)}) = p_n, \qquad n = 1, \ldots, N'.$$

Note: When $N' > N$, the plotting position p_n takes on a different value in each function.

Step 2. (Normal quantile transforms). Read Exercises 2 and 3 of Chapter 3 to learn the definition and the properties of the *normal quantile transform* (NQT). Define the variate Z as the NQT of estimator Y, and define the variate V as the NQT of predictand W:

$$Z = Q^{-1}(H(Y)),$$
$$V = Q^{-1}(G(W)).$$

Note: Each variate, Z and V, is NM(0, 1). Exercise 4 asks you to prove this.

Step 3. (Transformed realizations). Apply the *empirical* NQT of each variate to transform (i) every realization of Y into a realization of Z, and (ii) every realization of W into a realization of V:

$$z_{(n)} = Q^{-1}(\check{H}(y_{(n)})) = Q^{-1}(p_n), \qquad n = 1, \ldots, N;$$
$$v_{(n)} = Q^{-1}(\check{G}(w_{(n)})) = Q^{-1}(p_n), \qquad n = 1, \ldots, N'.$$

Note: Employ the rational function approximation to Q^{-1} (Table 2.3).

Step 4. (Transformed joint sample). Each sequence calculated above forms an **ordered** marginal sample of Z and V, respectively:

$$\{z_{(n)} : n = 1, \ldots, N\},$$
$$\{v_{(n)} : n = 1, \ldots, N'\}.$$

For each joint realization $(y(n), w(n))$ of the original variates (Y, W) identify the matching joint realization $(z(n), v(n))$ of the transformed variates (Z, V). Collate them into a *joint sample* of (Z, V):

$$\{(z(n), v(n)) : n = 1, \ldots, N\}. \tag{11.12}$$

Note: When $N' > N$, the variate V has N' realizations $v_{(n)}$ in the ordered sample, but only N of them have matching realizations $z_{(n)}$; ignore the unmatched realizations.

Step 5. (Likelihood parameters). The *normal-linear model* (10.5)–(10.7) is adapted to quantify the stochastic dependence between the transformed variates Z and V. Retaining all other symbols for the sake of analogy,

$$E(Z|V = v) = av + b,$$
$$Var(Z|V = v) = \sigma^2. \tag{11.13}$$

Estimate the likelihood parameters a, b, σ from the joint sample (11.12) by applying the procedures from Section 10.3.2.

Step 6. (Validation of assumptions). Apply the procedures from Section 10.3.3 to validate assumptions 2, 3, 4 of the normal-linear model: normality, linearity, homoscedasticity. *Note*: Compare these results with those from Step 2 of Section 11.3.3 (if it was executed earlier), to judge the effectiveness of the NQTs on the validity of the normal-linear likelihood model in the sample space of Z and V.

Step 7. (Informativeness score). Calculate the *informativeness score* of Z with respect to V:

$$IS = \left[\left(\frac{a}{\sigma} \right)^{-2} + 1 \right]^{-\frac{1}{2}}. \tag{11.14}$$

The properties of this *IS* are the same as those listed below equation (10.26) in Section 10.4.4. Three additional properties and explanations are essential.

1. This *IS* is also a measure of *informativeness* of estimator Y with respect to predictand W. This is so because the order of forecasters induced by *IS* is invariant under any strictly monotone transformation of each variate, Y and W; and the NQT is such a transformation.
2. The standard deviation S of V does not appear in formula (11.14) because the empirical NQT using the meta-Gaussian plotting positions ensures that the sample mean of V is zero and the sample variance of V is approximately 1. In short, $V \sim NM(0, 1)$; see Exercise 3 in Chapter 3.
3. While *IS* calculated from equation (11.14) is interpretable as the *Bayesian estimator* of the absolute value of the correlation between Z and V (more specifically, Pearson's product-moment correlation coefficient between Z and V), it can be mapped into the Bayesian estimators of the absolute values of the rank correlations between the original estimator Y and the original predictand W:
   - Spearman's rank correlation coefficient,

$$|\rho| = (6/\pi) \ \arcsin (IS/2); \tag{11.15}$$

   - Kendall's alternative rank correlation coefficient,

$$|\tau| = (2/\pi) \ \arcsin (IS). \tag{11.16}$$

The three statistics IS, $|\rho|$, $|\tau|$ are equivalent measures of informativeness of estimator Y with respect to predictand W.

Example 11.2   *(Meta-Gaussian model)*

Let $N = 5$ and $N' = 7$. Table 11.3 shows the joint and prior samples of the original variates (Y, W) collated in Step 0, and the matching samples of the transformed variates (Z, V) outputted from Step 4. Table 11.4 shows the empirical distribution functions of Y and W constructed in Step 1, and the transformed realizations obtained via the empirical NQT in Step 3. The visualization of the NQT is deferred to Exercise 5.

Table 11.3   Samples of (Y, W) and matching samples of (Z, V); joint sample size $N = 5$; prior sample size $N' = 7$.

	Original		Transformed	
n	*y(n)*	*w(n)*	*z(n)*	*v(n)*
7		110		1.556
6		20		−0.935
5	20	10	−1.435	−1.556
4	30	40	−0.660	0.000
3	60	90	0.660	0.935
2	50	30	0.000	−0.446
1	70	50	1.435	0.446

Table 11.4 Empirical distribution functions: $\{(y_{(n)}, p_n) : n = 1, \ldots, 5\}$ and $\{(w_{(n)}, p_n) : n = 1, \ldots, 7\}$ of Y and W, respectively; and the transformed realizations $\{z_{(n)} : n = 1, \ldots, 5\}$ and $\{v_{(n)} : n = 1, \ldots, 7\}$ obtained via the empirical NQT.

		Estimator					Predictand		
n	(n)	$y_{(n)}$	p_n	$z_{(n)}$	n	(n)	$w_{(n)}$	p_n	$v_{(n)}$
1	5	20	0.0756	−1.435	1	5	10	0.0598	−1.556
2	4	30	0.2538	−0.660	2	6	20	0.1743	−0.935
3	2	50	0.5	0.000	3	2	30	0.3267	−0.446
4	3	60	0.7462	0.660	4	4	40	0.5	0.000
5	1	70	0.9244	1.435	5	1	50	0.6733	0.446
					6	3	90	0.8257	0.935
					7	7	110	0.9402	1.556

The results produced by the Gaussian model are:

$$a = 0.47727, \quad b = 25, \quad \sigma = 13.5512, \quad S = 34.2261;$$

$$IS = 0.7696.$$

The results produced by the meta-Gaussian model are:

$$a = 0.96502, \quad b = 0.1198, \quad \sigma = 0.56922;$$
$$IS = 0.8613, \quad |\rho| = 0.8503, \quad |\tau| = 0.6607.$$

Clearly, the likelihood parameters a, b, σ and the informativeness score IS have different values. The results of Exercise 5 offer explanations: (i) the empirical distribution functions reveal that Y and W are nonnormal variates, and (ii) the scatterplot of the joint sample of (Y, W) reveals the heteroscedasticity. Both violations of the properties of the normal-linear model are almost eliminated by the NQT.

11.3.5 Mean Squared Error

An estimate y of realization w has error $\xi = y - w$. The random error is $\Xi = Y - W$. The mean squared error is defined by

$$MSE = E(\Xi^2) = E\left[(Y - W)^2\right]. \tag{11.17}$$

Inasmuch as $Var(\Xi) = E(\Xi^2) - E^2(\Xi)$,

$$MSE = Var(\Xi) + E^2(\Xi). \tag{11.18}$$

To relate MSE to the verification algorithm in Section 11.3.3, let us express Y in terms of the normal-linear model (10.5):

$$\Xi = Y - W$$
$$= aW + b + \Theta - W$$
$$= (a - 1)W + b + \Theta.$$

Now let us exploit the analogy between the above equation for the variate Ξ and equation (10.5) for the variate X, and use expressions (10.9) for the mean and variance of X as the analogs for the mean and variance of Ξ:

$$E(\Xi) = (a-1)M + b,$$
$$Var(\Xi) = (a-1)^2 S^2 + \sigma^2. \qquad (11.19)$$

In conclusion, *MSE* can be calculated via equation (11.18) from the parameters of the normal-linear model (a, b, σ) and the prior moments of the predictand: $E(W) = M$ and $Var(W) = S^2$.

While popular, the mean squared error is a dubious measure for empirical verification of forecasts. Except for a special case (see the note below), it measures neither the calibration of Y nor the informativeness of Y. Instead, it commingles the two attributes of the forecaster. (Exercise 20 illustrates the pitfall of ranking the forecasters in terms of *MSE*.)

Note on optimal estimators

The special case arises when the estimator Y equals a mean of W — either unconditional $E(W)$ or conditional $E(W|X=x)$, given the realization x of some predictor X. In such a case, Y is an optimal estimator in that it minimizes *MSE*; this is demonstrated in Section 12.4. The posterior mean $E(W|X=x)$ outputted from the BGF is such an optimal estimator of W, often called the *minimum mean squared error estimator*; it yields $E(\Xi) = 0$ and $Var(\Xi) = T^2$; thus $MSE = T^2$, the posterior variance of W. But this theoretic equality may not hold for forecasts being verified based on a joint sample. And it should not be assumed because it is the purpose of the verification to ascertain the degree to which a set of forecasts has been calibrated and informative. This purpose requires two separate verification measures.

11.4   Verification of Bayesian Forecaster

An advantage of the forecaster developed from a theory, such as the Bayesian Gaussian forecaster (BGF), is the possibility of characterizing its performance before deploying it in real-time forecasting. This should be qualified as the **potential performance**. Here are three theoretic properties of the BGF that characterize its potential performance.

11.4.1   Informativeness

When developing the BGF, the *IS* is calculated from equation (10.26) using values of the parameters a, σ, S gotten from the *estimation samples* — joint and prior. It is qualified as the informativeness score of the predictor X; it may also be qualified as the informativeness score of the BGF (which uses predictor X). This *IS* characterizes the potential informativeness of the BGF.

11.4.2   Calibration in-the-Large

This calibration property is specific to the calibration measure — the distribution calibration function (DCF) defined in Section 11.2.3. Suppose a Bayesian forecaster of a continuous predictand W supplies the posterior distribution function $\Phi(\cdot|x)$ repeatedly on different forecasting occasions; that is, given different predictor realizations $x \in \mathcal{X}$. Can anything be said about the potential calibration in-the-large of this forecaster?

Theorem (*Calibration in-the-large*)   *The Bayesian forecaster (10.1)–(10.2) is well calibrated in-the-large on every set of forecast probabilities $P \subseteq (0, 1)$.*

The proof is deferred to Section 11.7.2. The corollary, that the BGF is well calibrated in-the-large, is the subject of Exercise 7.

11.4.3 Calibration in-the-Margin

This calibration property is general: (i) it does not involve any calibration measure; (ii) it reveals the role of the prior density function g of W, which calibrates the posterior density functions in the following sense.

Definition The forecaster is said to be *well calibrated in-the-margin* if the expectation of the conditional density function $\phi(\cdot|x)$ of W, viewed as a function of $x \in \mathcal{X}$, equals the marginal density function g of W; that is, if for every $w \in \mathcal{W}$,

$$E[\phi(w|X)] = g(w).$$

Theorem *(Calibration in-the-margin)* *The Bayesian forecaster (10.1)–(10.2) is well calibrated in-the-margin.*

Proof: Take the expectation with respect to X on both sides of equation (10.2):

$$E[\phi(w|X)] = \int_{\mathcal{X}} \frac{f(x|w)g(w)}{\kappa(x)} \kappa(x)\, dx$$

$$= g(w) \int_{\mathcal{X}} f(x|w)\, dx$$

$$= g(w);$$

the last equality holds because the integral of density function $f(\cdot|w)$ over sample space $\mathcal{X}$ equals one for every $w \in \mathcal{W}$. QED.

Exercise 8 asks you to demonstrate that the above definition and theorem can be restated in terms of the distribution functions: for every $w \in \mathcal{W}$,

$$E[\Phi(w|X)] = G(w). \tag{11.20}$$

The equality between the expected posterior distribution function and the prior distribution function is a necessary as well as a desirable property. For it means that after making many forecasts, the average conditional probability $\Phi(w|x)$, with which the nonexceedance event $W \leq w$ has been forecasted, equals the marginal probability $G(w)$ of the event. Hence, if $G(w)$ represents the relative frequency of event $W \leq w$ occurring in "nature", then the Bayesian forecaster simulates "nature" faithfully: it predicts event $W \leq w$ with the same relative frequency $G(w)$, on average; and this statement holds for every $w \in \mathcal{W}$.

11.4.4 Marginal Calibration Score

Equation (11.20) can be verified empirically for any forecaster who outputs a continuous and strictly increasing distribution function of predictand W, whose prior distribution function G is continuous and strictly increasing, and constitutes the standard against which the forecaster should be calibrated (as discussed in Section 6.2.3).

The *empirical method* of verifying the calibration in-the-margin is detailed for the BGF, for the sake of specificity, but can be readily adapted to any forecaster. (See the notes below the algorithm.)

Step 0. Given is a joint sample of the predictor–predictand pair (X, W), on which the verification of forecasts is to be performed:

$$\{(x(n),\, w(n)) : n = 1, \ldots, N\}.$$

Step 1. From the BGF with specified values of the prior and posterior parameters $(M, S; A, B, T)$, get expression (10.15) for $G(w)$ and expression (10.16) for $\Phi(w|x)$.

Step 2. Arrange the realizations of W from the joint sample in ascending order of numerical values:

$$w_{(1)} \leq w_{(2)} \leq \cdots \leq w_{(N)},$$

where $w_{(n)}$ denotes the nth realization and $(n) \in \{1, \ldots, N\}$. When two or more realizations have an identical value, their order is irrelevant.

Step 3. For every $n \in \{1, \ldots, N\}$, calculate the prior nonexceedance probability

$$p_n = G(w_{(n)}),$$

and the sample mean (across all realizations of X) of the posterior nonexceedance probability

$$q_n = \frac{1}{N} \sum_{j=1}^{N} \Phi(w_{(n)} | x(j)).$$

Step 4. Treat $\{(w_{(n)}, p_n) : n = 1, \ldots, N\}$ and $\{(w_{(n)}, q_n) : n = 1, \ldots, N\}$ as two empirical distribution functions of W. Evaluate the goodness of fit of the mean posterior distribution function $\{(w_{(n)}, q_n)\}$ to the prior distribution function $\{(w_{(n)}, p_n)\}$. Any of the three procedures described in Section 3.6 may be applied. (i) The graphical comparison. (ii) The K-S comparison-of-distributions test detailed below. (iii) The uniform distance measure: relabel MAD as *MCS* — the *marginal calibration score*; calculate it:

$$MCS = \max_{1 \leq n \leq N} |p_n - q_n|. \tag{11.21}$$

The *MCS* is bounded, $0 \leq MCS < 1$, with $MCS = 0$ implying perfect calibration in-the-margin and *MCS* near 1 implying the worst possible calibration in-the-margin.

For a Bayesian forecaster, the *MCS* reveals the degree to which the property of calibration in-the-margin has been maintained in a set of forecasts being verified. For any other forecaster, the *MCS* reveals the degree to which his forecasts could be considered as posterior distribution functions relative to the specified prior distribution function — the standard for calibration relevant to a forecast user (as discussed in Section 6.2.3).

Notes.

(i) The algorithm can be adapted to forecasters other than the BGF by suitably modifying the steps before Step 4. Three particular modifications are sketched.

(ii) When a parametric prior distribution function G is unavailable, the marginal sample $\{w(n) : n = 1, \ldots, N\}$ may be used to construct the empirical distribution function of W in the form $\{(w_{(n)}, p_n) : n = 1, \ldots, N\}$, as described in Section 3.2.

(iii) When a forecaster provides a parametric distribution function of W sans explicit conditioning on a predictor realization, the symbol $\Phi(\cdot | x(j))$ should be replaced by the symbol $\Phi(\cdot | j)$, which denotes the distribution function of W on the jth forecasting occasion, $j \in \{1, \ldots, N\}$.

(iv) When an expert provides a judgmental forecast sans mathematical expression, a parametric distribution function should be fitted to the pointwise representation of the judgmental distribution function, as described in Section 9.3; and the symbol $\Phi(\cdot | x(j))$ should be replaced by the symbol $\Phi(\cdot | j)$, which denotes the parametric distribution function of W on the jth forecasting occasion, $j \in \{1, \ldots, N\}$.

Kolmogorov-Smirnov comparison-of-distributions test

This is a version of the goodness-of-fit test described in Section 3.6.3; it is usually presented as the two-sample K-S test. Here, it is specialized further to the purpose of verifying the calibration in-the-margin.

Null hypothesis: the mean posterior distribution function and the prior distribution function are identical.
Alternative hypothesis: the two functions differ in any way they can.
Test statistic: $D = MCS$.

Critical value.

(i) Approximate value for small sample size N and the significance level $\alpha = 0.05$:

$$C = \frac{1}{10}\left(\frac{25}{N} + 3\right), \qquad 5 \le N \le 20.$$

(ii) Asymptotic value for large sample size N and any significance level $\alpha > 0$:

$$C = \sqrt{-\frac{1}{N}\ln\frac{\alpha}{2}}, \qquad 20 < N.$$

P-value. For the given D and any $N > 20$:

$$\text{P-value} = 2\exp(-ND^2).$$

Example 11.3 (*BGF of snowmelt runoff — Example 11.1 continued*)

The Bayesian Gaussian forecaster described in Section 10.6 had its uniform calibration verified in Section 11.2.8. Here, its calibration in-the-margin is verified on the same joint sample (Table 10.1) of the predictor–predictand pair (X, W) from $N = 7$ years (1977–1983). The values of its prior and posterior parameters are listed in Table 10.2. The output from the verification algorithm is reported in Table 11.5. With sample size $N = 7$ being small and $MCS = 0.083$ implying a good fit of the mean posterior distribution function $\{(w_{(n)}, q_n)\}$ to the prior distribution function $\{(w_{(n)}, p_n)\}$, one may conclude that this forecaster is approximately well calibrated in-the-margin. Exercise 9 asks you to execute the Kolmogorov-Smirnov test and to determine the significance level of this conclusion.

The interpretation of one row of (rounded) numbers from Table 11.5 is this. In 1978, the seasonal snowmelt runoff volume had rank 5 (among $N = 7$ realizations) and was $w_5 = 473 \times 10^3$ ac-ft. The prior probability of this volume being not exceeded in any year is

$$p_5 = P(W \le 473) = G(473) = 0.63.$$

In the seven forecasts of W, the mean probability of this event was $q_5 = 0.66$. The absolute difference between the two probabilities is $|0.63 - 0.66| = 0.03$; it implies that, practically, the seven forecasts of the event $\{W \le 473\}$ have been well calibrated in-the-margin.

The calibration in-the-margin may also be visualized (Figure 11.3) by graphing the mean posterior probability q_n versus the prior probability p_n for $n = 1, \ldots, N$. Each point (p_n, q_n) of the graph is associated with the event $\{W \le w_{(n)}\}$, where $w_{(n)}$ is the p_n-probability quantile of W under the prior distribution function.

Table 11.5 Output from the in-the-margin calibration algorithm applied to the joint sample of (X, W) from Table 10.1; Bayesian Gaussian forecaster of the snowmelt runoff; January forecast; Weiser, Idaho; runoff volume $w_{(n)}$ is in 10^3 ac-ft; $MCS = 0.083$.

Year	n	(n)	$w_{(n)}$	p_n	q_n	$|p_n - q_n|$
1977	7	1	38.8	0.0173	0.0115	0.0058
1979	5	2	275.8	0.2192	0.2986	0.0794
1981	3	3	283.8	0.2327	0.3156	**0.0829**
1980	4	4	407.6	0.4874	0.5561	0.0687
1978	6	5	472.9	0.6318	0.6578	0.0260
1982	2	6	589.1	0.8390	0.8300	0.0090
1983	1	7	608.4	0.8641	0.8568	0.0073

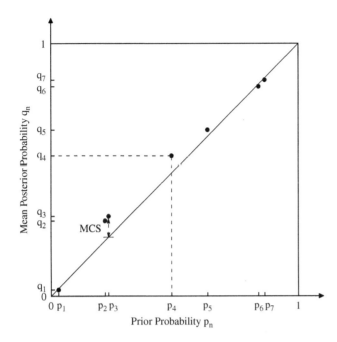

Figure 11.3 The *marginal calibration function* of the forecaster: a graph of the mean posterior probability q_n versus the prior probability p_n of the nonexceedance event $\{W \leq w_{(n)}\}$ for $n = 1, \ldots, 7$; the values are calculated in Table 11.5; they yield $MCS = 0.083$; Bayesian Gaussian forecaster of the snowmelt runoff; January forecast; Weiser, Idaho.

- A point on the main diagonal, $q_n = p_n$, implies perfect calibration of the forecast probabilities assigned to the event $\{W \leq w_{(n)}\}$.
- A point above the main diagonal, $q_n > p_n$, implies a bias toward *overestimating* the probabilities of the event $\{W \leq w_{(n)}\}$.
- A point below the main diagonal, $q_n < p_n$, implies a bias toward *underestimating* the probabilities of the event $\{W \leq w_{(n)}\}$.

Hence the graph, and the table, provide feedback to the forecaster regarding the two types of biases. When the biases are large, and the sample size N is large enough, the forecaster may consider improving the calibration of the posterior distributions, as discussed in Section 11.2.5.

11.5 Analysis of Judgmental Task

11.5.1 Calibration Report

The DCF defined in Section 11.2.3 characterizes the forecaster's calibration in a graphical form, displaying the order probabilities. Experimental results are often reported in a tabular form, showing the interval probabilities; this format is followed here.

Table 11.6 reports the calibrations pertaining to Task 9.1, in which the set of forecast probabilities was

$$\mathcal{P} = \{p_i : i = 1, \ldots, 5\} = \{0.01, 0.25, 0.5, 0.75, 0.99\}.$$

The first two columns of the table identify the group of forecasters (or the predictand) and the joint sample size N.

Table 11.6 Calibration of four groups of forecasters: each forecaster assessed 3 distribution functions of 3 different predictands in Task 9.1 (type 4 verification). For each group, the statistical probabilities of 9 events should be compared with the forecast probabilities of these events. The last row reports statistical probabilities for the HB group from an earlier, independent, but similar experiment.

Group	N	\multicolumn{6}{Simple events}						Median	50% CCI	98% CCI	CS
		Ω_1	Ω_2	Ω_3	Ω_4	Ω_5	Ω_6	Median	CCI	CCI	CS
Forecast P		0.01	0.24	0.25	0.25	0.24	0.01	0.5	0.5	0.98	
VE	3675	0.09	0.13	**0.22**	0.16	0.10	0.30	0.44	0.38	0.61	0.28
C1	270	0.07	0.14	**0.20**	0.16	0.12	0.31	0.41	0.36	0.62	0.29
C2	264	**0.05**	0.16	**0.29**	0.17	0.13	0.20	**0.50**	**0.46**	0.75	0.18
PF	36	**0.03**	**0.22**	**0.22**	**0.22**	0.03	0.28	**0.47**	0.44	0.69	0.26
HB	8000	0.16	0.12	0.16	0.17	0.12	0.27	0.44	0.33	0.57	0.26

VE — Virginia engineering students (1225) in 18 years.
C1 — Class 1 of students (90) in one year.
C2 — Class 2 of students (88) in another year.
PF — Professional forecasters (12) of the National Weather Service.
HB — Harvard business students (800) in one year; 10 predictands. Source: Alpert and Raiffa (1982, Table 2, Total — rounded to two decimal places).
Boldface statistical probability means an approximately well-calibrated credible interval.

The next six columns pertain to the *simple events* $\{\Omega_i : i = 1, \ldots, 6\}$. These are the credible intervals formed by the adjacent quantiles; they partition the sample space $\mathcal{W}$. Their forecast probabilities are:

$$P(\Omega_1) = P(W \leq Y_{0.01}) = 0.01,$$
$$P(\Omega_2) = P(Y_{0.01} < W \leq Y_{0.25}) = 0.24,$$
$$P(\Omega_3) = P(Y_{0.25} < W \leq Y_{0.5}) = 0.25,$$
$$P(\Omega_4) = P(Y_{0.5} < W \leq Y_{0.75}) = 0.25,$$
$$P(\Omega_5) = P(Y_{0.75} < W \leq Y_{0.99}) = 0.24,$$
$$P(\Omega_6) = P(Y_{0.99} < W) = 0.01.$$

The table reports their statistical probabilities $\{r_i : i = 1, \ldots, 6\}$. They can be either estimated from the joint sample (11.2) or calculated from the DCF, respectively:

$$r_1 = R(p_1);$$
$$r_i = R(p_i) - R(p_{i-1}), \qquad i = 2, \ldots, 5;$$
$$r_6 = 1 - R(p_5).$$

These adjacent credible intervals are well calibrated in-the-large if $r_i = P(\Omega_i)$ for $i = 1, \ldots, 6$.

The next three columns pertain to the *compound events*. These are the median, the 50% central credible interval (CCI), and the 98% CCI. Their forecast probabilities are:

$$P(W \leq Y_{0.5}) = P(\Omega_1) + P(\Omega_2) + P(\Omega_3) = 0.5,$$
$$P(Y_{0.25} < W \leq Y_{0.75}) = P(\Omega_3) + P(\Omega_4) = 0.5,$$
$$P(Y_{0.01} < W \leq Y_{0.99}) = 1 - P(\Omega_1) - P(\Omega_6) = 0.98.$$

The table reports their statistical probabilities, which are, respectively:

$$R(0.5) = r_1 + r_2 + r_3,$$
$$R(0.75) - R(0.25) = r_3 + r_4,$$
$$R(0.99) - R(0.01) = 1 - r_1 - r_6.$$

The median and the two CCIs are well calibrated in-the-large if for each the statistical probability equals the forecast probability.

The last column of the table reports the calibration score (*CS*).

11.5.2 Calibration of Forecasters: Interpretations

Task 9.1 was performed by 1225 students of engineering and sciences who took this course at the University of Virginia over 18 years, ending in 2020. In addition, 12 professional forecasters from the National Weather Service completed this task during introduction to a training workshop on probabilistic forecasting (see Sections 9.6.3 and 11.6.2).

Table 11.6 reports type 4 verifications for four groups; each individual forecasted three predictands. The Virginia engineering (VE) group is the aggregate of 18 classes; two of them constitute groups C1 and C2. Perusing the table, one notices immediately the pattern formed by the nine statistical probabilities: it is remarkably consistent across the groups despite hugely different sample sizes. Likewise for 16 classes not included in the table.

To learn the usefulness of the calibration table, let us interpret all the numbers for group VE, as if we were providing feedback to the forecasters regarding their "collective" performance.

Simple events: middle credible intervals
For interval Ω_3, the statistical probability $r_3 = 0.22$ is close to the forecast probability $P(\Omega_3) = 0.25$. Thus, this 25% credible interval is *approximately well calibrated*; this means that its width was about right: 22% of predictand realizations w fell inside the interval. For interval Ω_4, the statistical probability $r_4 = 0.16 < 0.25 = P(\Omega_4)$. Thus, this 25% credible interval is poorly calibrated; it was too narrow: only 16% of realizations w fell inside it instead of 25%. And too narrow a credible interval is indicative of *overconfidence* on the part of the forecaster. Likewise, the two 24% credible intervals, Ω_2 and Ω_5, were too narrow: $r_2 = 0.13 < 0.24 = P(\Omega_2)$ and $r_5 = 0.10 < 0.24 = P(\Omega_5)$, which affirms the overconfidence bias.

Simple events: tail credible intervals
Each 1% tail credible interval (TCI), Ω_1 and Ω_6, may be viewed as an extreme event (a rare event): only 1% of realizations w should fall inside it. Each is poorly calibrated, as $r_1 = 0.09 > 0.01 = P(\Omega_1)$ and $r_6 = 0.30 > 0.01 = P(\Omega_6)$. Thus, each was too wide, especially the upper 1% TCI: 30% of realizations w fell inside it instead of 1%. Of course, this is the inevitable consequence of the middle intervals being too narrow.

Compound events
The median is the union of the simple events $\Omega_1, \Omega_2, \Omega_3$. Its statistical probability is $r_1 + r_2 + r_3 = 0.44 < 0.5$. Thus, the median is not well calibrated; it was too low — 44% of realizations w fell below it instead of 50%. For the 50% CCI, being the union of events Ω_3 and Ω_4, the statistical probability is $r_3 + r_4 = 0.38 < 0.5$. Thus, this CCI is poorly calibrated; it was too narrow — only 38% of realizations w fell inside it instead of 50%. This affirms the overconfidence bias. Likewise for the 98% CCI, being the union of events $\Omega_2, \Omega_3, \Omega_4, \Omega_5$, whose statistical probability is $1 - r_1 - r_6 = 0.61 < 0.98$.

Partially well-calibrated forecaster
This term is convenient in practice. It may be applied if some quantiles (say, at least two out of five) and some credible intervals (say, at least two out of six adjacent credible intervals or one CCI) are approximately well calibrated. Accordingly, forecasters in groups C2 and PF may be characterized as partially well calibrated.

11.5.3   Calibration of Forecasters: Biases

To characterize succinctly the pattern of calibration of forecasters across the four groups, we focus on the median, the two CCIs, and the two TCIs.

Median

The median is underestimated or approximately well calibrated: $R(0.5) = 0.50, 0.47$ for groups C2, PF.

Central credible intervals

- The 50% CCI tends to be too narrow ($r_3 + r_4 < 0.5$) and skew ($r_3 \neq r_4$), though sometimes it is approximately well calibrated ($r_3 + r_4 = 0.46$ for group C2) or symmetric ($r_3 = r_4 = 0.22$ for group PF).
- The 98% CCI is much too narrow ($1 - r_1 - r_6 < 0.98$) and skew ($r_1 \neq r_6$); the statistical probability $1 - r_1 - r_6 = 0.75$ for group C2 is the closest to 0.98.
- Conclusion: The judgments of uncertainty exhibit a bias: *overconfidence*, which manifests itself in too narrow central credible intervals.

Tail credible intervals

- Each 1% TCI defines an *extreme event* (a *rare event*) — one might say a "surprise": after 100 forecasts only one predictand realization should fall inside this interval, on average.
- The lower 1% TCI tends to be too wide ($r_1 > 0.01$). In other words, the 0.01-probability quantile was not extreme enough. The best group, PF, recorded 3 "surprises" per 100 forecasts, on average, instead of one.
- The upper 1% TCI is much too wide ($r_6 > 0.01$). In other words, the 0.99-probability quantile was not extreme enough. The best group, C2, recorded 20 "surprises" per 100 forecasts, on average, instead of one.
- Conclusion: The judgments of uncertainty exhibit a bias: *underestimation of the magnitudes of extreme events* (*rare events*).

Reproducible calibration pattern

Task 9.1 parallels the experiment of Alpert and Raiffa (1982) conducted during the academic year 1968–1969 with 800 first-year students in the Master of Business Administration program at Harvard University. Their experiment comprised 10 predictands, but one was the same as ours (eggs). The statistical probabilities they reported are those for the Harvard business (HB) group in Table 11.6. The patterns of calibrations of the two groups, VE and HB, are strikingly similar. This *reproducibility* of results by experiments conducted nearly 50 years apart, with students representing different professions (and different generations), lends support for the following advice.

> **Advice** (Training for calibration). To become well calibrated, learners of the quantile technique, and aspiring forecasters of continuous predictands, should be open to these recommendations.
> 1. Hone the assessment of a well-calibrated median because it anchors the assessments of other quantiles.
> 2. Widen the central credible interval to overcome the overconfidence bias.
> 3. Push the extreme quantiles outward to counteract the underestimation bias in the magnitudes of extreme events (rare events).
> 4. Do not assess quantiles with extreme probabilities, such as $(0.01, 0.99)$ or $(0.001, 0.999)$, because they define rare events which the human mind is usually unprepared, or unexperienced, to envisage. Instead, assess quantiles with moderate probabilities (per the guideline in Section 9.2.2), and let a parametric distribution function extrapolate them into the tails.

11.5.4   Calibrations for Predictands

Table 11.7 reports type 3 verifications for group VE with respect to each of the three variates; specifically, the 3675 forecasts verified in Table 11.6 are disaggregated into 3 subsets of 1225 forecasts of each variate. Each variate is

Table 11.7 Calibration of group VE of forecasters (see Table 11.6) with respect to each variate (W_E, eggs; W_P, precipitation; W_Y, year) in Task 9.1 (type 3 verification). For each variate, the statistical probabilities of 9 events should be compared with the forecast probabilities of these events.

| Variate | N | Simple events | | | | | | 50% | 98% | | |
| | | Ω_1 | Ω_2 | Ω_3 | Ω_4 | Ω_5 | Ω_6 | Median | CCI | CCI | CS |
Forecast P		0.01	0.24	0.25	0.25	0.24	0.01	0.5	0.5	0.98	
W_E	1225	**0.02**	0.03	0.05	0.06	0.13	0.71	0.10	0.11	0.27	0.70
W_P	1225	0.12	0.17	**0.26**	0.19	0.11	0.15	0.55	0.45	0.73	0.14
W_Y	1225	0.12	**0.20**	0.36	**0.22**	0.07	**0.03**	0.68	0.58	0.85	0.18

Boldface statistical probability means an approximately well-calibrated credible interval.

factual (i.e., having a known realization, which is revealed in Exercise 4 of Chapter 9), and had been selected based on a different hypothesis, which was affirmed by many students in later discussions.

Variate W_E — eggs. The hypothesis was that the student's mind would seek a **conceptual model** (e.g., beginning with the US population, estimating the number of eggs a person consumes per day, etc.). A concomitant hypothesis was that the uncertainty about this variate is large because of the many ways the eggs are used, and that, therefore, W_E is a "difficult" predictand.

Variate W_P — precipitation. The hypothesis was that the student would have 2–3 years of **experience** with weather in Charlottesville, perhaps supplemented with some knowledge of US geography, as a basis for quantifying uncertainty.

Variate W_Y — year. The hypothesis was that the uncertainty quantification would start from an **anchor** — a year known from US history (e.g., 1773, Boston Tea Party; 1776, Declaration of Independence; 1787, Constitutional Convention).

The calibration of students with respect to variate W_E reveals that, with the exception of the 0.01 quantile which is approximately well calibrated, all other quantiles were grossly underestimated. Each CCI, the 50% and the 98%, was much too narrow — revealing large bias of overconfidence. The upper 1% TCI, having $r_6 = 0.71$ instead of 0.01, reveals a gross underestimation of the magnitude of the extreme event. These findings confirm the hypothesis that W_E is a "difficult" predictand.

The calibration of the same students with respect to the variates W_P and W_Y appears considerably better. The interpretation of the calibration report for each of these variates is left to the reader as Exercise 14.

Difficulty of predictand

The results from Task 9.1, the large verification samples, and the consistent calibration patterns across 18 groups of engineering students, prompt four general reflections.

1. The innate ability of an individual — technically educated but untrained in uncertainty quantification — to assess judgmentally a distribution function of a continuous predictand depends on the predictand's "difficulty".
2. A predictand with which the individual has some experience (like W_P), or for which knowledge can supply an anchor (like W_Y), is "easy". Judgmental distribution functions may be at least partially well calibrated.
3. A predictand for which a conceptual model needs to be formulated is "difficult". Judgmental distribution functions are likely to be miscalibrated — useless.
4. To be potentially useful, the quantification of uncertainty about a "difficult" predictand requires a professional approach: context-specific training and slow-methodical reasoning, aided by the tools for structuring and analyzing the predictive information (as recommended in Section 9.2.3, and as done in applications described in Section 9.6 and verified in Section 11.6).

11.5.5 Overconfidence and Self-Confidence

As overconfidence is a frequent bias in judgmental forecasts, it is apt to juxtapose it with a frequent personality trait of self-confidence.

Confidence in one's abilities is generally a positive trait, especially when implementing an action already decided, leading an organization toward a goal already set, or interacting with people — to teach, to convince, or to inspire them. But in forecasting, self-confidence is generally a negative trait because you do not interact with people. As a forecaster, you need to contemplate the future trajectory of a process beyond your influence — a physical process (as in forecasting weather), an economic process (as in forecasting prices), a socio-psychological process (as in forecasting consumer demand). Self-confidence tends to mislead the mind toward a higher degree of certainty than is justified by information and knowledge actually possessed and relevant to the predictand at hand. The effect shows up in the verification of judgmental forecasts as overconfidence. How to arrest this potential source of bias?

> **Advice** (Mind-set). Lock away the self-confidence and poise. Adopt a humble mind-set; realize a limit on predictability; judge the predictive strength of information; and filter opinions through skepticism. A heuristic: the greater the skepticism, the higher the degree of uncertainty; push the quantiles y_p and y_{1-p} outward.

11.6 Applications

This section presents the calibration reports from two applications of the judgmental forecasting procedure described in Section 9.6.

11.6.1 Auditing Financial Statements

A subset of the calibration report by Solomon (1982) is reproduced in Table 11.8. It shows type 3 verification: for a group of auditors with respect to one predictand (the "true" dollar amount of an account balance) forecasted on 6 different occasions (6 audit cases). Group IN comprises 26 individual auditors who assessed 149 distribution functions. Group TE comprises 13 audit teams that assessed 78 distribution functions. (Albeit assessed, the 0.1-quantile and 0.9-quantile are excluded in order to make Table 11.8 directly comparable with Tables 11.6 and 11.7.)

Table 11.8 Calibration of two groups of auditors with respect to a predictand: each individual auditor and each audit team assessed 6 distribution functions of the same predictand on 6 different occasions (type 3 verification). For each group, the statistical probabilities of 9 events should be compared with the forecast probabilities of these events.

Group	N	Simple events						Median	50% CCI	98% CCI	CS
		Ω_1	Ω_2	Ω_3	Ω_4	Ω_5	Ω_6				
Forecast P		0.01	0.24	0.25	0.25	0.24	0.01	0.5	0.5	0.98	
IN	149	**0.05**	0.07	**0.27**	**0.28**	**0.27**	0.06	0.39	0.55	0.89	0.13
TE	78	0.09	0.08	**0.24**	0.40	0.15	**0.04**	0.41	0.64	0.87	0.09

IN — Individual auditors (26). TE — Teams of three auditors (13) following the nominal group process (see Section 9.4). Boldface statistical probability means an approximately well-calibrated credible interval.
Source: Solomon (1982, Table 6, condition numbers 1, 2).

The interpretation of this calibration report is left to the reader as Exercise 16. But two insights from the author of the study are noteworthy.

Effect of training and experience

First, each group underestimated the median of the "true" account balance. This bias could be due to the nature of the training an auditor receives and of the experience an auditor accumulates, both of which foster "conservative" judgment. Second, each group assessed a too wide 50% CCI. Again, this bias toward underconfidence could be due to "caution" — a trait of the accounting profession.

Effect of extreme probabilities

Inasmuch as the auditors did assess quantiles for probabilities 0.1 and 0.9, we may compare the calibration of two central credible intervals. For the 98% CCI, the statistical probabilities reported in Table 11.8 are 0.89 and 0.87. For the 80% CCI, the statistical probabilities reported by the author of the study are 0.78 and 0.78. Clearly, the 98% CCI was too narrow, whereas the width of the 80% CCI was about right. This is an additional empirical evidence that supports the fourth piece of advice in Section 11.5.3.

11.6.2 Forecasting Precipitation Amount

The calibration report in Table 11.9 comes from the article by Krzysztofowicz and Sigrest (1999). It shows type 3 verifications: for a group of meteorologists with respect to each predictand (the average precipitation amount over a river basin accumulated during 24 hours), forecasted on different days (occasions). In order to make this table easily comparable with Tables 11.6–11.8, the simple events $\{\Omega_i : i = 2, 3, 4, 5\}$ and their forecast probabilities are relabeled as follows:

$$P(\Omega_2) = P(W \leq Y_{0.25}) = 0.25,$$
$$P(\Omega_3) = P(Y_{0.25} < W \leq Y_{0.5}) = 0.25,$$
$$P(\Omega_4) = P(Y_{0.5} < W \leq Y_{0.75}) = 0.25,$$
$$P(\Omega_5) = P(Y_{0.75} < W) = 0.25.$$

Their statistical probabilities are relabeled accordingly as r_2, r_3, r_4, r_5.

To provide systematic feedback to the forecasters during the 5-year experiment, the statistical probabilities for each river basin were estimated monthly from the joint sample accumulated during the preceding 3 months. The average sample size was 153. After 5 years, each r_i $(i = 2, \ldots, 5)$ had a time series of 58 values. These values fluctuated, of course. Table 11.9 reports the arithmetic average of 58 values of each r_i $(i = 2, \ldots, 5)$.

Table 11.9 Calibration of a group of meteorologists with respect to each predictand (river basin): each meteorologist assessed the distribution function of each predictand on many occasions (days, type 3 verification). For each predictand, the statistical probabilities of 6 events should be compared with the forecast probabilities of these events.

Basin	N	Simple events				50%		CS
		Ω_2	Ω_3	Ω_4	Ω_5	Median	CCI	
Forecast P		0.25	0.25	0.25	0.25	0.5	0.5	
MN	3090	0.42	**0.23**	0.13	**0.22**	0.65	0.36	0.17
AL	3069	0.38	**0.26**	0.14	**0.22**	0.64	0.40	0.14

MN — Monongahela river basin. AL — Allegheny river basin. Boldface statistical probability means an approximately well-calibrated credible interval.
Source: Krzysztofowicz and Sigrest (1999, Table 3).

The interpretation of this calibration report is left to the reader as Exercise 17. But three insights from the authors are worth mentioning.

Effect of mission

For each river basin, the 0.75-quantile was approximately well calibrated, whereas the 0.25-quantile and the median were overestimated. This calibration pattern has an explanation: meteorologists focus on forecasting extreme events — in this case, on large precipitation amounts that cause floods — because warning the public (to reduce property damage, injuries, death) is their mission. The bias toward overestimation of $Y_{0.25}$ and $Y_{0.5}$ can be attributed, at least in part, to this "intense focus" on extreme events — a trait of the weather forecasting profession.

Dynamic recalibration

For each forecast probability $p \in \{0.25, 0.5, 0.75\}$, the time series of 58 order probabilities $R(p)$ exhibited trends lasting several months. This suggested the possibility of improving the calibration of quantiles via *dynamic recalibration* as follows. During every month, each stated forecast probability p was replaced by the order probability $R(p)$ estimated from the joint sample accumulated during the preceding 3 months. After 58 months of such dynamic recalibration, the average CS dropped from 0.17 to 0.07 for the MN basin, and from 0.14 to 0.08 for the AL basin. Hence, the dynamic recalibration achieved its purpose: it brought the statistical probabilities closer to the face values of the forecast probabilities.

Calibration potential

The calibration of a group depends upon the metacognitive skill (recall Section 6.5) of individual forecasters. There were 9 meteorologists who participated in the experiment during the 5 years and prepared 87% of all forecasts. The calibration of each of them was verified monthly based on the joint sample accumulated during the preceding 6 months for both basins combined (type 2 verification). Then 9 time series of individual calibration scores CS were graphed. The scores varied among the forecasters and across time, but the minima reached below 0.06, to as low as 0.03. This showed the potential for mastering the metacognitive skill by individual forecasters and for improving the calibration of a group through individualized feedback.

11.7 Concepts and Proofs

11.7.1 General Definition of Calibration

If the Bayesian processor of forecast (BPF), formulated in Chapter 6 for a binary predictand, were reformulated for a continuous predictand W, it would revise a prior distribution function G of W into a posterior distribution function of W. For every $w \in \mathcal{W}$, this function would specify $P(W \le w | \mathbf{Y} = \mathbf{y})$ — the probability of the nonexceedance event $W \le w$, conditional on the vector of quantiles $\mathbf{Y} = \mathbf{y}$ provided by the forecaster on a particular occasion. Hence the following definition.

Definition   The forecaster of a continuous predictand is said to be *well calibrated* **against** the prior distribution function G if for every $\mathbf{y}$ and all $p \in \mathcal{P}$,

$$P(W \le y_p | \mathbf{Y} = \mathbf{y}) = p.$$

When $\mathcal{P} = \{p_i : i = 1, \dots, I\}$, there are I equalities which must be satisfied for every realization $\mathbf{y}$ of the I-dimensional vector of continuous variables, $\mathbf{y} = (y_{p_1}, \dots, y_{p_I})$. Clearly, this general definition is involved.

11.7.2 Calibration of Bayesian Forecaster

Theorem (*Calibration in-the-large*)   *The Bayesian forecaster (10.1)–(10.2) is well calibrated in-the-large on every set of forecast probabilities $\mathcal{P} \subseteq (0, 1)$.*

Proof: It proceeds in three steps.

1. For any $p \in P$, define $\Delta_p = W - Y_p$. The order probability (11.3) may now be written as

$$P(W \le Y_p) = P(W - Y_p \le 0) = P(\Delta_p \le 0).$$

Given $x \in \mathcal{X}$ on a particular forecasting occasion, the p-probability posterior quantile of W is $y_p = \Phi^{-1}(p|x)$; having observed $W = w$, the realization of Δ_p is

$$\delta_p = w - y_p = w - \Phi^{-1}(p|x).$$

2. Conditional on $X = x$ but before observing W,

$$(\Delta_p|X = x) = W - y_p = W - \Phi^{-1}(p|x).$$

Consequently,

$$\begin{aligned}
P(\Delta_p \le 0|X = x) &= P(W - y_p \le 0|X = x) \\
&= P(W \le y_p|X = x) \\
&= \Phi(y_p|x) \\
&= \Phi(\Phi^{-1}(p|x)|x) \\
&= p.
\end{aligned}$$

3. Because y_p is a function of x, the randomness of Y_p is due to the randomness of X, whose realization varies from one forecasting occasion to the next. Taking the expectation with respect to X yields the unconditional probability, which is the order probability:

$$\begin{aligned}
R(p) &= P(\Delta_p \le 0) \\
&= E[P(\Delta_p \le 0|X)] \\
&= \int_{\mathcal{X}} p\kappa(x)\, dx = p.
\end{aligned}$$

The equality $R(p) = p$ for all $p \in P$ defines the well-calibrated in-the-large forecaster. QED.

Note. At first, the proof may appear to hold for any statistical forecaster, but it does not. For suppose some non-Bayesian forecaster uses the same predictor X and outputs a conditional distribution function $\Psi(\cdot|x)$ of W. Thus in Step 1, $y_p = \Psi^{-1}(p|x)$. But in Step 2, executed from the viewpoint of a Bayesian decider, the posterior distribution function of W is $\Phi(\cdot|x)$. Hence,

$$\begin{aligned}
P(\Delta_p \le 0|X = x) &= P(W \le y_p|X = x) \\
&= \Phi(y_p|x) \\
&= \Phi(\Psi^{-1}(p|x)|x),
\end{aligned}$$

and

$$P(\Delta_p \le 0) = \int_{\mathcal{X}} \Phi(\Psi^{-1}(p|x)|x)\kappa(x)\, dx,$$

which cannot be proven to be equal to p for all $p \in P$.

11.7.3 Equivalent Calibrations

Theorem (*Equivalent calibrations*) *The forecaster of a continuous predictand is well calibrated in-the-large if and only if he is uniformly well calibrated.*

Proof: It consists of three steps.

1. *Preliminaries.* Let us reinterpret $\{\Phi(\cdot|x) : x \in \mathcal{X}\}$ as a family of functions from which the forecaster (any forecaster, not necessarily Bayesian) draws one function on each forecasting occasion, given information x available on that occasion. For every $x \in \mathcal{X}$, the function $\Phi(\cdot|x)$ possesses the three properties of the distribution function of a continuous predictand W: it is continuous and strictly increasing on the sample space $\mathcal{W}$, and its range is the open interval $(0, 1)$. For every $x \in \mathcal{X}$, the inverse function $\Phi^{-1}(\cdot|x)$ is such that

$$U = \Phi(W|x) \quad \Longleftrightarrow \quad W = \Phi^{-1}(U|x);$$

and for any $p \in (0, 1)$ and any $y_p \in \mathcal{W}$,

$$p = \Phi(y_p|x) \quad \Longleftrightarrow \quad y_p = \Phi^{-1}(p|x).$$

2. *Proof that* {well calibrated in-the-large} $\Longrightarrow$ {uniformly well calibrated}. The value $\Phi(w|x)$ is not presumed to be interpretable as the probability $P(W \leq w|X = x)$. Hence the derived variate U is not presumed to be independent of X and uniform — this is to be proven. Specifically, we must prove that for every probability $p \in (0, 1)$,

$$R(p) = p \quad \Longrightarrow \quad P(U \leq p) = p.$$

Reusing the notation and reasoning from Step 1 in Section 11.7.2, we begin with the second equation from Step 2, reusing its first two lines:

$$
\begin{aligned}
P(\Delta_p \leq 0|X = x) &= P(W - y_p \leq 0|X = x) \\
&= P(W \leq y_p|X = x) \\
&= P[\Phi^{-1}(U|x) \leq \Phi^{-1}(p|x)|X = x] \\
&= P[\Phi(\Phi^{-1}(U|x)|x) \leq \Phi(\Phi^{-1}(p|x)|x)|X = x] \\
&= P(U \leq p|X = x).
\end{aligned}
$$

The last probability is not necessarily equal to p because U is not necessarily the uniform variate. Therefore, we continue like in Step 3 of Section 11.7.2 by taking the expectation with respect to X; this yields the unconditional probability, which is the order probability:

$$
\begin{aligned}
R(p) &= P(\Delta_p \leq 0) \\
&= E[P(\Delta_p \leq 0|X)] \\
&= E[P(U \leq p|X)].
\end{aligned}
$$

Now, if $R(p) = p$ for every $p \in (0, 1)$, then

$$E[P(U \leq p|X)] = p.$$

This implies that U is stochastically independent of X and uniformly distributed: $P(U \leq p) = p$.

3. *Proof that* {uniformly well calibrated} $\Longrightarrow$ {well calibrated in-the-large}. On each forecasting occasion, given $x \in \mathcal{X}$, the uniformly well-calibrated forecaster draws one function from a family $\{\Phi(\cdot|x) : x \in \mathcal{X}\}$ having the property that, for every $x \in \mathcal{X}$, the variate $U = \Phi(W|x)$ is stochastically independent of X and has the uniform distribution function: $P(U \leq p) = p$. Now, we must prove that for every probability $p \in (0, 1)$,

$$P(U \leq p) = p \quad \Longrightarrow \quad R(p) = p.$$

In Step 2 above, it was shown that

$$P(U \leq p|X = x) = P(\Delta_p \leq 0|X = x).$$

Taking the expectation with respect to X on the left side,

$$E[P(U \leq p|X)] = P(U \leq p) = p,$$

by the property of PIT; and on the right side,

$$E[P(\Delta_p \leq 0|X)] = R(p),$$

by the equation in Step 3 of Section 11.7.2. Via the transitivity property of the equality, $R(p) = p$. QED.

Bibliographical Notes

Verification of uniform calibration via the Kolmogorov-Smirnov goodness-of-fit test was proposed by Wlper et al. (1994). The K-S comparison-of-distributions test for marginal calibration is an adaptation of the two-sample K-S test from Lindgren (1993).

The meta-Gaussian distribution (Kelly and Krzysztofowicz, 1995, 1997) is a bivariate (or multivariate) distribution constructed from any specified marginal distributions via a density weighting function (or copula); it allows pairwise nonlinear and heteroscedastic dependence structure (of a particular type); it has been used for probabilistic forecasting in hydrology (e.g., Krzysztofowicz and Kelly, 2000) and in meteorology (e.g., Krzysztofowicz and Evans, 2008).

The NQT was proposed by Krzysztofowicz (1992) for calculation of the informativeness score when both the estimator and the predictand are non-Gaussian. The formulae for Spearman's and Kendall's rank correlation coefficients come from Kruskal (1958).

The impact of different probabilities on the calibration of the tail credible interval (TCI) was investigated experimentally by Selvidge (1980) and by Alpert and Raiffa (1982). Overall, extreme probabilities (0.01, 0.99) or (0.001, 0.999) led to grossly miscalibrated 1% TCIs or 0.1% TCIs. The role of mind-set in performing various tasks successfully, including judgments and decisions, was researched by Dweck (2008).

The necessity of two attributes for verification of probabilistic forecasts of a continuous predictand has been rationalized within two paradigms. (i) The paradigm of optimal decision making under uncertainty adopted in this book. (ii) The paradigm of statistical optimization proposed by Gneiting et al. (2007): the forecaster should maximize the sharpness of the density function subject to a calibration constraint. In terms of their measures, the attributes (calibration, sharpness) are similar to, but not necessarily equivalent to, the attributes (calibration, informativeness) defined in this chapter.

Exercises

1 **Probability integral transform**. Reread Section 11.2.7. Apply the theory of Section 3.8.4 to derive the conditional density function $h(\cdot|x)$ of the variate U, for any $x \in \mathcal{X}$. Interpret the result.

2 **Probability integral transform for BGF**. Let $U = \Phi(W|w)$, as explained in Section 11.2.7, and let $\Phi(\cdot|x)$ be the distribution function of W outputted by the Bayesian Gaussian forecaster (BGF) and specified by expression (10.16). Prove that U is the uniform variate, which is stochastically independent of X.

3 **Kolmogorov-Smirnov test of uniform calibration**. Given the sample of the variate U from Table 11.2, perform the K-S test of the null hypothesis: U has the uniform distribution UN(0, 1). State the conclusion from this test in a way that answers two questions: Is this forecaster uniformly well calibrated? At what level of significance?

4 **Normal quantile transform**. Reread Step 2 of the algorithm in Section 11.3.4. Prove that the variate V, obtained through the NQT of variate W, has the standard normal distribution NM(0, 1).

5 **Visualization of NQT**. Reread Section 11.3.4. To get a sense of the empirical NQT, redo the calculations that transform the sample $\{y(n) : n = 1, \dots, 5\}$ of the estimator Y into the sample $\{z(n) : n = 1, \dots, 5\}$ of the variate Z; both samples are listed in Table 11.3. Likewise for samples of W and V. Next, complete the following exercises.

    5.1 Validate the sample moments after NQT: for each Z and V, calculate (i) the sample mean (is it 0?), and (ii) the sample standard deviation (is it close to 1?).

    5.2 Visualize the working of the NQT. (i) Graph the empirical distribution functions of Y and Z; compare them and judge the effect of the NQT. (ii) Graph the empirical distribution functions of W and V; compare them and judge the effect of the NQT. (iii) Make the scatterplots of the joint samples of (Y, W) and (Z, V); compare them and judge the effects of the NQTs.

6 **Mean squared error of BGF**. Let Y be an *estimator* of the predictand W. On every forecasting occasion, its realization y, called the *estimate* of W, is set to the posterior mean of W outputted by the Bayesian Gaussian forecaster and specified by equation (10.13). Three properties of this Bayesian estimator Y of the predictand W are mentioned in the note to Section 11.3.5. Prove them.

    6.1 The mean error of Y is zero.

    6.2 The variance of the error equals T^2.

    6.3 The MSE equals T^2.

7 **Calibration in-the-large of BGF**. The theorem stating that a Bayesian forecaster is well calibrated in-the-large on every set $\mathcal{P}$ is proven in Section 11.7.2. Prove the corollary: that a Bayesian Gaussian forecaster is well calibrated in-the-large on every set $\mathcal{P}$. *Hint*: Parallel the proof from Section 11.7.2, but employ expressions (10.16) and (10.18) for the functions Φ and Φ^{-1}, respectively.

8 **Calibration in-the-margin: proof**. The definition of this property is stated in Section 11.4.3.

    8.1 Restate it in terms of the prior distribution function G of W and the posterior distribution function $\Phi(\cdot|x)$ of W, given $x \in \mathcal{X}$.

    8.2 Prove that the Bayesian forecaster (10.1)–(10.2) satisfies the restated definition.

9 **Kolmogorov-Smirnov test of calibration in-the-margin**. Given the two empirical distribution functions of W from Table 11.5, perform the K-S test of the null hypothesis: the mean posterior distribution function of W and the prior distribution function of W are identical. State the conclusion from this test in a way that answers two questions: Is this forecaster well calibrated in-the-margin? At what level of significance?

10 **Equivalent calibrations of BGF**. Reread Section 11.7.3. Redo Step 1 and Step 2 of the proof for the Bayesian Gaussian forecaster; that is, employ expressions (10.16) and (10.18) for the functions Φ and Φ^{-1}, respectively.

11 **Uniform calibration: verification for BGF**. Consider the Bayesian Gaussian forecaster of the seasonal snowmelt runoff volume at Weiser, Idaho, which is described in Section 10.6. Its posterior parameter values are specified in the option. Its uniform calibration is to be verified on the joint sample of (X, W) from the years specified in the option.

    11.1 Collate the joint sample of (X, W) from the data given in Tables 10.6 and 10.7.

    11.2 Apply the algorithm for verification of the uniform calibration; report the output in the format of Table 11.2.

    11.3 Graph the uniform calibration function of the forecaster. (*Reminder*: The domain and the range must form a unit square.)

    11.4 Interpret the results of this verification.

Options for the joint sample and parameter values:

(75) Joint sample for verification: $N = 9$ (1975–1983).

Posterior parameter values: $A = 1.488$, $B = -139.34$, $T = 100.47$.

(They were obtained from the prior sample (1973–1983) and the joint sample (1977–1983).)

(73) Joint sample for verification: $N = 11$ (1973–1983).

Posterior parameter values as in option (75).

(71) Joint sample for verification: $N = 13$ (1971–1983).

Posterior parameter values: $A = 1.052$, $B = 7.70$, $T = 109.11$.

(They were obtained from the prior sample (1963–1983) and the joint sample (1971–1983).)

12 **Calibration in-the-margin: verification for BGF**. Consider the Bayesian Gaussian forecaster of the seasonal snowmelt runoff volume at Weiser, Idaho, which is described in Section 10.6. Its prior and likelihood parameter values are specified in the option. Its calibration in-the-margin is to be verified on the joint sample of (X, W) from the years specified in the option.

12.1 Collate the joint sample of (X, W) from the data given in Tables 10.6 and 10.7.

12.2 Apply the algorithm for verification of the calibration in-the-margin; report the output in the format of Table 11.5.

12.3 Graph the marginal calibration function of the forecaster. (*Reminder*: The domain and the range must form a unit square.)

12.4 Perform the K-S comparison-of-distributions test.

12.5 Interpret the results of this verification, and offer feedback to the analyst who developed this BGF.

Options for the joint sample and parameter values:

(75) Joint sample for verification: $N = 9$ (1975–1983).

Prior parameter values: $M = 413.25$, $S = 177.39$.

Likelihood parameter values: $a = 0.456$, $b = 182.70$, $\sigma = 55.64$.

(They were estimated from the prior sample (1973–1983) and the joint sample (1977–1983).)

(73) Joint sample for verification: $N = 11$ (1973–1983).

Parameter values as in option (75).

(71) Joint sample for verification: $N = 13$ (1971–1983).

Prior parameter values: $M = 421.07$, $S = 162.47$.

Likelihood parameter values: $a = 0.522$, $b = 173.26$, $\sigma = 76.86$.

(They were estimated from the prior sample (1963–1983) and the joint sample (1971–1983).)

13 **Actual versus potential performance**. As explained in Section 11.4, the BGF is potentially well calibrated in-the-large; this implies the calibration score $CS = 0$. Suppose the informativeness score calculated from equation (10.26) based on the values of parameters a, σ, S is $IS = 0.58$.

13.1 The verification based on a sample of actual forecasts yielded $CS = 0.15$ and $IS = 0.41$. Hypothesize the causes of the discrepancy between the potential performance and the actual performance of this BGF.

13.2 The verification based on another sample of actual forecasts yielded $CS = 0.04$ and $IS = 0.72$. Someone opines that these scores must be erroneous in light of the previous two sets of scores. Offer your opinion. Justify it by hypothesizing the causes of the differences between the three sets of scores.

14 **Interpretation of a calibration report**. Retrieve the calibration report from Task 9.1 specified in the option. (i) Write a detailed interpretation of this report as if you were providing feedback to a forecaster. (ii) Offer advice regarding future assessments of quantiles.

Options for the calibration report:

(C1) Table 11.6, group C1.	(HB) Table 11.6, group HB.
(C2) Table 11.6, group C2.	(WP) Table 11.7, variate W_P.
(PF) Table 11.6, group PF.	(WY) Table 11.7, variate W_Y.

15 **Distribution calibration function**. Retrieve the calibration report from Task 9.1 specified in the option, and complete the following exercises.

15.1 Construct the distribution calibration function, DCF.

15.2 Show the calculation of the calibration score, *CS*.

15.3 Graph the DCF. (*Reminder*: The domain and the range must form a unit square.)

15.4 Interpret the DCF as if you were providing feedback to a forecaster; identify judgmental biases; offer advice regarding future assessments of quantiles.

Options for the calibration report:

(C1) Table 11.6, group C1.	(WE) Table 11.7, variate W_E.
(C2) Table 11.6, group C2.	(WP) Table 11.7, variate W_P.
(PF) Table 11.6, group PF.	(WY) Table 11.7, variate W_Y.

16 **Calibration of auditors**. Retrieve the calibration report from Table 11.8 specified in the option, and complete the following exercises.

16.1 Write a detailed interpretation of this report as if you were providing feedback to an auditor.

16.2 Compare and contrast the judgmental biases of the auditors, A, who forecasted the "true" dollar amount of an account balance, with the judgmental biases of the students, S, who forecasted the variate specified in the option for which the calibration report is in Table 11.7.

16.3 Suppose the calibration report S came not from students but from a group of auditors who forecasted the "true" dollar amount, W_P or W_Y. You are asked to recommend one group of external auditors, A or S, to a company. Which group do you recommend and why?

Options for the auditors (A) *and students* (S):

(IP) Group IN, variate W_P.	(TP) Group TE, variate W_P.
(IY) Group IN, variate W_Y.	(TY) Group TE, variate W_Y.

17 **Calibration of meteorologists**. Retrieve the calibration report from Table 11.9 specified in the option, and complete the following exercises.

17.1 Construct the distribution calibration function, DCF.

17.2 Show the calculation of the calibration score, *CS*.

17.3 Graph the DCF. (*Reminder*: The domain and the range must form a unit square.)

17.4 Interpret the DCF as if you were providing feedback to a meteorologist; identify judgmental biases; offer advice regarding future assessments of quantiles.

Options for the river basin:

(MN) Monongahela;   (AL) Allegheny.

Mini-Projects

18 **Flood crest forecasts: Bayesian verification**. At the onset of a heavy thunderstorm over the mountains, the crest W of a flash flood at the river gauge in the piedmont is forecasted with the expected lead time of

4 hours. Each of the two hydrologist-interns, A and B, prepares a forecast of W relying on different rainfall and runoff models, and adjusting judgmentally the model outputs based on the latest hydrometeorological observations sent by sensors and spotters. Each forecast of W states an estimate y, measured in inches above the *flood stage* — the elevation of water surface at which property damage begins. The estimate y should be interpretable as the median of W. At the conclusion of the summer thunderstorm season, the realizations of Y and W were compiled:

	A	B	
n	y	y	w
7			110
6			20
5	20	40	10
4	30	50	40
3	60	100	90
2	50	60	30
1	70	80	50

18.1 Determine which hydrologist is better calibrated in-the-large.

18.2 Determine if one hydrologist is more informative than the other. (For each hydrologist, calculate the informativeness score based on the Gaussian model; graph the likelihood regression superposed on the scatterplot of the joint sample. Use the graph to explain the difference between the two scores.)

18.3 Overall, which hydrologist-intern would you recommend for hiring? Justify your recommendation.

19 **Flood crest forecasts** — *continued*: **Bayesian processor**. An emergency manager in a town located on the river bank near the gauge wants to use the flash-flood crest estimate to decide whether to issue or not to issue a warning to the town dwellers. When making this decision, she wants to account for the uncertainty about the actual crest, given its estimate $Y = y$ provided by a hydrologist. Quantify this uncertainty by applying the BGF using Y as the predictor. To wit, for each hydrologist, A and B, take the results from Exercise 18 and complete the following exercises.

19.1 Calculate the values of the posterior parameters A, B, T.

19.2 Graph the posterior mean $E(W|Y = y)$ as a function of y, superposed on the scatterplot of the joint sample, $w(n)$ versus $y(n)$ for $n = 1, \ldots, 5$.

19.3 Graph on the same figure (i) the posterior density function of W, conditional on the crest estimate $Y = 45$ inches, and (ii) the prior density function of W.

19.4 Given $Y = 45$, calculate the posterior probability of W exceeding 50 inches — the water level that triggers the *evacuation advisory* to town dwellers. Compare this probability with the prior probability of the same event.

19.5 Compare the above results for the two hydrologists in a way that supports the recommendation for hiring you made in Exercise 18.3.

20 **Flood crest forecasts** — *continued*: **mean squared error.** A supervisor in the forecast office wants to boil down the comparison of the two hydrologist-interns to a comparison of two numbers — the mean squared errors.

20.1 Calculate *MSE* of each hydrologist using the parameters estimated in Exercise 18. Which hydrologist would be hired? Compare this decision with your recommendation in Exercise 18.3; draw a conclusion.

20.2 For each hydrologist, take the BGF developed in Exercise 19 and recalibrate his every estimate $y(n)$ into the estimate $y^*(n) = Ay(n) + B$, which is the posterior median of W. Given the joint samples $\{(y^*(n), w(n))\}$, determine which hydrologist would be better calibrated if he disseminated $y^*(n)$ instead of $y(n)$.

20.3 Based on the theoretic facts (see Exercise 6 and the note to Section 11.3.5), the recalibrated estimator Y^* has $MSE = T^2$, calculated in Exercise 19.1. Given these MSE values, which hydrologist would be hired?

20.4 Argue for or against the use of MSE as a performance measure of a hydrologist (in particular) and a forecaster (in general).

Note. The quantity $\sqrt{MSE}$ is called the *standard error* of an estimator. It is preferred for discussion because its unit is identical to the unit of W, here inches, whereas the unit of MSE is [unit of $W]^2$, here inches squared.

21 **Judgmental forecasts of temperature**. Allan H. Murphy, a meteorologist, and Robert L. Winkler, a statistician, teamed up to design and execute several pioneering experiments in judgmental forecasting of weather variates. Their grand objectives were (i) to evaluate the ability of professional weather forecasters to quantify the uncertainty, and (ii) to promote the concept of probabilistic weather forecasting. In one of the experiments (Murphy and Winkler, 1974), there were two continuous predictands: the daytime maximum temperature, W_M, and the overnight minimum temperature, W_m, in Denver, Colorado. Two professional forecasters of the National Weather Service, having bachelor degrees in meteorology and over 18 years of experience, prepared probabilistic forecasts when on duty. Forecasts prepared on the day shift had the lead time (LT) of 12 h for W_m (tonight's low) and 24 h for W_M (tomorrow's high). Forecasts prepared on the midnight shift had LT of 12 h for W_M (today's high) and 24 h for W_m (tonight's low). Each forecast comprised 5 judgmentally assessed quantiles corresponding to the probabilities 0.125, 0.25, 0.5, 0.75, 0.875. A total of 132 probabilistic forecasts were prepared between August 1972 and March 1973. From the sample of joint realizations of the quantile vector and the predictand, calibration reports for six sets of forecasts were collated (Table 11.10). Use the two reports specified in the option.

21.1 Identify the type of verification to be performed based on the specified calibration reports.

Table 11.10 Calibration reports from an experiment on probabilistic forecasting of the daily maximum and minimum temperatures, with lead times of 12 h and 24 h, by two professional forecasters in Denver, Colorado.

Set	N	Simple events					
		Ω_1	Ω_2	Ω_3	Ω_4	Ω_5	Ω_6
Forecast P		0.125	0.125	0.25	0.25	0.125	0.125
Max. temperature	66	0.15	0.14	0.22	0.29	0.11	0.09
Min. temperature	66	0.06	0.07	0.14	0.25	0.15	0.23
Lead time 12 h	66	0.09	0.14	0.25	0.26	0.15	0.11
Lead time 24 h	66	0.12	0.17	0.15	0.24	0.11	0.21
Forecaster A	64	0.09	0.21	0.21	0.16	0.19	0.14
Forecaster B	68	0.12	0.10	0.29	0.24	0.07	0.18

Source: Murphy and Winkler (1974, Tables 1, 3); statistical probabilities rounded to two decimal places; the union $\Omega_3 \cup \Omega_4$ partitioned approximately into simple events.

21.2 Define the simple events, the compound events, their forecast probabilities, and their statistical probabilities — in parallel to the definitions in Section 11.5.1.

Note. Perform Exercises 21.3–21.6 for each of the two sets of forecasts.

21.3 Expand the calibration report to include the three compound events and the calibration score, *CS*.

21.4 Write a detailed interpretation of the expanded report as if you were providing feedback to the two weather forecasters; identify judgmental biases; offer advice regarding future assessments of quantiles.

21.5 Construct and graph the distribution calibration function, DCF. (*Reminder*: The domain and the range must form a unit square.)

21.6 Interpret concisely the DCF.

21.7 Compare and contrast the calibrations of the two sets of forecasts. Overall, which set of forecasts is better calibrated, in your opinion?

21.8 For options (MM) and (LT) only: Hypothesize an explanation for the distinct calibrations of the two sets of forecasts (especially for the distinct judgmental biases).

Options for the calibration reports:

(MM) Predictands (maximum temperature and minimum temperature).
(LT)   Lead times (12 h and 24 h).
(FF)   Forecasters (A and B).

22 Commodity futures forecasts: Gaussian model. An investor trades commodity futures contracts on the New York Mercantile Exchange (NYMEX). One of the commodities is platinum — a noble metal utilized in making jewelry, catalytic converters, electrical devices, chemotherapy drugs, etc. Its futures contract is for 50 troy ounces (1.555 kg); thus, a price change of $1 per troy ounce yields a gain or a loss of $50 per contract. A contract has an expiration date by which (i) the holder of a *long position* (who bought a contract) must either offset it by selling a contract or accept delivery of the metal, whereas (ii) the holder of a *short position* (who sold a contract) must either offset it by buying a contract or deliver the metal to a warehouse. The investor trades the contract with the largest *open interest* — the total number of outstanding long contracts (which equals the total number of outstanding short contracts); this is usually one that expires in the nearest January, April, July, or October. Every Monday, he decides to buy, or to sell, or to stay out of the market based on a forecast of the change W from the settlement price on Friday to the settlement price on the next Friday. The forecast specifies three quantiles of W: $\mathbf{y} = (y_{0.2}, y_{0.5}, y_{0.8})$. It is based on fundamental analysis (e.g., projections of supply and demand, economic growth, political developments) and on technical analysis of the price time series (patterns, statistics, and heuristics such as buy and sell "signals"). The forecast is prepared on Saturday and is updated on Monday morning, before the NYMEX opens, based on the world news and the opening prices at exchanges in Asia and Europe. Thus the forecast lead time is 5 days. Use the record of $\mathbf{Y} = (Y_{0.2}, Y_{0.5}, Y_{0.8})$ and W compiled in Table 11.11 to perform a complete verification analysis of these forecasts.

22.1 *Calibration in-the-large.* (i) Construct and graph the distribution calibration function. (ii) Calculate the calibration score. (iii) Interpret the results, identify biases, provide feedback to the forecaster.

22.2 *Uniform calibration.* (i) Construct and graph the uniform calibration function. (ii) Calculate the uniform calibration score. (iii) Perform the K-S goodness-of-fit test. (iv) Interpret the results.

Hint. Given the quantile vector $\mathbf{y}$, apply the quantiles–moments method (Section 9.3.2) to estimate the parameters of the $NM(\mu, \sigma)$ distribution function of W; use this function in lieu of $\Phi(\cdot|x)$ in the algorithm of Section 11.2.8.

22.3 *Informativeness.* (i) Calculate the informativeness score based on the Gaussian model. (ii) Graph the likelihood regression superposed on the scatterplot of the joint sample. (iii) Determine whether or not the median $Y_{0.5}$ constitutes (at least approximately) a single sufficient statistic of informativeness of the forecast; qualify the interpretation of *IS* accordingly.

Table 11.11 Record of $(y_{0.2}, y_{0.5}, y_{0.8})$ — the forecast quantiles, and of w — the weekly change of the Friday's settlement price [$ per troy oz] of the platinum futures contract (for 50 troy ounces). Quantiles are hypothetical; price change is actual in 2021, as reported by *The Wall Street Journal*. Month designates the contract expiration. Date (day.month) designates Friday on which w is determined.

Month	Date	$y_{0.2}$	$y_{0.5}$	$y_{0.8}$	w	Month	Date	$y_{0.2}$	$y_{0.5}$	$y_{0.8}$	w
April	8.01	−8	9	24	−7.9	October	2.07	−27	−13	0	−18.0
	15.01	14	30	45	18.6		9.07	7	24	50	8.0
	22.01	−1	10	21	21.7		16.07	5	21	35	12.8
	29.01	−33	−15	0	−32.4		23.07	−35	−19	0	−47.1
	5.02				53.8		30.07	−14	2	17	−13.0
	12.02				126.0		6.08	−50	−34	−20	−76.2
	19.02	40	54	70	34.1		13.08	−20	35	50	53.8
	26.02	−67	−50	−35	−107.8		20.08	−35	−22	−5	−31.8
	5.03	−56	−40	−25	−57.0		27.08	−10	5	20	12.3
	12.03	28	43	58	72.0		3.09	−5	11	26	15.1
	19.03	−2	14	27	−0.2		10.09	−45	−29	−13	−65.1
							17.09	−27	−10	7	−25.9
July	1.04	2	17	32	26.9	January	1.10	−10	6	20	−2.7
	9.04				0.7		8.10	10	23	35	54.6
	16.04				−0.6		15.10	0	13	28	30.7
	23.04	30	47	61	24.4		5.11				15.1
	30.04	−57	−42	−28	−27.9		12.11	27	40	53	53.4
	7.05	10	26	42	49.3		19.11				−53.2
	14.05				−31.7		26.11	−46	−31	16	−81.7
	21.05	−40	−26	−11	−53.4		3.12	−30	−17	0	−28.1
	28.05	−20	−5	10	13.0		10.12				8.0
	4.06	−23	−7	15	−18.0						
	11.06	−39	−23	7	−14.3						
	18.06				−110.1						

23 **Snowmelt runoff forecasts: BGF** — Chapter 10, Exercise 18 *continued*. Retrieve the Bayesian Gaussian forecaster of the seasonal snowmelt runoff volume which you developed in that exercise. Retrieve also the joint sample of the predictor–predictand pair (X, W) from 13 years (1971–1983); this is the sample which you used to estimate the likelihood parameters of your BGF. Use this joint sample (i) to produce the forecasts of W retrospectively, and (ii) to verify the produced forecasts.

23.1 *Calibration in-the-large.* (i) Construct and graph the uniform calibration function. (ii) Calculate the uniform calibration score. (iii) Interpret the results.

23.2 *Calibration in-the-margin.* (i) Construct and graph the marginal calibration function. (ii) Calculate the marginal calibration score. (iii) Perform the K-S comparison-of-distributions test. (iv) Interpret the results, identify biases, provide feedback to the analyst who developed the forecasting model.

23.3 *Informativeness.* (i) Calculate the informativeness score based on the Gaussian model. (ii) Graph the likelihood regression superposed on the scatterplot of the joint sample. (iii) Validate the assumptions of the normal-linear model.

24 **Recalibration based on uniform calibration function**. Let $\Phi(\cdot|j)$ be a continuous and strictly increasing distribution function of predictand W on the jth forecasting occasion, $j \in \{1, 2, \dots\}$. It is produced consistently on each occasion by the same forecasting method. Suppose a joint sample of past distribution functions and predictand realizations was inputted to the algorithm of Section 11.2.8 and the empirical distribution function $\{(u_{(n)}, p_n) : n = 1, \dots, N\}$ was constructed. Suppose, furthermore, that this function departs significantly from the uniform distribution function, but its graph appears "smooth", not "jerky", so that it can be interpolated by a continuous and strictly increasing function $T : (0, 1) \rightarrow (0, 1)$. Let T be named the *uniform calibration function*.

24.1 Invent a method for recalibration of every future forecast $\Phi(\cdot|j)$ in order to obtain distribution functions of W that are uniformly well calibrated.

24.2 Create a numerical example demonstrating the working of your recalibration method.

Hints.

(i) Create an empirical distribution function $\{(u_{(n)}, p_n)\}$ consisting of a few points and reflecting particular biases; fit to it one of the parametric distribution functions on the bounded open interval $(0, 1)$ to obtain T.

(ii) Suppose $\Phi(\cdot|j)$ comes from a family of parametric distribution functions, and its parameter values vary from one forecasting occasion to the next; for example, on occasion j, the function $\Phi(\cdot|j)$ comes from model $\mathrm{WB}(\alpha_j, \beta_j, \eta)$.

(iii) Consider three very different forecasting occasions, say $j = 1, 2, 3$. For each j, choose judiciously the parameter values; graph the forecast distribution function $\Phi(\cdot|j)$ and the recalibrated distribution function.

(iv) Explain how the biases quantified by T are removed through recalibration.

12

Target-Decision Theory

There is a class of decision problems under uncertainty wherein the optimal value of a continuous decision variable a would be set to the realization w of the random input variable W, if only one knew that realization with certainty at the time the decision must be made. This class includes problems of statistical estimation and problems of setting targets in the context of management, planning, and control. The term *target* may have various connotations: commitment level, contract magnitude, production goal, inventory level, stock size, benchmark return, and so on.

This chapter presents the fundamental theory and the analytic solutions to the target-setting problem with three forms of the criterion function: the two-piece linear opportunity loss function, the quadratic difference opportunity loss function, and the impulse utility function. Applications follow later in this chapter and in Chapter 13.

12.1 Target-Setting Problem

12.1.1 Elements

The general target-setting problem is a decision problem under uncertainty with the following elements:

a — decision variable (continuous), whose feasible values form a decision space $\mathcal{A}$;
W — input variate (continuous), whose possible realizations form a sample space $\mathcal{W}$;
G — distribution function of W such that for any realization $w \in \mathcal{W}$,

$$G(w) = P(W \leq w);$$

g — density function of W corresponding to G;
u — *utility function* that quantifies the strength of preference for outcomes of decision–input pairs, to wit,

$$u : \mathcal{A} \times \mathcal{W} \to \mathfrak{R},$$

such that $u(a, w)$ denotes the utility of outcome resulting from the decision $a \in \mathcal{A}$ and the input $w \in \mathcal{W}$. This utility is in a monetary unit when it represents profit, or it is dimensionless when it represents a subjective valuation of the desirability of an outcome.

The defining properties of the *target-setting problem* are (i) that the decision space $\mathcal{A}$ and the sample space $\mathcal{W}$ coincide: $\mathcal{A} = \mathcal{W}$; and (ii) that if input w were known with certainty, then the optimal decision would be $a = w$. Ergo, the utility function u is such that for every $w \in \mathcal{W}$,

$$\max_{a \in \mathcal{A}} u(a, w) = u(w, w). \tag{12.1}$$

When input W is uncertain, the problem is to find the value $a^* \in \mathcal{A}$ of the decision variable which (i) constitutes an estimate of the unknown realization of W, and (ii) is optimal with respect to the specified criterion function.

Probabilistic Forecasts and Optimal Decisions, First Edition. Roman Krzysztofowicz.
© 2025 John Wiley & Sons Ltd. Published 2025 by John Wiley & Sons Ltd.
Companion website: www.wiley.com/go/ProbabilisticForecastsandOptimalDecisions1e

Example 12.1 (i) A school principal needs an estimate, a, of the number of children, W, who will seek enrollment in her school next year. (ii) A civil engineer must decide the height of a flood levee, a, built to protect a town against the maximum flood crest, W, during the design lifetime, say the next 50 years. (iii) A nurseryman must decide the number of Christmas trees to stock before December, a, while the number of trees the customers will actually want to purchase, W, is uncertain. In each of these decision problems, a misestimation of the input has negative consequences.

12.1.2 Opportunity Loss Function

Oftentimes, it is more natural or convenient to frame the criterion function not as a utility function u, but as an *opportunity loss function*

$$l : \mathcal{A} \times \mathcal{W} \rightarrow \mathfrak{R},$$

such that $l(a, w)$ denotes the opportunity loss resulting from the decision $a \in \mathcal{A}$ and the input $w \in \mathcal{W}$. This opportunity loss is defined as the difference between the utility of the optimal decision and the utility of a given decision, when w is fixed (as if it were known):

$$l(a, w) = \max_{a \in \mathcal{A}} u(a, w) - u(a, w)$$
$$= u(w, w) - u(a, w); \tag{12.2}$$

the second equality results from property (12.1). It follows that l is a nonnegative function with the minimum of zero at $a = w$; specifically, for every $(a, w) \in \mathcal{A} \times \mathcal{W}$,

$$l(a, w) \geq 0 \quad \text{and} \quad l(w, w) = 0.$$

12.1.3 Decision Criteria

In the presence of uncertainty about the input W, the decision $a^* \in \mathcal{A}$ is optimal if it *maximizes the expected utility*:

$$U^* = E[u(a^*, W)] = \max_{a \in \mathcal{A}} E[u(a, W)]. \tag{12.3}$$

Equivalently, the decision $a^* \in \mathcal{A}$ is optimal if it *minimizes the expected opportunity loss*:

$$L^* = E[l(a^*, W)] = \min_{a \in \mathcal{A}} E[l(a, W)]. \tag{12.4}$$

The two decision criteria are equivalent because, employing relationship (12.2),

$$\min_{a \in \mathcal{A}} E[l(a, W)] = \min_{a \in \mathcal{A}} \{E[u(W, W) - u(a, W)]\}$$
$$= E[u(W, W)] - \max_{a \in \mathcal{A}} E[u(a, W)];$$

in short,

$$L^* = E[u(W, W)] - U^*. \tag{12.5}$$

To sum up, the criterion function for a target-setting problem can be framed either as a utility function or as an opportunity loss function. Then the optimal decision a^* under uncertainty about the input W can be found either by maximizing the expected utility or, equivalently, by minimizing the expected opportunity loss.

12.1.4 Value of Perfect Forecaster

The ideal decision situation would be one without uncertainty. A question thus arises: How much is one losing, in terms of utility, because of the presence of uncertainty? Alternatively: How much, at most, should one be willing to pay an expert for a perfect forecast of W?

Imagine a clairvoyant who reveals the input w. Per expression (12.1), the optimal decision is $a = w$ and its utility is $u(w, w)$. Of course, a clairvoyant is not actually available, but is only contemplated. While the true input w remains unknown, one can ask the question: What is the distribution function of inputs that the clairvoyant could state? It is, naturally, the same as the distribution function G of the random input W. Therefore, the optimal decision with a perfect forecast can be evaluated *a priori* in terms of the *expected maximum utility*:

$$U^{**} = E[\max_a u(a, W)]$$
$$= E[u(W, W)]. \tag{12.6}$$

The *value of the perfect forecaster* of the input W is defined as the difference between the expected maximum utility U^{**} and the maximum expected utility U^*:

$$VPF = U^{**} - U^*$$
$$= L^*, \tag{12.7}$$

where the last equality is obtained by replacing U^{**} with expression (12.6), and recognizing the resultant as expression (12.5). It turns out that VPF equals the minimum expected opportunity loss L^*, as defined by equation (12.4). Because l is a nonnegative function, $L^* \geq 0$. Hence VPF is never negative.

When L^* is in monetary unit, the VPF admits two useful interpretations. First, it is the maximum price the decider should be willing to pay for one perfect forecast of the input. Second, it is an upper bound on the value of any imperfect forecast.

Having defined the target-setting problem, we can focus on the criterion function. A general modeling methodology is outlined in Section 12.9. But for four particular mathematical forms of the criterion function there exist elegant analytic solutions. We shall derive them and study their properties. Next, we shall use theses theoretic results in several applied decision models, in this chapter and in Chapter 13.

12.2 Two-Piece Linear Opportunity Loss

12.2.1 Criterion Function

The consequences of misestimation of the input are quantified in terms of two constants, which may be in monetary units or may represent subjective valuations:

λ_o — marginal opportunity loss from overestimation of the input [\$/(unit of w)];
λ_u — marginal opportunity loss from underestimation of the input [\$/(unit of w)].

The opportunity loss resulting from the decision a and the input w is proportional to the absolute difference between a and w, but its magnitude depends on the direction of misestimation:

$$l(a, w) = \begin{cases} \lambda_o(a - w) & \text{if } w \leq a, \\ \lambda_u(w - a) & \text{if } a \leq w. \end{cases} \tag{12.8}$$

For a fixed decision a, this opportunity loss is a two-piece linear function of the input w (Figure 12.1(a)), with slope coefficients $\lambda_o > 0$ and $\lambda_u > 0$.

12.2.2 Optimal Decision Procedure

The optimal decision procedure consists of three steps. First, any feasible decision $a \in \mathcal{A}$ is evaluated in terms of the *expected opportunity loss* resulting from that decision:

$$L(a) = E[l(a, W)]. \tag{12.9}$$

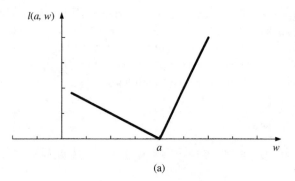

(a)

Figure 12.1 Special criterion functions for an estimation, or a target-setting, problem: (a) two-piece linear opportunity loss with $\lambda_o = 1/2$ and $\lambda_u = 2$ (a special case of which is absolute difference opportunity loss, when $\lambda_o = \lambda_u$); (b) quadratic difference opportunity loss with $\lambda = 1$; and (c) impulse utility.

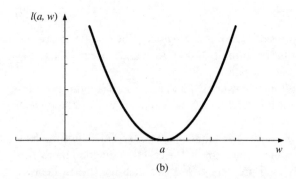

(b)

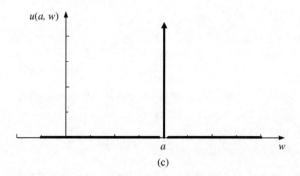

(c)

Next, the *optimal decision* a^* is found, which minimizes the expected opportunity loss:

$$\min_a L(a). \tag{12.10}$$

Finally, the *minimum expected opportunity loss* is determined:

$$
\begin{aligned}
L^* &= L(a^*) \\
&= \min_a L(a) \\
&= \min_a E[l(a, W)].
\end{aligned} \tag{12.11}
$$

12.2.3 Analytic Solution

The objective now is to derive expressions for L, a^*, and L^* in terms of the input elements λ_o, λ_u, G, and g. For derivations, it is assumed that the sample space of the random input W is bounded: $\mathcal{W} = (\eta_L, \eta_U)$. When $\mathcal{W}$ is of different type, one or both integration limits should be replaced: η_L by $-\infty$, η_U by ∞. With l given by expression (12.8) and the density function g of W specified, the *expected opportunity loss* can be expressed as follows:

$$L(a) = \int_{\eta_L}^{\eta_U} l(a, w)g(w)\, dw$$

$$= \lambda_o \int_{\eta_L}^{a} (a - w)g(w)\, dw + \lambda_u \int_{a}^{\eta_U} (w - a)g(w)\, dw$$

$$= \lambda_o a \int_{\eta_L}^{a} g(w)\, dw - \lambda_o \int_{\eta_L}^{a} wg(w)\, dw + \lambda_u \int_{a}^{\eta_U} wg(w)\, dw - \lambda_u a \int_{a}^{\eta_U} g(w)\, dw$$

$$= \lambda_o a G(a) - \lambda_o \int_{\eta_L}^{a} wg(w)\, dw + \lambda_u \left[\int_{\eta_L}^{\eta_U} wg(w)\, dw - \int_{\eta_L}^{a} wg(w)\, dw \right] - \lambda_u a[1 - G(a)],$$

where the last step employs the relationship between the density function g and the distribution function G. The final expression takes the form

$$L(a) = (\lambda_u + \lambda_o)aG(a) - \lambda_u a + \lambda_u E(W) - (\lambda_u + \lambda_o)\int_{\eta_L}^{a} wg(w)\, dw. \tag{12.12}$$

If the integral in the last summand had η_U as its upper limit, it would give the expectation of W. Because the integration is truncated at a, the integral is called the *incomplete expectation* of W. Methods of solving this integral are covered in Section 12.3.

In order to find the optimal decision, the first derivative of $L(a)$ with respect to a must be obtained:

$$\frac{dL(a)}{da} = (\lambda_u + \lambda_o)G(a) + (\lambda_u + \lambda_o)ag(a) - \lambda_u - (\lambda_u + \lambda_o)ag(a)$$

$$= (\lambda_u + \lambda_o)G(a) - \lambda_u.$$

The necessary condition for a^* to be the optimal decision is that $dL(a^*)/da = 0$. Hence

$$(\lambda_u + \lambda_o)G(a^*) - \lambda_u = 0.$$

It follows that a^* is the *optimal decision* if

$$G(a^*) = \frac{\lambda_u}{\lambda_u + \lambda_o} = \frac{1}{1 + \lambda_o/\lambda_u}. \tag{12.13}$$

Because W is a continuous variate, its distribution function G, which is continuous and strictly increasing, has the inverse G^{-1}, called the *quantile function* of W. Thus,

$$a^* = G^{-1}\left(\frac{\lambda_u}{\lambda_u + \lambda_o}\right) = G^{-1}\left(\frac{1}{1 + \lambda_o/\lambda_u}\right). \tag{12.14}$$

Finally, the expression for the *minimum expected opportunity loss* is derived by inserting solution (12.13) into expression (12.12):

$$L^* = L(a^*)$$

$$= (\lambda_u + \lambda_o)a^* \frac{\lambda_u}{\lambda_u + \lambda_o} - \lambda_u a^* + \lambda_u E(W) - (\lambda_u + \lambda_o)\int_{\eta_L}^{a^*} wg(w)\, dw,$$

wherefrom

$$L^* = \lambda_u E(W) - (\lambda_u + \lambda_o) \int_{\eta_L}^{a^*} wg(w)\,dw. \tag{12.15}$$

The effective calculation of L^* requires the calculation of the expectation of W and the incomplete expectation of W. These tasks are detailed in Section 12.3.

12.2.4 Interpretation of the Solution

The optimal solution to a target-setting problem with a two-piece linear opportunity loss function has three basic properties, illustrated in Figure 12.2.

1. In order to find the optimal decision a^*, one need only know
   (a) the distribution function G of the random input W, and
   (b) the ratio λ_o/λ_u of the marginal opportunity losses ($0 < \lambda_o/\lambda_u < \infty$).

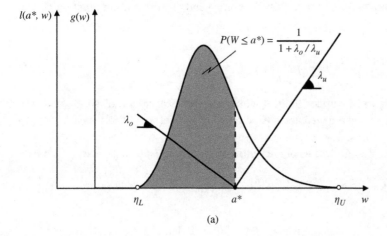

(a)

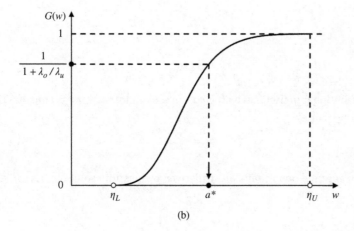

(b)

Figure 12.2 Optimal solution to a target-setting problem: (a) the density function g of the random input W on the sample space (η_L, η_U), and the two-piece linear opportunity loss function $l(a^*, \cdot)$ for the optimal decision a^*, with the marginal opportunity losses λ_o (from overestimation) and λ_u (from underestimation); (b) the distribution function G of W, and the determination of the optimal decision a^* from the ratio λ_o/λ_u of the marginal opportunity losses.

Table 12.1 Under the two-piece linear opportunity loss function, the ratio λ_o/λ_u of the marginal opportunity losses determines the optimal decision a^* and the associated optimal probability of overestimation, $P(W < a^*) = G(a^*)$.

λ_o/λ_u	5/1	3/1	1/1	1/3	1/5
$G(a^*)$	1/6	1/4	1/2	3/4	5/6

2. The optimal decision a^* equals the *quantile* of the random input W corresponding to the probability

$$P(W \leq a^*) = G(a^*) = \frac{1}{1 + \lambda_o/\lambda_u}. \qquad (12.16)$$

3. Associated with the optimal decision a^* are the probability of overestimation of the input, $P(W < a^*) = G(a^*)$, and the probability of underestimation of the input, $P(W > a^*) = 1 - G(a^*)$. (Note that $P(W \leq a^*) = P(W < a^*)$ and $P(W = a^*) = 0$ because W is a continuous variate.)

Table 12.1 illustrates the essence of this solution. For instance, when a decider judges the marginal opportunity loss from underestimation to be three times as large as that from overestimation, $\lambda_o/\lambda_u = 1/3$, then $G(a^*) = 3/4$; that is, the optimal decision a^* equals the third quartile of W under G.

12.2.5 Absolute Difference Opportunity Loss

A special case of the two-piece linear opportunity loss function arises when the marginal opportunity losses are identical: $\lambda_o = \lambda_u = \lambda$. The single constant has the following interpretation:

λ — marginal opportunity loss from misestimation of the input, regardless of the direction of misestimation [$/(unit of w)].

The opportunity loss function (12.8) is now symmetric about decision a,

$$l(a, w) = \begin{cases} \lambda(a - w) & \text{if } w \leq a, \\ \lambda(w - a) & \text{if } a \leq w, \end{cases}$$

and can be rewritten concisely as

$$l(a, w) = \lambda|a - w|. \qquad (12.17)$$

In other words, the opportunity loss is proportional to the absolute difference between the decision a (the estimate) and the realization w of the input variate W. From expression (12.14), it follows that the optimal decision a^* equals the *median* of the random input W:

$$a^* = G^{-1}(1/2). \qquad (12.18)$$

Thus in order to find the optimal decision a^*, one need not know λ and G; it suffices to know only the median of W.

12.3 Incomplete Expectations

For some families of distributions, the incomplete expectation of the variate W has a nice expression, as the integral which defines it can be solved analytically. Six such distributions are: the uniform (UN), the normal (NM), the exponential (EX), the Weibull (WB), the power type I (P1), the power type II (P2). For the UN and NM distributions, first the expectation of the variate W is derived in order to provide a background and contrast for the subsequent derivation of the incomplete expectation of W. For the EX, WB, P1, and P2 distributions, only expressions are given, while some derivations are deferred to exercises. For other distributions, numerical integration may be necessary.

12.3.1 The Uniform Variate

The uniform distribution $UN(\eta_L, \eta_U)$ is defined in Section 2.5. When the variate W has the UN distribution on an open interval (η_L, η_U), where $-\infty < \eta_L < \eta_U < \infty$, its density function is

$$g(w) = \frac{1}{\eta_U - \eta_L}, \qquad \eta_L < w < \eta_U. \tag{12.19}$$

The expectation (the mean) of the uniform variate is

$$E(W) = \int_{\eta_L}^{\eta_U} w g(w) \, dw$$

$$= \frac{1}{\eta_U - \eta_L} \int_{\eta_L}^{\eta_U} w \, dw$$

$$= \frac{1}{\eta_U - \eta_L} \left[\frac{1}{2} w^2 \right]_{\eta_L}^{\eta_U}$$

$$= \frac{\eta_U + \eta_L}{2}. \tag{12.20}$$

Given any constant a such that $\eta_L < a < \eta_U$, the incomplete expectation of the uniform variate W is

$$\int_{\eta_L}^{a} w g(w) \, dw = \frac{1}{\eta_U - \eta_L} \int_{\eta_L}^{a} w \, dw$$

$$= \frac{1}{\eta_U - \eta_L} \left[\frac{1}{2} w^2 \right]_{\eta_L}^{a}$$

$$= \frac{a^2 - \eta_L^2}{2(\eta_U - \eta_L)}. \tag{12.21}$$

12.3.2 The Normal Variate

The normal distribution $NM(\mu, \sigma)$ is defined in Section 2.6.2. When the variate W has the NM distribution with mean $E(W) = \mu$ and variance $Var(W) = \sigma^2$, its density function is

$$g(w) = \frac{1}{\sigma} q\left(\frac{w - \mu}{\sigma} \right), \qquad -\infty < w < \infty, \tag{12.22}$$

where q is the standard normal density function given by

$$q(z) = \frac{1}{\sqrt{2\pi}} e^{-\frac{1}{2} z^2}, \qquad -\infty < z < \infty. \tag{12.23}$$

The relationship between a quantile w of the normal variate W (having mean μ and variance σ^2) and the corresponding quantile z of the standard normal variate Z (having mean 0 and variance 1) is

$$w = \mu + \sigma z. \tag{12.24}$$

The expectation (the mean) of the normal variate W is

$$E(W) = \int_{-\infty}^{\infty} w g(w) \, dw$$

$$= \int_{-\infty}^{\infty} (\mu + \sigma z) q(z) \, dz$$

$$= \mu \int_{-\infty}^{\infty} q(z) \, dz + \frac{\sigma}{\sqrt{2\pi}} \int_{-\infty}^{\infty} z e^{-\frac{1}{2}z^2} \, dz$$

$$= \mu + \frac{\sigma}{\sqrt{2\pi}} \left[-e^{-\frac{1}{2}z^2} \right]_{-\infty}^{\infty}$$

$$= \mu. \tag{12.25}$$

Given any constant a such that $-\infty < a < \infty$, the incomplete expectation of the normal variate W is derived in terms of the standard normal density function q and the standard normal distribution function Q. In addition, the substitution $w = \mu + \sigma z$ in the integrand entails a change of the upper limit of integration from a to $b = (a - \mu)/\sigma$. Thus,

$$\int_{-\infty}^{a} w g(w) \, dw = \int_{-\infty}^{b} (\mu + \sigma z) q(z) \, dz$$

$$= \mu \int_{-\infty}^{b} q(z) \, dz + \frac{\sigma}{\sqrt{2\pi}} \int_{-\infty}^{b} z e^{-\frac{1}{2}z^2} \, dz$$

$$= \mu Q(b) + \frac{\sigma}{\sqrt{2\pi}} \left[-e^{-\frac{1}{2}z^2} \right]_{-\infty}^{b}$$

$$= \mu Q(b) - \frac{\sigma}{\sqrt{2\pi}} e^{-\frac{1}{2}b^2}$$

$$= \mu Q(b) - \sigma q(b)$$

$$= \mu Q \left(\frac{a - \mu}{\sigma} \right) - \sigma q \left(\frac{a - \mu}{\sigma} \right). \tag{12.26}$$

12.3.3 The Exponential Variate

The exponential distribution $EX(\alpha, \eta)$ is defined in Section C.3.1; it has the scale parameter $\alpha > 0$ and the shift parameter η $(-\infty < \eta < \infty)$. Given any constant $a \in (\eta, \infty)$, the incomplete expectation of the exponential variate W is

$$\int_{\eta}^{a} w g(w) \, dw = \alpha + \eta - (\alpha + a) \exp \left(-\frac{a - \eta}{\alpha} \right). \tag{12.27}$$

12.3.4 The Weibull Variate

The Weibull distribution $WB(\alpha, \beta, \eta)$ is defined in Section C.3.2; it has the scale parameter $\alpha > 0$, the shape parameter $\beta > 0$, and the shift parameter η $(-\infty < \eta < \infty)$. Given any constant $a \in (\eta, \infty)$, the incomplete expectation of the Weibull variate W is

$$\int_{\eta}^{a} w g(w) \, dw = \frac{\alpha}{\beta} \gamma \left(\frac{1}{\beta}, \left(\frac{a - \eta}{\alpha} \right)^{\beta} \right) - a \exp \left[-\left(\frac{a - \eta}{\alpha} \right)^{\beta} \right] + \eta, \tag{12.28}$$

where γ is the incomplete gamma function (see Section 2.8).

12.3.5 The Power Type Variates

The power type I distribution $P1(\beta, \eta_L, \eta_U)$, and the power type II distribution $P2(\beta, \eta_L, \eta_U)$, are defined in Sections C.4.1 and C.4.2, respectively. Each distribution has the shape parameter $\beta > 0$ and the shift parameters η_L, η_U, where $-\infty < \eta_L < \eta_U < \infty$. Given any constant $a \in (\eta_L, \eta_U)$, the incomplete expectation of the variate W is: when W is power type I,

$$\int_{\eta_L}^{a} wg(w)\, dw = \frac{\beta a + \eta_L}{\beta + 1} \left(\frac{a - \eta_L}{\eta_U - \eta_L} \right)^{\beta}; \tag{12.29}$$

when W is power type II,

$$\int_{\eta_L}^{a} wg(w)\, dw = \frac{\eta_U + \beta \eta_L}{\beta + 1} - \frac{\eta_U + \beta a}{\beta + 1} \left(\frac{\eta_U - a}{\eta_U - \eta_L} \right)^{\beta}. \tag{12.30}$$

12.4 Quadratic Difference Opportunity Loss

12.4.1 Criterion Function

The consequences of misestimation of the input are quantified in terms of one constant, which may be in monetary units or may represent subjective valuation:

λ — marginal opportunity loss from misestimation of the input, regardless of the direction of misestimation [$/(unit of w)^2].

The opportunity loss resulting from the decision a and the input w takes the form

$$l(a, w) = \lambda(a - w)^2. \tag{12.31}$$

In other words, the opportunity loss is proportional to the quadratic difference between the decision a (the estimate) and the realization w of the input variate W (Figure 12.1(b)).

12.4.2 Optimal Decision Procedure

The optimal decision procedure outlined in Section 12.2.2 is followed in order to find the analytic solution. First, the *expected opportunity loss* resulting from decision a is found:

$$\begin{aligned} L(a) &= E[\lambda(a - W)^2] \\ &= \lambda E[(a - W)^2] \\ &= \lambda[a^2 - 2aE(W) + E(W^2)]. \end{aligned} \tag{12.32}$$

Toward finding the optimal decision, the first derivative of $L(a)$ with respect to a is obtained:

$$\frac{dL(a)}{da} = \lambda[2a - 2E(W)].$$

The necessary condition for a^* to be the optimal decision is that $dL(a^*)/da = 0$. Hence

$$2\lambda[a^* - E(W)] = 0.$$

It follows that the *optimal decision* a^* equals the expectation (*the mean*) of the random input W:

$$a^* = E(W). \tag{12.33}$$

Finally, the expression for the *minimum expected opportunity loss* is derived by inserting solution (12.33) into expression (12.32):

$$L^* = L(a^*)$$
$$= \lambda E[(a^* - W)^2]$$
$$= \lambda E[(E(W) - W)^2].$$

Recognizing that $E[(E(W) - W)^2] = Var(W)$, we can write

$$L^* = \lambda Var(W). \tag{12.34}$$

12.4.3 Interpretation of the Solution

The optimal solution to a target-setting problem with a quadratic difference opportunity loss function has two basic properties.

1. In order to find the optimal decision a^*, one need only know the mean $E(W)$ of the random input W.
2. In order to evaluate the minimum expected opportunity loss L^*, one need only know the variance $Var(W)$ of the random input W and the marginal opportunity loss λ.

In other words, the optimal solution is specified in terms of the first two moments of the random input W, the mean $E(W)$ and the variance $Var(W)$. The simplicity of this solution is one of the reasons for the popularity of the quadratic loss criterion in applied decision models.

12.5 Impulse Utility

12.5.1 Criterion Function

Suppose the opportunity loss from misestimation of the input is infinite, regardless of the direction and magnitude of misestimation. An equivalent way of modeling this strength of preference is in terms of a *utility function u* such that

$$u(a, w) = \begin{cases} \infty & \text{if } a = w, \\ 0 & \text{if } a \neq w. \end{cases} \tag{12.35}$$

That is, when the decision a is a perfect estimate of the input w, the resulting utility is infinite; and when the decision a misestimates the input w, the resulting utility is zero, regardless of the direction and magnitude of misestimation. It follows that the utility function u can be expressed in terms of the *impulse function* δ shifted to position a (Figure 12.1(c)):

$$u(a, w) = \delta(w - a). \tag{12.36}$$

The *impulse function*, also called the *Dirac function*, and widely used in control theory, is defined by

$$\delta(w) = \begin{cases} \infty & \text{if } w = 0, \\ 0 & \text{if } w \neq 0, \end{cases} \tag{12.37}$$

for all $w \in \Re$. It is a constant, zero-valued function, except at the origin where it is infinite. Shifting function (12.37) by a yields function (12.35); hence equality (12.36).

With that criterion function, the density function g of the input variate W on the sample space $\mathcal{W}$ must be bounded: $g(w) < \infty$ for all $w \in \mathcal{W}$. This requirement constrains the shape parameter β for some distribution families (see graphs of density functions in Appendix C).

12.5.2 Optimal Decision Procedure

First, any feasible decision $a \in \mathcal{A} = \mathcal{W}$ is evaluated in terms of its utility. The *utility of decision*, $U(a)$, is defined as the *expected utility of outcome* resulting from that decision:

$$U(a) = E[u(a, W)]$$

$$= \int_{\mathcal{W}} u(a, w)g(w)\, dw$$

$$= \int_{\mathcal{W}} \delta(w - a)g(w)\, dw$$

$$= g(a). \tag{12.38}$$

(The last equality is the *sifting property* of the impulse function: when the integrand is a product of a continuous function g on $\mathcal{W}$ and the impulse function δ shifted to position $a \in \mathcal{W}$, the integral retrieves the value of the function g at point a.)

Next, the *optimal decision* a^* is found, which maximizes the expected utility of outcome:

$$U(a^*) = \max_a U(a)$$

$$= \max_a g(a). \tag{12.39}$$

The last equality implies that the optimal decision a^* is a point in the sample space $\mathcal{W}$ of the input variate W at which the density function g attains the maximum. In other words, the optimal decision a^* equals the *mode* of the random input W.

Finally, the *maximum expected utility of outcome* (equivalently, the *maximum utility of decision*) equals the ordinate of the density function g at the mode:

$$U^* = U(a^*) = g(a^*). \tag{12.40}$$

12.5.3 Value of Perfect Forecaster

To determine the value of the perfect forecaster, *VPF*, per defining equation (12.7), we must find U^{**}, the expected maximum utility. Per equation (12.6), with u given by expression (12.36),

$$U^{**} = E[u(W, W)]$$

$$= \int_{\mathcal{W}} u(w, w)g(w)\, dw$$

$$= \delta(0) \int_{\mathcal{W}} g(w)\, dw$$

$$= \delta(0) = \infty. \tag{12.41}$$

Hence,

$$VPF = U^{**} - U^* = \infty. \tag{12.42}$$

This says that there is no upper bound on the amount which the decider may want to pay for one perfect forecast of the input.

12.6 Implications for Analysts

12.6.1 Summary of Solutions

We have studied models for a class of decision problems known as estimation problems or as target-setting problems. Four models have been formulated, each with a special criterion function, and analytic solutions have been derived. It turns out that each of the four estimators of W (a quantile, the median, the mean, the mode) is optimal with respect to some criterion function (Table 12.2).

12.6.2 Impact of Uncertainty

Suppose W has a Weibull distribution WB(α, β, η) with scale parameter $\alpha = 1$, shift parameter $\eta = 0$, and shape parameter $\beta = 1$, or $\beta = 2$, or $\beta = 6$. Figure C.3.2 shows graphs of the three density functions and their corresponding distribution functions. Section C.3.2 presents the formulae for the four estimators. Table 12.3 lists the optimal decisions a^*.

Table 12.2 Analytic solutions to estimation, or target-setting, problems with special criterion functions; W is the variate whose realization is being estimated.

Criterion function	Optimal decision a^*
Two-piece linear opportunity loss	Quantile of W
Absolute difference opportunity loss	Median of W
Quadratic difference opportunity loss	Mean of W
Impulse utility	Mode of W

Table 12.3 The optimal decision a^* in an estimation, or target-setting, problem depends on the criterion function (see Table 12.2) and the distribution function of the input variate W — here, the Weibull distribution having scale parameter $\alpha = 1$, shift parameter $\eta = 0$, and shape parameter β as specified.

Optimal decision a^*	Shape parameter β		
	1	2	6
0.2-quantile of W	0.22	0.47	0.78
0.6321-quantile of W	1.00	1.00	1.00
0.8-quantile of W	1.61	1.27	1.08
Median of W	0.69	0.83	0.94
Mean of W	1.00	0.89	0.93
Mode of W	0	0.71	0.97

The p-quantile of W corresponds to the ratio λ_o/λ_u of the marginal opportunity losses as follows:

$$p = 0.2 \quad \Leftrightarrow \lambda_o/\lambda_u = 4/1,$$
$$p = 0.6321 \Leftrightarrow \lambda_o/\lambda_u = 291/500,$$
$$p = 0.8 \quad \Leftrightarrow \lambda_o/\lambda_u = 1/4.$$

Let us first look at the graphs in Figure C.3.2. When β increases, the uncertainty about W decreases, as measured by $Var(W)$. This is manifested in the density functions, which become narrower and taller, and in the distribution functions, which become steeper in the center. Both visual effects imply an increased concentration of the probability mass under the density function.

The range of optimal decisions under different criterion functions (Table 12.3) is the widest $[0, 1.61]$ when uncertainty is the largest ($\beta = 1$), and decreases to $[0.78, 1.08]$ when uncertainty is the smallest ($\beta = 6$).

Conclusion

The larger the uncertainty about the input variate W, the more important it is for the decider to know his preferences over outcomes, and for the analyst to choose an appropriate form of the criterion function (to capture the decider's preferences).

12.6.3 Impact of Skewness

When $\beta = 1$ or $\beta = 2$, the Weibull density function (Figure C.3.2) is skew to the right (i.e., has a long right tail); and the order of the optimal decisions under the three symmetric criterion functions is the mode, the median, the mean (which is a reversal of their alphabetical order — recall Section 2.6.4):

$$w_M < w_{0.5} < E(W).$$

When $\beta = 6$, the Weibull density function (Figure C.3.2) is skew to the left (slightly); and the order of the optimal decisions under the three symmetric criterion functions is reversed (recall Section 2.6.4):

$$E(W) < w_{0.5} < w_M.$$

Under a symmetric density function (which the Weibull density function is approximately when $\beta = 3.6$), the three optimal decisions would be equal: $a^* = E(W) = w_{0.5} = w_M$, ($w_{0.5} = 0.90$ under the Weibull density function with $\beta = 3.6$).

Conclusion

When the density function of the input variate W is symmetric, the optimal decision a^* is identical under every criterion function which is symmetric in w.

12.6.4 Impact of Asymmetry

The three symmetric criterion functions yield optimal decisions (the median, the mean, the mode), which are measures of the central tendency of a variate. Only the asymmetric criterion function (the two-piece linear opportunity loss function) can yield the optimal decision a^* which is not a measure of the central tendency but can, in fact, be any quantile of W (i.e., any point in the sample space of W).

Conclusion

The asymmetry is a consequential property of the criterion function. When it is present in the real world, it should be retained in a decision model.

12.6.5 Meta-Decision Problem

Sometimes, especially in the context of statistical estimation problems, when no direct connection is discernible between the misestimation and its consequences in the real world, you may be left free to choose the criterion function. This is a *meta-decision problem*: to decide how to decide.

It should be understood thusly: the choice of an estimator of W (a quantile, the median, the mean, the mode) is tantamount to accepting (if only implicitly) the corresponding criterion function as representative of your preferences over outcomes; and vice versa, the choice (explicit) of the criterion function prescribes the type of estimator which is optimal.

The apparent propensity of many analysts to prefer the mean (as the estimator) and to deplore the so-called "bias" (any deviation from the mean) finds no justification in the theory of rational decision making as applied to real-world problems: ample empirical evidence from cost–benefit analyses and preference-elicitation studies indicates that opportunity loss functions are asymmetric more often than not. Chapters 13 and 15 provide some of this evidence.

12.6.6 Decisions with Forecasts

In every one of the four decision models, the distribution function G of the input variate W may constitute either a judgmental forecast (as defined in Chapter 9) or a statistical forecast (as defined in Chapter 10). In the latter case: (i) the symbol G should be replaced by the symbol $\Phi(\cdot|x)$ — the posterior distribution function; and (ii) the optimal solution should be made a function of x — the realization of the predictor. A general decision model with forecasts is formulated in Section 12.10.

12.6.7 Applications

The decision model with a two-piece linear opportunity loss is applied to inventory management and capacity planning — the two core problems for businesses — in Chapter 13; then it is adapted to yield control problems for airline reservations and college admissions in Chapter 15.

An application of the decision model with an impulse utility is presented in Section 12.7; then it is extended, in Section 12.8, to a decision problem which calls for a double-impulse utility.

12.7 Weapon-Aiming Model

This decision model is dedicated to warriors. It attempts to capture a situation in which the decision and the state variables are continuous, but the outcome variable is binary and extreme — in that the decider's strength of preference for one outcome over the other is indefinitely great.

12.7.1 Decision Situation

This is the situation of a warrior in a battle: kill the enemy or be killed. Here are two such situations in the warriors' own words.

Situation 1 *(Ground battle)* Fallujah, Iraq, November 2004, US Navy SEAL Chris Kyle (2012, pp. 188–189):

> I went back up on the roofs and started doing overwatches again .... There was so much dust and grit in the air, I kept my goggles on ... my helmet cinched tight, wary of the chips and cement frags that flew from the battered masonry during a firefight .... Within seconds, we were fully engaged ... Insurgents kept popping up from behind the stones ... I carefully put my scope on a target, steadied the aim on center mass, then squeezed ever so smoothly ... the bullet leapt from the barrel .... My target fell.

In Situation 1, the target is nearly stationary, but its exact position may be uncertain because blowing dust and grit blur the distinction between the insurgent and the surrounding stones.

Situation 2 *(Aerial battle)* Over North Vietnam, January 1966, US Brigadier General Robin Olds (2010, p. 281):

> [A Soviet-made MiG-21 fighter jet] has popped up out of the undercast beneath us .... We're set up for radar missiles and he's right at min range .... [My crewman] aims the radar that way and is locked on almost immediately. I squeeze the trigger and launch one, then another Sparrow. They zoom away but apparently aren't guiding ... I've got another 21 engaged ... I shoot again. Splash! The MiG's right wing comes off and he snaps right and down.

In Situation 2, the target is in motion; it is tracked by radar, which repeatedly estimates its position, velocity, and acceleration rate in order to predict its position at the time it is reached by the missile. The prediction uncertainty may arise from several sources; for instance, the noise in the radar system (see Chapter 5, Exercise 6), the imperfections of the tracking and prediction algorithms.

Situation 3 *(A hunt)* The buck being hunted (for trophy antlers, venison, deer population control, or all of these) is moving cautiously through the landscape. The rising morning fog blurs the distinction between its antlers and the surrounding thicket, creating uncertainty about its exact position. The hunter scans the horizon and judges the probability of the buck hiding within each narrow interval. What is the best aiming point?

12.7.2 Mathematical Formulation

The target (an enemy, a buck) is located at point $w \in \mathcal{W}$ of a line segment $\mathcal{W} = (\eta_L, \eta_U)$. But to the decider (a warrior, a hunter), the *target position* is uncertain; hence it is a variate W with sample space $\mathcal{W}$.

The uncertainty about W is assessed (visually or algorithmically) in terms of a density function g; it may be multimodal (Figure 12.3), with the modes centered at the prominent silhouettes or the strongest reflected signals (e.g., when multiple radars track the target, g is a mixture of the density functions from all radars, and may be multimodal). What should the *aiming point* $a \in \mathcal{W}$ be? Should it be the median $w_{0.5}$, or the mean $E(W)$ — as the advocates of "unbiased" estimates might argue — or some other point? Look at Figure 12.3 and decide.

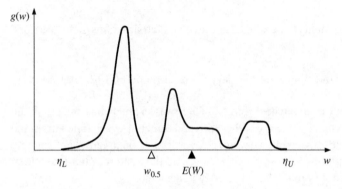

Figure 12.3 A trimodal density function g of the target position W on a line segment $\mathcal{W} = (\eta_L, \eta_U)$, with median $w_{0.5}$, and mean $E(W)$. What is the optimal aiming point?

Now, consider the two outcomes {hitting the target, missing the target}, and ponder the difference between their utilities: (i) to a warrior — in terms of life or death, winning or losing a battle; (ii) to a hunter — in terms of elation or disappointment.

Suppose the utility difference (i) is not a function of the distance $|a - w|$ by which the bullet misses the target, and (ii) is indefinitely great. Then both aspects of this preference are captured mathematically in the utility function (12.35).

12.7.3 Model and Solution

In summary, the model for aiming a weapon at a target whose position is uncertain has four elements:

a — aiming point, a continuous decision variable, $a \in \mathcal{W}$;
W — target position, a continuous variate, with realization $w \in \mathcal{W}$, and sample space $\mathcal{W} = (\eta_L, \eta_U)$;
g — density function of W, bounded on $\mathcal{W}$;
u — utility function that quantifies the strength of preference for two outcomes resulting from a hit ($a = w$) or a miss ($a \neq w$), and that is specified by equation (12.36).

The optimal decision procedure, detailed in Section 12.5.2, prescribes the *optimal aiming point $a^* = w_M$*, which is the mode of W. This solution basically confirms the intuition of many sharpshooters: aim at the "most likely" target position.

In general, the mode $a^* = w_M$ is the optimal estimate of a continuous variate W whenever predicting exactly the realization of W matters enormously (infinitely). But, the *maximum expected utility of outcome* remains finite: $U^* = g(a^*)$, by equation (12.40).

This is instructive: U^* can be increased only if the uncertainty about W can be reduced in a way that raises the peak of the density function g. The potential increase of U^* is unbounded. Ergo, a reduction of uncertainty about the target position may have enormous utility. (It is infinite in the limit, when g converges to an impulse function and uncertainty vanishes; consistently, *VPF* $= \infty$, per equation (12.42).) In a battle, learning the true target position to achieve victory may be worth any cost.

This is congruent with the strategic preference in time of war. "You ask what is our aim? I can answer in one word: victory. Victory at all costs" (Winston Churchill).

Bibliographical Notes

The quotations describing the two battle situations come from the autobiographies of two veterans: Chris Kyle (2012), US Navy SEAL, an elite soldier and record-holding lethal sniper who served four combat tours (1999–2009) on the battlefields of Operation Iraqi Freedom; and Robin Olds (2010), US Brigadier General, an ace fighter pilot and audacious wartime commander, who flew 107 combat missions in the Second World War (1944–1945) and 152 missions in the Vietnam War (1965–1968), and then served as the commandant of the US Air Force Academy (1968–1971).

The statement by Winston Churchill comes from his first speech to the House of Commons on 13 May 1940, after he became prime minister of Britain 3 days earlier — on the day the Germans invaded the Netherlands, Belgium, and Luxembourg, initiating World War II in Western Europe (Olson, 2020).

12.8 Weapon-Aiming-with-Friend Model

This decision model is dedicated to military cadets. One of them, who had studied Section 12.7, brought up the danger of "friendly fire", which inspired this model.

12.8.1 Decision Situation

This danger is real, as the histories of many military operations attest. Here is one of them from the First World War: the battle between American and German troops at Belleau Wood — a dense, almost impenetrable forest, 80 km east of Paris toward Reims.

Situation 4 *(Friend in battle)* Belleau Wood, France, June 1918, Paul Dowswell (2004):

> Fighting for possession of Belleau Wood took on a grisly, claustrophobic quality ... trees were close together and it was constantly dark .... The entire battle was fought in an atmosphere of great confusion. So dense were the woods, it was possible for enemies to pass within a few feet of each other and not see their opponents. In such a place, edgy soldiers had to exercise great care not to shoot their own comrades.

Our objective is plain: to extend the model of Section 12.7, so that the account is taken of a friend whose position is within the line segment that contains the target. Each position is distinct, uncertain, and stochastically independent of the other.

12.8.2 Mathematical Formulation

This extended model comprises the following elements:

a — aiming point, a continuous decision variable, $a \in \mathcal{W}$;
W — target position, a continuous variate, with realization $w \in \mathcal{W}$, and sample space $\mathcal{W} = (\eta_L, \eta_U)$;
V — friend position, a continuous variate with realization $v \in \mathcal{W}$, such that $v \neq w$;
g — density function of W, bounded on $\mathcal{W}$;
h — density function of V, bounded on $\mathcal{W}$;
u — utility function that quantifies the strength of preference for three outcomes resulting from hitting the target ($a = w$), or hitting the friend ($a = v$), or missing each ($a \neq w, a \neq v$); to wit,

$$u(a, w, v) = \begin{cases} \infty & \text{if } a = w, \\ 0 & \text{if } a \neq w, a \neq v, \\ -\infty & \text{if } a = v. \end{cases} \tag{12.43}$$

The outcome variable is now trinary, with two opposite extremes. Like outcome $a = w$, outcome $a = v$ is extreme, in that the decider's strength of preference for missing the friend ($a \neq v$) over hitting the friend ($a = v$) is indefinitely great; this is modeled by the utility difference being $0 - (-\infty) = \infty$. Employing the impulse function (12.37) to model function (12.43), we obtain

$$u(a, w, v) = \delta(w - a) - \delta(v - a), \qquad w \neq v. \tag{12.44}$$

As a function of a, for fixed w and v ($w \neq v$), this is a *double impulse* function (Figure 12.4), with the positive impulse (∞) at the target position, $a = w$, and the negative impulse ($-\infty$) at the friend position, $a = v$.

The variates W and V are assumed (i) to have non-coinciding realizations, $w \neq v$; (ii) to have bounded and nonidentical density functions: $g(w) < \infty$, $h(w) < \infty$ for all $w \in \mathcal{W}$, and $g(w) \neq h(w)$ for some $w \in \mathcal{W}$; (iii) to be stochastically independent of each other, so that their joint density function p on the sample space $\mathcal{W} \times \mathcal{W}$ is given by

$$p(w, v) = \begin{cases} g(w)h(v) & \text{if } w \neq v, \\ 0 & \text{if } w = v. \end{cases} \tag{12.45}$$

Figure 12.4 Special criterion function for weapon-aiming-with-friend model: double impulse utility $u(a, w, v)$ as a function of the aiming point a, for fixed positions of the target, w, and the friend, v.

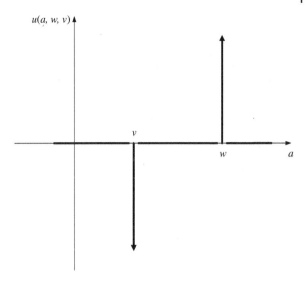

12.8.3 Optimal Decision Procedure

In parallel to Section 12.5.2, the procedure consists of three steps. First, the *utility of decision* $a \in \mathcal{W}$ is defined as the *expected utility of outcome* resulting from that decision:

$$
\begin{aligned}
U(a) &= E[u(a, W, V)] \\
&= \int_{\mathcal{W}} \int_{\mathcal{W}} u(a, w, v) p(w, v) \, dw \, dv \\
&= \int_{\mathcal{W}} \int_{\substack{\mathcal{W} \\ \{w \neq v\}}} [\delta(w - a) - \delta(v - a)] g(w) h(v) \, dw \, dv \\
&= \int_{\mathcal{W}} h(v) \left[\int_{\substack{\mathcal{W} \\ \{w \neq v\}}} \delta(w - a) g(w) \, dw \right] dv - \int_{\mathcal{W}} g(w) \left[\int_{\substack{\mathcal{W} \\ \{v \neq w\}}} \delta(v - a) h(v) \, dv \right] dw \\
&= \int_{\mathcal{W}} h(v) g(a) \, dv - \int_{\mathcal{W}} g(w) h(a) \, dw \\
&= g(a) \int_{\mathcal{W}} h(v) \, dv - h(a) \int_{\mathcal{W}} g(w) \, dw \\
&= g(a) - h(a).
\end{aligned} \tag{12.46}
$$

The equality between the fourth line and the fifth line results from the sifting property of the impulse function, which is explained in Section 12.5.2.

Next, the *optimal decision* a^* is found, which maximizes the expected utility of outcome:

$$
\begin{aligned}
U(a^*) &= \max_a U(a) \\
&= \max_a \{g(a) - h(a)\}.
\end{aligned} \tag{12.47}
$$

Thus, the *optimal aiming point* $a^* \in \mathcal{W}$ is one at which the difference between the target density function and the friend density function is maximal; in short, a^* is the mode under the difference function $g - h$.

Finally, the *maximum expected utility of outcome* (equivalently, the *maximum utility of decision*) is

$$U^* = U(a^*) = g(a^*) - h(a^*).$$ (12.48)

The assumptions imposed on W and V, and the optimal decision procedure, guarantee that the maximum utility of decision is positive and finite: $0 < U^* < \infty$.

Figure 12.5 shows an example with piecewise-linear density functions g and h on a bounded interval $\mathcal{W} = (\eta_L, \eta_U)$, the resultant piecewise-linear utility function $U = g - h$ of decision, and the optimal solution (a^*, U^*): the optimal aiming point a^* and its utility U^*. Figure 12.6 shows an example with continuous density functions.

12.8.4 Value of Perfect Forecaster

Although there are two variates, W and V, a clairvoyant who reveals the target position $W = w$ suffices to achieve the objective: hit the target $(a = w)$ and miss the friend $(a \neq v)$ because $v \neq w$, by assumption. The maximum utility of this outcome is

$$
\begin{aligned}
\max_a u(a, w, v) &= u(w, w, v) \\
&= \delta(w - w) - \delta(v - w) \\
&= \delta(0) - 0 = \infty;
\end{aligned}
$$

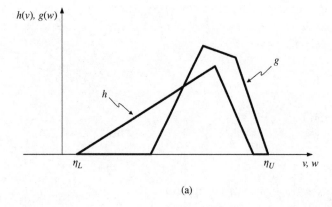

(a)

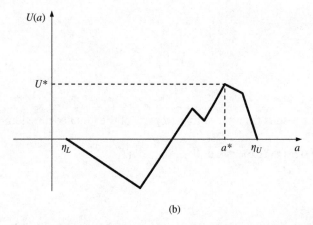

(b)

Figure 12.5 Elements of the weapon-aiming-with-friend model: example with piecewise-linear density functions g and h of target position W and friend position V, respectively, on a line segment $\mathcal{W} = (\eta_L, \eta_U)$.

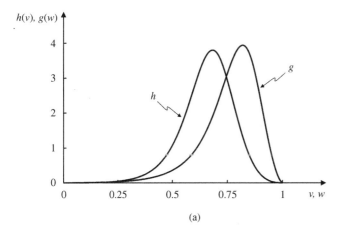

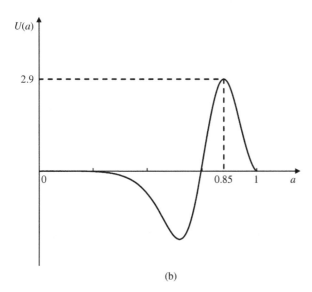

Figure 12.6 Elements of the weapon-aiming-with-friend model: example with log-ratio logistic density functions; $W \sim$ LR1$(0, 1)$-LG$(1.3, 0.4)$, and $V \sim$ LR1$(0, 1)$-LG$(0.7, 0.3)$.

and the *expected maximum utility* is

$$U^{**} = E[\max_a u(a, W, V)]$$
$$= \int_w \int_{\substack{w \\ \{w \neq v\}}} \delta(0) g(w) h(v) \, dw \, dv$$
$$= \delta(0) \int_w g(w) \, dw \int_w h(v) \, dv = \delta(0) = \infty. \tag{12.49}$$

Hence, the value of perfect forecaster, *VPF*, per defining equation (12.7), is

$$VPF = U^{**} - U^* = \infty. \tag{12.50}$$

Plainly: in a battle, learning the true target position to achieve victory may be worth any cost.

12.9   General Modeling Methodology

A general modeling methodology is outlined. It couples (i) the theory of decision making under uncertainty from Section 12.1 with (ii) an output function and (iii) a utility function, or a multi-attribute utility function. It offers a *mathematical structure* for modeling decision systems more complex than the ones presented in this chapter.

   The reader's objective may be merely to realize the existence of such a mathematical structure, or to actually employ it in modeling — but this objective may require further study, beyond the scope of this book.

12.9.1   System with Single Outcome Variable

A *decision system* with a single outcome variable is a 6-tuple:

$$\{\mathcal{A}, \mathcal{W}, \mathcal{Z}, g, h, v\}. \tag{12.51}$$

It consists of three sets and three functions having the following meanings.

1. A set of decisions that are feasible (the decision space): $\mathcal{A} = \{a\}$.
2. A set of inputs that are possible (the sample space of the input variate W): $\mathcal{W} = \{w\}$.
3. A set of outcomes (outputs) that are induced (the outcome space): $\mathcal{Z} = \{z\}$.
4. A density function g on the sample space $\mathcal{W}$, which quantifies the uncertainty about the input variate W.
5. An outcome function (output function),

$$h : \mathcal{A} \times \mathcal{W} \to \mathcal{Z},$$

   such that $z = h(a, w)$ is the outcome resulting from the decision $a \in \mathcal{A}$ and the input $w \in \mathcal{W}$. (It is the model of a relationship or a process (e.g., physical, chemical, biological, economic, social), which the decision a is intended to affect.)
6. A utility function,

$$v : \mathcal{Z} \to \mathfrak{R},$$

   which quantifies the decider's strength of preference for outcomes in $\mathcal{Z}$. (It is a cardinal scale, defined formally in Section 14.3. Oftentimes, z is called an attribute, and v is called a single-attribute utility function.)

   The *structure* of this decision system is depicted by a block diagram in Figure 12.7(a). It may be regarded as a recipe for simulation: The decision a is made, the function g generates the input w, the function h maps the decision–input pair into the outcome z, whose utility to the decider is $v(z)$. (This structure is employed in the asking price model of Section 15.1.)

12.9.2   System with Multiple Outcome Variables

A *decision system* with multiple outcome variables is a 6-tuple:

$$\{\mathcal{A}, \mathcal{W}, \mathcal{Z}, g, \mathbf{h}, v\}. \tag{12.52}$$

It is an extension of system (12.51) to a vector of N ($N \geq 2$) outcome variables, obtained by redefining three elements as follows.

3. A set of outcomes $\mathcal{Z} = \{\mathbf{z}\}$; it is a Cartesian product:

$$\mathcal{Z} = \mathcal{Z}_1 \times \cdots \times \mathcal{Z}_N,$$
$$\mathbf{z} = (z_1, \ldots, z_N),$$
$$\mathcal{Z}_n = \{z_n\}, \qquad n = 1, \ldots, N.$$

Figure 12.7 Mathematical structure of the decision system: (a) with single-outcome variable; (b) with multiple-outcome variables; (c) after reduction.

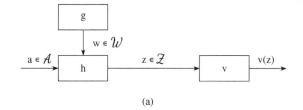

(a)

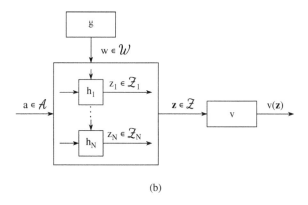

(b)

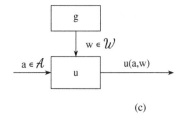

(c)

5. An outcome function **h**; it is a vector of functions:

$$\mathbf{h} = (h_1, \dots, h_N),$$
$$h_n : \mathcal{A} \times \mathcal{W} \to \mathcal{Z}_n, \qquad n = 1, \dots, N;$$
$$z_n = h_n(a, w).$$

6. A multi-outcome utility function v (oftentimes called a multi-attribute utility function):

$$v : \mathcal{Z} \to \mathfrak{R},$$

such that $v(\mathbf{z}) = v(z_1, \dots, z_N)$ is the utility of the outcome vector $\mathbf{z}$ to the decider.

The *structure* of this decision system is depicted by a block diagram in Figure 12.7(b).

12.9.3 General Decision System

The purpose, and the art, of modeling is to transform a verbal description of a decision problem into a mathematical formulation, (12.51) or (12.52), which consists of three well-defined sets and three well-defined functions. Once the modeling has been accomplished, the optimal decision procedure needs to be formulated. For this purpose, a parsimonious formulation suffices. Called a *general decision system*, it is a 4-tuple:

$$\{\mathcal{A}, \mathcal{W}, g, u\}, \tag{12.53}$$

with

$$u : \mathcal{A} \times \mathcal{W} \to \mathfrak{R}.$$

The *structure* of this decision system is depicted by a block diagram in Figure 12.7(c).

In the system with a single outcome variable, the function u is the composition of functions v and h:

$$u = v \circ h,$$
$$u(a, w) = v(h(a, w)).$$

In the system with multiple outcome variables, function u is the composition of functions v and $\mathbf{h}$:

$$u = v \circ \mathbf{h},$$
$$u(a, w) = v(h_1(a, w), \dots, h_N(a, w)).$$

The $u(a, w)$ represents the utility of the outcome z, or the outcome vector $(z_1, \dots, z_N)$, resulting from decision $a \in \mathcal{A}$ and input $w \in \mathcal{W}$. The actual value of the outcome variable, or outcome vector, is irrelevant for decision making; hence $\mathcal{Z}$ does not appear in system (12.53).

An equivalent *general decision system* is a 4-tuple:

$$\{\mathcal{A}, \mathcal{W}, g, l\}, \tag{12.54}$$

with l being the opportunity loss function derived from u according to the relationship (12.2).

When function l has one of the mathematical forms studied in this chapter, the analytic solution to the decision problem is available. When this is not the case, the solution needs to be found analytically or numerically. Or one may try to approximate the given function l locally, in the space $\mathcal{A} \times \mathcal{W}$, in terms of a two-piece linear function or a quadratic difference function, and thereby to obtain an approximate solution.

In summary, the theory of decision making under uncertainty presented in Section 12.1 is applicable to any static decision system, (12.51) or (12.52), having any outcome function and any utility function (single-attribute or multi-attribute).

12.10   General Forecast–Decision System

12.10.1   System Structure

The general decision system (12.53) coupled with the Bayesian forecaster (10.1)–(10.2) is a 6-tuple:

$$\{\mathcal{A}, \mathcal{W}, \mathcal{X}, g, f, u\}. \tag{12.55}$$

It is called the *forecast–decision system* (F–D system).

The *structure* of this F–D system is depicted by a block diagram in Figure 12.8. It may be regarded as a recipe for simulation. In the *forecast system*, the function g generates the input w that selects the function $f(\cdot|w)$ from which predictor realization x is generated. In the *decision system*, the optimal decision function α^* (to be defined in the next section) maps x into the decision a; then the function u maps the decision–input pair into the utility $u(a, w)$ of outcome to the decider.

Henceforth, our objective is fourfold. (i) To formulate the optimal decision procedure that uses the probabilistic forecasts. (ii) To define the (economic) value of the forecaster. (iii) To define mathematically the informativeness relation, which was defined verbally in Sections 10.4.1 and 11.3.1. (iv) To prove the comparison theorem stated in Section 10.4.5 and relied upon in Sections 11.3.3 and 11.3.4.

Figure 12.8 Mathematical structure of the general forecast–decision system $\{\mathcal{A}, \mathcal{W}, \mathcal{X}, g, f, u\}$ with the decision function α^* deduced through the optimal decision procedure.

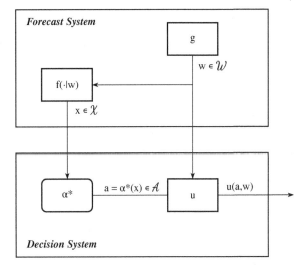

12.10.2 Optimal Decision Procedure

Given predictor realization $x \in \mathcal{X}$, the Bayesian forecaster outputs the posterior density function $\phi(\cdot|x)$ of the predictand W. On this occasion, the *utility of decision* $a \in \mathcal{A}$ is defined as the *expected utility of outcome* resulting from that decision, *conditional on* $X = x$:

$$U(a, x) = E[u(a, W)|X = x]$$
$$= \int_{\mathcal{W}} u(a, w)\phi(w|x)\, dw. \tag{12.56}$$

The *maximum utility of decision* is

$$U^*(x) = \max_{a \in \mathcal{A}} U(a, x). \tag{12.57}$$

When executed for every $x \in \mathcal{X}$, the maximization yields the *optimal decision function*

$$\alpha^* : \mathcal{X} \to \mathcal{A}, \tag{12.58}$$

such that $\alpha^*(x) \in \mathcal{A}$ is the optimal decision when $X = x$, and

$$U^*(x) = U(\alpha^*(x), x). \tag{12.59}$$

To obtain an overall evaluation of the forecast–decision system across all possible occasions, the *integrated maximum utility* is calculated by taking the expectation of $U^*(X)$ with respect to the predictor X:

$$UF^* = E[U^*(X)]$$
$$= \int_{\mathcal{X}} U^*(x)\kappa(x)\, dx. \tag{12.60}$$

The total expression for UF^* in terms of the six elements (12.55) defining the F–D system is

$$UF^* = \int_{\mathcal{X}} \left[\max_{a \in \mathcal{A}} \int_{\mathcal{W}} u(a, w)\phi(w|x)\, dw \right] \kappa(x)\, dx$$
$$= \int_{\mathcal{X}} \left[\max_{a \in \mathcal{A}} \int_{\mathcal{W}} u(a, w)f(x|w)g(w)\, dw \right] dx, \tag{12.61}$$

where the second equality is obtained by replacing $\phi(w|x)$ by its defining expression (10.2).

This optimal decision procedure with forecasts has an essential property, whose proof is deferred to Exercise 12: in terms of its utility, UF^*, it is preferred at least as the procedure with a fixed prior density function g of W, whose utility U^* is given by equation (12.3); and it is preferred no more than the procedure with a perfect forecaster, whose utility U^{**} is given by equation (12.6).

Theorem *(Order of utilities)* *For any utility function u, any prior density function g, and any family of likelihood functions f of the Bayesian forecaster,*

$$U^* \leq UF^* \leq U^{**}.$$

12.10.3 Value of Forecaster

The *value of the forecaster* of the input W is defined as the difference between the evaluation with the forecaster and the evaluation without the forecaster:

$$VF = UF^* - U^*. \tag{12.62}$$

When the utilities are in monetary units, the VF is the maximum price the decider should be willing to pay for one forecast of the input. The VF is never negative and is never greater than the value of the perfect forecaster, defined by equation (12.7):

$$0 \leq VF \leq VPF. \tag{12.63}$$

The *efficiency of the forecaster* is defined as the quotient

$$EF = \frac{VF}{VPF} = \frac{UF^* - U^*}{U^{**} - U^*}. \tag{12.64}$$

The EF is bounded, $0 \leq EF \leq 1$. Its interpretation and usage are the same as those discussed in Section 7.4.6.

When the criterion function is framed as an opportunity loss function (12.2), the objects (12.56)–(12.61) may be readily redefined by replacing u with l, U with L, and max with min. Relationships (12.2) and (12.6) imply that the *integrated minimum opportunity loss* is

$$LF^* = U^{**} - UF^*.$$

Therefore,

$$VF = L^* - LF^*.$$

The first relationship, together with the order-of-utilities theorem and equation (12.7), implies that $0 \leq LF^* \leq L^*$.

12.10.4 Informativeness Relation

In the definition (12.55) of the F–D system, the sets $\{\mathcal{A}, \mathcal{W}\}$ specify a class of decision problems, the functions $\{g, u\}$ characterize a decider, and the pair $\{\mathcal{X}, f\}$ characterizes a forecaster. It follows that the value of the forecaster VF is a function of these input elements; this can be indicated explicitly by rewriting equation (12.62) in the form

$$VF(u, g; f) = UF^*(u, g; f) - U^*(u, g). \tag{12.65}$$

Now suppose there are two forecasters F_i $(i = 1, 2)$, each using a different predictor X_i and therefore characterized by different $\{\mathcal{X}_i, f_i\}$. Furthermore, suppose there are multiple deciders, each having different u and g. Thus to say "for every rational decider", as the verbal definitions of Sections 10.4.1 and 11.3.1 say, is equivalent to saying "for every u and g". Hence the following mathematical definition.

Definition   For all deciders facing decision problems from the class $\{A, W\}$, forecaster F_1 is said to be *more informative than* forecaster F_2 with respect to predictand W if, for every utility function u and every prior density function g,

$$VF(u, g; f_1) \geq VF(u, g; f_2).$$

The order of forecasters in terms of the *informativeness relation* does not always exist. Given two forecasters, it is possible that neither F_1 is more informative than F_2, nor F_2 is more informative than F_1.

Note on the class

The class $\{A, W\}$ of decision problems may be large, especially when the predictand W is relevant to many kinds of decisions with identical A; perhaps it may include almost all decision problems affected by W. In that case, F_1 may be said to be more informative than F_2 with respect to W for **all** rational deciders — those who make decisions based on the rationality postulates stated in Appendix A. This interpretation, together with the principles of utilitarian ethics (see Section A.2.2), makes it imperative to verify the informativeness of forecasts serving multiple users. A company or an agency producing such forecasts (e.g., the National Weather Service producing weather forecasts for the general public) should ensure that any intended improvement to its forecast system increases, or at least does not decrease, the economic value of forecasts to every user.

12.10.5   Sufficiency Relation

To establish the existence of the informativeness relation directly based on its definition, one would have to determine the economic values of forecasters F_1 and F_2 for every pair of feasible functions u and g — a gigantic task! Remarkably, the Bayesian verification theory offers a statistical procedure for inferring the informativeness relation, if it exists, based solely on a comparison of the two families of likelihood functions, f_1 and f_2. This procedure rests on two definitions and one theorem.

Definition   A nonnegative function $\psi : \mathcal{X}_1 \times \mathcal{X}_2 \to \mathfrak{R}$, whose value at point (x_1, x_2) is denoted $\psi(x_2 | x_1)$, is called a *stochastic transformation* from $\mathcal{X}_1$ onto $\mathcal{X}_2$ if

$$\int_{\mathcal{X}_2} \psi(x_2 | x_1) \, dx_2 = 1, \qquad x_1 \in \mathcal{X}_1;$$

$$0 < \int_{\mathcal{X}_1} \psi(x_2 | x_1) \, dx_1 < \infty, \qquad x_2 \in \mathcal{X}_2.$$

Definition   Forecaster F_1 is said to be *sufficient* for forecaster F_2, with respect to predictand W, if there exists a stochastic transformation ψ from $\mathcal{X}_1$ onto $\mathcal{X}_2$ such that

$$f_2(x_2 | w) = \int_{\mathcal{X}_1} \psi(x_2 | x_1) f_1(x_1 | w) \, dx_1, \qquad x_2 \in \mathcal{X}_2, \quad w \in \mathcal{W}.$$

The first definition establishes a bivariate function ψ such that for every $x_1 \in \mathcal{X}_1$, the univariate function $\psi(\cdot | x_1)$ is interpretable as a density function of X_2, conditional on $X_1 = x_1$.

The second definition (which applies the total probability law) formalizes the intuitive notion of **randomness** as a comparative attribute of predictors. To explain it, let forecaster F_2 be the component of the F–D system depicted in Figure 12.8. When ψ exists, there are two recipes for simulating X_2, as diagrammed in Figure 12.9.

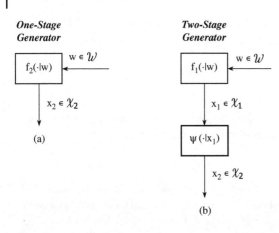

One-Stage Generator

Two-Stage Generator

(a)

(b)

Figure 12.9 Two generators of a realization of predictor X_2, conditional on predictand realization $W = w$: (a) forecaster F_2 generates x_2 directly; (b) forecaster F_1, which is sufficient for forecaster F_2, generates x_1, from which the stochastic transformation ψ generates x_2.

- In the *one-stage generator*, the given input w selects the function $f_2(\cdot|w)$ from which the predictor realization x_2 is generated.
- In the *two-stage generator*, the given input w selects the function $f_1(\cdot|w)$ from which the predictor realization x_1 is generated; next, this x_1 selects the function $\psi(\cdot|x_1)$ from which the predictor realization x_2 is generated.

The two-stage generator offers two insights. First, the function $f_2(\cdot|w)$ is not involved. Thus, if ψ exists, then forecaster F_1 is sufficient for simulating forecaster F_2. Second, to output a realization of X_2, the stochastic transformation ψ performs an *auxiliary randomization* of a realization of X_1. In that sense, predictor X_2 is "more random" than predictor X_1, conditional on every predictand realization $W = w$ (i.e., on every forecasting occasion). Now, if you had to select a forecaster of W, which one would it be: F_2 using predictor X_2, or F_1 using predictor X_1?

Theorem *(Sufficiency–Informativeness)* *If forecaster F_1 is sufficient for forecaster F_2, then F_1 is more informative than F_2.*

Proof: It proceeds in two steps.

1. *Value of forecaster.* As is to be proven in Exercise 13, the inequality in the definition of "more informative than" is equivalent to $UF_1^* \geq UF_2^*$, where for any fixed (u, g) and $i = 1, 2$,

$$UF_i^* = UF^*(u, g; f_i) = \int_{\mathcal{X}_i} \left[\max_{a \in A} \int_W u(a, w) f_i(x_i|w) g(w) \, dw \right] dx_i,$$

with the right side being equation (12.61) rewritten for F_i.
2. *Order of integrated maximum utilities.* It is established by (i) replacing $f_2(x_2|w)$ with the right side of the sufficiency relation, (ii) introducing $\kappa_1(x_1)$ derived from $f_1(x_1|w)$ and $g(w)$ via equation (10.1), (iii) recognizing expression (10.2) for $\phi_1(w|x_1)$, and expression (12.56) for $U_1(a, x_1)$, (iv) relocating the max operator, and (v) exploiting the property of $\psi(\cdot|x_1)$ as a density function. Specifically,

$$UF_2^* = \int_{\mathcal{X}_2} \left[\max_{a \in A} \int_W u(a, w) f_2(x_2|w) g(w) \, dw \right] dx_2$$

$$= \int_{\mathcal{X}_2} \left\{ \max_{a \in A} \int_W u(a, w) \left[\int_{\mathcal{X}_1} \psi(x_2|x_1) f_1(x_1|w) \, dx_1 \right] g(w) \, dw \right\} dx_2$$

$$= \int_{\mathcal{X}_2} \left\{ \max_{a \in A} \int_{\mathcal{X}_1} \psi(x_2|x_1) \left[\int_w u(a,w) \frac{f_1(x_1|w)g(w)}{\kappa_1(x_1)} dw \right] \kappa_1(x_1) dx_1 \right\} dx_2$$

$$= \int_{\mathcal{X}_2} \left\{ \max_{a \in A} \int_{\mathcal{X}_1} \psi(x_2|x_1) \left[\int_w u(a,w)\phi_1(w|x_1) dw \right] \kappa_1(x_1) dx_1 \right\} dx_2$$

$$= \int_{\mathcal{X}_2} \left[\max_{a \in A} \int_{\mathcal{X}_1} \psi(x_2|x_1) U_1(a,x_1) \kappa_1(x_1) dx_1 \right] dx_2$$

$$\leq \int_{\mathcal{X}_2} \int_{\mathcal{X}_1} \left[\max_{a \in A} \psi(x_2|x_1) U_1(a,x_1) \kappa_1(x_1) dx_1 \right] dx_2$$

$$= \int_{\mathcal{X}_1} \int_{\mathcal{X}_2} \psi(x_2|x_1) dx_2 \left[\max_{a \in A} U_1(a,x_1) \right] \kappa_1(x_1) dx_1$$

$$= \int_{\mathcal{X}_1} U_1^*(x_1) \kappa_1(x_1) dx_1$$

$$= UF_1^*,$$

where the inequality arises because the expectation (with respect to X_1) of the maximum of a function is not smaller than the maximum of the expectation of that function. Inasmuch as the inequality $UF_1^* \geq UF_2^*$ holds for any fixed (u, g), it holds for every u and every g:

$$UF^*(u,g;f_1) \geq UF^*(u,g;f_2).$$

QED.

12.10.6 Sufficiency Characteristic of the BGF

Suppose each F_i ($i = 1, 2$) is a Bayesian Gaussian forecaster (BGF) of the predictand W, has likelihood parameters (a_i, b_i, σ_i) and the sufficiency characteristic $SC = |a_i|/\sigma_i$, with $|a_i| > 0$ and $\sigma_i > 0$, as defined in Section 10.4.

Theorem (*Sufficient comparison*) *If $SC_1 > SC_2$, then forecaster F_1 is sufficient for forecaster F_2 with respect to predictand W.*

Proof: It proceeds in five steps.

1. *Likelihood functions*. With $\equiv$ meaning the identity of objects written in different symbols, the family of likelihood functions of forecaster F_i ($i = 1, 2$) is specified as

$$f_i(x_i|w) \equiv NM(a_i w + b_i, \sigma_i), \qquad a_i \neq 0, \, \sigma_i > 0.$$

2. *Hypothesis*: the stochastic transformation ψ exists and takes the form

$$\psi(x_2|x_1) \equiv NM(Cx_1 + D, \tau), \qquad C \neq 0, \, \tau > 0.$$

3. *Derivation via pattern*, as explained in Step 2 of Section 10.9.4. Given f_1 and ψ specified above, the total probability law (which defines the sufficiency relation) yields a derived f_2. Symbolically,

$$NM\left(C(a_1 w + b_1) + D, \sqrt{C^2 \sigma_1^2 + \tau^2} \right) \equiv \int_{\mathcal{X}_1} NM(Cx_1 + D, \tau) \cdot NM(a_1 w + b_1, \sigma_1).$$

4. The specified f_2 and the derived f_2 are identical if, under the NM models, the means are equal and the variances are equal:

$$a_2 w + b_2 = C(a_1 w + b_1) + D,$$

$$\sigma_2^2 = C^2 \sigma_1^2 + \tau^2.$$

The solution of the two equations for the parameters of ψ is

$$C = \frac{a_2}{a_1}, \qquad D = b_2 - \frac{a_2}{a_1} b_1;$$

$$\tau^2 = \sigma_2^2 - \frac{a_2^2}{a_1^2} \sigma_1^2.$$

5. These parameters are valid for the NM model of ψ if

$$a_1 \neq 0, \qquad \sigma_2^2 - \frac{a_2^2}{a_1^2} \sigma_1^2 > 0.$$

The first condition is true by assumption. The second condition holds if and only if the following inequality holds:

$$\frac{|a_1|}{\sigma_1} > \frac{|a_2|}{\sigma_2}.$$

Ergo, the hypothesized stochastic transformation ψ exists if $SC_1 > SC_2$. QED.

Note 1. If $|a_1|/\sigma_1 = |a_2|/\sigma_2$, then $\tau_2 = 0$ and ψ degenerates to an impulse function. In such a case, F_1 is sufficient for F_2, and F_2 is sufficient for F_1; in other words, F_1 and F_2 are *mutually sufficient*.

Note 2. The first part of the theorem on comparison of predictors for the BGF in Section 10.4.5 is actually a corollary of the above two theorems — on sufficient comparison and on sufficiency–informativeness inference.

Bibliographical Notes

Decision models for statistical estimation with two-piece linear and quadratic difference opportunity loss functions are well known (e.g., Raiffa and Schlaifer, 1961; DeGroot, 1970; Berger, 1985; Pratt et al., 1995). The model with an impulse utility function has an equivalent in statistics, dubbed the "zero–one" loss function (Bernardo and Smith, 1994). The target-setting model of Section 12.2 has two extensions with utility functions (Krzysztofowicz, 1986a, 1990) and a comprehensive application to decision making with forecasts for optimal water supply planning (Krzysztofowicz, 1986b, 1986c; Krzysztofowicz and Watada, 1986).

The sufficiency relation for comparing forecasters parallels the concept of sufficient comparison of experiments originated by Blackwell (1951, 1953). The concept has been further elaborated in various contexts by Blackwell and Girshick (1954), DeGroot (1962, 1970), Le Cam (1964) and others. The sufficiency characteristic of the Bayesian Gaussian forecaster comes from Krzysztofowicz (1987).

Exercises

1 **Incomplete expectation of uniform variate**. Prove that, for any constant $a \in (\eta_L, \eta_U)$, expression (12.21) can be interpreted as $E(W|W \leq a)P(W \leq a)$.

2 **Incomplete expectation of exponential variate**. Derive expression (12.27). Next prove that, for any constant $a \in (\eta, \infty)$, this expression can be interpreted as $E(W) - E(W|W > a)P(W > a)$.

3 **Incomplete expectation of power type I variate**. Derive expression (12.29). Next prove that, for any constant $a \in (\eta_L, \eta_U)$, this expression can be interpreted as $E(W|W \leq a)P(W \leq a)$.

4 **Incomplete expectation of power type II variate**. Derive expression (12.30). Next prove that, for any constant $a \in (\eta_L, \eta_U)$, this expression can be interpreted as $E(W) - E(W|W > a)P(W > a)$.

5 **Impact of criterion function on decision**. Create an example parallel to the ones discussed in Section 12.6. The input variate W has distribution $WB(1, \beta, 0)$, with the value of the shape parameter β specified in the option. The main tasks are as follows.

    5.1 Calculate each optimal decision a^*, as it is defined in Table 12.3. (A statistician might say: calculate each optimal estimate a^* of the variate W.)

    5.2 Graph the density function of W.

    5.3 Interpret each optimal decision (each optimal estimate of W), and relate it to the criterion function; compare all six decisions, and relate them to the shape of the density function of W.

    Options: (A) $\beta = 0.5$,   (D) $\beta = 4.5$,
                     (B) $\beta = 1.5$,   (E) $\beta = 5.5$,
                     (C) $\beta = 2.5$,   (F) $\beta = 6.5$.

6 **Medication dose**. A medication for treating a contagious, rapid-onset (a few days), short-duration (a few weeks), deadly disease had been developed, and then clinical trials were conducted to determine the daily dose W [mg]. Too small a dose does not cure the disease, and the patient dies. Too large a dose is toxic, and the patient dies. The trials found that the "optimal" dose varies from patient to patient. The variation is partly explained by the patient's weight X [kg], and partly remains unexplained (i.e., is random). A Bayesian Gaussian forecaster (Chapter 10) applied to data from the clinical trials has these parameter estimates:

    $M = 137$ mg, $S = 13$ mg;
    $a = 0.49$ kg/mg, $b = 5.94$ kg, $\sigma = 3.72$ kg.

Determine the optimal daily dose for each of the three patients whose weights are specified in the option.

    Options: (A) 43, 57, 70;   (D) 49, 62, 87;
                     (B) 48, 55, 82;   (E) 53, 71, 90;
                     (C) 46, 51, 79;   (F) 45, 68, 84.

7 **Two tracking radars**. Two radars track an airplane that entered a no-fly zone. The radars, indexed by j ($j = 1, 2$), are synchronized, and each outputs a density function g_j of the airplane position at a given instant. Then the forecast of the position is made in the form of a *mixed density function* $g = g_1 p_1 + g_2 p_2$, where p_j is the probability weight attached to radar j, with $0 < p_j < 1$ and $p_1 + p_2 = 1$. For each of the two cases, $p_1 = 0.3$ and $p_1 = 0.7$, complete the following exercises.

    7.1 Graph the density functions g_1, g_2, specified in the option, and the mixture g.

    7.2 Determine the optimal aiming point a^* for a surface-to-air missile that must take down the invading airplane; calculate its utility U^*; indicate (a^*, U^*) on the graph of function g.

    Options for density functions g_1, g_2:

        (A)   LG(1.1, 0.4), LG(3.3, 0.7);   (F)   LG(1.1, 0.4), LP(3.2, 0.7);
        (B)   LP(1.2, 0.5), LP(3.4, 0.8);   (G)   LG(1.1, 0.4), GB(3.3, 0.7);
        (C)   GB(1.3, 0.6), GB(3.5, 0.9);   (H)   LG(1.1, 0.4), RG(3.4, 0.7);
        (D)   RG(1.4, 0.7), RG(3.6, 0.4);   (I)   GB(1.3, 0.6), LP(3.2, 0.3);
        (E)   NM(1.5, 0.8), NM(3.7, 0.5);   (J)   GB(1.5, 0.6), RG(3.6, 0.9).

8 **Foe and friend**. Consider the decision situation modeled in Section 12.8, with the density functions g and h specified in the option.

    8.1 Graph the functions g and h, and the function $U = g - h$.

    8.2 Determine the optimal aiming point a^* and its utility U^*; indicate (a^*, U^*) on the graph of function U.
    Options for density functions g, h:

        (A) LG(2.1, 0.4), LG(4.3, 0.7);   (F) LG(4.1, 0.4), LP(6.3, 0.7);
        (B)   LP(2.2, 0.5), LP(4.4, 0.8);   (G) LG(4.1, 0.4), GB(6.4, 0.7);
        (C) GB(2.3, 0.6), GB(4.5, 0.9);   (H) LG(4.1, 0.4), RG(6.5, 0.7);
        (D) RG(2.4, 0.7), RG(4.6, 0.4);   (I) GB(2.3, 0.6), LP(4.1, 0.3);
        (E) NM(2.5, 0.8), NM(4.7, 0.5);   (J) GB(2.5, 0.6), RG(4.7, 0.9).

9 **Example of extreme outcome**. Come up with an example of the decision problem (apart from the weapon-aiming one) wherein the decision and the state variables are continuous, but the outcome variable is binary or trinary and extreme — to the point that the utility function could be represented by an impulse function or a double-impulse function. Describe this problem concisely.

10 **Electricity generation for cooling** — Chapter 3, Exercise 9 *continued*. The manager of an electric power plant faces the decision problem described in Exercise 9 of Chapter 3. She must decide the optimal maximum power to be generated on every day on which the maximum power load Y [MW] due to cooling of buildings is uncertain and has the distribution function you derived in Exercise 9 of Chapter 3 (with the options remaining the same as in that exercise). The plant incurs an economic loss whenever it generates more or less power than the actual power load.

    10.1 Formulate a mathematical model of this decision problem (i.e., identify and interpret every element); in particular, identify the sample space $\mathcal{Y}$ of the power load Y, and write the expression for the distribution function of Y on $\mathcal{Y}$.

    10.2 Suppose the criterion function has the form of the two-piece linear opportunity loss function with the marginal opportunity losses λ_o, λ_u [$/MW] specified in the option. (i) Find the optimal maximum power to be generated. (ii) Find the probability of power shortage (which occurs when the actual power load exceeds the generated power).

    10.3 Imagine you live in a city serviced by that power plant, and on every day the power shortage occurs the electricity is turned off for 12 hours. (i) What is the smallest reliability (the probability of no power shortage in a day) that you would consider acceptable? (ii) Find the imputed ratio of the marginal opportunity losses. (iii) Find the power that should be generated in order to satisfy your preference for the smallest acceptable reliability.

    Options: (A) $\lambda_o = 7$, $\lambda_u = 45$;   (D) $\lambda_o = 11$, $\lambda_u = 72$;
                     (B) $\lambda_o = 8$, $\lambda_u = 54$;   (E) $\lambda_o = 12$, $\lambda_u = 83$;
                     (C) $\lambda_o = 9$, $\lambda_u = 71$;   (F) $\lambda_o = 13$, $\lambda_u = 97$.

11 **Electricity generation for heating** — Chapter 3, Exercise 10 *continued*. Complete Exercise 10 above when the maximum power load Y [MW] due to heating of buildings is uncertain and has the distribution function you derived in Exercise 10 of Chapter 3 (with the options remaining the same as in that exercise). [Organize your report into sections 11.1, 11.2, 11.3, paralleling the tasks specified in Exercise 10.]

12 **Order of utilities**. Prove the theorem stating the order of utilities, $U^* \leq UF^* \leq U^{**}$, at the conclusion of Section 12.10.2. *Hint*: Prove each inequality separately; begin each proof with the total expression for UF^*.

13 **Informativeness relation**. Consider two inequalities, the elements of which are defined in Section 12.10.4:

$$VF(u,g;f_1) \geq VF(u,g;f_2),$$
$$UF^*(u,g;f_1) \geq UF^*(u,g;f_2).$$

13.1 Prove that the two inequalities are equivalent (i.e., that one inequality holds if, and only if, the other inequality holds).

13.2 The definition of "more informative than" is stated in terms of the first inequality. Restate it in terms of the second inequality.

13

Inventory and Capacity Models

The target-decision theory presented in Chapter 12 is applied in this chapter to formulate decision models for two classes of problems: inventory management and capacity expansion. Both classes of problems are ubiquitous in the manufacturing, retail, and service sectors of the economy. Excess inventory or capacity constitutes frozen capital; insufficient inventory or capacity translates into lost revenue. The objective of a decision model is to minimize the expected opportunity loss based on a probabilistic forecast of the uncertain demand; the model prescribes the optimal decision rule for replenishing inventory or expanding capacity (when and by how much?), and determines the (economic) value of the perfect forecaster, which constitutes an upper bound on the value of any probabilistic forecaster.

13.1 Inventory Systems

13.1.1 Inventory Management Problem

Decisions about the level of inventory are critical to the profitability of a business, be it manufacturing, retailing, or service, be it small (e.g., a family-run restaurant) or large (e.g., a global manufacturer of airplanes). When the demand for a commodity (part, product, or service) during a specified period cannot be predicted with certainty, the inventory turns out to be too small or too large almost surely. What matters, of course, is the size of shortage or surplus because either one results in an *opportunity loss*: a shortage translates into lost revenue; a surplus translates into frozen capital (nonworking capital) or a loss (partial or total), depending on the value retained by an item stored beyond the specified period.

The search for the preferred *trade-off* between these two opportunity losses is reflected in the slogans that have been coined to promote various management policies:

1. "Just-in-case" inventory management focuses on avoiding the opportunity loss from shortage (by keeping a large inventory of items all the time).
2. "Just-in-time" delivery management focuses on avoiding the opportunity loss from surplus (by manufacturing or purchasing items only when they are demanded).

Such one-sided policies are unlikely to be economically optimal over the long term. Rather, the search for an optimal trade-off requires a balanced approach — one that takes all uncertainties and opportunity losses into account in a coherent manner, and can adapt to changing uncertainties and economic conditions.

Yet adaptability remains a recurring challenge to many businesses. As the US economy was recovering from the disruption caused by the Covid-19 pandemic to supply–demand relations, *The Wall Street Journal* on 8 November 2021 summed up the situation: "Companies are wrestling with how big their inventories should be, since the pandemic highlighted the dangers of having too much or too little stored away."

In large corporations, inventory management has evolved into *supply chain management* — in essence, a systems approach to managing the entire process, from purchasing raw materials, to shipping, to manufacturing, to distributing finished products in a coordinated manner.

Nonetheless, the basic inventory models remain essential because they prescribe the fundamental rules for managing inventories in a rational manner, because they can be adapted as components of supply chain management systems, and because they are all that many small businesses need to manage their inventories in a near-optimal manner. (Think of a restaurant, where the refrigerator and the freezer contain most of the operating inventory.)

13.1.2 Basic Inventory System

The basic inventory system which we shall study is depicted in Figure 13.1. There is a storage facility (a lot, a silo, a warehouse) in which a commodity (e.g., coal, wheat, bananas) or a part (e.g., tires, bearings, computer chips), or a product (e.g., cars, shoes, cell phones) is stored. To model this system, we fix the *decision period* (e.g., a week, a month, a season). The *state* of the system at the beginning of the period is the level of inventory (the stock level); the *input* to the system during the period is the number of ordered (received and paid-for) items; the *output* from the system during the period is the number of demanded (sold and shipped) items.

The decider is a manufacturer, a wholesaler, or a retailer. The inventory is replenished by placing an order, either to produce or to purchase the items. An *inventory policy* is a rule for deciding: When to order? How much to order?

There exist many models for managing inventory. They can be classified according to the features of an inventory system which they incorporate. Here is a partial list of such features.

1. Operating horizon: finite or infinite (as an approximation to an unspecified lifetime of the system).
2. Inventory review (update of the state and decision): continuous (in time) or periodic (e.g., once a week).
3. Demand during a period: known or uncertain.
4. Delivery of orders: instantaneous (for all practical purposes) or with a lag time.
5. Shortages: not allowed or allowed.
6. Backlogging of unsatisfied demand (shifting the demand from the current period to a future period): not allowed or allowed.
7. Quantity discount on orders: not available or available.
8. Setup cost incurred with each order: absent or present.

A model incorporating all these features as options would be complicated indeed. For this reason, numerous models have been developed, each incorporating some of the features. In effect, the choice of a decision model for a given inventory system amounts to matching the dominant features of the system with the features offered by a model, and then accepting the model as a mathematical approximation to reality.

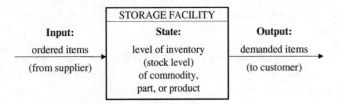

Figure 13.1 Basic inventory system.

13.2 Basic Inventory Model

13.2.1 Purpose and Assumptions

The purpose of the basic inventory model is to determine the optimal *stock level* of a product at the beginning of the period during which the product should be sold, consumed, or used. Among the products of this kind are those which

1. become obsolete (e.g., a daily newspaper, a weekly magazine, a monthly periodical, a calendar for a year);
2. are perishable (e.g., milk, butter, eggs, fruit, vegetables, bread);
3. have seasonal demand (e.g., Christmas decorations, Valentine's greeting cards, skis, snow shovels, swim suits, heating oil).

The decider for whom the basic inventory model is formulated may be a manufacturer who must decide the size of the production for the given period, or a wholesaler who must decide the size of the purchase order to be placed with a manufacturer, or a retailer who, likewise, must decide the size of the purchase order to be placed with a wholesaler.

There are five main assumptions from which the model is formulated. First, the period is fixed beforehand. Second, the demand during the period is uncertain and is forecasted probabilistically. Third, a shortage of the product is allowed. Fourth, the unsatisfied demand cannot be backlogged and satisfied later. Fifth, at the end of the period, the remaining stock of the product is either worthless (as is the case with the daily newspaper), or may be disposed of at a reduced price (as might be the case with milk whose expiration date approaches), or may be held in storage until the next season when it is sold either at a reduced price (as might be the case with skis — the previous year's model can be offered at the next pre-season sale) or at the prevailing price (as might be the case with heating oil).

13.2.2 Mathematical Formulation

This basic inventory model comprises seven elements:

a — stock level of the product at the beginning of the period; it is the *decision variable*, assumed to be continuous and positive, $a > 0$, and taking on values in the decision space $\mathcal{A}$ such that $\mathcal{A} = \mathcal{W}$; see the next element.

W — demand for the product during the period; it is a *random variable*, assumed to be continuous and positive, with the sample space $\mathcal{W}$ being either $(0, \infty)$; or (η, ∞) when it is certain that the demand will exceed level η, where $\eta > 0$; or (η_L, η_U) when it is certain that the demand will exceed level η_L and not exceed level η_U, where $0 < \eta_L < \eta_U$.

G — distribution function of the demand W such that for any realization $w \in \mathcal{W}$,

$$G(w) = P(W \le w).$$

c_p — unit purchase cost (or unit production cost).

c_h — unit holding cost of unsold product until the next season (if applicable); for example, the cost of storing a pair of skis from March to October.

p_s — unit sale price.

p_r — unit reduced sale price (if applicable); for example, the price at which a pair of one-year old skis will be offered at the next pre-season sale.

The costs and prices must satisfy three nontriviality conditions:

$p_s > c_p$ because otherwise the operation is not economical;

$p_r > c_h$ because otherwise holding the product beyond the period is not economical;

$c_p > p_r - c_h$ because otherwise it would be optimal to purchase (or to produce) an infinite quantity of the product and to hold the unsold items.

The decision problem is to find the optimal stock level a^* that maximizes the expected profit from the operation during a single period.

13.2.3 Criterion Function

The *profit per period* $r(a, w)$ depends upon the stock level a at the beginning of the period and the demand w during the period. If $w \leq a$, then the quantity w is sold and the quantity $a - w$ remains unsold. If $a \leq w$, then the entire stock level a is sold. Thus

$$r(a, w) = \begin{cases} p_s w + (p_r - c_h)(a - w) - c_p a & \text{if } w \leq a, \\ p_s a - c_p a & \text{if } a \leq w, \end{cases}$$

$$= \begin{cases} (p_s - p_r + c_h)w + (p_r - c_h - c_p)a & \text{if } w \leq a, \\ (p_s - c_p)a & \text{if } a \leq w. \end{cases} \tag{13.1}$$

For a fixed demand w, the profit per period is a two-piece linear function of the stock level a; this function attains the maximum at $a = w$. Thus if the demand were known (were revealed by a clairvoyant), then (i) the optimal stock level would be $a = w$, and (ii) the *maximum profit per period* would be

$$\max_a r(a, w) = r(w, w)$$

$$= (p_s - c_p)w. \tag{13.2}$$

In order to find the optimal stock level when the demand is uncertain, it is advantageous to transform the profit function (13.1) into an opportunity loss function.

For a fixed demand w, *the opportunity loss* $l(a, w)$ is defined as the difference between the profit that would be made if the stock level were w and the profit that is made if the stock level is a. That is,

$$l(a, w) = \max_a r(a, w) - r(a, w)$$

$$= r(w, w) - r(a, w). \tag{13.3}$$

When expressions (13.1) and (13.2) are inserted into equation (13.3), one obtains

$$l(a, w) = \begin{cases} (c_p + c_h - p_r)(a - w) & \text{if } w \leq a, \\ (p_s - c_p)(w - a) & \text{if } a \leq w. \end{cases}$$

Denoting

$$\lambda_o = c_p + c_h - p_r,$$
$$\lambda_u = p_s - c_p, \tag{13.4}$$

the expression for the opportunity loss takes the form

$$l(a, w) = \begin{cases} \lambda_o(a - w) & \text{if } w \leq a, \\ \lambda_u(w - a) & \text{if } a \leq w, \end{cases} \tag{13.5}$$

where λ_o is the *marginal opportunity loss from overstocking* the product (or overestimating the demand), and λ_u is the *marginal opportunity loss from understocking* the product (or underestimating the demand). For a fixed stock level a, the opportunity loss is a two-piece linear function of the demand w; this function is nonnegative (as $\lambda_o > 0$, $\lambda_u > 0$) and attains the minimum of zero at $a = w$.

The profit per period $r(a, w)$ can now be expressed in terms of the opportunity loss $l(a, w)$. From equation (13.3),

$$
\begin{aligned}
r(a, w) &= r(w, w) - l(a, w) \\
&= (p_s - c_p)w - l(a, w) \\
&= \lambda_u w - l(a, w).
\end{aligned}
\tag{13.6}
$$

The advantage of this reformulation of the profit function r is that the decision variable a appears only as an argument of the opportunity loss function l. Moreover, the structure of this opportunity loss function, which is given by equation (13.5), is identical to the structure of the opportunity loss function (12.8) for which the optimal solution has already been derived. Consequently, we can use the existing analytic solution (from Section 12.2.3) and its interpretation (from Section 12.2.4) to establish the solution to the basic inventory problem.

13.2.4 Optimal Decision Procedure

For a given stock level a, the *expected profit* is given by

$$
\begin{aligned}
R(a) &= E[r(a, W)] \\
&= E[\lambda_u W - l(a, W)] \\
&= \lambda_u E(W) - E[l(a, W)] \\
&= \lambda_u E(W) - L(a),
\end{aligned}
\tag{13.7}
$$

where $L(a) = E[l(a, W)]$. The decision problem is to find the optimal stock level a^* which maximizes the expected profit:

$$
\begin{aligned}
R(a^*) &= \max_a R(a) \\
&= \max_a \{\lambda_u E(W) - L(a)\} \\
&= \lambda_u E(W) - \min_a L(a) \\
&= \lambda_u E(W) - L(a^*).
\end{aligned}
\tag{13.8}
$$

This relationship has a usable implication: the decision a^* that maximizes the expected profit is the same decision a^* that minimizes the expected opportunity loss. Because the solution to the latter optimization problem is already derived in Section 12.2.3, it can be employed here — specifically, its expressions (12.14) and (12.15).

The *optimal stock level a^** at the beginning of the period is given by

$$
a^* = G^{-1}\left(\frac{1}{1 + \lambda_o/\lambda_u}\right).
\tag{13.9}
$$

The associated *maximum expected profit $R^* = R(a^*)$* is given by

$$
\begin{aligned}
R^* &= \lambda_u E(W) - L^* \\
&= (\lambda_u + \lambda_o)\int_{\eta_L}^{a^*} wg(w)\, dw,
\end{aligned}
\tag{13.10}
$$

and the calculation of the integral is explained below expression (12.15) for $L^* = L(a^*)$.

13.2.5 Properties of Optimal Solution

This optimal solution to the basic inventory problem has three important properties, illustrated in Figure 13.2.

1. In order to find the optimal stock level a^* at the beginning of the period, one need only know

(a) the distribution function G of the uncertain demand W, and

(b) the ratio λ_o/λ_u of the marginal opportunity losses from overstocking to understocking $(0 < \lambda_o/\lambda_u < \infty)$.

2. The *optimal stock level* a^* at the beginning of the period equals the *quantile* of the uncertain demand W corresponding to the probability

$$P(W \leq a^*) = G(a^*) = \frac{1}{1 + \lambda_o/\lambda_u}. \tag{13.11}$$

3. Associated with the optimal stock level a^* is the *optimal reliability of supply* — the probability that the demand will be met (equivalently, the probability that the demand will not exceed the supply): $P(W \leq a^*) = G(a^*)$.

The last property, illustrated in Table 13.1, deserves an additional commentary. First, the optimal reliability depends on the economic factors — costs and prices that yield the ratio λ_o/λ_u of the marginal opportunity losses. Second, this optimal reliability is specified solely by the degree of asymmetry in the marginal opportunity losses; in particular,

$$\frac{\lambda_o}{\lambda_u} \begin{Bmatrix} > \\ = \\ < \end{Bmatrix} 1 \Rightarrow P(W \leq a^*) \begin{Bmatrix} < \\ = \\ > \end{Bmatrix} \frac{1}{2}.$$

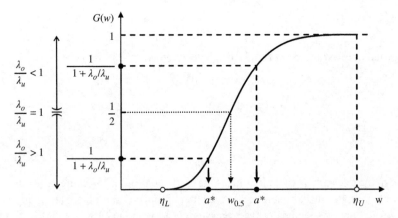

Figure 13.2 Optimal solution to the basic inventory problem: the distribution function G of the demand W on the sample space (η_L, η_U); the ratio λ_o/λ_u of the marginal opportunity losses; the optimal stock level a^*, which is the p^*-probability quantile of W, where $p^* = 1/(1 + \lambda_o/\lambda_u)$. In this graph, G is log-ratio logistic LR1(η_L, η_U)-LG(β, α), with $\eta_L = 1.5$, $\eta_U = 8.5$, $\beta = -0.6$, $\alpha = 0.4$, while a^* is for $p^* = 0.2$ (when $\lambda_o/\lambda_u = 4$) and $p^* = 0.8$ (when $\lambda_o/\lambda_u = 0.25$).

Table 13.1 Optimal reliability of supply $P(W \leq a^*)$, and optimal probability of shortage $P(W > a^*)$; each probability is a function of the ratio λ_o/λ_u of the marginal opportunity losses.

Ratio λ_o/λ_u	Reliability $P(W \leq a^*)$	Probability of shortage $P(W > a^*)$
5	$\frac{1}{6} = 0.167$	0.833
2	$\frac{1}{3} = 0.333$	0.667
1	$\frac{1}{2} = 0.5$	0.5
$\frac{1}{2}$	$\frac{2}{3} = 0.667$	0.333
$\frac{1}{5}$	$\frac{5}{6} = 0.833$	0.167

Third, the above interpretation can be framed in terms of the probability of shortage:

$$P(W > a^*) = 1 - P(W \leq a^*).$$

In essence, the optimal solution to the basic inventory problem states that it is optimal to allow for shortage with a probability specified by the ratio λ_o/λ_u of the marginal opportunity losses.

13.2.6 Value of Perfect Forecaster

How much at most should the decider be willing to pay for a perfect forecast of the demand? To answer this question, imagine a clairvoyant who states a value w such that $P(W = w) = 1$. Given such a value, the decision problem reduces to deterministic optimization: find the stock level a that maximizes the profit per period. From expression (13.2), it is already known that the optimal solution to this problem is $a = w$ and that, with $\lambda_u = p_s - c_p$,

$$\max_a r(a, w) = \lambda_u w. \tag{13.12}$$

Of course, a clairvoyant is not actually available, but is only contemplated. Thus the single value w is unknown; however, one can ask the question: What is the distribution function of the values of the demand that the clairvoyant could state? It is, naturally, the same as the distribution function G of the random demand W. Therefore, the optimal decision with a perfect forecast can be evaluated *a priori* in terms of the *expected maximum profit*:

$$
\begin{aligned}
R^{**} &= E[\max_a r(a, W)] \\
&= E(\lambda_u W) \\
&= \lambda_u E(W). \tag{13.13}
\end{aligned}
$$

The *value of the perfect forecaster* of the demand W is defined as the difference between the expected maximum profit R^{**} and the maximum expected profit R^*:

$$
\begin{aligned}
VPF &= R^{**} - R^* \\
&= L^*, \tag{13.14}
\end{aligned}
$$

where the last equality is obtained by replacing R^{**} with expression (13.13) and R^* with expression (13.10). It turns out that *VPF* equals the minimum expected opportunity loss L^*, which can be calculated from the input elements via equation (12.15).

When R^{**} and R^* are in monetary units, the *VPF* admits two useful interpretations. First, it is the maximum price the manager of the inventory system should be willing to pay for one perfect forecast of the demand during a period. Second, it is an upper bound on the value of any imperfect forecast.

13.2.7 Impact of Loss Asymmetry

That the asymmetry of the criterion function impacts the optimal decision a^* was demonstrated in Section 12.6. Here, we ask a parallel question: How does the asymmetry of the marginal opportunity losses, λ_o and λ_u, impact the *VPF*? It turns out — profoundly.

Theorem (Graph of VPF versus λ_o/λ_u)   *Let G be strictly increasing on $\mathcal{W} = (\eta_L, \eta_U)$, with $0 < \eta_L < \eta_U$, and let $\lambda = \max\{\lambda_o, \lambda_u\}$, with $\lambda_o > 0, \lambda_u > 0$. There exists a continuous graph of VPF versus λ_o/λ_u on the interval $(0, 1] \cup [1, \infty)$, which has these properties.*

1. If $\lambda_o = \lambda_u = \lambda$ (so that $\lambda_o/\lambda_u = 1$ and $a^* = G^{-1}(1/2) = w_{0.5}$, the median of W), then VPF attains the maximum, which is

$$VPF^* = \lambda\left[E(W) - 2\int_{\eta_L}^{w_{0.5}} wg(w)\,dw\right].$$

2. If $\lambda_u = \lambda$ is held fixed, and λ_o decreases from λ to 0 (so that λ_o/λ_u decreases from 1 to 0), then VPF decreases strictly from VPF^* to 0, with

$$\lim_{\lambda_o \to 0} a^* = \eta_U, \qquad \lim_{\lambda_o \to 0} VPF = 0.$$

3. If $\lambda_o = \lambda$ is held fixed, and λ_u decreases from λ to 0 (so that λ_o/λ_u increases from 1 to ∞), then VPF decreases strictly from VPF^* to 0, with

$$\lim_{\lambda_u \to 0} a^* = \eta_L, \qquad \lim_{\lambda_u \to 0} VPF = 0.$$

The proof is presented in Section 13.6; the properties are illustrated in Figure 13.3; and exploration of these properties is the subject of Exercises 1 and 2.

Interpretations

For fixed stock level a and variable demand w, the two-piece linear opportunity loss function $l(a, \cdot)$ has a V-shape with the vertex at $w = a$. Let us examine three cases (Figure 13.4).

1. *Symmetry.* When the marginal opportunity losses from overstocking and from understocking are identical, $\lambda_o/\lambda_u = 1$, the V is symmetrical (Figure 13.4(a)); a^* equals the median of the demand W, and VPF is maximal.
2. *Left skew.* When $\lambda_o/\lambda_u < 1$, the V is skew leftward (Figure 13.4(b)); and if the marginal opportunity loss from overstocking vanishes $(\lambda_o \to 0)$, then $a^* \to \eta_U$, the upper bound of the demand W. In other words, stocking "just-in-case" becomes economically preferred, irrespective of the distribution function G of W. Consequently, a perfect forecaster of W has no economic value: $VPF \to 0$.

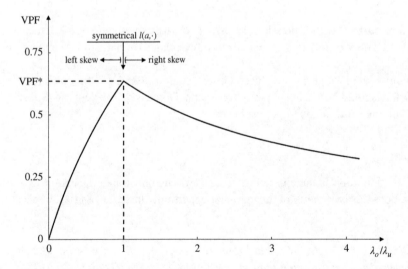

Figure 13.3 Graph of the value of a perfect forecast, *VPF*, versus the ratio of marginal opportunity losses, λ_o/λ_u, constructed per the theorem of Section 13.2.7, when $W \sim P1(4, 0, 1)$ and $\lambda = 5$.

Figure 13.4 Two-piece linear opportunity loss function $l(a, \cdot)$ of the demand w, when the stock level a is fixed: (a) symmetrical, $\lambda_o/\lambda_u = 1$; (b) asymmetrical with left skew, $\lambda_o/\lambda_u < 1$; (c) asymmetrical with right skew, $\lambda_o/\lambda_u > 1$.

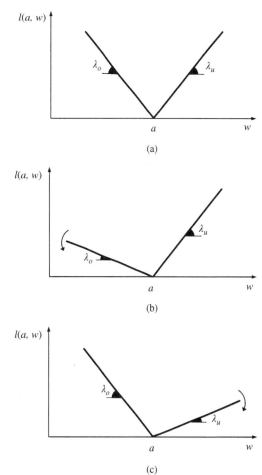

3. *Right skew.* When $\lambda_o/\lambda_u > 1$, the V is skew rightward (Figure 13.4(c)); and if the marginal opportunity loss from understocking vanishes $(\lambda_u \to 0)$, then $a^* \to \eta_L$, the lower bound of the demand W. In other words, stocking "just-in-time" becomes economically preferred, irrespective of the distribution function G of W. Consequently, a perfect forecaster of W has no economic value: $VPF \to 0$.

Overall conclusion

As the opportunity loss function $l(a, \cdot)$ becomes extremely asymmetric (skew): (i) the optimal stock level a^* approaches one of the endpoints of the sample space (η_L, η_U) of the demand; (ii) the value of a perfect forecaster of the demand, VPF, approaches zero; and, by implication, (iii) the value of any imperfect forecaster approaches zero.

A precise characterization of this behavior may be obtained for a specified distribution function G of the demand by graphing VPF versus the ratio λ_o/λ_u. This task is the subject of Exercise 2.

13.2.8 Subjective Assessment of the Ratio

When there are no economic data for estimation of costs and prices from which the marginal opportunity losses could be calculated according to equation (13.4), the manager of the inventory system may assess the ratio $r = \lambda_o/\lambda_u$ of the marginal opportunity losses subjectively. In this way, the optimal stock level (equation (13.9)) and the optimal reliability of supply (equation (13.11)) can be calculated, and the inventory system can be operated consistently with the manager's preferences over consequences.

To assess the ratio $r = \lambda_o / \lambda_u$ subjectively, the manager should follow a four-step procedure.

Step 1. Identify and describe the consequences. To wit, write down two lists: (i) all consequences of overstocking the product (when a quantity remains unsold at the end of the period), and (ii) all consequences of understocking the product (when an additional quantity could have been sold during the period). The contemplated quantity should be large enough to cause significant consequences, or painful regret from having ordered too much or too little.

Step 2. Order the two scenarios in terms of their undesirability. For this purpose, imagine the end of the period; imagine it twice, each time with a different scenario. Next, answer the question: Is it worse to overstock or to understock? If you assess the two scenarios to be about equally undesirable, then set $r = 1$, and go to Step 4.

Step 3. Assess the degree of relative undesirability of the consequences. Toward this end:

    3.1 If in Step 2 you responded "to overstock", then ask yourself: How much worse is it to overstock than to understock? Come up with a number r greater than 1 which reflects the degree to which overstocking by an amount appears to you worse than understocking by the same amount.

    3.2 If in Step 2 you responded "to understock", then ask yourself: How much worse is it to understock than to overstock? Come up with a number s greater than 1 which reflects the degree to which understocking by an amount appears to you worse than overstocking by the same amount.

Step 4. If you assessed $r = 1$ in Step 2, or $r > 1$ in Step 3.1, then set $\lambda_o / \lambda_u = r$. If you assessed $s > 1$ in Step 3.2, then set $\lambda_o / \lambda_u = 1/s$.

13.2.9 Applying the Model

To apply the basic inventory model to a problem, follow these four steps.

Step 1. Map the problem description into the elements of the mathematical formulation (Section 13.2.2) and judge whether or not this formulation, as well as its purpose and assumptions (Section 13.2.1), match the problem, at least approximately. Some approximations may be inevitable: (i) When the product is countable, the continuous variables a and w are adequate provided that rounding the optimal solution a^* to the nearest integer does not alter $L(a^*)$ significantly. (ii) When there are multiple products to manage, the model may be applied independently to every product.

Step 2. Quantify the uncertainty about the demand W by preparing a probabilistic forecast that specifies a distribution function G of W. Such a forecast may be prepared either (i) judgmentally (Chapter 9), or (ii) statistically (Chapter 10).

Step 3. Evaluate the consequences of overstocking and understocking in terms of the marginal opportunity losses, λ_o and λ_u, or their ratio, λ_o / λ_u. This may be accomplished through either (i) economic analysis of costs and prices (Section 13.2.3), or (ii) subjective assessment of the degree of undesirability (Section 13.2.8).

Step 4. Calculate the optimal stock level a^* (formula (13.9)), the optimal reliability of supply $G(a^*)$ (formula (13.11)), and, if desirable and feasible, the maximum expected profit R^* (formula (13.10)), and the value of the perfect forecaster VPF (formula (13.14)).

13.3 Model with Initial Stock Level

13.3.1 Purpose and Assumptions

This model is an extension of the basic inventory model (Section 13.2): it allows for an initial stock level, already paid for, to exist at the beginning of the period. In addition to being handy, this model serves as the foundation for a capacity planning model (Section 13.4).

13.3.2 Mathematical Formulation

The extension involves two elements:

s — initial stock level of the product at the beginning of the period; it is the *state variable*, assumed to be continuous and nonnegative, $s \geq 0$, and taking on values in the decision space $\mathcal{A}$.

a — target stock level of the product at the beginning of the period (after replenishment if ordering); it is the *decision variable*, assumed to be continuous and positive, $a > 0$, and taking on values in the decision space $\mathcal{A}$ such that $\mathcal{A} = \mathcal{W}$.

All other elements of the basic inventory model are retained. Thus, the extension amounts to defining the state variable s and redefining the decision variable a. The new definitions imply that (i) an order to replenish the inventory should be placed only if $a > s$, and (ii) the quantity to be ordered is $a - s$.

13.3.3 Decision Tree

The decision problem can be structured graphically as a *decision tree* (Figure 13.5). Given the initial stock level s at the beginning of the period, the first decision is to order or not to order an additional quantity; each alternative is represented by a *branch* emanating from the first *decision node* (a square).

When *no order* is placed, s is the stock level available for the period. The demand $W = w$, generated randomly from a distribution function G of W, represented by the *chance node* (a circle), is observed during the period; and the profit $r_0(s, w)$ is realized at the end of the period.

When an *order* is to be placed, the second decision is to set the target stock level a; obviously $a > s$, for otherwise no order would be needed. The amount $a - s$ is ordered, whereupon the stock is replenished to the target level a; the demand $W = w$ is observed during the period; and the profit $r_1(s, a, w)$ is realized at the end of the period.

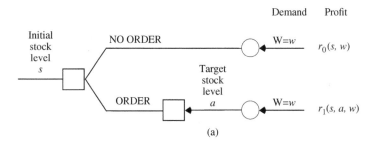

(a)

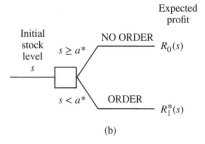

(b)

Figure 13.5 Inventory model with initial stock level: (a) the decision tree; (b) the choice of the branch. *Note*: The black sector initiating a branch signifies a continuous variable (whose possible values are uncountable).

The decision analysis proceeds in three steps. (i) In the no-order branch, we must derive an expression for the expected profit. (ii) In the order branch, we must derive expressions for the optimal target stock level and for the maximum expected profit. (iii) At the first decision node, we must establish the optimal ordering rule.

13.3.4   The No-Order Branch

Given the initial stock level $s > 0$, and supposing demand w during the period, the *profit per period* is

$$r_0(s, w) = \begin{cases} p_s w + (p_r - c_h)(s - w) & \text{if } w \leq s, \\ p_s s & \text{if } s \leq w. \end{cases} \tag{13.15}$$

Because $p_s > p_r - c_h$, maximum profit is realized when, by luck, $w = s$:

$$\max_w r_0(s, w) = r_0(s, s) = p_s s. \tag{13.16}$$

Thus, the *opportunity loss* from being unlucky, when $w \neq s$, can be defined as

$$\begin{aligned} l_0(s, w) &= \max_w r_0(s, w) - r_0(s, w) \\ &= r_0(s, s) - r_0(s, w). \end{aligned} \tag{13.17}$$

After inserting expressions (13.15) and (13.16) into equation (13.17), one obtains

$$l_0(s, w) = \begin{cases} (\lambda_u + \lambda_o)(s - w) & \text{if } w \leq s, \\ 0 & \text{if } s \leq w, \end{cases} \tag{13.18}$$

where λ_o and λ_u are the marginal opportunity losses in the basic model, defined by expression (13.4). For a fixed initial stock level s, this opportunity loss is a two-piece linear function of the demand w.

Expression (13.18) for l_0 is analogous to expression (13.5) for l in the basic model. Like in that model, it is advantageous to express the profit per period in terms of the opportunity loss:

$$\begin{aligned} r_0(s, w) &= r_0(s, s) - l_0(s, w) \\ &= p_s s - l_0(s, w). \end{aligned} \tag{13.19}$$

Given state s, the expected profit is

$$\begin{aligned} R_0(s) &= E[r_0(s, W)] \\ &= E[p_s s - l_0(s, W)] \\ &= p_s s - E[l_0(s, W)] \\ &= p_s s - L_0(s), \end{aligned} \tag{13.20}$$

where $L_0(s) = E[l_0(s, W)]$. Inasmuch as $l_0(s, w)$, specified by equation (13.18), has the same form as $l(a, w)$, specified by equation (13.5), we can look at the solution (12.12) in Section 12.2.3 for $L(a) = E[l(a, W)]$, replace in it a by s, λ_o by $(\lambda_u + \lambda_o)$, λ_u by 0, and thereby obtain the solution for the *expected opportunity loss*:

$$L_0(s) = (\lambda_u + \lambda_o)sG(s) - (\lambda_u + \lambda_o) \int_{\eta_L}^{s} wg(w)\, dw. \tag{13.21}$$

When solution (13.21) is inserted into expression (13.20), the final formula is obtained for the *expected profit when not ordering*:

$$R_0(s) = p_s s - (\lambda_u + \lambda_o)sG(s) + (\lambda_u + \lambda_o) \int_{\eta_L}^{s} wg(w)\, dw. \tag{13.22}$$

13.3.5 The Order Branch

Given the initial stock level s, the decision to replenish the stock to level a ($a > s$) by purchasing quantity $a - s > 0$, and supposing demand w during the period, the *profit per period* is

$$r_1(s, a, w) = \begin{cases} p_s w + (p_r - c_h)(a - w) - c_p(a - s) & \text{if } w \leq a, \\ p_s a - c_p(a - s) & \text{if } a \leq w, \end{cases}$$
$$= c_p s + r(a, w),$$

(13.23)

where $r(a, w)$ is specified by expression (13.1) or, equivalently, by expression (13.6).

Given state s and decision a, the expected profit is

$$R_1(s, a) = E[r_1(s, a, W)]$$
$$= c_p s + E[r(a, W)]$$
$$= c_p s + R(a),$$

(13.24)

where $R(a)$ is identical with the expected profit (13.7) in the basic model.

The decision problem is to find the optimal target stock level a^*, which maximizes the expected profit:

$$R_1(s, a^*) = \max_a R_1(s, a)$$
$$= \max_a \{c_p s + R(a)\}$$
$$= c_p s + \max_a R(a)$$
$$= c_p s + R(a^*)$$
$$= c_p s + R^*,$$

(13.25)

where R^* is identical with the maximum expected profit (13.10) in the basic model. Denoting $R_1^*(s) = R_1(s, a^*)$, and substituting R^* with an explicit expression, yields the final formula for the *maximum expected profit when ordering*:

$$R_1^*(s) = c_p s + (\lambda_u + \lambda_o) \int_{\eta_L}^{a^*} w g(w) \, dw.$$

(13.26)

The term $R(a)$ maximized in equation (13.25) is identical with the term maximized in equation (13.8) of the basic model. Therefore, solution (13.9) is valid here as well: the *optimal target stock level when ordering* is

$$a^* = G^{-1}\left(\frac{1}{1 + \lambda_o/\lambda_u}\right).$$

(13.27)

Interestingly, a^* is independent of s.

13.3.6 Choice of the Branch

Given the initial stock level s, the choice between the no-order branch and the order branch is made by comparing the expected profit $R_0(s)$ with the maximum expected profit $R_1^*(s)$, as depicted in Figure 13.5. The result is this theorem.

Theorem *(Branch choice)* *In the inventory model with the initial stock level s and the optimal target stock level a^*:*

> *If $s = a^*$, then $R_0(s) = R_1^*(s)$.*
>
> *If $s < a^*$, then $R_0(s) < R_1^*(s)$.*

The proof is left as Exercise 3, while we turn to the practical implication of this result.

13.3.7 Optimal Ordering Rule

The analytic solution for a^* and the branch choice theorem offer a theoretic basis for managing inventory of a product as follows. At the beginning of the period:

Step 1. Observe the initial stock level.
Step 2. Update the costs and prices, and calculate the marginal opportunity losses λ_o and λ_u from equation (13.4), and their ratio λ_o / λ_u.
Step 3. Forecast the demand for the product during the period in terms of a distribution function G; techniques described in Chapters 9 and 10 may be used for this purpose.
Step 4. Calculate the optimal target stock level a^* from formula (13.27).
Step 5. Make a decision according to the following *optimal ordering rule*:

$$\text{If } s \geq a^*, \text{ then do not order.}$$

$$\text{If } s < a^*, \text{ then order quantity } a^* - s.$$

The above decision algorithm is optimal for a single-period inventory problem. Of course, it could be applied at the beginning of each new period in a multi-period inventory problem. Such an application (i) may be convenient, especially in a nonstationary environment, when costs, prices, and the distribution function of the demand change from period to period, but (ii) is *myopic* in that it ignores all period-to-period dependence that may exist. Consequently, it may be *suboptimal*.

13.4 Capacity Planning Model

13.4.1 Capacity Planning Problem

Every production, service, or transport system has a maximum capacity at which it can operate. Here are a few examples.

> Number of airplanes an assembly line can produce per month.
> Number of megawatts a hydroelectric power plant can generate.
> Number of seats in a theater (restaurant, arena, stadium).
> Number of telephone calls a service can answer simultaneously.
> Number of skiers a lift can ferry to the peak per hour.
> Volume of potable water a pipeline can deliver per day.

To build a capacity requires a capital-intensive, long-term investment; and then to operate a system at that capacity requires a cash flow. Hence two basic questions with significant economic consequences to the owner of the system: (i) Should the capacity of a system be expanded or shrunk? (ii) If so, then by how much?

These questions are recurrent because the economy is dynamic — it expands or contracts. Consequently, the demand for a product, service, or transport changes over time, and so should the system capacity in order to minimize the opportunity loss from overcapacity or undercapacity. Not infrequently, decisions about the system capacity determine the success or failure of an enterprise — be it a small business or a large corporation.

Example 13.1 *(Small business)*
Shelby American designs and produces performance automobiles (e.g., the Shelby 1000 Mustang) and performance parts. In 2012, in the midst of a recession, it celebrated its 50th anniversary, and on that occasion *Road & Track* (May 2012) interviewed the company's president John Luft:

R & T: How's business these days?

J. L.: It's always challenging. The auto industry, no surprise, has taken a beating. Shelby American has weathered the storm, but we've had to swim real hard. But it's really about keeping your business properly sized for the amount of business you do. That's fundamental.

Example 13.2 *(Large corporations)*

The onset of the recession coincided with rapid and radical changes in the production capacities of many large corporations. As reported by *The Wall Street Journal* (3 February 2010), during 2009 production capacities in some sectors expanded while in other sectors contracted, for example, semiconductors (+10.4%), computers (+5.5%), oil and gas (+1.2%); plastic and rubber products (−3.1%), furniture (−5.4%), textile mills (−7.1%). Overall, the US production capacity contracted by about 1%. Whereas some of these adjustments were the result of changing economic climate or restructuring aimed at increasing efficiency, others were a direct response to changing demand; for example, the total US auto and truck sales reached 17.5 million in 2005, but were forecasted to be only between 11.5 million and 12.5 million in 2010.

13.4.2 Purpose and Assumptions

This model is a reinterpretation of the inventory model with initial stock level (Section 13.3). Its purpose is to determine the optimal *capacity* of a system to be operated over a *planning period* (e.g., 5 years) during which the demand in every *operation cycle* (e.g., 1 day) is stationary, and economic coefficients remain constant.

13.4.3 Mathematical Formulation

This capacity planning model comprises the following elements:

s — current capacity, already paid for, per operation cycle; it is the *state variable*, assumed to be continuous and nonnegative, $s \geq 0$, and taking on values in the decision space $\mathcal{A}$.

a — target capacity per operation cycle; it is the *decision variable*, assumed to be continuous and positive, $a > 0$, and taking on values in the decision space $\mathcal{A}$ such that $\mathcal{A} = \mathcal{W}$; see the next element.

W — demand for a product, service, or transport during the operation cycle; it is a *random variable*, assumed to be continuous and positive, with the sample space $\mathcal{W}$ being either $(0, \infty)$; or (η, ∞) when it is certain that the demand will exceed level η, where $\eta > 0$; or (η_L, η_U) when it is certain that the demand will exceed level η_L and not exceed level η_U, where $0 < \eta_L < \eta_U$.

G — distribution function of the demand W such that for any realization $w \in \mathcal{W}$,

$$G(w) = P(W \leq w).$$

c_p — cost of building one unit of capacity, amortized over the planning period and scaled down to the operation cycle.

c_h — cost of operating one unit of capacity, per operation cycle.

p_s — profit from utilizing one unit of capacity, per operation cycle.

$p_r = 0$ reflects an assumption that unutilized capacity cannot be diverted to another profitable activity.

$p_s > c_p$ is the nontriviality condition because otherwise the operation is not economical.

An example of costs and profit is provided in Exercise 10. The decision problem is to find the optimal target capacity a^* that maximizes the expected profit from the operation cycle.

13.4.4 Optimal Capacity Adaptation Rule

With the above reinterpretation of costs and profit, the marginal opportunity losses, defined first in expression (13.4), are redefined thusly:

$$\lambda_o = c_p + c_h,$$
$$\lambda_u = p_s - c_p, \tag{13.28}$$

where λ_o is the *marginal opportunity loss from overcapacity* (unit building cost plus unit operating cost), and λ_u is the *marginal opportunity loss from undercapacity* (unit utilization profit minus unit building cost).

The ratio λ_o/λ_u of the marginal opportunity losses from overcapacity to undercapacity, and the distribution function G of the demand W, specify the optimal solution (equation (13.27)). It is reinterpreted for the current decision problem as the *optimal target capacity when expanding*:

$$a^* = G^{-1}\left(\frac{1}{1 + \lambda_o/\lambda_u}\right). \tag{13.29}$$

Again, a^* is independent of s, the current capacity.

Finally, the branch choice theorem (Section 13.3.6) implies the following *optimal capacity adaptation rule*:

$$\text{If } s \begin{Bmatrix} < \\ = \\ > \end{Bmatrix} a^*, \text{ then } \begin{cases} \text{expand by } a^* - s \text{ units,} \\ \text{maintain current capacity } s, \\ \text{shrink by } s - a^* \text{ units.} \end{cases}$$

13.4.5 Optimal Reliability

The optimal solution to the capacity planning problem has the same three important properties which the optimal solution to the basic inventory problem has (Section 13.2.5). In particular, associated with the optimal capacity a^* is the optimal reliability — the probability that the demand will be met (equivalently, the probability that the demand for product, service, or transport will not exceed the capacity of the system):

$$P(W \le a^*) = G(a^*) = \frac{1}{1 + \lambda_o/\lambda_u}. \tag{13.30}$$

Again, the optimal reliability depends on the economic factors — costs and profit that yield the ratio λ_o/λ_u of the marginal opportunity losses.

13.5 Inventory and Macroeconomy

"Warehouses offer insight into business sentiment". This headline in *The Wall Street Journal* (10 April 2012) underscores the significance of inventory management and capacity planning beyond an individual company. To wit, the sum of inventories across all companies in a sector (e.g., manufacturing of home appliances, production of crude oil) is a proxy measure of that sector's economic state — "a barometer of economic health". But to understand this state, which may be a precursor of the economy's *expansion* or *contraction*, one must diagnose its cause. That task in itself involves many uncertainties.

A buildup of inventories may occur because (i) the demand in the recent past has been weaker than planned for, or (ii) the demand is anticipated to increase in the near future, or (iii) the prices of raw materials are expected to rise, or (iv) the companies managed their inventories nonoptimally, and so on.

A reduction of inventories may occur because (i) the demand in the recent past has been stronger than planned for, or (ii) the demand is anticipated to decrease in the near future, or (iii) the prices of raw materials are expected to decline, or (iv) the companies have increased sales before replenishing inventories in order to raise cash (to improve the balance sheet before quarter's end), and so on.

Admittedly, inventory management and capacity planning are dynamic decision problems with many uncertainties. The decision models which we have studied offer fundamental concepts for framing the decisions that a company must make in the face of uncertainty about the future demand.

13.6 Concepts and Proofs

Expressions needed for proving the theorem ***Graph of VPF versus*** λ_o/λ_u of Section 13.2.7 are: equation (13.9),

$$a^* = G^{-1}\left(\frac{\lambda_u}{\lambda_u + \lambda_o}\right) = G^{-1}\left(\frac{1}{1 + \lambda_o/\lambda_u}\right);$$

equation (13.11) in three alternative forms,

$$G(a^*) = \frac{\lambda_u}{\lambda_u + \lambda_o}; \qquad \lambda_u + \lambda_o = \frac{\lambda_u}{G(a^*)}; \qquad \lambda_u = \lambda_o\frac{G(a^*)}{1 - G(a^*)};$$

and equation (13.14) with L^* replaced by expression (12.15),

$$VPF = \lambda_u E(W) - (\lambda_u + \lambda_o)\int_{\eta_L}^{a^*} wg(w)\,dw.$$

Proof of Part 2. To prove the limits, observe that $\lambda_o \to 0$ implies

$$\frac{\lambda_u}{\lambda_u + \lambda_o} \to 1,\ G(a^*) \to 1,\ a^* \to \eta_U,\ \int_{\eta_L}^{a^*} wg(w)\,dw \to E(W);$$

consequently, in the limit, $VPF = \lambda_u E(W) - (\lambda_u + 0)\,E(W) = 0$.

To prove that VPF increases strictly on the interval $(0, 1]$ of λ_o/λ_u, observe that a^* is a strictly decreasing function of λ_o/λ_u, with the range $[w_{0.5}, \eta_U)$. Therefore, it suffices to show that VPF is a strictly decreasing function of a^* on $[w_{0.5}, \eta_U)$. This is done in three steps. (i) Fix λ_u, and express VPF as a function of a^*:

$$VPF = \lambda_u\left[E(W) - \frac{1}{G(a^*)}\int_{\eta_L}^{a^*} wg(w)\,dw\right]. \tag{13.31}$$

(ii) Find the first derivative:

$$\frac{dVPF}{da^*} = \frac{\lambda_u\,g(a^*)}{G(a^*)}\left[\frac{1}{G(a^*)}\int_{\eta_L}^{a^*} wg(w)\,dw - a^*\right]. \tag{13.32}$$

(iii) Show that $dVPF/da^* < 0$ for every $a^* \in [w_{0.5}, \eta_U)$. This is true if the term within the brackets is negative; equivalently, if

$$\int_{\eta_L}^{a^*} wg(w)\,dw < a^*\,G(a^*) = \int_{\eta_L}^{a^*} a^*g(w)\,dw;$$

the inequality holds because $w < a^*$ for every $w \in (\eta_L, a^*)$. QED.

Proof of Part 3. To prove the limits, observe that $\lambda_u \to 0$ implies

$$\frac{\lambda_u}{\lambda_u + \lambda_o} \to 0, \; G(a^*) \to 0, \; a^* \to \eta_L, \; \int_{\eta_L}^{a^*} wg(w)\,dw \to 0;$$

consequently, in the limit, $VPF = 0 \cdot E(W) - (0 + \lambda_o) \cdot 0 = 0$.

To prove that VPF decreases strictly on the interval $[1, \infty)$ of λ_o/λ_u, observe that a^* is a strictly decreasing function of λ_o/λ_u, with the range $(\eta_L, w_{0.5}]$. Therefore, it suffices to show that VPF is a strictly increasing function of a^* on $(\eta_L, w_{0.5}]$. This is done in three steps. (i) Fix λ_o, and express VPF as a function of a^*:

$$VPF = \lambda_o \frac{G(a^*)}{1 - G(a^*)} \left[E(W) - \frac{1}{G(a^*)} \int_{\eta_L}^{a^*} wg(w)\,dw \right]$$

$$= \frac{\lambda_o}{1 - G(a^*)} \left[E(W)G(a^*) - \int_{\eta_L}^{a^*} wg(w)\,dw \right]. \tag{13.33}$$

(ii) Find the first derivative:

$$\frac{dVPF}{da^*} = \frac{\lambda_o\, g(a^*)}{(1 - G(a^*))^2} \left[E(W) - a^*(1 - G(a^*)) - \int_{\eta_L}^{a^*} wg(w)\,dw \right]. \tag{13.34}$$

(iii) Show that $dVPF/da^* > 0$ for every $a^* \in (\eta_L, w_{0.5}]$. This is true if the term within the brackets is positive. Expressing $E(W)$ as a sum of two integrals on $(\eta_L, a^*]$ and (a^*, η_U), and $1 - G(a^*)$ as an integral on (a^*, η_U), yields the inequality

$$\int_{\eta_L}^{a^*} wg(w)\,dw + \int_{a^*}^{\eta_U} wg(w)\,dw - a^* \int_{a^*}^{\eta_U} g(w)\,dw - \int_{\eta_L}^{a^*} wg(w)\,dw > 0,$$

which simplifies to

$$\int_{a^*}^{\eta_U} wg(w)\,dw > \int_{a^*}^{\eta_U} a^* g(w)\,dw;$$

this inequality holds because $w > a^*$ for every $w \in (a^*, \eta_U)$. QED.

Proof of Part 1. First, when $\lambda_o = \lambda_u = \lambda$, equation (13.9) yields $a^* = G^{-1}(1/2) = w_{0.5}$. At this point, the left derivative (13.34) is positive, and the right derivative (13.32) is negative. Hence, VPF attains a maximum at $w_{0.5}$. To prove that this maximum is unique, show that VPF is continuous at $w_{0.5}$ by setting $\lambda_o = \lambda_u = \lambda, a^* = w_{0.5}$, and $G(a^*) = 1/2$ in each of the two pieces of VPF: piece (13.33) on $(\eta_L, w_{0.5}]$ is continuous from the left at $w_{0.5}$, with the value

$$VPF^- = 2\lambda \left[E(W)(1/2) - \int_{\eta_L}^{w_{0.5}} wg(w)\,dw \right];$$

piece (13.31) on $[w_{0.5}, \eta_U)$ is continuous from the right at $w_{0.5}$, with the value

$$VPF^+ = \lambda \left[E(W) - 2 \int_{\eta_L}^{w_{0.5}} wg(w)\,dw \right].$$

Hence, $VPF^- = VPF^+ = VPF^*$, which is the unique global maximum of VPF on the interval (η_L, η_U) of the optimal decision a^*; or, equivalently, on the interval $(0, \infty)$ of the ratio λ_o/λ_u. QED.

Exercises

1 **Value of perfect forecaster**. Consider the value of the perfect forecaster, VPF, of the demand W in the basic inventory model.

    1.1 Starting from equation (13.14), derive the equation

$$VPF = \lambda_u \left[E(W) - \left(1 + \frac{\lambda_o}{\lambda_u}\right) \int_{\eta_L}^{a^*} w g(w)\, dw \right].$$

    1.2 Show every step of differentiation $dVPF/da^*$ that leads to equation (13.32) for $a^* \in [w_{0.5}, \eta_U)$ and to equation (13.34) for $a^* \in (\eta_L, w_{0.5}]$.

2 **Sensitivity analysis**. Consider the above expression for the value of the perfect forecaster, VPF, of the demand W in the basic inventory model. Suppose W is in thousands of units per week.

    2.1 Given the distribution of W specified in the option, write the expression for a^*, and derive the expression for VPF in the simplest parametric form. *Hint*: make use of the solution for the incomplete expectation, Section 12.3.

    2.2 Given the above expressions, prove the limits of a^* and of VPF, as they are stated in parts 2 and 3 of the theorem of Section 13.2.7.

    2.3 Given the parameter value of the distribution of W specified in the option, graph VPF as a function of the ratio of the marginal opportunity losses λ_o/λ_u, when $\lambda = \$3/\text{unit}$.

    Hint. First, set $\lambda_u = 3$, and vary λ_o in small increments over the interval $(0, 3]$; second, set $\lambda_o = 3$, and vary λ_u in small increments over the interval $(0, 3]$. Choose an interval for the domain of the graph so that the convergence of VPF to zero is apparent at both ends.

    2.4 Summarize the sensitivity of VPF to the ratio λ_o/λ_u. In particular: (i) characterize in words the shape of the graph; (ii) explain the relationship between VPF and the asymmetry, or skew, of the two-piece linear opportunity loss function.

Options for the distribution:		*Options for the parameter*:	
(UN)	$W \sim \text{UN}(0, 1)$.	(A)	$\alpha = 0.5$, or $\beta = 0.5$.
(EX)	$W \sim \text{EX}(\alpha, 0)$.	(B)	$\alpha = 1$, or $\beta = 2$.
(P1)	$W \sim \text{P1}(\beta, 0, 1)$.	(C)	$\alpha = 4$, or $\beta = 6$.
(P2)	$W \sim \text{P2}(\beta, 0, 1)$.		

    Note. The graphs of the distribution functions and the density functions of type EX, P1, P2 with the above parameter values are shown in Appendix C.

3 **Branch choice theorem**. Prove this theorem. (It provides the basis for the optimal ordering rule in the single-period inventory model with initial stock level.)

4 **Imputed loss ratio**. Having received a probabilistic forecast of the demand W for the beach umbrellas popular during summer, the manager of an oceanfront store examined the distribution function of W, which is specified in the option, and decided intuitively to order 570 umbrellas. Assuming that the profit function is two-piece linear, find the ratio of the marginal opportunity losses λ_o/λ_u under which the manager's order would be optimal in the sense of maximizing the expected profit from umbrella sales.

$$
\begin{aligned}
\text{Options: (WB)} \quad & W \sim \text{WB}\,(241, 3.2,\ 340). \\
\text{(IW)} \quad & W \sim \text{IW}(196, 2.8,\ 280). \\
\text{(LW)} \quad & W \sim \text{LW}\,(5.9,\ 4.6,\ 410). \\
\text{(LL)} \quad & W \sim \text{LL}\,(263, 1.7,\ 230).
\end{aligned}
$$

5 Daily batch size. A bakery has a contract to deliver 250 baguettes to a supermarket each morning. It also sells the baguettes in its own store, where the demand varies from day to day. This demand, between the time the store opens and the mid-afternoon, is never smaller than 30 and never greater than 90. Baguettes not sold by the mid-afternoon are processed into bread crumbs. A baguette costs $0.98 to produce, sells for $1.37 when delivered on contract, and retails for $1.69; it yields 0.2 kg of crumbs which sell for $2.75/kg; the cost of processing and packaging the crumbs is $0.35/kg. The uncertain part of the demand, W, has a distribution as specified in the option.

Options for the distribution: *Options for the parameter*:

(UN)	$W \sim \text{UN}\left(\eta_L, \eta_U\right).$	(A)	$\beta = 0.5,$
(P1)	$W \sim \text{P1}\left(\beta, \eta_L, \eta_U\right).$	(B)	$\beta = 2,$
(P2)	$W \sim \text{P2}\left(\beta, \eta_L, \eta_U\right).$		

(C)	$\beta = 4,$
(D)	$\beta = 5.$

Note. The graphs of the distribution functions and the density functions of type P1, P2 with the above parameter values are shown in Section C.4.

5.1 How many baguettes should be baked each morning in order to maximize the expected profit? (Illustrate the answer on a graph of the distribution function of the demand.)

5.2 What is the maximum expected profit per day? *Hint*: Make use of an expression for the incomplete expectation, Section 12.3.

5.3 How much at most should the bakery be willing to pay, each morning, for a perfect forecast of the daily demand?

5.4 What is the probability that a customer can find a baguette in the bakery store just before the mid-afternoon?

6 Daily newspaper order. A bookstore also sells a daily newspaper, for which the demand varies from day to day, but is never smaller than 230 and never greater than 390. The bookstore orders the newspaper from the publisher at $1.15 per copy, sells it for $1.50 per copy, and sends the unsold copies to a recycling center, which pays $0.02 per pound of paper. On average, 27 newspaper copies make 1 pound.

Options: the same as in Exercise 5.

6.1 How many newspaper copies should be ordered each day in order to maximize the expected profit? (Illustrate the answer on a graph of the distribution function of the demand.)

6.2 What is the maximum expected profit per day? *Hint*: Make use of the expression for the incomplete expectation, Section 12.3.

6.3 How much at most should the bookstore be willing to pay, each day, for a perfect forecast of the daily demand?

6.4 What is the probability that a customer can find a newspaper in the bookstore just before it closes in the evening?

7 Restaurant operation planning. A chef-owner of a 200-seat restaurant in a seaside resort wants to determine the number of diners he should plan for each evening. The restaurant is always full during the summer months when he hires staff to help him, and is always near empty during the winter months when he and his family can handle all tasks. These are the off-season holidays (Labor Day, Columbus Day, Thanksgiving Day,

New Year's Eve, President's Day, Easter Sunday, Mother's Day) that are the most difficult to prepare for. While a fair weather forecast usually increases the demand, the uncertainty remains huge. You have been hired to structure this decision problem.

7.1 List the key decisions the owner must make and the lead time needed to implement each.

7.2 List the major consequences of overpreparing and underpreparing.

7.3 Define the costs and prices. Write down the criterion function (Section 13.2.3), first, as the profit per evening and, second, as the opportunity loss per evening.

7.4 List the items of information that you would collect in order to produce a probabilistic forecast of the number of diners on a particular holiday; consider making both, a judgmental forecast and a statistical forecast. What should be the lead time of the forecast?

7.5 Estimate the time you would need to perform the analysis that would determine the optimal number of diners the restaurant owner should plan for on a particular holiday.

8 **Restaurant operation planning** — *continued.* An analysis performed in Exercise 7 produced the following estimates per evening of restaurant operation during an off-season holiday: $48 is the average bill paid per diner (it includes a tip); $19 is the average cost of making one seat serviceable and servicing it if occupied (it includes wages of the cook, server, kitchen and janitorial staff, etc.); $21 is the cost of preparing a meal per diner and cleaning afterwards (it includes the costs of ingredients, energy for cooking a meal and washing dishes, etc.). On a holiday evening, a seat is planned for one diner only. Ten days before the approaching Mother's Day, the restaurateur prepared judgmentally a forecast of the number of diners. The forecast is in terms of quantiles specified in the option, and is based on his experience from years past, the state of regional economy, and the weather forecast for this seaside resort retrieved from the National Weather Service website. He has hired you to complete the decision analysis.

Hints. (i) Formulate the criterion function: first, as the profit per evening; second, as the opportunity loss per evening.

  (ii) Estimate the distribution parameters via the LS method.

  (iii) Calculate the incomplete expectation via a formula for the specified power type distribution, or via a numerical method for a log-ratio type distribution.

Options for the distribution:

(P1) Power Type I.   (LR-LG) Log-ratio logistic.
(P2) Power Type II.   (LR-LP) Log-ratio Laplace.
                         (LR-GB) Log-ratio Gumbel.
                         (LR-RG) Log-ratio reflected Gumbel.

Options for the quantiles $(y_{0.25}, y_{0.5}, y_{0.75})$ *under a P1 distribution*:

(A) (14, 50, 109);   (B) (80, 123, 164);   (C) (118, 150, 177).

Options for the quantiles $(y_{0.25}, y_{0.5}, y_{0.75})$ *under a P2 distribution*:

(D) (14, 32, 57);   (E) (27, 58, 100);   (F) (90, 148, 186).

Options for the quantiles $(y_{0.2}, y_{0.5}, y_{0.8})$ *under a log-ratio distribution*:

(K) (43, 127, 156);   (L) (45, 74, 108);
(M) (27, 145, 192);   (N) (8, 55, 173).

8.1 For how many diners should the restaurateur plan on Mother's Day evening in order to maximize the expected profit? (Illustrate the answer on a graph of the distribution function of the demand.)

8.2 What is the maximum expected profit per evening?

8.3 How much at most should the restaurateur be willing to pay for a perfect forecast of the number of diners that evening?

8.4 What is the probability that the last customer calling on Mother's Day evening will have her request for a reservation accepted? Rationalize this probability on economic grounds. (Why is it high, or why is it low?)

9 **Parcel delivery capacity**. During the nonpeak weeks, the average on-time delivery rate for packages shipped by ground via United Parcel Service (UPS) is 0.97. During the week beginning on Thanksgiving Day in 2015, the on-time delivery rate dropped to 0.91 because the online sales of merchandise surged unexpectedly and, consequently, the volume of packages shipped by retailers to consumers was higher than the volume UPS had planned for. While managers from the headquarters in Atlanta were dispatched to the processing centers afield to help man the operations, UPS still lacked the capacity (i.e., staff and equipment) to deliver all packages on time (*The Wall Street Journal*, 11 December 2015).

Note. The capacity can be expanded permanently by investing in infrastructure (e.g., larger centers, modernized equipment, more trucks). It can be expanded temporarily during peak periods by hiring seasonal workers, extending delivery hours, or increasing the transit time promised to retailers, say, from 3 to 4 days; however, even a temporary expansion must be planned with lead time sufficient for implementation.

9.1 Consider the task of determining the optimal capacity (in terms of the number of packages that can be delivered on time) of a parcel service during the Thanksgiving holiday week. Specifically: (i) Interpret all elements of the capacity planning model for this decision problem. (ii) Sketch a distribution function of the demand and indicate on it the capacity a_1^* that ensures the reliability (the on-time delivery rate) of 0.97. (iii) Infer the implied ratio λ_o/λ_u of the marginal opportunity losses and interpret it.

9.2 Explain the situation faced by UPS during Thanksgiving in 2015. Specifically: (i) Keeping a_1^* as determined in Exercise 9.1, sketch a new distribution function of the demand on which this a_1^* yields the reliability of 0.91. (ii) If this were the preferred reliability, then what would be the implied ratio λ_o/λ_u of the marginal opportunity losses? (iii) If the new distribution function of the demand were used in the analysis, what would be the capacity a_2^* needed to ensure the reliability of 0.97? (Indicate a_2^* on the sketch.) (iv) If the goal of the analysis in Exercise 9.1 was to find a^* that ensures the reliability of 0.97, but the actual reliability turned out to be 0.91, then what went wrong in that analysis? (Offer an explanation, supposing that the distribution function of the demand was assessed judgmentally by a forecaster using the methodology of Chapter 9.)

10 **Cinema capacity**. The owner of a 40-seat cinema has just learned that a large corporation will relocate into town, doubling the population in 2 years. Should she expand the cinema's capacity to profit from the boom? If so, then a builder would complete the design and construction within 2 years, and she would want to recoup the investment in the following 8 years (planning period) while running three screenings per day (operation cycle), 300 days per year ($8 \times 300 = 2400$ operation cycles in the planning period). To apply the capacity planning model you need to estimate: c_p — the cost of design and construction per seat-day (if amortized linearly, it is the total cost per seat divided by 2400); c_h — the cost of daily operation per seat (e.g., the costs of heating and cooling, janitor, box office cashier, projector operator, plus movie rental fee); and p_s — the profit from a seat occupied during every screening per day (e.g., the revenue from three tickets minus the cost of daily operation per seat). The owner supplied these estimates for a 50-seat addition:

design and construction	$345 000;
heating and cooling	$4200/year;
janitor (wage plus benefits)	$19/h × 3 h/day;

box office cashier	\$32/h × 8 h/day;
projector operator	\$36/h × 8 h/day;
movie rental fee	\$200/screening;
price of a ticket	\$14.50.

A demographer analyzed the record of ticket sales in years past and concluded that the distribution function of the daily demand for tickets is of the type specified in the option. Next, taking into account the projected growth of the town's population, he followed the methodology of Chapter 9 and assessed judgmentally the lower bound $\eta = 45$, the median $y_{0.5} = 225$, and the upper quartile $v_{0.75} = 375$.

Assumption. The demand for tickets on any day is divided equally among three screenings.

Hints. (i) Use the estimates supplied by the owner to calculate c_p, c_h, p_s. (ii) Ensure that the units of these parameters and of all variables are consistent. (iii) Consult Section 12.3 for the incomplete expectation.

Options: (WB) Weibull.
            (IW) Inverted Weibull.
            (LW) Log-Weibull.
            (LL) Log-logistic.

10.1 How many seats should be added to the cinema in order to maximize the expected profit? (Illustrate the answer on a graph of the distribution function of the demand.)

10.2 What is the maximum expected profit per day?

10.3 What is the probability that the last cinemagoer of the day will find a ticket still available?

10.4 How much at most should the cinema owner be willing to pay for a perfect forecast of the daily demand? *Note*: The benefit of this forecast will accrue over the 2400 operation cycles in the planning period.

11 **Airplane capacity**. An airline wants to determine the optimal capacity of an airplane for a particular 5-hour flight operated daily on a fixed route between two cities. The demand for tickets is variable (from day to day), but its distribution is stationary (across the days). The economic parameters are: c_p — the capital cost of providing a seat for one 5-hour flight, calculated from the airplane purchase price plus the lifetime maintenance cost, amortized over the airplane lifetime (the total number of flying hours), and scaled to 5 hours; c_h — the operating cost per seat of one flight (e.g., the costs of fuel, crew, janitors, passenger services); p_t — the price of a ticket for one flight; and p_s — the operating profit from an occupied seat on one flight $(p_s = p_t - c_h)$. The parameter values, the assessed quantiles of the daily demand for tickets, and the distribution type are specified in the options.

Hints. (i) Estimate the distribution parameters via the methodology of Chapter 9. (ii) Consult Section 12.3 for the incomplete expectation.

Options for the costs:

(C1) $c_p = \$441$, $c_h = \$307$.
(C2) $c_p = \$516$, $c_h = \$298$.
(C3) $c_p = \$495$, $c_h = \$392$.

Options for the demand quantiles:

(D1) $\eta = 30$, $y_{0.5} = 75$, $y_{0.9} = 135$.
(D2) $\eta = 50$, $y_{0.5} = 85$, $y_{0.9} = 155$.
(D3) $\eta = 70$, $y_{0.5} = 95$, $y_{0.9} = 175$.

Options for the distribution:

(WB)   Weibull.
(IW)   Inverted Weibull.
(LW)   Log-Weibull.
(LL)   Log-logistic.

11.1 What is the threshold ticket price below which the flight on a particular route is not economical? Set the ticket price p_t to $200 above that threshold.

11.2 How many passenger seats should the airplane have in order to maximize the expected profit per flight? (Illustrate the answer on a graph of the distribution function of the demand.)

11.3 What is the maximum expected profit per flight?

11.4 What is the probability that a traveler will find a ticket still available just before the ticket counter closes for a flight on a particular day?

11.5 If before each flight it were feasible to choose an airplane whose capacity matches the demand for tickets on that day, how much at most should the airline be willing to pay for a perfect forecast of the daily demand?

11.6 Suppose an airplane can be purchased with the number of passenger seats equal to any positive integer divisible by 30. (i) Recommend the capacity of the airplane to be purchased. (ii) Calculate the expected profit per flight from the operation of this airplane. (iii) Determine the expected opportunity loss due to the infeasibility of purchasing an airplane with the optimal capacity. (iv) Is this opportunity loss significant to an airline that operates 500 similar airplanes on different routes? Support your answer with an argument. *Note*: The list price of a new Boeing 737 Max jet in 2018 was about $96 million.

Mini-Projects

12 **Strategic production planning**. A company plans a 2-year production of a new model of the personal computer (PC) which will be sold on the domestic market and on a foreign market. The demand in each of the markets during the 2-year period is uncertain. An expert forecasts this demand in terms of quantiles as follows.

	Domestic demand	Foreign demand
median	37 000	54 000
0.25-quantile	33 000	45 000
0.75-quantile	41 000	62 000
0.125-quantile	—	42 000
0.875-quantile	—	69 000

The production cost is estimated to be $490 per PC, which will retail for $990. A PC will be shipped from the storage facility directly to the buyer upon receipt of an order. Any PC unsold by the end of the 2-year period (when a new model will be introduced) will be advertised for $290, a price that surely will attract a buyer.

Assumption. The demand in each of the markets has a normal distribution and is stochastically independent of the demand in the other market.

12.1 For each market, estimate the parameters of the distribution function of the demand; graph the assessed quantiles and the estimated distribution function; appraise the pros and cons of the assumption that this distribution function is normal.

12.2 Consider a *decentralized* production planning wherein each market is analyzed separately. For each market, determine the values of the three planning variables:

12.2.1 the optimal production size;

12.2.2 the maximum expected profit;

12.2.3 the value of a perfect forecast of the demand.

Illustrate the solution for the optimal production size on the graph of the distribution function of the demand.

12.3 Aggregate the two separate production plans. To wit, for each of the three planning variables listed in Exercise 12.2, obtain the total value by summing the values from the two plans.

12.4 Consider a *centralized* production planning wherein the total demand equals the sum of the demands in the two markets.

12.4.1 Estimate the parameters of the distribution function of the total demand; graph this distribution function.

12.4.2 Determine the values of the three planning variables listed in Exercise 12.2.

Illustrate the solution for the optimal total production size on the graph of the distribution function of the total demand.

12.5 Compare the two approaches to production planning, the one followed in Exercises 12.2–12.3 where the central plan is obtained through the *aggregation of optimal production plans* developed separately for each market, and the one followed in Exercise 12.4 where a *single optimal production plan* is developed centrally for both markets.

12.5.1 Contrast the results and explain the differences.

12.5.2 Draw conclusions.

13 **Strategic production planning** — *continued*: **correlated demands**. Revisit Exercise 12 above. Suppose that based on past experience with computer products in the two markets, the company anticipates that the demand for a PC in the foreign market will be strongly correlated with the demand for the same PC in the domestic market.

Modified assumption. The demands in the two markets have a joint distribution which is bivariate normal.

13.1 An expert assessed the correlation coefficient between the demands in the two markets to be 0.8.

13.1.1 Repeat Exercise 12.4 under the modified assumption.

13.1.2 Compare the results obtained here with the results of Exercise 12.4.

13.2 An expert assessed the correlation coefficient between the demands in the two markets to be −0.8.

13.2.1 Repeat Exercise 12.4 under the modified assumption.

13.2.2 Compare the results obtained here with the results of Exercise 12.4.

13.3 Compare the results of Exercises 12.4, 13.1, and 13.2; draw conclusions.

14 **Seasonal irrigation planning**. A farmer from Idaho hires you as a consultant. He wants to plan crop production so as to maximize his expected profit. Essential to the crop production is irrigation water that he takes from a tributary of the Snake River. Crops must be irrigated from the beginning of April to the end of July. The farmer has been recording the volume of water flowing through the river gauging station during this 4-month period in years past. Under his riparian right, he can take from the river 1% of the actual water volume. Also from past experience he knows that if the volume of water he can actually take from the river equals exactly the volume he planned for, then his profit from crop production is $150 per acre-foot of water. If the actual water supply is larger than the volume he planned for, then each acre-foot of surplus brings him $80 of profit. If the actual supply is smaller than the volume he planned for, then each acre-foot of deficit reduces his profit by $30, in relation to the profit he would have made if neither deficit nor surplus occurred.

Options for station and record length (Tables 10.6 and 10.8):

(W)   Weiser on the Weiser River.
      (13) Record from 13 years, 1971–1983.
      (21) Record from 21 years, 1963–1983.
      (31) Record from 31 years, 1953–1983.

(T)   Twin Springs on the Boise River.
      (13) Record from 13 years, 1971–1983.
      (21) Record from 21 years, 1963–1983.
      (33) Record from 33 years, 1951–1983.
      (72) Record from 72 years, 1912–1983.

Assumption. The seasonal water volume has (approximately) a normal distribution. If you estimated this distribution in Exercise 18 of Chapter 10, then use it here (with the same options for the station and the record length).

14.1 Formulate a mathematical model of this decision problem.

14.2 Determine the optimal volume of water the farmer should use in crop production planning (which he does in the fall of the preceding year), and the maximum expected profit per year.

14.3 Determine how much, at most, the farmer should be willing to pay annually for the construction of a storage reservoir that would provide him every year with the same volume of water, equal to the 1% of mean flow in the river during April–July. *Note*: Be mindful of the units.

15 **Seasonal irrigation planning with forecast** — Chapter 10, Exercise 18 *continued*. A farmer faces the decision problem described in Exercise 14 above; however, instead of using a distribution function of the seasonal water volume estimated from measurements recorded in years past, he wants to use a probabilistic forecast of the seasonal snowmelt runoff volume prepared by the Bayesian Gaussian forecaster you developed in Exercise 18 of Chapter 10 (with the same options for the station, the prior sample, and the forecast lead time).

15.1 Formulate a mathematical model of this decision problem.

15.2 Determine the optimal volume of water the farmer should use in crop production planning, and the maximum expected profit, in a year in which the predictor realization (an estimate of the runoff volume produced by hydrologists) is (i) $x_1 = M - 100$, and (ii) $x_2 = M + 100$, where M is the prior mean of the runoff volume.

15.3 Determine the value of the perfect forecast of the runoff volume in a year in which the predictor realization is (i) $x_1 = M - 100$, and (ii) $x_2 = M + 100$, where M is the prior mean of the runoff volume.

16 **Seasonal irrigation planning, farmer B**. Complete Exercise 14 with the following modification to the profit function. If the volume of water actually available to the farmer under his riparian right equals exactly the volume of water he planned for, then his profit from crop production is $250 per acre-foot of water. Otherwise, his profit is reduced by $40 for every [ac-ft]^2 of the difference between the actually available volume and the planned volume. (Organize your report into Sections 16.1, 16.2, 16.3, paralleling the tasks specified in Exercise 14.)

17 **Seasonal irrigation planning with forecast, farmer B** — Chapter 10, Exercise 18 *continued*. Complete Exercise 15 with the modified profit function detailed in Exercise 16. (Organize your report into Sections 17.1, 17.2, 17.3, paralleling the tasks specified in Exercise 15.)

14

Investment Models

Among an individual's unalienable rights (to life, liberty, and the pursuit of happiness), the right to liberty implies, *inter alia*, the right to spend time as one wishes, and this includes doing work whose fruits can be exchanged for money. Money saved and accumulated forms capital. Hence, capital is accumulated work. Hence, one has the right to keep it, to spend it, or to risk multiplying it through investment that, in turn, creates opportunities for others to work and, thereby, to fulfill their rights to liberty. In a nutshell, it is moral to make money.

Rational investment decisions should consider (i) uncertainties about outcomes (losses or gains), which should be identified and quantified, and (ii) preferences for outcomes, which should be uncovered and assessed. This chapter presents fundamentals of building mathematical models to guide investment decisions.

14.1   Investment Choice Problem

14.1.1   Elements

The basic investment choice problem is a decision problem under uncertainty with the following elements.

a_k — investment alternative, one of K alternatives available from the set

$$\mathcal{A} = \{a_k : k = 1, \dots, K\}.$$

Z — outcome variate, discrete or continuous, with the sample space $\mathcal{Z}$ being either finite or an interval (bounded or unbounded); its realization $z \in \mathcal{Z}$ is a monetary amount representing *loss* or *gain* relative to the *status quo* (the current assets of the decider):

$$z < 0 \qquad \text{loss},$$

$$z = 0 \qquad \text{status quo},$$

$$z > 0 \qquad \text{gain}.$$

G_k — distribution function of outcome Z from alternative a_k, such that for any realization $z \in \mathcal{Z}$,

$$G_k(z) = P(Z \le z | a_k);$$

there are K distribution functions, which form a set

$$\mathcal{G} = \{G_k : k = 1, \dots, K\}.$$

Probabilistic Forecasts and Optimal Decisions, First Edition. Roman Krzysztofowicz.
© 2025 John Wiley & Sons Ltd. Published 2025 by John Wiley & Sons Ltd.
Companion website: www.wiley.com/go/ProbabilisticForecastsandOptimalDecisions1e

The decision problem is to rank the investment alternatives according to one's preferences, and to choose the most preferred alternative, when the outcome Z from each alternative a_k is uncertain, with the uncertainty being quantified in terms of a distribution function G_k. Inasmuch as there is a one-to-one correspondence between alternative a_k and distribution function G_k, choosing an alternative from the set $\mathcal{A}$ is equivalent to choosing a distribution function of outcome from the set $\mathcal{G}$.

14.1.2 Binary Relations

To talk about, and to model, the preferences of a decider, it is advantageous to introduce several symbols. First, with a, b being any two statements, define two *sentential connective* symbols:

$\implies$ conditional symbol:
$a \implies b$ reads "if a, then b "or" a implies b";

$\iff$ biconditional symbol:
$a \iff b$ reads "a if and only if b".

Second, with $z_1, z_2 \in \mathcal{Z}$ being any two outcomes, define three *binary relation* symbols:

$\succ$ preference relation on $\mathcal{Z}$:
$z_1 \succ z_2$ reads "z_1 is preferred to z_2";

$\sim$ indifference relation on $\mathcal{Z}$:
$z_1 \sim z_2$ reads "z_1 is indifferent to z_2";

$\succsim$ preference–indifference relation on $\mathcal{Z}$:
$z_1 \succsim z_2 \iff z_1 \succ z_2$ or $z_1 \sim z_2$;
reads "z_1 is preferred or indifferent to z_2", or
"z_1 is at least as preferred as z_2", or
"z_2 is not preferred to z_1".

The three binary relations apply also to comparisons of alternatives from the set $\mathcal{A}$, distribution functions from the set $\mathcal{G}$, and special cases of distributions called gambles.

14.1.3 Gamble

When Z is a discrete variate, its probability function may be presented as a gamble. For instance, a two-outcome *gamble G* is depicted as

$$\left\langle \begin{array}{c} p_1, z_1 \\ p_2, z_2 \end{array} \right\rangle$$

and reads: you receive outcome z_1 with probability p_1 or outcome z_2 with probability p_2 ($p_1 + p_2 = 1$). The symbol G also denotes the distribution function of outcome Z. If $z_1 < z_2$, then for any $z \in (-\infty, \infty)$, the nonexceedance probability $G(z) = P(Z \leq z)$ is given by

$$G(z) = \begin{cases} 0 & \text{if } z < z_1, \\ p_1 & \text{if } z_1 \leq z < z_2, \\ 1 & \text{if } z_2 \leq z. \end{cases}$$

14.2 Stochastic Dominance Relation

14.2.1 Absolute Preference

Task 14.1 (***Absolute preference***).   You are offered an opportunity to play one of the two gambles. Which one do you choose? Circle your choice.

$$\begin{matrix} G_1 \\ \left\langle \begin{matrix} 0.3, \$800 \\ 0.3, \$300 \\ 0.4, \$400 \end{matrix} \right\rangle \end{matrix} \qquad \begin{matrix} G_2 \\ \left\langle \begin{matrix} 0.1, \$350 \\ 0.4, \$900 \\ 0.5, \$500 \end{matrix} \right\rangle \end{matrix}$$

Let us compare the two alternatives analytically. The order of the expected gains,

$$E(Z|a_1) = \$490 < E(Z|a_2) = \$645,$$

might explain the preference $G_1 \prec G_2$. But a stronger explanation can be inferred from the graphs of the distribution functions (Figure 14.1). They reveal that for every z,

$$P(Z > z|a_1) \leq P(Z > z|a_2)$$
$$1 - G_1(z) \leq 1 - G_2(z)$$
$$G_1(z) \geq G_2(z).$$

In other words, for every z, the probability of winning an amount larger than z is greater, or at least not smaller, under gamble G_2 than it is under gamble G_1. Visually, the distribution function G_2 lies to the right of, or overlaps, the distribution function G_1. This explains the *absolute preference*: upon reflection, **everyone** should prefer gamble G_2 over gamble G_1. A situation of this kind has a formal characterization.

Definition   Suppose the preference order increases on $\mathcal{Z}$ (i.e., a larger z is preferred to a smaller z). The distribution function G_2 *stochastically dominates* the distribution function G_1 if for every $z \in \mathcal{Z}$,

$$G_1(z) \geq G_2(z),$$

with strict inequality holding for some $z \in \mathcal{Z}$.

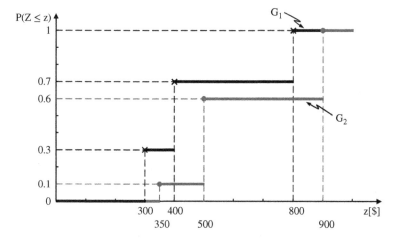

Figure 14.1   Distribution functions offered in Task 14.1; G_2 stochastically dominates G_1.

That stochastic dominance implies the absolute preference can be proven mathematically when an individual's strength of preferences for outcomes in $\mathcal{Z}$ is represented by a cardinal utility function v on $\mathcal{Z}$. Such a function is defined in Section 14.3. In the meantime let us state the theorem:

Theorem *(First-degree stochastic dominance)* *If the distribution function G_2 stochastically dominates the distribution function G_1, then G_2 has greater utility than G_1 has,*

$$U(G_1) < U(G_2),$$

*for **every** cardinal utility function v which is continuous, strictly increasing, and bounded on $\mathcal{Z}$, and which yields*

$$U(G_k) = U(a_k) = E[v(Z)|a_k], \quad k = 1, 2.$$

The proof is given in Section 14.7.3. The statement "for every cardinal utility function" is practically equivalent to the statement "for every rational decider". This makes the stochastic dominance relation useful for *screening* the investment alternatives. Given a set $\mathcal{G}$, every distribution function that is stochastically dominated by some other distribution function can be identified and discarded. This task can be performed by an analyst without asking the decider for his preferences. The set $\mathcal{G}^*$ of the remaining distribution functions is a subset of $\mathcal{G}$ and has a name.

Definition The *set of stochastically nondominated distributions* of outcome Z is a subset $\mathcal{G}^* \subseteq \mathcal{G}$ containing every distribution function that is not stochastically dominated by some other distribution function from $\mathcal{G}$.

To sum up, the set $\mathcal{G}$ can be dichotomized into a subset $\mathcal{G}^*$ of stochastically nondominated distributions and a subset $\mathcal{G} - \mathcal{G}^*$ of stochastically dominated distributions (the subset $\mathcal{G} - \mathcal{G}^*$ is the complement of $\mathcal{G}^*$ relative to $\mathcal{G}$). Further analysis toward ranking the investment alternatives and choosing the most preferred alternative needs to consider only the subset $\mathcal{A}^* \subseteq \mathcal{A}$ of alternatives whose distributions of outcome belong to the subset $\mathcal{G}^*$. But the analyst needs also to ask the decider for his preferences.

14.2.2 Individual Preference

Task 14.2 *(**Individual preference**).* You are offered an opportunity to play one of the two gambles. Which one do you choose? Circle your choice.

$$
\begin{matrix}
G_3 & & G_4 \\
\left\langle \begin{matrix} 0.5, \$250 \\ 0.1, \$950 \\ 0.4, \$700 \end{matrix} \right\rangle & & \left\langle \begin{matrix} 0.2, \$550 \\ 0.6, \$450 \\ 0.2, \$600 \end{matrix} \right\rangle
\end{matrix}
$$

Let us compare the two alternatives analytically. The equality of the expected gains,

$$E(Z|a_3) = \$500 = E(Z|a_4) = \$500,$$

does not explain the preference between the gambles, which may be $G_3 \prec G_4$ or $G_3 \succ G_4$. The graphs of the distribution functions (Figure 14.2) cross each other, implying that neither distribution function stochastically dominates the other. This is the situation in which the choice is a matter of *individual preference*.

When the four gambles from Tasks 14.1 and 14.2 are considered as a set of distribution functions

$$\mathcal{G} = \{G_1, G_2, G_3, G_4\}$$

offered by four investment alternatives $\mathcal{A} = \{a_1, a_2, a_3, a_4\}$, the simultaneous comparison of the four graphs (Figures 14.1–14.2) leads to the conclusion that (i) G_1 is stochastically dominated by G_2, and (ii) no stochastic

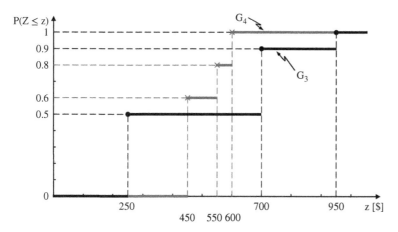

Figure 14.2 Distribution functions offered in Task 14.2; G_3 and G_4 cross each other.

dominance relation exists between any other two distribution functions; hence

$$G^* = \{G_2, G_3, G_4\}.$$

In general, the choice between any two alternatives from the set G^* of stochastically nondominated distributions requires an expression of individual preference. How to model an individual's preferences between distributions of outcomes is the next topic.

14.3 Utility Function

14.3.1 Purpose and Approach

The purpose of utility theory, a branch of decision theory, is to elicit and to model a decider's preferences in order to obtain a criterion for making decisions under uncertainty — decisions which are rational and optimal for that decider. To grasp the gist of the approach, place yourself in the position of a decider (an investor) and an analyst (a modeler) at the same time, and follow these three steps.

1. Contemplate several choices between **simple investment alternatives** (two-outcome gambles) and express your preference in several **easy choices** (between a sure outcome and a gamble). The expressed preferences between alternatives manifest your strength of preferences for outcomes from these alternatives.
2. Encode the implied *strength of preferences* for outcomes in a cardinal utility function v, which extrapolates a few measurements into the entire outcome space $\mathcal{Z}$.
3. Use the utility function v in the decision criterion of a mathematical model to make **difficult choices**: to rank many **complex investment alternatives** (continuous distribution functions of outcomes) and to choose the most preferred alternative (which is optimal for you), in a manner *consistent* with the easy choices you made earlier.

This approach is *normative* for it prescribes (i) how the strength of preferences expressed in easy choices with a few discrete outcomes should be extrapolated to the entire outcome space $\mathcal{Z}$, and (ii) how decisions under uncertainty should be made in accordance with the postulates of rationality (Appendix A).

The normative approach of decision theory must be distinguished from the *descriptive* approach of behavioral decision theory, a branch of cognitive psychology, whose purpose is to uncover, understand, and describe how individuals or groups do make decisions intuitively. Inasmuch as intuitive decisions frequently violate the postulates of rationality, utility theory and normative models should not be employed to predict human intuitive decisions.

14.3.2 Utility Theory

Utility theory derives from postulates of rationality (axioms of Bayesian decision theory — see Appendix A). The consequence of accepting the rationality postulates as principles for making decisions under uncertainty is two theorems. In them, $\mathcal{Z}$ is a bounded interval and $\mathcal{G}$ is a set of distribution functions on $\mathcal{Z}$.

Theorem (*Existence of utility*) *For any outcomes $z, z_1, z_2 \in \mathcal{Z}$ such that $z_1 \precsim z \precsim z_2$, (i) there exists a unique probability p ($0 \le p \le 1$), such that*

$$z \sim \left\langle \begin{matrix} p, & z_2 \\ 1-p, & z_1 \end{matrix} \right\rangle ; \tag{14.1}$$

(ii) there exists a utility function $v : \mathcal{Z} \to \mathfrak{R}$ such that if the above indifference relation holds, then

$$v(z) = pv(z_2) + (1-p)v(z_1); \tag{14.2}$$

(iii) the utility function v is a cardinal scale. (This scale is defined in Section 14.3.5. That this scale is sufficient is proven in Section 14.7.2.)

Definition Outcome z that satisfies relation (14.1) and equality (14.2) is called the *cash equivalent* of the gamble.

Relation (14.1) is a qualitative statement: an individual is indifferent between receiving outcome z with certainty and receiving outcome generated from the gamble, which yields outcome z_1 with probability $1 - p$ or outcome z_2 with probability p. Equality (14.2) is a quantitative model of relation (14.1): the utility of the cash equivalent equals the expected utility of the outcome generated from the gamble. Remarkably, this model is sufficient to establish a general criterion for decision making under uncertainty.

Theorem (*Expected utility criterion*) *For any distribution functions $G_1, G_2 \in \mathcal{G}$,*

$$G_1 \precsim G_2 \iff U(G_1) \le U(G_2), \tag{14.3}$$

where the utility of the distribution function is

$$U(G_k) = E[v(Z)|a_k], \qquad k = 1, 2. \tag{14.4}$$

Relations (14.1)–(14.2) provide the basis for assessing the utility function v, the subject of the following subsections. Relations (14.3)–(14.4) supply the criterion for making optimal decisions, the subject of Section 14.4.

14.3.3 Contextual Conditioning

A rational decider evaluates possible outcomes in a context (Figure 14.3). At a given time, he considers a decision problem (D), such as choosing an investment alternative. His preferences may be affected by a reference point (R), such as current assets or wealth (the *status quo*) and current circumstances (such as age, income, job security); aspirations (A), such as purchasing a new car, financing college education, or planning retirement annuities; and experience (E), such as past investments and their outcomes. As he considers the next investment, he should contemplate possible outcomes, form preferences, and express their strength.

The strength of preference is a primitive notion — undefined but assumed to be formable in one's mind, and conditional on (D, R, A, E). Consequently, a measurement of the strength of preferences for outcomes $z \in \mathcal{Z}$ is also conditional; this can be indicated by writing $v(z|D, R, A, E)$. Henceforth, the abbreviated notation will be employed,

$$v(z) = v(z|D, R, A, E),$$

remembering that the utility function v is always conditional on the current (D, R, A, E), and that a change in this context may necessitate a reassessment of v.

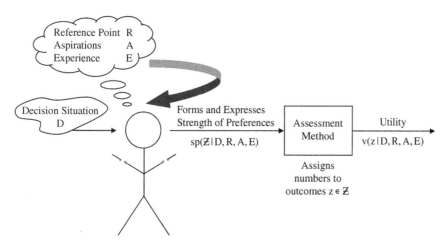

Figure 14.3 Judgmental task of assessing utility.

14.3.4 Strength of Preference

The preference is a *binary relation*. You compare two outcomes, say z_1 with z_2, and conclude that you prefer z_2 to z_1; in symbols, $z_1 \prec z_2$; in utilities, $v(z_1) < v(z_2)$.

The strength of preference is a *quaternary relation*. You compare two pairs of outcomes, say (z_1, z) with (z, z_2), such that $z_1 \prec z \prec z_2$. The task is facilitated by contemplating two *exchanges*. (i) Having currently at hand outcome z_1, you are given an opportunity to exchange z_1 for z. (ii) Having currently at hand outcome z, you are given an opportunity to exchange z for z_2. (iii) After which exchange would you be more elated? Which exchange would bring you greater satisfaction or pleasure? If it is the second exchange, then write

$$(z_1, z) \prec (z, z_2), \tag{14.5}$$

and interpret this quaternary relation thusly: your strength of preference for z_2 over z is greater than your strength of preference for z over z_1.

The strength of preference is modeled mathematically in terms of the difference of utilities. Hence, the above qualitative relation is modeled quantitatively as the inequality between utility differences:

$$v(z) - v(z_1) < v(z_2) - v(z). \tag{14.6}$$

Example 14.1 *(Mid-morning snack)*
Your mid-morning snack is a piece of fruit: an apple (A), or a banana (B), or a tangerine (T). Normally, your preference order is $B \prec T \prec A$. But today, only B was left in the refrigerator. (i) As you are about to eat B, a friend proposes to exchange her T for your B. Are you pleased? How much? (ii) As you are getting ready to eat T, another friend arrives and proposes to exchange her A for your T. Are you pleased? How much? (iii) Which exchange brought you greater pleasure? If it is the second one, then your strength of preference for A over T is greater than your strength of preference for T over B; in symbols,

$$(B, T) \prec (T, A) \iff v(T) - v(B) < v(A) - v(T).$$

Example 14.2 *(Scholarship)*
You applied for several scholarships (D), and while awaiting decisions, you identified the assets currently available to you (R) to finance the next year in college (A), and also reviewed the financial aid you had received last year (E). Under college rules, you may accept only one scholarship per year. Now contemplate this situation.

(i) A letter arrives with a scholarship offer of $5000 for the next year. Are you elated? How strongly? (ii) The next day, another letter arrives with a scholarship offer of $8000. Are you elated? How strongly? (iii) Which letter brought you a greater degree of elatedness? If it is the first one, then your strength of preference for $5000 over $0 (the status quo) is greater than your strength of preference for $8000 over $5000; in symbols,

$$(0, 5000) > (5000, 8000) \iff v(5000) - v(0) > v(8000) - v(5000).$$

14.3.5 Measurement of Utility

Definition A measurement scale is called a *cardinal scale* (or an *interval scale*) if its admissible transformation is a *positive linear transformation*,

$$y = \alpha x + \beta,$$

where x and y are the alternative scales, and $\alpha > 0$ and β are constants.

The cardinal scale is defined by fixing either (i) the origin and the unit, or (ii) two reference points. For example, a temperature scale is defined by assigning numbers to the ice point and the steam point of water at atmospheric pressure. Daniel G. Fahrenheit (1686–1736) assigned numbers 32 and 212, whereas Anders Celsius (1701–1744) assigned numbers 0 and 100. The mapping of a measurement t_C in degrees Celsius into the measurement t_F in degrees Fahrenheit is a positive linear transformation: $t_F = 1.8t_C + 32$.

The physical quantity "temperature" can be thought of as an analog of the psychological quantity "strength of preference". Hence, you may extend your intuition about temperature measurements to utility measurements.

To assess the utility function v on the set $\mathcal{Z}$ of losses and gains from an investment, three outcomes are distinguished:

z — any outcome representing loss ($z < 0$) or gain ($z > 0$) relative to the status quo ($z = 0$);
z^e — a least preferred outcome, fixed as a loss if loss from any investment alternative is possible;
z^s — a most preferred outcome, assessed as a gain.

The scale of the utility function is defined here by (i) fixing the origin $v(0) = 0$, which implies $v(z) < 0$ if $z < 0$, and $v(z) > 0$ if $z > 0$, and (ii) requiring the unit to satisfy equation (14.8) of the following methodology.

14.3.6 Minimal Modeling Methodology

A minimal methodology for assessing and modeling the decider's utility function v on a set $\mathcal{Z}$ of losses and gains from an investment consists of four steps. This methodology is minimal because (i) it requires assessing one outcome only, and (ii) it assumes an exponential model for v.

Step 1. Fix probability p ($0 < p < 1$) and loss z^e, and assess gain z^s such that you are indifferent between the status quo and the gamble:

$$0 \sim \left\langle \begin{matrix} p, & z^s \\ 1-p, & z^e \end{matrix} \right\rangle. \tag{14.7}$$

In the first pass through the methodology, set $p = 0.5$. In subsequent passes, you may consider different probabilities, as explained in Section 14.3.7. To assess z^s, apply the *bounding technique*, which is as follows.

1.1 Create two choices between the status quo and the gamble: one with a small gain, so that the status quo is preferred, and one with a large gain, so that the gamble is preferred. For example:

$$0 > \left\langle \begin{matrix} 0.5, & \$300 \\ 0.5, & -\$300 \end{matrix} \right\rangle, \quad 0 < \left\langle \begin{matrix} 0.5, & \$900 \\ 0.5, & -\$300 \end{matrix} \right\rangle.$$

The implication of these preferences is that $300 < z^s < 900$. The goal now is to narrow this interval.

1.2 Increase the gain from the left gamble, decrease the gain from the right gamble, and reassess the preferences. For example:

$$0 > \left\langle \begin{matrix} 0.5, & \$350 \\ 0.5, & -\$300 \end{matrix} \right\rangle, \quad 0 < \left\langle \begin{matrix} 0.5, & \$800 \\ 0.5, & -\$300 \end{matrix} \right\rangle.$$

1.3 Continue in this way until deciding the preferences becomes difficult and the interval becomes narrow, so that its midpoint is an acceptable approximation to z^s. For example:

$$0 > \left\langle \begin{matrix} 0.5, & \$500 \\ 0.5, & \$300 \end{matrix} \right\rangle, \quad 0 < \left\langle \begin{matrix} 0.5, & \$600 \\ 0.5, & -\$300 \end{matrix} \right\rangle.$$

Now $z^s = (500 + 600)/2 = 550$ is accepted as the gain from the gamble which makes this particular decider indifferent:

$$0 \sim \left\langle \begin{matrix} 0.5, & \$550 \\ 0.5, & -\$300 \end{matrix} \right\rangle.$$

Step 2. Write the *measurement equation* corresponding to the indifference relation assessed in Step 1:

$$v(0) = p\,v(z^s) + (1 - p)\,v(z^e).$$

Applying the scaling condition $v(0) = 0$ yields the equation

$$v(z^s) = -\frac{1-p}{p}v(z^e). \tag{14.8}$$

When $p = 0.5$, this equation becomes $v(z^s) = -v(z^e)$.

Step 3. Examine the result.

3.1 Consider three points on the graph of v: $(z^e, v(z^e))$, $(0, 0)$, $(z^s, v(z^s))$. Calculate the abscissa:

$$z_L^s = -\frac{1-p}{p}\,z^e.$$

When $p = 0.5$, this abscissa is $z_L^s = -z^e$. The location of z^s relative to z_L^s implies the global shape of v that passes through the three points (Figure 14.4).

3.2 Infer the global shape of v:

$$z^s > z_L^s \implies \text{concave } v,$$
$$z^s < z_L^s \implies \text{convex } v,$$
$$z^s = z_L^s \implies \text{linear } v.$$

Most often, function v is (i) concave, and (ii) steeper for losses ($z < 0$) than for gains ($0 < z$).

3.3 When $p = 0.5$, the three points are such that

$$v(0) - v(z^e) = v(z^s) - v(0).$$

This equality of the utility differences has an implication.

3.3.1 Check your understanding of the implication, which is this: Your strength of preference for recovering a loss z^e equals your strength of preference for realizing a gain z^s. To probe deeper, imagine two scenarios at separate times. (i) Yesterday, your investment showed a loss z^e. Today, it recovered to the status quo $z = \$0$. Are you elated? How strongly? (ii) Yesterday, your investment was at the status quo $z = \$0$. Today, it shows a gain z^s. Are you elated? How strongly?

3.3.2 Was the degree of your elatedness (satisfaction) in each scenario about equal? If yes, then continue to Step 4. If no, then adjust z^s to achieve equality.

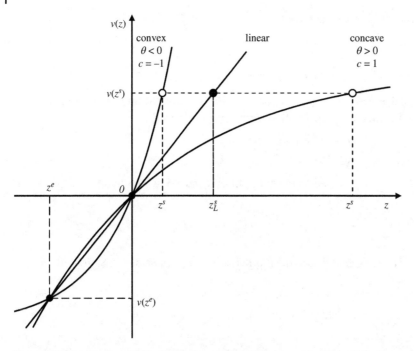

Figure 14.4 Inference of the global shape of the utility function v based on three points: $(z^e, v(z^e))$, $(0, 0)$, $(z^s, v(z^s))$.

Step 4. Select and estimate a parametric model for v.

    4.1 The model selected here is an exponential function:

$$v(z) = c\left(1 - e^{-\frac{z}{\theta}}\right), \qquad -\infty < z < \infty. \tag{14.9}$$

It satisfies the scaling condition $v(0) = 0$. It has a single parameter $\theta \neq 0$, whose dimension matches the dimension of z (e.g., dollars), and whose sign controls the shape as follows:

$$\theta > 0 \iff c = 1, \text{ concave } v;$$
$$\theta < 0 \iff c = -1, \text{ convex } v.$$

If the conclusion from Step 3.2 was that v is linear, then set $v(z) = z$, and skip Steps 4.2–4.3.

    4.2 To estimate parameter θ, function v is inserted into the measurement equation (14.8):

$$c\left(1 - e^{-\frac{z^s}{\theta}}\right) = -\frac{1-p}{p}\, c\left(1 - e^{-\frac{z^e}{\theta}}\right). \tag{14.10}$$

This equation is next rewritten in the form of a contraction operator, which allows for an *iteration method*. (i) If v is concave ($\theta > 0$), then the equation is solved for θ on the right side:

$$\theta = -z^e\left[\ln\left(\frac{1}{1-p}\left(1 - pe^{-\frac{z^s}{\theta}}\right)\right)\right]^{-1}. \tag{14.11}$$

(ii) If v is convex ($\theta < 0$), then the equation is solved for θ on the left side:

$$\theta = -z^s\left[\ln\left(\frac{1}{p}\left(1 - (1-p)e^{-\frac{z^e}{\theta}}\right)\right)\right]^{-1}. \tag{14.12}$$

    4.3 Each of the two equations is now in the form of a *contraction operator* ξ, such that $\theta = \xi(\theta)$. An arbitrary initial point θ_0 is chosen (say, $\theta_0 = z^s$ or $\theta_0 = z^e$), and a sequence $\{\theta_1, \theta_2, \theta_3, \ldots\}$ is calculated recursively as $\theta_{i+1} = \xi(\theta_i)$ for $i = 0, 1, 2, \ldots$ until $|\theta_{i+1} - \theta_i| < \epsilon$ for some small $\epsilon > 0$. Then θ_{i+1} is taken as an estimate of θ.

Example 14.3   Let $p = 0.5$, $z^e = -300$, $z^s = 550$ so that the three points examined in Step 3 have utilities such that

$$v(0) - v(-300) = v(550) - v(0).$$

The implication: This decider's strength of preference for recovering the loss of −$300 equals her strength of preference for realizing the gain of $550. The utility function passing through the three points is concave ($\theta > 0$). Therefore, equation (14.11) is employed in the iteration method. Starting with the initial point $\theta_0 = 500$, the calculated sequence is $\theta_1 = 587$, $\theta_2 = 631.4, \ldots, \theta_{10} = 679.5$, $\theta_{11} = 679.7$, $\theta_{12} = 679.7$. We take $\theta = 680$ as the solution. To check it, we take model (14.9) with $c = 1$; calculate $v(-300) = -0.5545$, $v(550) = 0.5546$; and conclude that the equality $v(z^e) = -v(z^s)$ holds approximately.

14.3.7   Coherence and Validation

The minimal modeling methodology does not reveal any errors that may trickle into v because (i) it lacks any coherence checks in the assessment of z^s, and (ii) it lacks validation of the exponential model. The joint effect of the assessment and model errors can be uncovered by repeating the methodology several times.

Each time: (i) Fix the loss z^e at a significantly different value. Possibly, fix the probability p at a different value, remembering that $p \neq 0.5$ makes the task of intuitive evaluation of the gamble more difficult. (ii) Assess z^s. (iii) Estimate the parameter θ.

Compare all estimates of θ. A small dispersion is to be expected; if this is the case, then taking the geometric mean of all estimates yields a utility function that interpolates between random errors. On the other hand, a large dispersion of the estimates indicates either incoherence among the several assessments, or the inadequacy of the exponential model, or both.

In summary, an exponential utility function estimated from just one assessed point is an approximation, which may capture your strength of preferences over losses and gains globally, but not necessarily locally. Empirical utility functions estimated from many assessed points tend to be steeper for losses than for gains, and may exhibit a noticeable change of the slope in the vicinity of the status quo ($z = \$0$). In such cases, the exponential function, which rises continuously from the subdomain of losses into the subdomain of gains, is usually too steep over small gains. This caveat notwithstanding, it offers a convenient representation of the decider's strength of preferences on $\mathcal{Z}$; and it admits an easy behavioral interpretation, whose impact can be easily traced into the investment choices.

14.3.8   Inference of Risk Attitude

Task 14.3   (*Risk attitude*).   Your investment is showing a gain (sure outcome z). You may take it and liquidate the investment, or you may continue to hold the investment for another period at the end of which you will realize either a loss or a gain, as quantified by gamble G. Circle your preference or indifference.

$$
\begin{array}{ccc}
z & & G \\
& \succ & \\
900 & \lessgtr & \left(\begin{array}{cc} 0.5, & 2100 \\ 0.5, & -300 \end{array} \right) \\
& \sim &
\end{array}
$$

The sure outcome z happens to be equal to the expected outcome $E(Z)$ from gamble G. This is the kind of choice situation through which one's risk attitude is characterized.

Definition   The type of *global risk attitude* on the outcome space $\mathcal{Z}$ is revealed by consistent choice between (i) an outcome to be generated randomly from distribution function G, and (ii) a sure outcome equal to the expected outcome $E(Z)$ from G. The decider is said to be *globally*

risk averse (RA)	if	$E(Z) \succ G,$
risk seeking (RS)	if	$E(Z) \prec G,$
risk neutral (RN)	if	$E(Z) \sim G,$

and if the relation $\succ$ or $\prec$ or $\sim$ holds consistently for every G on $\mathcal{Z}$.

The requirement, that the preference or indifference relation between $E(Z)$ and G be consistent across **all** distribution functions G on $\mathcal{Z}$, makes it infeasible to uncover one's risk attitude through direct questioning (*à la* Task 14.3). However, as proven in Section 14.7.4, there is a way of inferring the type of risk attitude from the shape of the utility function v.

Theorem *(Inference of risk attitude)*   *Let the outcome space* $\mathcal{Z} \subseteq \mathfrak{R}$, *and let the decider's strength of preferences on* $\mathcal{Z}$ *be represented by a continuous utility function* v. *Then*

strictly concave v	$\implies$	RA *attitude,*
strictly convex v	$\implies$	RS *attitude,*
linear v	$\implies$	RN *attitude.*

Corollary *(Exponential model and risk attitude)*   *If function* v *is exponential with parameter* $\theta \neq 0$, *then*

$$\theta > 0 \implies \text{RA } attitude,$$
$$\theta < 0 \implies \text{RS } attitude.$$

Moreover, parameter θ may be thought of as a measure of the *degree of risk aversion* (if $\theta > 0$) or the *degree of risk seeking* (if $\theta < 0$). This measure is comparable across deciders: if each of two deciders has an exponential utility function (14.9) on $\mathcal{Z}$, then the one having smaller $|\theta|$ is more risk averse (if $\theta > 0$) or more risk seeking (if $\theta < 0$) than the other.

In a nutshell: $0 < |\theta| < \infty$; decreasing $|\theta|$ implies strengthening of RA or RS attitude; increasing $|\theta|$ implies convergence toward RN attitude (because for large $|\theta|$, the exponential utility function approaches a linear function).

14.3.9 Recapitulation of Utility Measurement

The role of a cardinal utility function v in an investment choice problem can be summarized thusly.

1. The function v measures an individual's strength of preferences for outcomes (losses, gains) relative to the status quo (current assets). Any shape of v is admissible, and the distinctiveness of the shapes explains the differences in preferences of individuals facing identical choice problems.
2. The shape of v (if it is concave, or convex, or linear) also allows one to infer the type of global risk attitude. But the reverse is not true: the type of global risk attitude (even if it were identified via the definition) does not imply the shape of v.
3. The function v is always conditional on the current (D, R, A, E) — the decision problem D, the reference point R, the aspirations A, and the experience E of the decider. Together, these factors define the context within which the utility function v is assessed. When this context changes, the function v may have to be reassessed.

14.3.10 Calculation of Expected Utility

Having determined the utility function v on the outcome space $\mathcal{Z}$, equation (14.9), we can return to the expected utility criterion (14.3)–(14.4) for operational calculations.

Discrete distribution

Generalizing the definition of a two-outcome gamble (Section 14.1.3), let $G = \langle p_1, z_1; p_2, z_2; \ldots; p_J, z_J \rangle$ represent a J-outcome gamble that yields outcome z_j with probability p_j for $j = 1, \ldots, J$ ($p_1 + p_2 + \cdots + p_J = 1$). Per equation (14.4), the utility of gamble G is calculated as the expected utility of outcome Z generated from G:

$$U(G) = \sum_{j=1}^{J} v(z_j) \, p_j. \tag{14.13}$$

Continuous distribution

Let G be a distribution function on the outcome space $\mathcal{Z}$, and let g be the corresponding density function. Per equation (14.4), the utility of distribution function G is calculated as the expected utility of outcome Z generated from G:

$$U(G) = \int_{\mathcal{Z}} v(z) g(z) \, dz. \tag{14.14}$$

Cash equivalent

Generalizing relations (14.1)–(14.2), let z^* be the cash equivalent of the distribution function G, either discrete or continuous. By definition, $z^* \sim G$ and $v(z^*) = U(G)$. Hence, having calculated $U(G)$, one can find

$$z^* = v^{-1}(U(G)). \tag{14.15}$$

14.4 Investment Choice Model

14.4.1 Elements

The basic investment choice model consists of three sets $\mathcal{A}, \mathcal{Z}, \mathcal{G}$, which define the choice problem (Section 14.1), plus the cardinal utility function v of the decider (Section 14.3), which encodes his strength of preferences over outcomes in $\mathcal{Z}$. It is assumed that $\mathcal{Z} \subseteq \mathfrak{R}$ is an interval (bounded or unbounded), and that each distribution function $G_k \in \mathcal{G}$ has a corresponding density function g_k. The model should order the investment alternatives consistently with the decider's preferences. Imagine performing this task sans model.

> **Task 14.4** (*Intuitive ordering*). You are offered four investment alternatives; the outcome from each is forecasted in terms of the distribution function graphed in Figure 14.5. Order G_1, G_2, G_3, G_4 in terms of your preference.
>
> _____ _____ _____ _____
>
>   least preferred   most preferred

14.4.2 Optimal Decision Procedure

First, each alternative a_k ($k = 1, \ldots, K$) is evaluated in terms of its utility. The *utility of an alternative*, $U_k = U(a_k) = U(G_k) = E\left[v(Z)|a_k\right]$, is defined as the *expected utility of outcome* resulting from that alternative:

$$U_k = \int_{\mathcal{Z}} v(z) g_k(z) \, dz. \tag{14.16}$$

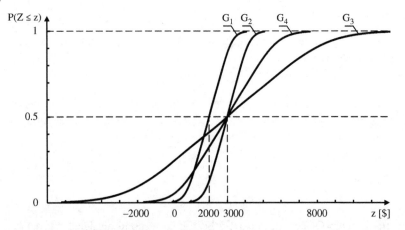

Figure 14.5 Normal distribution functions G_k ($k = 1, 2, 3, 4$) of outcome variate Z in the investment choice problem tackled intuitively in Task 14.4 and solved analytically in Table 14.1.

Second, the alternatives are arranged in ascending order of their utilities,

$$U_{(1)} \leq U_{(2)} \leq \cdots \leq U_{(K)},$$ (14.17)

where $U_{(k)}$ denotes the kth utility, and $(k) \in \{1, \ldots, K\}$. This order of utilities implies the order of alternatives: $a_{(1)}$ is the *least preferred* alternative, and $a_{(K)}$ is the *most preferred* alternative. When two or more alternatives have identical utilities, their order is irrelevant, and they are considered to be indifferent.

Third, the *most preferred* alternative $a^* \in \{a_1, a_2, \ldots, a_K\}$ is that which ensures maximum utility:

$$U_{(K)} = \max\{U_1, U_2, \ldots, U_K\},$$ (14.18)

$$a^* = a_{(K)}.$$ (14.19)

14.4.3 Gaussian–Exponential Model

An implementation of the optimal decision procedure requires models for the density functions g_k ($k = 1, \ldots, K$) and the utility function v. One useful and convenient, albeit approximate, pair of models is Gaussian–exponential, which arises from three assumptions.

1. The outcome space $\mathcal{Z}$ is the unbounded interval $\mathcal{Z} = (-\infty, \infty)$.
2. The outcome variate Z resulting from alternative a_k has a normal distribution: $(Z|a_k) \sim \text{NM}(\mu_k, \sigma_k)$. Thus, its density function is

$$g_k(z) = \frac{1}{\sigma_k \sqrt{2\pi}} \; e^{-\frac{1}{2}\left(\frac{z-\mu_k}{\sigma_k}\right)^2}, \quad -\infty < z < \infty,$$ (14.20)

with the parameters (μ_k, σ_k) determined by forecasting outcome Z from alternative a_k. The forecasting method may be judgmental, as described in Chapter 9, or statistical, as described in Chapter 10. In the latter case, μ_k is the posterior mean of Z, and σ_k^2 is the posterior variance of Z, given a_k.
3. The utility function is exponential:

$$v(z) = c(1 - e^{-\frac{z}{\theta}}), \quad -\infty < z < \infty,$$ (14.21)

with the parameters $\theta \neq 0$ and $c \in \{-1, 1\}$ determined as described in Section 14.3.6.

14.4.4 Analytic Solution

With functions g_k and v specified by expressions (14.20) and (14.21), the integral (14.16), which defines the utility of alternative a_k, can be solved analytically:

$$U_k = \int_{-\infty}^{\infty} v(z) g_k(z)\, dz$$

$$= \int_{-\infty}^{\infty} c(1 - e^{-\frac{z}{\theta}}) \frac{1}{\sigma_k \sqrt{2\pi}}\, e^{-\frac{1}{2}\left(\frac{z-\mu_k}{\sigma_k}\right)^2} dz$$

$$= c - \frac{c}{\sigma_k \sqrt{2\pi}} \int_{-\infty}^{\infty} e^{-\left[\frac{z}{\theta} + \frac{1}{2}\left(\frac{z-\mu_k}{\sigma_k}\right)^2\right]} dz$$

$$= c - \frac{c}{\sigma_k \sqrt{2\pi}} \left\{ \sigma_k \sqrt{2\pi} e^{-\frac{1}{\theta}\left[\mu_k - \frac{1}{2\theta}\sigma_k^2\right]} \right\}$$

$$= c \left(1 - e^{-\frac{1}{\theta}\left[\mu_k - \frac{1}{2\theta}\sigma_k^2\right]} \right), \tag{14.22}$$

where the solution to the integral in the third line appears within the braces in the fourth line.

14.4.5 Cash Equivalent of Alternative

The remarkable feature of the above solution is that its form matches the form of the exponential utility function (14.21), except that argument z is replaced by the expression within the brackets. Denoting

$$z_k^* = \mu_k - \frac{1}{2\theta}\sigma_k^2, \tag{14.23}$$

we find that the utility of alternative a_k equals the utility of outcome z_k^*:

$$U_k = c\left(1 - e^{-\frac{z_k^*}{\theta}}\right)$$

$$= v\left(z_k^*\right). \tag{14.24}$$

Definition An outcome z_k^* is called the *cash equivalent* of the distribution function G_k (or of alternative a_k) if $U_k = v(z_k^*)$.

Inasmuch as this equality of utilities holds if, and only if, the indifference relation $G_k \sim z_k^*$ holds, the solution (14.23)–(14.24) may be interpreted thusly. Given (i) the parameters (μ_k, σ_k) of the normal distribution function of the outcome variate and (ii) the parameter θ of the exponential utility function on the outcome space, we can calculate via expression (14.23) a cash equivalent z_k^* of the distribution function G_k, which has the following property: the decider would be indifferent between an investment alternative from which the outcome is uncertain, and is generated randomly according to distribution function G_k, and an investment alternative that yields sure outcome z_k^*.

Equality (14.24) also implies that the alternatives can be ordered either in terms of their utilities, as in relation (14.17), or in terms of their cash equivalents:

$$z_{(1)}^* \leq z_{(2)}^* \leq \cdots \leq z_{(K)}^*, \tag{14.25}$$

where $z_{(k)}^*$ denotes the kth cash equivalent in ascending order of numerical values, and $(k) \in \{1, \ldots, K\}$.

Example 14.4   (Four distributions)

Let $K = 4$. For each alternative a_k ($k = 1, \ldots, 4$), Table 14.1 lists values of the parameters (μ_k, σ_k), and reports two solutions (z_k^*, U_k): one for $\theta = \$5555$, which implies a RA decider, and one for $\theta = -\$12\,500$, which implies a RS decider. Let us analyze these solutions.

1. Across all alternatives, $\mu_k > z_k^*$ for the RA decider, whereas $\mu_k < z_k^*$ for the RS decider. These inequalities are consistent with the definition of the type of risk attitude because $z_k^* \sim G_k$; consequently, $\mu_k > G_k$ by the RA decider, whereas $\mu_k < G_k$ by the RS decider.

2. Because $\mu_1 < \mu_2$ and $\sigma_1 = \sigma_2$, distribution G_1 is stochastically dominated by distribution G_2, as can be seen in Figure 14.5; consequently $G_1 \prec G_2$ by each decider.

3. The rankings of the cash equivalents, per relation (14.25), are:

$$z_3^* < z_1^* < z_4^* < z_2^* \qquad \text{for RA decider,}$$

$$z_1^* < z_2^* < z_4^* < z_3^* \qquad \text{for RS decider.}$$

The implied preference orders of the distributions are:

$$G_3 \prec G_1 \prec G_4 \prec G_2 \qquad \text{by RA decider,}$$

$$G_1 \prec G_2 \prec G_4 \prec G_3 \qquad \text{by RS decider.}$$

Conspicuous is the reversal of the rank of G_3: it is the most preferred distribution by the RS decider, and the least preferred distribution by the RA decider. Moreover, the RA decider prefers G_1 to G_3 even though G_1 is stochastically dominated by G_2. Thus, a distribution which is stochastically dominated by another distribution is not necessarily the least preferred among all distributions being ranked.

4. Given the three distributions with identical means, $\mu_2 = \mu_3 = \mu_4$, the preference order between the distributions corresponds to the following order of standard deviations:

$$\sigma_3 > \sigma_4 > \sigma_2 \qquad \text{for RA decider,}$$

$$\sigma_2 < \sigma_4 < \sigma_3 \qquad \text{for RS decider.}$$

In other words, the RA decider prefers the distribution with a smaller variability of outcome, whereas the RS decider prefers the distribution with a larger variability of outcome.

5. The order of the four distributions in terms of the probability of loss $P(Z < 0 | a_k)$, assuming the smaller probability is preferred, is

$$G_3 \prec G_4 \prec G_1 \prec G_2.$$

Table 14.1   Solution to an investment choice problem with four normal distributions $\{NM(\mu_k, \sigma_k): k = 1, 2, 3, 4\}$ for two deciders, each having an exponential utility function with parameter θ; μ_k, σ_k, z_k^* are in dollars.

	Alternative		RA decider $\theta = \$5555$		RS decider $\theta = -\$12\,500$		Probability of loss
k	μ_k	σ_k	z_k^*	U_k	z_k^*	U_k	$P(Z < 0 | a_k)$
1	2000	900	1927	0.29	2032	0.18	0.013
2	3000	900	2927	0.41	3032	0.27	0.000
3	3000	4000	1560	0.24	3640	0.34	0.227
4	3000	2000	2640	0.38	3160	0.29	0.067

This order, or its reversal, does not match any of the two preference orders established in point 3. Therefore we can conclude that, in general, minimization of the probability of loss is not consistent with maximization of the expected utility of outcome.

6. For a RN decider, $v(z) = z$ and therefore $U_k = E(Z|a_k) = \mu_k$ ($k = 1, \ldots, 4$). The implied preference order between the distributions is: $G_1 \prec G_k$ for $k = 2, 3, 4$, and

$$G_2 \sim G_3 \sim G_4.$$

In other words, the RN decider is indifferent between the three distributions that yield the identical mean outcome, $\mu_2 = \mu_3 = \mu_4$, because the uncertainty about outcome Z, as quantified here by the standard deviations, $\sigma_2 < \sigma_4 < \sigma_3$, does not affect such a decider's utility of the distribution, and hence her preference order between the distributions.

14.4.6 Mean–Variance Trade-off

In the Gaussian–exponential model, the uncertainty about outcome Z from an investment alternative a_k is quantified by distribution G_k, which is NM(μ_k, σ_k), and whose two parameters admit an intuitive interpretation (Chapter 9): μ_k — mean, median, mode, a measure of central tendency of Z; σ_k^2 — variance, a measure of uncertainty about Z. Therefore, the preference order between the distributions may be framed in terms of the preferred *trade-off* between the mean (a measure of central tendency) and the variance (a measure of uncertainty).

Example 14.5   (Four distributions — Example 14.4 continued)
The choice between G_1 and G_3 involves a trade-off between

$$[\mu_1 = \$2000, \ \sigma_1^2 = (\$900)^2] \quad \text{and} \quad [\mu_3 = \$3000, \ \sigma_3^2 = (\$4000)^2].$$

A preference between G_1 and G_3 may be interpreted thusly. (i) The RA decider, for whom $G_1 \succ G_3$, is willing to reduce the mean from \$3000 to \$2000 *in exchange for* a reduction of the variance (i.e., reduction of uncertainty) from $(\$4000)^2$ to $(\$900)^2$. (ii) The RS decider, for whom $G_1 \prec G_3$, is willing to increase the variance (i.e., increase the uncertainty) from $(\$900)^2$ to $(\$4000)^2$ *in exchange for* an increase of the mean from \$2000 to \$3000.

The notion of the trade-off exemplified above can be characterized precisely via equations (14.23)–(14.24). Consider two distributions, G_1 and G_2, which are indifferent:

$$G_1 \sim G_2 \iff U_1 = U_2 \iff z_1^* = z_2^*.$$

Via equation (14.23), the last equality takes the form

$$\mu_1 - \frac{1}{2\theta}\,\sigma_1^2 = \mu_2 - \frac{1}{2\theta}\,\sigma_2^2,$$

wherefrom

$$\frac{\Delta\sigma^2}{\Delta\mu} = \frac{\sigma_2^2 - \sigma_1^2}{\mu_2 - \mu_1} = 2\theta. \tag{14.26}$$

That is, the two distributions are indifferent if, and only if, the difference $\Delta\sigma^2$ of the variances divided by the difference $\Delta\mu$ of the means is constant and equals 2θ. Hence the following property.

> **Substitution property.** In the Gaussian–exponential model, the mean and the variance are substitutes, and 2θ is the *marginal rate of substitution*: a simultaneous change of the mean by one unit and of the variance by 2θ units leaves the utility unchanged ($U_1 = U_2$).

Recalling the interpretation of θ from Section 14.3.8, equation (14.26) says that the marginal rate of substitution between the variance and the mean (i) is positive for a RA decider, (ii) is negative for a RS decider, and (iii) decreases in magnitude $2|\theta|$ as the RA or RS attitude becomes stronger. In that sense, the parameter θ of the utility function encodes the decider's most preferred trade-off between the mean and the variance of the outcome. Note also the consistency of the units in equation (14.26): $[\$]^2/[\$] = [\$]$.

Example 14.6 *(Marginal rate of substitution)*
Let the outcome variate Z be in dollars, and let $\theta = \$55.55$, so that $2\theta = \$111.1$.

Case 1. If $\Delta\mu = 10$, then $\Delta\sigma^2 = 111.1(10) = 1111$. Interpretation: a $\$^2 1111$ increase of the variance of outcome must be accompanied by a $\$10$ increase of the mean outcome in order for this decider to be indifferent between the original distribution, $NM(\mu_1, \sigma_1)$, and the new one, $NM\left(\mu_1 + 10, \sqrt{\sigma_1^2 + 1111}\right)$, regardless of the values of μ_1 and σ_1^2.

Case 2. If $\Delta\mu = -10$, then $\Delta\sigma^2 = 111.1(-10) = -1111$. Interpretation: a $\$10$ decrease of the mean outcome must be accompanied by a $\$^2 1111$ decrease of the variance of outcome in order for this decider to be indifferent between the original distribution, $NM(\mu_1, \sigma_1)$, and the new one, $NM\left(\mu_1 - 10, \sqrt{\sigma_1^2 - 1111}\right)$, regardless of the values of μ_1 and σ_1^2, but provided that $\sigma_1^2 > 1111$.

Case 3. Let $\Delta\mu = -10$ and $\sigma_1 = 40$. Then equation (14.26) yields

$$\sigma_2^2 = \sigma_1^2 + 2\theta \cdot \Delta\mu$$
$$= (40)^2 + (111.1)(-10)$$
$$= 489,$$

giving $\sigma_2 = 22.11$ and $\Delta\sigma = 22.11 - 40 = -17.89$. Interpretation: a $\$10$ decrease of the mean outcome must be accompanied by a $\$17.89$ decrease of the standard deviation of outcome in order for this decider to be indifferent between the original distribution, $NM(\mu_1, 40)$, and the new one, $NM(\mu_1 - 10, 22.11)$, regardless of the value of μ_1.

14.5 Capital Allocation Model

14.5.1 Elements

The problem is to decide the amount of capital to be invested in an alternative from which the return rate over a specified period (e.g., 3 months, 6 months, 1 year, 3 years) is uncertain and is forecasted probabilistically. The alternative may be a company's stock, an investment fund, a futures contract (of a commodity, a currency, a stock index, Treasury bonds or notes). The investment may involve taking a long position (buying) or taking a short position (selling), though the latter is not feasible for every investment category (e.g., for managed investment funds). The basic capital allocation model comprises the following elements.

k — index of the investment alternative being considered, $k = 1, \ldots, K$.

a_k — amount of capital to be invested in alternative k; it is a continuous, nonnegative *decision variable*, $a_k \geq 0$.

W_k — return rate from investment alternative k, after fees and expenses (assuming reinvestment of dividends and capital gains, if applicable), over a specified period; it is a continuous variate, with the sample space $\mathcal{W}$, and the realized return rate $w \in \mathcal{W}$ being negative, or zero, or positive.

Z — outcome variate, continuous, with the sample space $\mathcal{Z}$, and the realized outcome $z \in \mathcal{Z}$ being *loss* ($z < 0$) or *gain* ($z > 0$), relative to the *status quo* ($z = 0$); given an amount a_k invested in alternative k,

$$Z = a_k W_k.$$

g_k — density function of outcome Z from investment alternative k.

v — cardinal utility function on $\mathcal{Z}$ of the decider (the investor).

The problem is to find the optimal amount of capital a_k^* to be invested that maximizes the utility of alternative $k \in \{1, \dots, K\}$, independently of the other alternatives being considered. The problem of investing simultaneously in two alternatives (a portfolio) is studied in Section 14.6.

14.5.2 Gaussian–Exponential Model

The adaptation of model (14.20) (14.21) to the capital allocation problem involves four assumptions and derivations.

1. The sample space $\mathcal{W}$ is the unbounded interval $\mathcal{W} = (-\infty, \infty)$.
2. The return rate W_k from alternative k has a normal distribution: $W_k \sim \text{NM}(\mu_k, \sigma_k)$, with the parameters (μ_k, σ_k) determined by forecasting W_k probabilistically. (The forecast may be visualized by interpreting μ_k as the median of W_k and σ_k as the width of the 38% central credible interval of W_k.)
3. Applying the derived distribution theory (Section 3.8), the outcome variate Z, given the invested amount a_k, is found to have a normal distribution: $(Z|a_k) \sim \text{NM}(a_k \mu_k, a_k \sigma_k)$. Thus, its density function is

$$g_k(z) = \frac{1}{a_k \sigma_k \sqrt{2\pi}} e^{-\frac{1}{2}\left(\frac{z - a_k \mu_k}{a_k \sigma_k}\right)^2}, \quad -\infty < z < \infty. \tag{14.27}$$

4. The utility function is exponential:

$$v(z) = c\left(1 - e^{-\frac{z}{\theta}}\right), \quad -\infty < z < \infty, \tag{14.28}$$

with the parameters $\theta \neq 0$ and $c \in \{-1, 1\}$ determined as described in Section 14.3.6.

Probability of loss

The normal distribution of the return rate W_k yields a simple formula for the probability of a loss. To wit, for any amount $a_k > 0$ of invested capital, $Z < 0$ if and only if $W_k < 0$, the event whose probability is

$$P(W_k < 0) = Q\left(\frac{0 - \mu_k}{\sigma_k}\right)$$
$$= 1 - Q\left(\frac{\mu_k}{\sigma_k}\right), \tag{14.29}$$

where the second equality results from equation (2.5) — a property of the standard normal distribution function Q.

Coefficients

A variate having $\mu_k > 0$ is sometimes characterized in terms of the *coefficient of variation*, σ_k / μ_k. Being dimensionless, it is a suitable measure for comparing the degrees of uncertainty associated with different variates W_k ($k = 1, \dots, K$). But equation (14.29) contains the reciprocal of this measure; hence the following definition.

Definition For an investment alternative k whose return rate W_k has a normal distribution with mean $\mu_k > 0$ and standard deviation $\sigma_k > 0$, the quotient μ_k / σ_k is called the *coefficient of surety*.

The larger the coefficient of surety, the surer you can be of realizing a gain (equivalently, of not suffering a loss). Here are a few values calculated from equation (14.29) to guide your intuition:

μ_k / σ_k	0.2	0.5	1	1.5	2	2.5	3
$P(W_k < 0)$	0.421	0.309	0.159	0.067	0.023	0.0062	0.0014

You may now create a simple heuristic for screening a large number of investment alternatives to a smaller number K for analytic evaluation. Choose a probability threshold p_o, such that any probability of loss $P(W_k < 0) < p_o$ is acceptable to you; find the corresponding coefficient of surety μ_o/σ_o; and consider only the alternatives having $\mu_k/\sigma_k > \mu_o/\sigma_o$. For example, $p_o = 0.023$ yields $\mu_o/\sigma_o = 2$.

Parameters

Financial advisors typically publish estimates of the mean annual return rate μ_k based on samples from the recent few years (e.g., 1, 3, 5, 10 years), as well as from the inception of an investment fund. However, the standard deviation σ of the annual return rate is rarely published — a glaring omission. Section 14.6.7 discusses the task of forecasting W_k. Exercises offer some examples and suggestions.

14.5.3 Optimal Decision Procedure

The utility of investing an amount a_k in alternative k ($k = 1, \ldots, K$) equals the *expected utility of outcome* resulting from that investment: $U_k(a_k) = E[v(Z)|a_k]$. In the Gaussian–exponential model,

$$
U_k(a_k) = \int_{-\infty}^{\infty} v(z) g_k(z)\, dz
$$

$$
= \int_{-\infty}^{\infty} c\left(1 - e^{-\frac{z}{\theta}}\right) \frac{1}{a_k \sigma_k \sqrt{2\pi}}\, e^{-\frac{1}{2}\left(\frac{z - a_k \mu_k}{a_k \sigma_k}\right)^2}\, dz.
\tag{14.30}
$$

Via analogy between this expression and expression (14.22), the solution to the integral takes the form:

$$
U_k(a_k) = c\left(1 - e^{-\frac{1}{\theta}\left[a_k \mu_k - \frac{1}{2\theta} a_k^2 \sigma_k^2\right]}\right)
$$

$$
= v\left(\zeta_k(a_k)\right),
\tag{14.31}
$$

with

$$
\zeta_k(a_k) = a_k \mu_k - \frac{1}{2\theta} a_k^2 \sigma_k^2.
\tag{14.32}
$$

Definition An outcome $\zeta_k(a_k)$ is called the *conditional cash equivalent* of the investment alternative k, in which an amount a_k is invested, if $U_k(a_k) = v(\zeta_k(a_k))$.

The optimal amount a_k^* should maximize the expected utility of outcome $U_k(a_k)$. Based on equality (14.31), this maximization is equivalent to the maximization of the conditional cash equivalent $\zeta_k(a_k)$ because the utility function v is strictly increasing. Toward finding a_k^*, the first derivative of $\zeta_k(a_k)$ with respect to a_k is taken:

$$
\frac{d\zeta_k(a_k)}{da_k} = \mu_k - \frac{1}{\theta} a_k \sigma_k^2.
$$

Next, setting $d\zeta_k(a_k^*)/da_k = 0$ as the necessary condition for a_k^* to be optimal, and solving the resultant equation, yields

$$
a_k^* = \theta \frac{\mu_k}{\sigma_k^2}, \qquad \theta \neq 0.
\tag{14.33}
$$

This *optimal amount of capital* to be invested in alternative k has four properties.

1. Because $\theta \neq 0$ and $\sigma_k^2 > 0$, the optimal amount is zero ($a_k^* = 0$) only if the mean return rate is zero ($\mu_k = 0$).

2. A decider having $\theta > 0$ (RA attitude) should invest a positive amount ($a_k^* > 0$) only if the mean return rate is positive ($\mu_k > 0$). Thereupon, a_k^* increases linearly with the mean μ_k and decreases hyperbolically with the variance σ_k^2 (i.e., the larger the uncertainty about the return rate, the smaller the optimal amount a_k^* to be invested).

3. The smaller θ ($\theta > 0$), the stronger the degree of decider's risk aversion (recall Section 14.3.8) and, consequently, the smaller the optimal amount a_k^* to be invested.

4. A decider having $\theta < 0$ (RS attitude) should invest a positive amount ($a_k^* > 0$) only if the mean return rate is negative ($\mu_k < 0$). Is this wise?

> **Advice** (Risk attitude for investing). Despite being generally allowed by the utility theory (as a natural consequence of individual freedom), a globally risk-seeking (RS) attitude toward money is foolish in the context of investment decisions whose outcomes are uncertain.

Adhering to this advice, we henceforth restrict the parameters:

$$\theta > 0, \qquad c = 1, \qquad \mu_k > 0 \text{ for } k = 1, \dots, K. \tag{14.34}$$

14.5.4 Preference Order of Alternatives

Let us return to functions (14.31) and (14.32). By replacing the argument a_k of each function with a_k^* from equation (14.33), two quantities are derived: the *maximum utility* of the investment alternative,

$$
\begin{aligned}
U_k^* &= \max_{a_k} U_k(a_k) \\
&= U_k(a_k^*) \\
&= 1 - e^{-\frac{1}{2}\left(\frac{\mu_k}{\sigma_k}\right)^2};
\end{aligned} \tag{14.35}
$$

and the *cash equivalent* of the investment alternative,

$$
\begin{aligned}
z_k^* &= \zeta(a_k^*) \\
&= \frac{1}{2} a_k^* \mu_k = \frac{1}{2}\theta\left(\frac{\mu_k}{\sigma_k}\right)^2.
\end{aligned} \tag{14.36}
$$

The maximum utility U_k^* of the investment alternative has three noteworthy properties.

1. It is independent of the utility parameter θ. Accordingly, U_k^* is identical for every rational investor.

2. It depends only upon, and increases strictly with, the coefficient of surety μ_k/σ_k. Accordingly, under the Gaussian–exponential model, the coefficient of surety is sufficient for establishing the preference order of investment alternatives ($k = 1, \dots, K$) that holds for every rational investor. Ergo, this is the *absolute preference order* (as defined in Section 14.2.1) — but only **in the context** of the Gaussian–exponential model.

3. Letting $(\mu/\sigma)_k = \mu_k/\sigma_k$, this absolute preference order can be written

$$\left(\frac{\mu}{\sigma}\right)_{(1)} \leq \left(\frac{\mu}{\sigma}\right)_{(2)} \leq \cdots \leq \left(\frac{\mu}{\sigma}\right)_{(K)}, \tag{14.37}$$

where $(\mu/\sigma)_{(k)}$ denotes the kth coefficient of surety in ascending order of numerical values, and $(k) \in \{1, \dots, K\}$. The cash equivalent z_k^* of the investment alternative has two noteworthy properties.

4. It increases linearly with the utility parameter θ (equivalently, with the optimal amount of capital a_k^* to be invested): the weaker the degree of the decider's risk aversion, the larger θ (recall Section 14.3.8) and, consequently, the larger the cash equivalent z_k^*. Hence, z_k^* is personal — it is the minimum sure gain the decider should be willing to accept in lieu of investing in a risky alternative which generates the outcome randomly according to the distribution $NM(a_k^*\mu_k, a_k^*\sigma_k)$.

5. It increases quadratically with the coefficient of surety μ_k/σ_k which, again, is sufficient for evaluating the desirability of the investment alternative.

Example 14.7 *(Four alternatives)*

Let $K = 4$. For each alternative k ($k = 1, 2, 3, 4$), Table 14.2 lists the values of the parameters (μ_k, σ_k), the coefficient of surety μ_k/σ_k, the maximum utility U_k^*, and reports the solution (a_k^*, z_k^*) for a RA decider having the utility parameter $\theta = \$1100$. Let us analyze the results.

1. The absolute preference order of the alternatives, written in terms of the coefficients of surety (14.37), is

$$(1.75)_2 < (2)_1 = (2)_4 < (2.05)_3.$$

It is consistent with the order of the maximum utilities: $U_2^* < U_1^* = U_4^* < U_3^*$. Hence, the absolute preference order of the alternatives is

$$2 < 1 \sim 4 < 3.$$

2. For a decider having the utility parameter $\theta = \$1100$, the optimal amount of capital to invest in the most preferred alternative is $a_3^* = \$57\,857$. Interestingly, it is neither the smallest nor the largest amount among $\{a_1^*, a_2^*, a_3^*, a_4^*\}$.

3. If alternative $k = 3$ were unavailable, then the choice would be between alternatives $k = 1$ and $k = 4$, which are indifferent. To choose one of them, the decider may employ an auxiliary criterion: When the cash equivalents are equal, $z_1^* = z_4^* = \$2200$, choose the alternative for which the optimal amount of capital to invest is smaller; here it is $a_4^* = \$48\,889$.

Table 14.2 Solution to a capital allocation problem with four normal distributions $\{NM(\mu_k, \sigma_k): k = 1, 2, 3, 4\}$ of the return rate: absolute measures and individual measures for a decider having an exponential utility function with parameter θ; a_k^*, z_k^* are in dollars.

Alternative			Absolute measures		RA decider $\theta = \$1100$		Probability of loss	Rank
k	μ_k	σ_k	$\dfrac{\mu_k}{\sigma_k}$	U_k^*	a_k^*	z_k^*	$P(Z < 0\|k)$	(k)
1	0.06	0.030	2.00	0.865	73 333	2200	0.023	(2)
2	0.07	0.040	1.75	0.784	48 125	1684	0.040	(1)
3	0.08	0.039	2.05	0.878	57 857	2314	0.020	(4)
4	0.09	0.045	2.00	0.865	48 889	2200	0.023	(3)

14.5.5 Constraint on Capital

If $a_{(K)}^*$ does not exceed your available capital, then invest $a_{(K)}^*$ in the most preferred alternative (K). If the maximum amount you can invest is b, which is less than the optimal amount $a_{(K)}^*$, and possibly less than some other amounts $a_{(k)}^*$, then proceed thusly.

Step 1. For every alternative $k \in \{1, \ldots, K\}$, set the investment amount:

$$a_k = \begin{cases} b & \text{if } b < a_k^*, \\ a_k^* & \text{if } a_k^* \le b. \end{cases} \tag{14.38}$$

Table 14.3 Solution to a capital allocation problem when the capital is constrained to amount $b = \$50\,000$; the same four distributions and the same decider as in Table 14.2.

k	a_k^*	a_k	$\zeta_k(a_k)$
1	73 333	50 000	1977
2	48 125	48 125	1684
3	57 857	50 000	1728
4	48 889	48 889	2200

Step 2. Evaluate each alternative in terms of its conditional cash equivalent $\zeta_k(a_k)$ specified by equation (14.32). Note that if $a_k = a_k^*$, then $\zeta_k(a_k) = z_k^*$.

Step 3. Rank the alternatives in the ascending order of their conditional cash equivalents; invest in the alternative having the highest $\zeta_k(a_k)$.

Example 14.8   (Four alternatives — Example 14.7 continued)
Suppose $b = \$50\,000$. For each alternative k, Table 14.3 reports the solution $(a_k, \zeta_k(a_k))$ for a RA decider having the utility parameter $\theta = \$1100$. The ascending order of the conditional cash equivalents implies the preference order of the alternatives:

$$2 < 3 < 1 < 4.$$

This preference order differs from the absolute preference order established using the optimal amounts a_k^*, which are unconstrained. Hence the following advice.

> **Advice** (Constrained capital for investing). When the capital at your disposal is constrained, do not invest in the absolutely preferred alternative. Find the most preferred alternative for you: it depends upon (i) the capital you have available, b, and (ii) the degree of your risk aversion, as encoded in the utility parameter, θ.

14.6   Portfolio Design Model

14.6.1   Elements

To "diversify" is a frequent investment advice. A vehicle for diversification is the *portfolio* — a combination (a subset) of several investment alternatives. Hence the portfolio design problem: Which alternatives to include? How much to invest in each?

To answer these questions, the Gaussian–exponential model of Section 14.5 is extended to a portfolio comprising two investment alternatives whose return rates are stochastically dependent on each other. To account for, and to understand the impact of, this stochastic dependence is our primary objective. The basic portfolio design model consists of all the elements defined in Section 14.5.1, plus the following.

k, l — indices of two investment alternatives comprising the portfolio, $k \neq l$; $k, l \in \{1, \dots, K\}$.

a — amount of capital to be invested in the portfolio; it is a continuous, nonnegative *decision variable*:

$$a = a_k + a_l, \qquad a \geq 0.$$

α — fraction of capital a to be allocated to alternative k when it is paired with alternative l; it is a continuous *decision variable*:

$$\alpha = \frac{a_k}{a}, \qquad 0 \le \alpha \le 1;$$

hence, $1 - \alpha = a_l/a$ is fraction of capital a to be allocated to alternative l.

W — return rate from the portfolio, after fees and expenses, over a specified period; it is a continuous variate, with the sample space $\mathcal{W}$, and the realized return rate $w \in \mathcal{W}$ being negative, or zero, or positive:

$$W = \alpha W_k + (1 - \alpha)W_l.$$

Z — outcome variate, continuous, with the sample space $\mathcal{Z}$, and the realized outcome $z \in \mathcal{Z}$ being *loss* $(z < 0)$ or *gain* $(z > 0)$, relative to the *status quo* $(z = 0)$; given amount a invested in the portfolio,

$$Z = aW$$
$$= a_k W_k + a_l W_l.$$

g — density function of outcome Z from the portfolio.

The expression for W constitutes a mathematical definition of the portfolio: it is an investment with the return rate being a *positive linear combination* of the return rates from the alternatives (two here, many in general).

14.6.2 Gaussian–Exponential Model

This model retains assumptions 1–4 listed in Section 14.5.2, to which the following two assumptions are appended.

5. The return rate W from the portfolio has a normal distribution: $W \sim \text{NM}(\mu, \sigma)$. Being a weighted sum of two normal variates, W_k and W_l, which are not identical and may be stochastically dependent on each other, the variate W has its mean and variance specified by the second theorem of Section 10.7.4, with the weights inserted accordingly:

$$\mu = \alpha \mu_k + (1 - \alpha)\mu_l, \tag{14.39a}$$
$$\sigma^2 = \alpha^2 \sigma_k^2 + (1 - \alpha)^2 \sigma_l^2 + 2\alpha(1 - \alpha)\gamma \sigma_k \sigma_l, \tag{14.39b}$$

where

$$\gamma = Cor(W_k, W_l)$$

is the correlation between W_k and W_l, as defined by expression (10.28), with $-1 < \gamma < 1$.

6. The outcome variate Z, given the invested amount a, is found to have a normal distribution: $(Z|a) \sim \text{NM}(a\mu, a\sigma)$. Thus, its density function is

$$g(z) = \frac{1}{a\sigma\sqrt{2\pi}} e^{-\frac{1}{2}\left(\frac{z-a\mu}{a\sigma}\right)^2}, \qquad -\infty < z < \infty. \tag{14.40}$$

Note. The elements $a, \alpha, W, Z, g, \mu, \sigma, \gamma$ depend on the alternatives comprising the portfolio. Hence, they could have the subscript kl; it is omitted for now to keep the notation simple.

14.6.3 Optimal Decision Procedure

For the portfolio comprising investment alternatives (k, l), the decision model is specified in terms of:

- the exponential utility function (14.28) of the decider on the outcome space $\mathcal{Z}$;
- the normal density function (14.40) of the outcome variate Z from the portfolio;

- the six parameters whose values must be known: θ; (μ_k, σ_k), (μ_l, σ_l), γ.
- the two decision variables, a and α, whose optimal values must be found.

Expression (14.40) includes explicitly the capital amount a to be invested in the portfolio, and implicitly (through μ and σ) the allocation fraction α. Keeping this in mind, the objective of the decision procedure is to find the optimal values a^* and α^* that maximize the *utility of the portfolio*. That utility equals the *expected utility of outcome* resulting from the portfolio: $U(a, \alpha) = E[v(Z)|a, \alpha]$. In the Gaussian–exponential model,

$$U(a, \alpha) = \int_{-\infty}^{\infty} v(z)g(z)\,dz$$

$$= \int_{-\infty}^{\infty} c(1 - e^{-\frac{z}{\theta}}) \frac{1}{a\sigma\sqrt{2\pi}}\, e^{-\frac{1}{2}\left(\frac{z-a\mu}{a\sigma}\right)^2} dz. \tag{14.41}$$

The optimization problem is to find the capital amount a^* and the allocation fraction α^* such that

$$U^* = U(a^*, \alpha^*) = \max_{a,\alpha} U(a, \alpha), \qquad a \geq 0,\, 0 \leq \alpha \leq 1. \tag{14.42}$$

The derivation of the solution is deferred to Section 14.7.5 and to exercises, while the next section presents, interprets, and illustrates the solution. It answers the question: How much to invest in each alternative of a given portfolio?

14.6.4 Solution to Portfolio Optimization

The solution to the portfolio optimization is presented in three parts: (i) the condition under which a pair of alternatives constitutes an *admissible* portfolio; (ii) the optimal solution framed as a *disaggregation* of the total amount into two subamounts; (iii) the optimal solution framed as an *aggregation* of two subamounts into the total amount.

Admissible portfolio

The optimal amount, a^*, to be invested in the portfolio equals the sum of the optimal subamounts, a_k^o and a_l^o, to be invested in the alternatives k and l which comprise the portfolio: $a^* = a_k^o + a_l^o$. Naturally, the portfolio exists only if each subamount is positive. Hence, the portfolio is said to be *admissible* if $a_k^o > 0$ and $a_l^o > 0$. It turns out that not every pair of alternatives constitutes an admissible portfolio.

Theorem (*Admissibility condition*) *The portfolio comprising two alternatives, k and l, characterized by the parameters $(\mu_k, \sigma_k; \mu_l, \sigma_l; \gamma)$, is admissible if and only if*

$$\gamma < \min \left\{ \frac{\mu_k/\sigma_k}{\mu_l/\sigma_l}, \frac{\mu_l/\sigma_l}{\mu_k/\sigma_k} \right\}.$$

The proof is left as Exercise 12. Inasmuch as each coefficient of surety, μ_k/σ_k and μ_l/σ_l, is positive, the smaller of the two ratios is a number from the interval $(0, 1]$. This number constrains the magnitude of positive correlation γ between the return rates. But the magnitude of negative correlation remains unconstrained.

Another way to interpret this condition: Every pair of alternatives with negative correlation γ constitutes an admissible portfolio; but a pair of alternatives with positive correlation γ is admissible only if γ is smaller than a bound.

Disaggregation framework

The optimal solution framed as a disaggregation of the optimal amount a^* into two subamounts, a_k^o and a_l^o, takes the following form.

Optimal allocation fraction:

$$\alpha^* = \frac{\sigma_l^2 \mu_k - \gamma \sigma_k \sigma_l \mu_l}{(\sigma_l^2 - \gamma \sigma_k \sigma_l)\mu_k + (\sigma_k^2 - \gamma \sigma_k \sigma_l)\mu_l}.$$ (14.43)

Optimal amount to be invested in the portfolio:

$$a^* = \theta \frac{(\sigma_l^2 - \gamma \sigma_k \sigma_l)\mu_k + (\sigma_k^2 - \gamma \sigma_k \sigma_l)\mu_l}{(1 - \gamma^2)\sigma_k^2 \sigma_l^2}.$$ (14.44)

Optimal subamounts to be invested in the alternatives:

$$a_k^o = \alpha^* a^*, \qquad a_l^o = (1 - \alpha^*)a^*.$$ (14.45)

Cash equivalent of the portfolio:

$$z^* = \frac{1}{2}a^*[\alpha^* \mu_k + (1 - \alpha^*)\mu_l].$$ (14.46)

This disaggregation framework reveals three noteworthy properties of the optimal solution.

1. The optimal allocation fraction α^* is independent of the utility parameter θ. Hence, α^* is identical for every rational investor. Hence, α^* is a property of the portfolio (precisely, of the two alternatives comprising the portfolio).
2. The optimal investment amount a^* for the portfolio has the structure parallel to that for a single alternative (equations (14.44) and (14.33)): a^* is proportional to the utility parameter θ; the smaller θ ($\theta > 0$), the stronger the degree of decider's risk aversion (recall Section 14.3.8) and, consequently, the smaller the optimal amount a^* to be invested in the portfolio.
3. The cash equivalent z^* of the portfolio likewise has the structure parallel to that for a single alternative (equations (14.46) and (14.36)): z^* increases linearly with a^* and with the mean return rate from the portfolio (which equals the weighted sum of μ_k and μ_l, with α^* and $1 - \alpha^*$ serving as the weights).

Aggregation framework

Let a_k^* be the optimal amount of capital to be invested in alternative k, specified by equation (14.33), independently of the other alternatives being considered. Let a_l^* be the like amount for alternative l ($k \neq l$). Then the solution for the portfolio comprising these two alternatives can be framed thusly.

Optimal subamounts to be invested in the alternatives:

$$a_k^o = \frac{1}{1 - \gamma^2}\left(a_k^* - a_l^* \gamma \frac{\sigma_l}{\sigma_k}\right),$$

$$a_l^o = \frac{1}{1 - \gamma^2}\left(a_l^* - a_k^* \gamma \frac{\sigma_k}{\sigma_l}\right).$$ (14.47)

Optimal amount to be invested in the portfolio:

$$a^* = a_k^o + a_l^o.$$ (14.48)

Optimal allocation fraction:

$$\alpha^* = \frac{a_k^o}{a^*}.$$ (14.49)

Cash equivalent of the portfolio:

$$z^* = \frac{1}{2}a^*[\alpha^* \mu_k + (1 - \alpha^*)\mu_l].$$ (14.50)

This aggregation framework reveals the transformation (14.47) between the investment amounts (a_k^*, a_l^*), which are optimal for each single alternative, and the investment amounts (a_k^o, a_l^o), which are optimal for the portfolio comprising the two alternatives. This transformation is determined by two characteristics of the portfolio: (i) the ratio σ_k/σ_l of the standard deviations of the two return rates — a measure of the diversity of uncertainties; and (ii) the correlation γ between the return rates — a measure of the association.

When the two return rates are equally uncertain, $\sigma_k/\sigma_l = 1$, the standard deviations have no influence on the transformation of (a_k^*, a_l^*) into (a_k^o, a_l^o). When the two return rates are uncorrelated, $\gamma = 0$, the transformation degenerates to the identity functions $a_k^o = a_k^*$ and $a_l^o = a_l^*$; in other words, the problem of portfolio optimization becomes vacuous — the two optimal investment amounts can be determined independently of each other by applying the capital allocation model of Section 14.5. The sensitivity of transformation (14.47) to the correlation is subject of Exercise 17.

Example 14.9 *(Influence of correlation)*

Consider the portfolio comprising alternatives $(k, l) = (1, 2)$ from Table 14.2. Its admissibility condition is

$$\gamma < \min \left\{ \frac{2.00}{1.75}, \frac{1.75}{2.00} \right\} = 0.875.$$

If $\gamma = 0.9$ were allowed, equations (14.47) would yield $a_1^o = \$82\,016$ and $a_2^o = -\$7236$, clearly an inadmissible solution. For five admissible γ values, Table 14.4 reports the optimal solution obtained via equations (14.47)–(14.50) for a RA decider having the utility parameter $\theta = \$1100$. Observe the pattern of numbers.

As the positive correlation weakens from 0.6 to 0, and then the negative correlation strengthens from 0 to -0.6: (i) the optimal allocation fraction α^* (which is independent of θ) converges toward 0.5, (ii) the optimal amount a^* to be invested increases from \$78\,000 to over \$301\,000, and (iii) the cash equivalent z^* of the portfolio rises from \$2460 to \$9679. In short, the utility of the portfolio to the investor increases substantially.

This phenomenal influence of the correlation provides a theoretic ground for "diversification" — for investing in a portfolio rather than in a single alternative.

14.6.5 Solution to Portfolio Choice

We are ready to answer the question: Which alternatives to include in the portfolio? This is the portfolio choice problem. Consistently with the procedure of Section 14.5.4, the measure for ordering the portfolios is either U^* — the maximum utility of the portfolio, expression (14.42), or z^* — the cash equivalent of the portfolio, expression (14.46). Each of these expressions can be rewritten in a form that contains the following measure.

Table 14.4 Impact of correlation γ between the return rates on the optimal solution for the portfolio comprising alternatives (1, 2) from Table 14.2 for a decider having the utility parameter $\theta = \$1100$; the amounts a_1^o, a_2^o, a^*, z^* are in dollars.

γ	a_1^o	a_2^o	a^*	α^*	z^*
0.6	54 427	23 633	78 060	0.6972	2460
0.3	59 432	34 753	94 185	0.6310	2999
0	73 333	48 125	121 458	0.6038	3884
−0.3	101 740	71 016	172 756	0.5889	5538
−0.6	174 739	126 758	301 497	0.5796	9679

Definition For a portfolio comprising two alternatives, k and l, characterized by the parameters (μ_k, σ_k; μ_l, σ_l; γ), let the mean μ^* and the standard deviation σ^* of the return rate W be specified by equations (14.39), with the allocation fraction $\alpha = \alpha^*$ being optimal and coming from equation (14.43). Then the quotient μ^*/σ^* is called the *coefficient of surety of the portfolio*.

Expressions can now be derived (see Exercise 15) for three quantities: the square of the coefficient of surety of the portfolio,

$$\left(\frac{\mu^*}{\sigma^*}\right)^2 = \frac{1}{1-\gamma^2}\left[\left(\frac{\mu_k}{\sigma_k}\right)^2 + \left(\frac{\mu_l}{\sigma_l}\right)^2 - 2\gamma\left(\frac{\mu_k}{\sigma_k}\right)\left(\frac{\mu_l}{\sigma_l}\right)\right]; \tag{14.51}$$

the maximum utility of the portfolio,

$$U^* = 1 - e^{-\frac{1}{2}\left(\frac{\mu^*}{\sigma^*}\right)^2}; \tag{14.52}$$

and the cash equivalent of the portfolio,

$$z^* = \frac{1}{2}\theta\left(\frac{\mu^*}{\sigma^*}\right)^2. \tag{14.53}$$

Expressions (14.52)–(14.53) parallel expressions (14.35)–(14.36); hence their properties parallel the five properties discussed in Section 14.5.4. But expression (14.51) offers new insight: it reveals the interaction between the coefficients of surety of the alternatives and the correlation — the interaction that determines the portfolio's coefficient of surety and, hence, the portfolio's utility U^* and cash equivalent z^*.

Inasmuch as U^*, as well as z^*, increases strictly with μ^*/σ^*, the coefficient of surety is sufficient for establishing the preference order of portfolios that holds for every rational investor. Ergo, this is the *absolute preference order* (as defined in Section 14.2.1) — but only **in the context** of the Gaussian–exponential model.

14.6.6 Portfolio Design: Synthesis

To synthesize the portfolio design procedure, a subscript identifying the portfolio is attached to every element of the optimal solution (14.43)–(14.53). For portfolio kl comprising alternatives (k, l), these elements are now γ_{kl}, μ_{kl}^*, σ_{kl}^*, U_{kl}^*, and a_{kl}^o, a_{lk}^o, a_{kl}^*, α_{kl}^*, z_{kl}^*. For the subamounts, the order of subscripts matters: a_{kl}^o is the optimal subamount to be invested in alternative k when it is paired with alternative l; and a_{lk}^o is the optimal subamount to be invested in alternative l when it is paired with alternative k.

The portfolio design procedure employs equations from Sections 14.5.3, 14.6.4, 14.6.5 within the following algorithm.

Step 0. Given are K investment alternatives, indexed by $k \in \{1, \dots, K\}$, and each characterized by (μ_k, σ_k), the mean and the standard deviation of the return rate. Given also is the utility parameter θ of the investor.

Step 1. For every alternative k, calculate the coefficient of surety μ_k/σ_k.

Step 2. There are at most $K(K-1)/2$ portfolios, each comprising two alternatives, k and l, such that

$$k < l, \qquad k \in \{1, \dots, K-1\}, \qquad l \in \{k+1, \dots, K\}.$$

Select the portfolios you wish to consider (possibly all of them), and for each: (i) estimate statistically or assess judgmentally the correlation γ_{kl}, and (ii) verify the admissibility condition. If the latter is violated, then discard this portfolio.

Step 3. For every admissible portfolio (k, l), calculate μ_{kl}^*/σ_{kl}^*.

Step 4. Order all admissible portfolios in terms of the numerical values of their coefficients of surety, $\{\mu_{kl}^*/\sigma_{kl}^*\}$, from the lowest (the least preferred portfolio) to the highest (the most preferred portfolio).

Step 5. For the most preferred portfolio (m, n), calculate a_{mn}^o, a_{nm}^o, a_{mn}^*, α_{mn}^*, z_{mn}^*.

Constraint on capital

Let b denote your available capital. If $b \geq a^*_{mn}$, then invest the amount a^*_{mn} in portfolio (m, n); specifically, invest the amount a^o_{mn} in alternative m and the amount a^o_{nm} in alternative n. If $b < a^*_{mn}$, then you need to adapt the above algorithm to account for the constraint on capital, analogously to the adaptation made in Section 14.5.5; this is the subject of Exercise 24.

Example 14.10   (Portfolio choice)

Consider three portfolios, each comprising two alternatives from Table 14.2: $(1, 2)$, $(2, 3)$, $(3, 4)$. As shown in Example 14.7, the absolute preference order of the alternatives is $2 \prec 1 \sim 4 \prec 3$. For each portfolio, Table 14.5 reports the correlation γ_{kl}, the coefficient of surety μ^*_{kl}/σ^*_{kl}, and the solution elements a^*_{kl}, α^*_{kl}, z^*_{kl}, as well as z^*_k, z^*_l for reference. There are two cases.

In case A, all correlations are positive. The order of the coefficients of surety (which matches the order of the cash equivalents, $z^*_{12} < z^*_{34} < z^*_{23}$) implies the absolute preference order of the portfolios:

$$(1, 2) \prec (3, 4) \prec (2, 3).$$

This shows that combining the worst and the best alternative may create a portfolio $(2, 3)$ which has higher utility than the portfolio created from the best two alternatives $(3, 4)$ has.

In case B, the correlations are negative, zero, positive. The order of the coefficients of surety (which matches the order of the cash equivalents, $z^*_{34} < z^*_{23} < z^*_{12}$) implies the absolute preference order of the portfolios:

$$(3, 4) \prec (2, 3) \prec (1, 2).$$

This shows that combining the worst two alternatives which are negatively correlated may create the most preferred portfolio $(1, 2)$, while combining the best two alternatives which are positively correlated may create the least preferred portfolio $(3, 4)$.

> **Advice** (Portfolio design). Do not discard any admissible portfolios before evaluating them because it is the interaction between the coefficients of surety and the correlation, per equation (14.51), that determines the portfolio's utility.

14.6.7   Investment Forecast–Decision Process

An investor often must be a forecaster and a decider. This requires two separate mind-sets to prevent spurious influences (e.g., wishful thinking). The model helps because it **structures** and **decomposes** the problem into (i) forecasting the return rates, and (ii) assessing the strength of preferences for outcomes (losses, gains).

Table 14.5   Two cases of portfolio choice: the alternatives $k, l \in \{1, 2, 3, 4\}$ come from Table 14.2; the decider's utility parameter is $\theta = \$1100$; the amounts $a^*_{kl}, z^*_{kl}, z^*_k, z^*_l$ are in dollars, rounded to the nearest 100.

Case	Portfolio		Amount	Fraction	Cash equivalents		
	(k, l)	γ_{kl}	μ^*_{kl}/σ^*_{kl}	a^*_{kl}	α^*_{kl}	z^*_{kl}	z^*_k	z^*_l
A	$(1, 2)$	0.3	2.335	94 200	0.63	3000	2200	1700
	$(2, 3)$	0.1	2.572	96 400	0.45	3600	1700	2300
	$(3, 4)$	0.3	2.512	82 200	0.55	3500	2300	2200
B	$(1, 2)$	−0.3	3.173	172 800	0.59	5500	2200	1700
	$(2, 3)$	0.0	2.658	106 000	0.45	4000	1700	2300
	$(3, 4)$	0.3	2.512	82 200	0.55	3500	2300	2200

For forecasting, a sample $\{w_k(n) : n = 1, \ldots, N'\}$ of return rates from investment alternative k over N' past periods can supply estimates of the prior mean and variance of W_k. Mutual funds publish mean return rates in sales brochures and disclosure statements (see Exercise 23). They invariably include a warning: "past performance is no guarantee of future results". Its translation to our terminology: for a future period, the reported statistic is just a prior mean of W_k. Ergo, W_k should be forecasted probabilistically, as we studied in Chapters 9 and 10, to get the posterior mean and variance of W_k.

Over the long term, investing should not be viewed as a one-time problem, but rather as a sequential forecast–decision process. The reason is twofold. (i) The contextual conditioning of the decider's strength of preferences for outcomes is **nonstationary** — it evolves in time (recall Sections 14.3.3 and 14.3.9). (ii) The investment market is **nonstationary** — it evolves in time (e.g., the uncertainty associated with the return rates decreases or increases, the correlation between the return rates weakens or strengthens, new investment alternatives emerge).

Practically, the decision analyses prescribed by the models of Sections 14.5 and 14.6 should be repeated whenever one or both of the following changes are detected.

Decider-induced changes

Whenever any of the four elements (D, R, A, E), which condition the utility function, changes, the value of the utility parameter θ may change as well. Per equations (14.44)–(14.46), this triggers a change in the values of a^*, a_k^o, a_l^o, z^* — a change which may call for reallocating the capital.

Market-induced changes

Whenever the forecast of the return rates and their correlations changes, the values of the distribution parameters $(\mu_k, \sigma_k; \mu_l, \sigma_l; \gamma)$ may change as well. Per equations (14.43)–(14.46) and (14.51), this triggers a change in the values of α^*, a^*, a_k^o, a_l^o, z^*, μ^*/σ^* — a change which may call for rebalancing the portfolio, reallocating the capital, or redesigning the portfolio.

14.7 Concepts and Proofs

14.7.1 Measurement Scales

Measurement theory is a branch of mathematics that deals rigorously with representations of qualitative relations in terms of quantitative relations. Such representations require measurement scales. There are five fundamental, progressively stronger, types of scales:

nominal scale (e.g., bar code of a product),
ordinal scale (e.g., shoe size),
cardinal scale or interval scale (e.g., temperature),
ratio scale (e.g., distance),
absolute scale (e.g., probability).

14.7.2 Cardinal Scale

Theorem (Cardinality of utility)   *The utility function v in expressions (14.2) and (14.4) is a cardinal scale.*

Proof: By definition of a cardinal scale (also called an interval scale) stated in Section 14.3.5, it is admissible to replace v by its positive linear transformation v' such that for any $z \in \mathcal{Z}$,

$$v'(z) = \alpha v(z) + \beta,$$

where $\alpha > 0$ and β are constants. (In brief, "v is unique up to a positive linear transformation".) To prove this, we must show that the preference order between distribution functions is not changed when v is replaced by v'. Starting with relation (14.4) for $k = 1, 2$, let

$$U'(G_k) = E[v'(Z)|a_k]$$

$$= E[\alpha v(Z) + \beta|a_k]$$

$$= \alpha E[v(Z)|a_k] + \beta$$

$$= \alpha U(G_k) + \beta.$$

Inserting this result into relation (14.3) yields

$$G_1 \precsim G_2 \iff U'(G_1) \le U'(G_2)$$

$$\iff \alpha U(G_1) + \beta \le \alpha U(G_2) + \beta$$

$$\iff \alpha U(G_1) \le \alpha U(G_2)$$

$$\iff U(G_1) \le U(G_2),$$

where dividing both sides by α does not change the direction of the inequality because $\alpha > 0$. Thus the preference order between distribution functions G_1 and G_2 is unaffected by the choice of scale, v or v'. QED.

14.7.3 Stochastic Dominance

Theorem (*First-degree stochastic dominance*) *If the distribution function G_2 stochastically dominates the distribution function G_1, then G_2 has greater utility than G_1 has,*

$$U(G_1) < U(G_2),$$

*for **every** cardinal utility function v which is continuous, strictly increasing, and bounded on $\mathcal{Z}$, and which yields*

$$U(G_k) = U(a_k) = E[v(Z)|a_k], \qquad k = 1, 2.$$

Proof: It proceeds in six steps.

1. *Rescaling of utility function.* Because v is a cardinal scale and is assumed to be bounded on $\mathcal{Z}$, it can always be rescaled to the range $(0, 1)$. Therefore, let $v : \mathcal{Z} \rightarrow (0, 1)$.
2. *Transformation of variate.* Because v is assumed to be continuous and strictly increasing, it has an inverse v^{-1} such that $y = v(z)$ if, and only if, $z = v^{-1}(y)$ for $z \in \mathcal{Z}$ and $y \in (0, 1)$. Consequently, any distribution function G of outcome variate Z induces a distribution function F of utility variate Y, which is (recall Section 3.8.3) $F(y) = G(v^{-1}(y))$, for any $y \in (0, 1)$.
3. *Equivalent relations.* The assumption that G_2 stochastically dominates G_1 implies that, for every $z \in \mathcal{Z}$,

$$G_1(z) \ge G_2(z) \iff G_1(v^{-1}(y)) \ge G_2(v^{-1}(y))$$

$$\iff F_1(y) \ge F_2(y).$$

4. *Integration by parts*, involving any two functions H and F, proceeds according to the rule:

$$\int H(y)\frac{dF(y)}{dy}\, dy = H(y)F(y) - \int F(y)\frac{dH(y)}{dy}\, dy.$$

5. *The utility of distribution function G can now be expressed by applying the above rule, with* $H(y) = y$:

$$U(G) = \int_{Z} v(z)g(z)\,dz$$

$$= \int_{0}^{1} yf(y)\,dy$$

$$= \int_{0}^{1} y\frac{dF(y)}{dy}\,dy$$

$$= yF(y)]_{0}^{1} - \int_{0}^{1} F(y)\,dy$$

$$= 1 - \int_{0}^{1} F(y)\,dy.$$

6. *The difference between the utilities of* G_2 *and* G_1 *thus becomes*

$$U(G_2) - U(G_1) = \left[1 - \int_{0}^{1} F_2(y)\,dy\right] - \left[1 - \int_{0}^{1} F_1(y)\,dy\right]$$

$$= \int_{0}^{1} [F_1(y) - F_2(y)]\,dy > 0.$$

That this difference is strictly positive follows from the inequality $F_1(y) \geq F_2(y)$ demonstrated in Step 3, which is strict for some y because the inequality $G_1(z) \geq G_2(z)$ is strict for some z, per the definition of stochastic dominance. Consequently, $U(G_1) < U(G_2)$. QED.

14.7.4 Risk Attitude

Theorem (*Inference of risk attitude*) *Let the outcome space be* $Z \subseteq \mathfrak{R}$, *and let the decider's strength of preferences on* Z *be represented by a continuous utility function* v. *Then*

strictly concave v	$\implies$	RA *attitude,*
strictly convex v	$\implies$	RS *attitude,*
linear v	$\implies$	RN *attitude.*

Proofs for a strictly convex v and a linear v are deferred to Exercises 8 and 9. The proof for a strictly concave v proceeds in two steps.

1. *Jensen's inequality.* If v is a strictly concave function on Z, then there exists a line $az + b$ that touches v at point $z = E(Z) \in Z$,

$$aE(Z) + b = v(E(Z)),$$

and that lies above v at every other point $z \in \mathcal{Z}$,

$$az + b > v(z).$$

Replacing the realization z by the variate Z, and taking the expectation on each side of the inequality, yields

$$aE(Z) + b > E[v(Z)].$$

By transitivity,

$$v(E(Z)) > E[v(Z)].$$

This is known as *Jensen's inequality*. In the context of decision theory, it says that the utility of the sure outcome, which equals the expected outcome $E(Z)$, is greater than the expected utility of the random outcome Z.

2. *Global risk attitude.* By equation (14.4), the utility of the distribution function G on $\mathcal{Z}$ is $U(G) = E[v(Z)]$. Hence for every G, Jensen's inequality is equivalent to

$$v(E(Z)) > U(G).$$

By statement (14.3) of the expected utility criterion, this inequality holds if and only if the following preference relation holds:

$$E(Z) \succ G,$$

which is the definition of the global RA attitude. QED.

14.7.5 Portfolio Optimization

The solution to the two-dimensional maximization problem (14.42) is found by solving two one-dimensional maximization problems in three steps, and then evaluating the solution in the fourth step. The specific tasks are: (i) Derive the expression for α^*, conditional on an arbitrary a. (ii) Derive the expression for a^*. (iii) Return to the expression for conditional α^* and replace a by a^*. (iv) Derive the expression for the cash equivalent of the portfolio z^*, such that $U^* = v(z^*)$. The optimal solution thus obtained is in the disaggregation framework (14.43)–(14.46).

Step 1. (Conditional allocation fraction). Integral (14.41) is identical to integral (14.30); hence the solutions are identical. We adapt expression (14.32) for the conditional cash equivalent:

$$\zeta(a) = a\mu - \frac{1}{2\theta} a^2 \sigma^2. \tag{14.54}$$

After replacing μ and σ by expressions (14.39), we treat α as the decision variable, conditional on an arbitrary investment amount a, and write

$$\zeta(\alpha|a) = a[\alpha\mu_k + (1-\alpha)\mu_l] - \frac{a^2}{2\theta} [\alpha^2\sigma_k^2 + (1-\alpha)^2\sigma_l^2 + 2\alpha(1-\alpha)\gamma\sigma_k\sigma_l]. \tag{14.55}$$

Now we can find the optimal allocation fraction α^* that maximizes $\zeta(\alpha|a)$ for an arbitrary amount a. Exercise 14 asks you to prove that

$$\alpha^* = \frac{\theta}{a} \frac{m}{s} + \frac{n}{s}, \tag{14.56}$$

with

$$m = \mu_k - \mu_l, \qquad n = \sigma_l^2 - \gamma\sigma_k\sigma_l, \tag{14.57a}$$

$$s = \sigma_k^2 + \sigma_l^2 - 2\gamma\sigma_k\sigma_l. \tag{14.57b}$$

Next, the above expression for α^* is inserted into expressions (14.39) for μ and σ^2, the terms are manipulated to obtain a common denominator, and the result is:

$$\mu = \frac{1}{a^2 s} \; [am(\theta m + an) + a^2 s \mu_l],$$

$$\sigma^2 = \frac{1}{a^2 s} \; [(\theta m + an)^2 - 2an(\theta m + an) + a^2 s \sigma_l^2].$$

These are the mean and the variance of the return rate W from the portfolio with the optimal allocation fraction α^* for an arbitrary investment amount a.

Step 2. (Optimal investment amount). The optimal amount a^* should maximize the conditional cash equivalent $\zeta(a)$. The expression for it is identical to expression (14.32); hence the solution (14.33) is identical. We adapt it here:

$$a = \theta \frac{\mu}{\sigma^2}. \tag{14.58}$$

It is not the expression for a^* yet because μ and σ derived in Step 1 are conditional on an arbitrary amount a. Replacing μ and σ by these derived expressions yields

$$a = \theta \frac{am(\theta m + an) + a^2 s \mu_l}{(\theta m + an)^2 - 2an(\theta m + an) + a^2 s \sigma_l^2}.$$

This is the equation that the optimal investment amount a^* must satisfy. After it is solved, and the parameters m, n, s are replaced by their expressions from Step 1, equation (14.44) for a^* is obtained.

Step 3. (Optimal allocation fraction). In the expression for α^*, derived in Step 1, the arbitrary amount a is replaced by the optimal amount a^* specified by expression (14.44), parameters m, n, s are replaced by their expressions, and the terms are manipulated until equation (14.43) for α^* is obtained.

Step 4. (Cash equivalent of the portfolio). Again, we adapt expression (14.36):

$$z^* = \frac{1}{2} a^* \mu. \tag{14.59}$$

Next, μ is replaced by the mixture (14.39a) in which α is substituted by α^*. The result is equation (14.46).

Bibliographical Notes

The utility function (as an analytic concept) and the maximization of expected utility of outcome (as the criterion for making personal decisions under uncertainty) were proposed by Daniel Bernoulli (1738). The notion of the strength of preference, measured in terms of utility difference, was advanced by economists in the 1920s. The measurement of utility based on preferences between gambles was first axiomatized by von Neumann and Morgenstern (1947). The mathematical characterization of risk attitude based on the curvature of a utility function was invented by Pratt (1964). For applications, a variety of assessment methods and models have been developed and documented, for instance by Keeney and Raiffa (1976) and Farquhar (1984).

The "level of aspiration" — as a motivator of the individual's behavior — was recognized by psychologists in the 1930s. That it may affect the utility function was theorized by Siegel (1957) and demonstrated empirically by Becker and Siegel (1962). That the "reference point" affects the utility of monetary outcomes was shown experimentally by Payne et al. (1980, 1981).

Classic applications of the utility theory in economics define the outcome as the total wealth of an individual. But Markowitz (1952) reasoned that utility should be defined over losses and gains, relative to the present wealth. The validity of this framing was demonstrated by Swalm (1966), who assessed utility functions of businessmen and found them to be two-piece. Empirical evidence was compiled by Kahneman and Tversky (1979) across many contexts, and by Fishburn and Kochenberger (1979) in the investment contexts. As proven by Fishburn (1977), this framing is congruent with maximization of expected utility. It also received a specialized axiomatization by Krzysztofowicz (1994).

[Daniel Kahneman, professor of psychology at Princeton University (who died in 2024), received the Nobel Prize in economics in 2002 for pioneering (with Amos Tversky, who died in 1996) the field of behavioral economics, which "integrated insights from psychological research into economic science, especially concerning human judgment and decision making".]

Exercises

1 **Stochastic dominance relation**. Given is the set $\mathcal{G} = \{G_k : k = 1, 2, 3, 4\}$ of gambles, the outcomes of which are in dollars:

$$G_1 = \langle 0.2, -350; 0.3, 150; 0.3, 350; 0.2, 650 \rangle,$$
$$G_2 = \langle 0.3, -150; 0.4, 200; 0.3, 550 \rangle,$$
$$G_3 = \langle 0.2, -350; 0.3, 100; 0.3, 300; 0.2, 600 \rangle,$$
$$G_4 = \langle 0.4, -150; 0.3, 200; 0.3, 500 \rangle.$$

1.1 Identify the set of stochastically nondominated gambles.

1.2 An individual chose G_3 to play: if the outcome Z turns out to be positive, the broker pays him; if the outcome Z turns out to be negative, he pays the broker. Is G_3 a rational choice? Support your answer with a cogent and concise argument.

2 **Ordering gambles**. Given is the set $\mathcal{A} = \{a_k : k = 1, 2, 3, 4\}$ of investment alternatives; the outcome Z in dollars from alternative a_k has a discrete distribution G_k, which is specified in Exercise 1.

2.1 Calculate the expected outcome $E(Z|a_k)$ from each alternative. Use this measure to order the alternatives.

2.2 Calculate the standard deviation of outcome $Var^{1/2}(Z|a_k)$ from each alternative. Use this measure to order the alternatives.

2.3 An investor's utility function v is exponential with parameter θ specified in the option. Calculate the utility U_k of each alternative. Use this measure to order the alternatives.

2.4 Find the cash equivalent z_k^* of each gamble G_k. Interpret z_k^*. Use this measure to order the alternatives. *Hint*: Revisit Section 14.3.10.

2.5 Compare the preference orders of alternatives; identify the input needed from the decider to establish each preference order; draw conclusions.

Options:			
(A)	$\theta = 680;$	(D)	$\theta = -991;$
(B)	$\theta = 580;$	(E)	$\theta = -691;$
(C)	$\theta = 480;$	(F)	$\theta = -491.$

3 **Utility parameter**. With outcome z in dollars, an assessment supplied p, z^e, z^s specified in the option. (i) Find the value of parameter θ of the exponential utility function. (ii) Infer the type of risk attitude implied by that utility function.

Options:			
(A)	$p = 0.5,$	$z^e = -300,$	$z^s = 600.$
(B)	$p = 0.5,$	$z^e = -300,$	$z^s = 250.$
(C)	$p = 0.25,$	$z^e = -300,$	$z^s = 2200.$
(D)	$p = 0.25,$	$z^e = -300,$	$z^s = 500.$
(E)	$p = 0.75,$	$z^e = -300,$	$z^s = 200.$
(F)	$p = 0.75,$	$z^e = -300,$	$z^s = 90.$

4 **Consistency**. An individual has an exponential utility function with parameter θ dollars. Given p and z^e, what value of z^s should this individual assess to be consistent with his utility function?

Options: (A) $\quad\theta = \quad 600, \qquad p = 0.5, \qquad z^e = -350.$
(B) $\quad\theta = -1400, \qquad p = 0.5, \qquad z^e = -300.$
(C) $\quad\theta = \quad 1600, \qquad p = 0.25, \qquad z^e = -400.$
(D) $\quad\theta = \quad -900, \qquad p = 0.25, \qquad z^e = -450.$
(E) $\quad\theta = \quad 800, \qquad p = 0.75, \qquad z^e = -500.$
(F) $\quad\theta = \quad -700, \qquad p = 0.75, \qquad z^e = -550.$

5 **Your utility function**. Consider the decision problem (D) of choosing an investment alternative for the next 12 months, given your current reference point (R), aspirations (A), and experience (E). (For instance, imagine you have saved a specific amount of cash. You want to invest it for a purpose (say, to increase savings, to pay college tuition, to travel in summer, to buy a car) — contemplate it! Then proceed with the exercise.)

5.1 Assess your utility function of outcome (monetary loss or gain to be realized in 12 months) by applying the minimal modeling methodology. Report the chosen probability p and outcome z^e; the assessed outcome z^s; the estimate of parameter θ.

Requirement. If $z^s = -z^e$, then for the purpose of this exercise set $z^s = -1.3z^e$.

5.2 Validate the result by repeating Exercise 5.1 with a different outcome z^e. Compare the results. If the two estimates of θ are significantly different, then ponder the source of the discrepancy, redo the assessment (either the first or the second, or both), and re-estimate θ. Take the geometric mean of the two estimates as the final value of θ.

5.3 Graph your utility function. Infer the type of your risk attitude, and characterize in words its degree.

6 **Investment choice**. Consider investment alternatives $\{a_k : k = 1, \ldots, 5\}$. The distribution function G_k of the outcome Z from alternative a_k is NM(μ_k, σ_k), with the parameter values specified in the option. The investor's utility function of outcome is exponential with parameter θ whose value is specified in the option.

6.1 Identify the set of stochastically nondominated alternatives.

6.2 Calculate the utility and the cash equivalent of each alternative. Rank all alternatives in terms of their utilities.

6.3 Interpret the cash equivalent of the most preferred alternative. Describe a decision situation in which knowing that cash equivalent would be useful.

6.4 Let $G_{(K)}$ be the most preferred distribution function of outcome, and let $(\mu_{(K)}, \sigma_{(K)})$ be its parameters. (i) Determine the values of the parameters (μ_k, σ_k) of two new distribution functions G_k ($k = 6, 7$), which have these properties: $G_6 \backsim G_{(K)} \backsim G_7$, and $\mu_6 = \mu_{(K)} - \Delta\mu$, $\mu_7 = \mu_{(K)} + \Delta\mu$. (ii) Graph the three distribution functions in one figure; then describe the trade-offs which make these distribution functions indifferent.

Options for distributions: (I) Table 14.6, Set I, $\Delta\mu = 5$.
(II) Table 14.6, Set II, $\Delta\mu = 50$.

Options for parameter: (Y) θ is yours, from Exercise 5.
(A) $\theta = \$680$.
(S) $\theta = -\$991$.

7 **Global risk attitude**. Let G be a distribution function on the outcome space $\mathcal{Z}$; let $E(Z)$ be the expected outcome under G; and let z^* be the cash equivalent of G assessed by a decider so that $z^* \backsim G$. Prove the following theorem.

Table 14.6 Parameter values of the distribution NM(μ_k, σ_k) of outcome from an investment alternative a_k ($k = 1, \ldots, 5$).

	Set I		Set II	
k	μ_k [\$]	σ_k [\$]	μ_k [\$]	σ_k [\$]
1	300	400	500	390
2	400	800	200	115
3	200	90	300	200
4	400	600	250	110
5	300	90	400	260

Theorem (*Global risk attitude*) *The decider is globally*

risk averse (RA)	*if*	$E(Z) > z^*$,
risk seeking (RS)	*if*	$E(Z) < z^*$,
risk neutral (RN)	*if*	$E(Z) = z^*$,

and if the relation > *or* < *or* = *holds consistently for every G on* $\mathcal{Z}$.

Note. Because each of the inequality (equality) relations in the above theorem is equivalent to the corresponding preference (indifference) relation in the definition of global risk attitude, the roles of the definition and this theorem are sometimes reversed.

8 **Inference of RS attitude**. Prove the second part of the theorem on inference of risk attitude: strictly convex v implies RS attitude (see Section 14.7.4).

9 **Inference of RN attitude**. Prove the third part of the theorem on inference of risk attitude: linear v implies RN attitude (see Section 14.7.4).

10 **Cash equivalent and maximum utility**. Derive expressions (14.36) and (14.35), in that order, by following the explanation that precedes them.

11 **Capital allocation**. Based on return rates realized in the past and the outlook for stock market performance in the next 12 months, an investor selected four mutual funds for detailed analysis. It began with probabilistic forecasting of the return rate W_k from each mutual fund k ($k = 1, 2, 3, 4$) over the next 12 months. Each forecast is in terms of five judgmentally assessed quantiles of W_k, expressed as percentages, and corresponding to probabilities 0.1, 0.25, 0.5, 0.75, 0.9.

Mutual fund 1:	(1.5, 3.1, 4.7, 6.3, 7.9).
Mutual fund 2:	(0.3, 2.8, 5.4, 8.1, 9.9).
Mutual fund 3:	(1.6, 2.7, 3.9, 5.2, 6.3).
Mutual fund 4:	(0.2, 1.8, 4.1, 6.5, 8.7).

11.1 For each mutual fund k, estimate μ_k and σ_k via the method of quantiles–moments (QM).

11.2 Perform a complete analysis of the capital allocation problem using the model of Section 14.5, with parameter θ specified in the option. In particular, for each mutual fund k, report: μ_k/σ_k, $P(W_k < 0)$, a_k^*, z_k^*, U_k^*.

11.3 Determine the absolute preference order of the mutual funds for investing during the next 12 months.

11.4 Having examined the results of your analysis, the investor notices that the capital she has available equals 70% of the amount that is optimal for investing in the most preferred mutual fund. Determine the preference order of the mutual funds for this investor.

Options for utility parameter θ [$]:

(Y) θ is yours, from Exercise 5.

(A)   900;   (B)   1100;   (C)   1300;   (D)   1500.

12   **Admissibility condition theorem**. Prove this theorem, which is stated in Section 14.6.4. *Hint*: employ equation (14.33) in equations (14.47) for a_k^o and a_l^o; then apply the definition of an admissible portfolio.

13   **Disaggregation–aggregation frameworks**. Given equations (14.43)–(14.44) for α^* and a^* in the disaggregation framework, derive equations (14.47) for a_k^o and a_l^o in the aggregation framework.

14   **Optimal conditional allocation fraction**. Prove that the allocation fraction α^*, specified by expression (14.56), maximizes $\zeta(\alpha|a)$ — the conditional cash equivalent (14.55) of the portfolio in which the amount a is invested.

15   **Portfolio's coefficient of surety**. Starting from equations (14.58)–(14.59), derive equations (14.51)–(14.53). *Hint*: Employ equations (14.39) and (14.43), or equations (14.39a) and (14.43)–(14.44).

16   **Effect of correlation on uncertainty**. For the portfolio specified in the option, perform the following sensitivity analysis.

16.1 Employ equation (14.39b) to graph the standard deviation σ of the return rate W from the portfolio as a function of the correlation γ on the interval $(-1, 1)$.

16.2 Explain the graph to a potential investor.

Options:	(A)	$\alpha = 0.7$,	$\sigma_1 = 0.09$,	$\sigma_2 = 0.04$.
	(B)	$\alpha = 0.8$,	$\sigma_3 = 0.07$,	$\sigma_4 = 0.03$.
	(C)	$\alpha = 0.2$,	$\sigma_5 = 0.05$,	$\sigma_6 = 0.09$.
	(D)	$\alpha = 0.3$,	$\sigma_7 = 0.03$,	$\sigma_8 = 0.08$.

17   **Effect of correlation on subamounts**. For the investor having the utility parameter $\theta = \$800$ and investing in the portfolio specified in the option, perform the following sensitivity analysis.

17.1 Calculate the optimal amounts, a_k^* and a_l^*, to be invested in each alternative, k and l, independently of the other alternative. Find the upper bound γ_U on the admissible correlation.

17.2 Employ equations (14.47) to graph each optimal subamount, a_k^o and a_l^o, as a function of the correlation γ on the interval $[-0.7, \gamma_U]$.

17.3 Explain the graph to the investor.

Options:	(A)	$\mu_1 = 0.04$,	$\sigma_1 = 0.03$;	$\mu_2 = 0.15$,	$\sigma_2 = 0.06$.
	(B)	$\mu_3 = 0.09$,	$\sigma_3 = 0.08$;	$\mu_4 = 0.14$,	$\sigma_4 = 0.07$.
	(C)	$\mu_5 = 0.07$,	$\sigma_5 = 0.06$;	$\mu_6 = 0.12$,	$\sigma_6 = 0.05$.
	(D)	$\mu_7 = 0.06$,	$\sigma_7 = 0.07$;	$\mu_8 = 0.10$,	$\sigma_8 = 0.04$.

18   **Portfolio optimization: three frameworks**. For the investor and the portfolio specified in the options, calculate all elements of the optimal solution $(a_k^o, a_l^o, a^*, \alpha^*, z^*)$ via three frameworks.

18.1 Aggregation framework: equations (14.47)–(14.50).

18.2 Disaggregation framework: equations (14.43)–(14.46).

18.3 Disaggregation framework with moments: (i) calculate α^* from equation (14.43); (ii) insert α^* into equations (14.39) and calculate μ^* and $(\sigma^*)^2$ — the mean and the variance of the return rate W from the portfolio with the optimal allocation fraction α^*; (iii) calculate the optimal amount of capital as $a^* = \theta\mu^*/(\sigma^*)^2$; (iv) complete the calculations using equations (14.45)–(14.46).

18.4 If you worked as an investment consultant, which framework would you choose for routine evaluations of the portfolios? Explain the reasons for your choice.

Options for utility parameter θ [\$]:

(Y) θ is yours, from Exercise 5.
(A) 800; (B) 1000; (C) 1200; (D) 1400.

Options for portfolio parameters $(\mu_k, \sigma_k; \mu_l, \sigma_l; \gamma)$:

(P) $(0.09, 0.06; 0.14, 0.07; -0.1)$.
(Q) $(0.07, 0.05; 0.12, 0.05; -0.2)$.
(R) $(0.06, 0.03; 0.10, 0.04; -0.3)$.

19 RN decider, single investment. Consider a decider having linear utility function: $v(z) = z$, for all $z \in \mathcal{Z}$.

19.1 Following the optimal decision procedure of Section 14.5.3, derive expressions for $U_k(a_k)$ — the utility of investing amount a_k in alternative k, and for $\zeta_k(a_k)$ — the conditional cash equivalent of the investment alternative k, in which the amount a_k is invested.

19.2 What is the optimal amount of capital to be invested in alternative k? Given three investment alternatives with $\{(\mu_k, \sigma_k) : k = 1, \ldots, 3\} = \{(0.05, 0.4), (0.09, 0.17), (0.07, 0.11)\}$, what is their preference order (for a fixed a_k)?

19.3 Draw conclusions.

20 RN decider, portfolio of investments. Consider a decider having linear utility function: $v(z) = z$, for all $z \in \mathcal{Z}$.

20.1 Following the optimal decision procedure of Section 14.6.3, derive expressions for $U(a, \alpha)$ — the utility of the portfolio with invested amount a and allocation fraction α, and for $\zeta(\alpha|a)$ — the conditional cash equivalent of the portfolio, given a and α. *Hint*: follow the logic of Step 1 in Section 14.7.5.

20.2 What is the optimal amount of capital to be invested in the portfolio? What is the optimal allocation fraction (for a fixed a)?

20.3 Draw conclusions.

21 Allocation between risky and riskless alternatives. Define b — the capital available for investment, a_k^* — the optimal amount for investing in a risky alternative (e.g., a stock fund) characterized by (μ_k, σ_k), and w_0 — the return rate from a riskless investment (e.g., insured certificate of deposit in a bank). Suppose $b > a_k^*$, and the decider plans to invest amounts a_k^* and $b - a_k^*$ in the risky and the riskless alternatives, respectively. Prove that this allocation is preferred to investing the entire capital b in the riskless alternative only if $\mu_k > 2w_0$. *Hint*: employ cash equivalents to establish the preference order.

Mini-Projects

22 Nonstationarity of allocations — Exercise 21 *continued*. Imagine you are the investor, and you plan to revisit the capital allocation problem once a year over three decades. Suppose only two parameters will be nonstationary. Firstly, your initial capital b (choose your own number) will increase from year to year (i.e., your wealth will grow). Secondly, your utility parameter θ (choose your own value, or use the estimate from Exercise 5 if you completed it) will evolve monotonically from year to year. Contemplate two scenarios: θ will decrease (i.e., you will become more RA with age); θ will increase (i.e., you will become less RA with age). Your objective is to investigate the effect of nonstationarity of (b, θ) on capital allocation decisions.

22.1 Write the formulae for (i) the optimal amount a_k^* to be invested in the risky alternative, (ii) the amount a_0 to be invested in the riskless alternative, and (iii) the fraction α of capital b to be allocated to the risky alternative. The formulae should be written in terms of the parameters characterizing this investment problem: $(\mu_k, \sigma_k; w_0; b, \theta)$.

22.2 Under each of the two scenarios, analyze the five trajectories over the three decades as follows. (i) Given the initial values of b and θ (year 0), change them to hypothetical values for years 10, 20, 30. Sketch the trajectories of b and θ on the interval $[0, 30]$ years by interpolating between the hypothetical values. (ii) Given these changing values of (b, θ) and given the constant values of $(\mu_k, \sigma_k; w_0)$ specified in the option, calculate the values of (a_k^*, a_0, α) for years 0, 10, 20, 30. Sketch the trajectories of a_k^*, a_0, α. (iii) Characterize the monotonicity of each trajectory. Write an interpretation of these results and the conclusions.

22.3 Which of the two scenarios regarding the nonstationarity of θ would you consider more likely to reflect the evolution of your strength of preferences for outcomes (losses, gains) over the next 30 years?

22.4 Some financial advisors offer a general suggestion: as the number of years to retirement decreases, the fraction of capital to be allocated to a risky alternative, α, should be decreasing as well. (i) Does the above investment model support this advice? If so, then what scenario do the financial advisors assume? (ii) How does your trajectory of α compare with the examples in Table 14.7?
Note. This suggestion is often phrased by retirement plans as a gradual reallocation of the capital "from more aggressive to more conservative" investment alternatives (e.g., from stocks to bonds to fixed interest accounts).

22.5 The "Target-Date Fund" applies a fixed trajectory of α (sometimes called a "glide path") that decreases with the investor's age (*The Wall Street Journal*, 10 January 2022). Can such a single trajectory be optimal for every person who wishes to maximize one's own utility of the invested capital at every age? Justify your answer by listing specific pros or cons.

22.6 When retirement looms on a person's planning horizon and the accumulated capital b appears insufficient to retire, some financial advisors recommend increasing α — "taking more risk", they say. Explain this recommendation using the above investment model. (Specifically: Which parameter value should be changed, in which direction, and why?)

Options:	(A)	$\mu_k = 0.059$,	$\sigma_k = 0.027$;	$w_0 = 0.025$.
	(B)	$\mu_k = 0.072$,	$\sigma_k = 0.034$;	$w_0 = 0.035$.
	(C)	$\mu_k = 0.066$,	$\sigma_k = 0.031$;	$w_0 = 0.030$.
	(D)	$\mu_k = 0.097$,	$\sigma_k = 0.038$;	$w_0 = 0.040$.

Table 14.7 Simplified example of a generally suggested interval for allocation fraction α (as defined in Exercise 22), based on age and "risk attitude" (as categorized by financial advisors).

Age	"Risk attitude"		
	Aggressive	Moderate	Conservative
31–40	(0.9, 1.0)	(0.6, 0.8)	(0.4, 0.6)
41–50	(0.8, 1.0)	(0.4, 0.6)	(0.3, 0.5)
51–60	(0.5, 0.8)	(0.3, 0.5)	(0.0, 0.3)

23 **Portfolio design.** For 12 mutual funds indexed by k ($k = 1, \ldots, 12$), Table 14.8 lists estimates of the mean annual return rates μ_k from four investment periods. For future investments, these estimates offer prior means of W_k. Here we supplement them with judgmental estimates of the standard deviations σ_k and correlations γ_{kl} in order to create a quasi-real portfolio design problem. It is specified by three options. (i) The utility parameter θ of the investor. (ii) The indices of four mutual funds from which portfolios are to be created. (iii) The investment period. The last two options designate the estimates of μ_k which need to be retrieved from Table 14.8. The second option alone designates the matrix with γ_{kl} and σ_l estimates; these estimates are assumed to be independent of the investment period.

23.1 Solve the portfolio design problem completely using the model of Section 14.6. Report the results in parallel to the steps of the algorithm from Section 14.6.6.

23.2 Discuss the results as if you were presenting them to a potential investor.

Note. Inasmuch as each investment alternative is a mutual fund, the portfolio being designed here is sometimes called a "fund of funds".

Options for utility parameter θ [$]:

(Y) θ is yours, from Exercise 5.
(A) 950; (B) 1150; (C) 1350; (D) 1550.

Options for mutual funds:

(P) $\{1, 2, 3, 4\}$; (Q) $\{5, 6, 7, 8\}$; (R) $\{9, 10, 11, 12\}$.

Options for investment period:

(01) 1 year; (03) 3 years; (05) 5 years; (10) 10 years.

Table 14.8 Mean annual return rates μ_k [%], as defined in Section 14.5.1, for 12 mutual funds. (These are approximate estimates calculated from realized return rates, minus reported fees and expenses, over the periods of 1 year, 3 years, 5 years, 10 years, ending in 2018.)

Mutual fund k	Investment period			
	1 year	3 years	5 years	10 years
1	2.38	8.67	6.05	12.25
2	4.87	11.76	8.24	14.52
3	−5.24	7.68	3.17	9.86
4	3.63	9.83	7.64	15.69
5	2.17	12.91	7.19	16.47
6	−0.19	8.44	7.32	18.28
7	7.15	15.27	12.69	17.11
8	5.89	11.48	8.92	16.56
9	−3.84	7.39	2.43	8.91
10	9.15	13.21	10.73	15.72
11	1.73	9.69	6.62	12.77
12	5.65	2.17	3.27	4.89

Source: an assortment of sales brochures and disclosure statements reporting fund performance statistics.

Options for correlation matrices and standard deviations [%]:

		l						l						l		
k	1	2	3	4		k	5	6	7	8		k	9	10	11	12
1		0.6	0.5	0.7		5		0.6	0.5	0.2		9		0.4	0.6	0.2
2			0.4	0.3		6			0.3	0.4		10			0.7	0.3
3				0.8		7				0.7		11				−0.2
σ_l	3.9	3.8	6.1	4.8		σ_l	5.7	7.2	4.1	4.3		σ_l	5.1	2.6	4.4	1.4

24 **Portfolio with constrained capital: algorithm**. Consider the portfolio design problem when the available capital b is less than the optimal amount a_{mn}^* needed for investing in the most preferred portfolio (m, n). Adapt the algorithm of Section 14.6.6 so that it accounts optimally for the constraint $b < a_{mn}^*$.

Hints. (i) The algorithm of Section 14.5.5 provides an analog for the logic: condition the solution to the portfolio optimization problem on the capital amount b. (ii) Study Section 14.7.5. (iii) Collect and interpret the relevant equations, among them (14.39), (14.51), (14.53), (14.54).

Note. The adapted algorithm will be the most general. It will fulfill the objective of designing the portfolio that is: (i) **admissible** (for any investor), (ii) **feasible** (for an investor whose capital is constrained to b), (iii) **optimal** (for an investor having utility parameter θ).

25 **Portfolio with constrained capital: application** — Exercises 23 and 24 *continued*. Apply the algorithm formulated in Exercise 24 to the portfolio design problem solved in Exercise 23. The specific tasks are as follows.

25.1 Summarize the solution to the portfolio design problem obtained in Exercise 23 sans constraint on the available capital.

25.2 Solve the same portfolio design problem but with the available capital b equal to 50% of the optimal amount a_{mn}^*, rounded to the nearest $100.

25.3 Compare the solutions from Exercises 25.1 and 25.2. Draw conclusions; write them in the form of an explanation for the investor who has capital b and wanted to invest it in portfolio (m, n).

26 **Distribution of utility parameter**. Exercise 5 was completed by 236 students who took this course at the University of Virginia in four consecutive years. The values of the utility parameter θ they reported are in Table 14.9. Imagine you are the manager of a stock fund (or a bond fund), and these students represent potential young investors in your fund. Hence, their values of θ constitute four samples of variate Θ — the utility parameter of a potential young investor.

26.1 Apply the distribution modeling methodology of Chapter 3 to the sample specified in the option to obtain a parametric model of variate Θ. Hypothesize as many models as necessary to find a good one. Estimate the parameters via the method of your choice. (i) Document all steps of the methodology; in particular, report the items listed in Exercises 4.1 and 4.2 in Chapter 3. (ii) Characterize the shapes of the distribution function and density function obtained. What might be their implication for the stock fund you are managing?

Note. This model of Θ may be viewed as the distribution of the degree of risk aversion (as interpreted in Section 14.3.8) in a pool of potential young investors.

26.2 Suppose you made a probabilistic forecast of the return rate W from your stock fund for the next 12 months; μ and σ are the posterior median and the posterior standard deviation of W; their values are specified in the option. You will disclose these values to each committed investor, who will then invest the amount a^* that maximizes his expected utility of outcome. For a potential investor, this amount is unknown; hence, it is a variate A^*. Derive the expressions for the distribution function and the density function of A^*; graph both functions. (*Reminder*: label the axes and show the units.) Are the distributions of Θ and A^* of the same type?

26.3 The distribution of A^* may be viewed as a probabilistic forecast of the amount of capital that will be invested in your stock fund by a potential young investor in the next 12 months. (i) Calculate the median and two quantiles of A^* corresponding to probabilities p and $1 - p$ specified in the option. (ii) Suppose you must communicate this information at a meeting of managers. Prepare a graphic and write an interpretation of the three quantiles to communicate forecast uncertainty, per the guidelines in Section 10.5.2.

Note. Each sample of Θ contains several realizations that are repeated two or more times. The explanation is twofold. First, almost all individuals began the assessment with the fixed loss $z^e = -\$300$; hence, the assessed gain z^s may fall within a narrow interval for several individuals. Second, the approximate nature of the assessment propels the individual to consider round numbers (e.g., \$350, \$400, \$500); hence, the number accepted as z^s may be identical for two or more individuals; consequently, their estimates of θ may be identical.

Options for the sample:

(1) from year 1;　(2) from year 2;　(3) from year 3;　(4) from year 4;
(9) pooled sample from any two, or any three, or all four years.

Options for posterior parameters (μ, σ):

(A)　$(0.15, 0.06)$;　(B)　$(0.14, 0.04)$;　(C)　$(0.13, 0.05)$;　(D)　$(0.12, 0.04)$.

Options for probability p:

(05)　0.05;　(10)　0.1;　(12)　0.125;　(20)　0.2;　(25)　0.25.

Table 14.9 Estimates of utility parameter θ for potential young investors obtained via the minimal modeling methodology in four years. When the same estimate was reported by two or more individuals, their number m ($m > 1$) is given; the sample size is N per year.

Year 1		Year 2		Year 3		Year 4	
θ	m	θ	m	θ	m	θ	m
457.14	4	432.81		443.80		326.80	
503.1		432.86		451.34		443.80	2
555.2		433.12		476.23	3	451.34	
584.0		433.79		492.31	3	492.31	4
614.5		457.14		503.1		549.7	
623.4	6	476.23		516.5	3	555.2	4
643.4		492.31	4	533.4		578.3	
680	3	503.1		555.2	3	623.4	6
766	8	516.5	3	623.4	9	654	
912	7	555.2	2	648.1	2	680	6
1030		614.5		680	7	766	7
1208	6	623.4	5	766	9	912	2
1506	2	680	6	828	2	1169	
2104	2	694		912	6	1208	9
3902	2	710		1030	2	2104	4
		766	8	1208	6		
		912	6	1506	6		
		1030		1805			
		1208	8	2104	4		
		1307		3902	3		
		1366					
		1591					
		1805					
		2104	8				
		2553					
		3902					
$N = 46$		$N = 67$		$N = 73$		$N = 50$	

15

Various Continuous Models

Three models are presented. First, an intuitively simple task of deciding the asking price for a property leads to a decision model whose analytic solution has a broad implication: it highlights the congruence between a principle of Bayesian rationality, a principle of utilitarian ethics, and a principle of the free market.

Next, a class of yield control problems is introduced via two paradigms: airline reservations and college admissions. They differ in two respects: the criterion function, and the nature of business — for profit versus nonprofit. But their decision models are similar: depending on the predictand distribution, the optimal solution can be derived analytically, or must be found numerically. The models help us to rationalize the over-booking and over-admitting tendencies observed in practice.

15.1 Asking Price Model

15.1.1 Decision Paradigm

You own some property (e.g., a rare book, a classic automobile, a piece of antique furniture, a studio apartment) and know its fair market value (as assessed by an appraiser and agreed upon by you). A prospective buyer has already placed a sealed bid for your property, but before the bid is open you must state your asking price. Then, if the bid exceeds your asking price, you will sell the property for the amount named in the bid. Otherwise, you keep the property. Here is a down-to-earth instance of this paradigm.

> **Task 15.1** (*Asking price*).   You get an envelope. You open it. It contains $100. You are told that someone else has already placed a written bid for the contents of this envelope. Before you can learn the value of the bid, however, you must state your asking price. Then, if the bid exceeds your asking price, you will exchange the $100 for the bid. Otherwise, you keep the $100. What is your asking price? Write it down:_____

15.1.2 Mathematical Formulation

We begin by extracting elements from the decision paradigm:

e — value of the property you own for which the bid has been placed.

a — asking price; it is the *decision variable*, assumed to be continuous and positive, $a > 0$, and taking on values in the decision space $\mathcal{A} = (0, \infty)$.

Probabilistic Forecasts and Optimal Decisions, First Edition. Roman Krzysztofowicz.
© 2025 John Wiley & Sons Ltd. Published 2025 by John Wiley & Sons Ltd.
Companion website: www.wiley.com/go/ProbabilisticForecastsandOptimalDecisions1e

w — bid. Before its value is disclosed, it is a continuous and positive *random variable W* with the sample space $\mathcal{W} = (0, \infty)$.

z — outcome; it is the amount of money you end up with, either in property value or in cash. Before the bid is open, it is a *random variable Z* with the sample space $\mathcal{Z} = (0, \infty)$; it is specified by an outcome function $h : \mathcal{A} \times \mathcal{W} \to \mathcal{Z}$ such that for any $a \in \mathcal{A}$ and $w \in \mathcal{W}$,

$$z = h(a, w) = \begin{cases} e & \text{if } w \leq a, \\ w & \text{if } a < w. \end{cases} \tag{15.1}$$

In order to complete the mathematical formulation, it is necessary to introduce two more elements — one that characterizes the uncertainty and one that characterizes the preference.

g — density function of random bid W; it characterizes your uncertainty about the bid at the time you must decide the asking price.

v — utility function of outcome variable z; it characterizes your subjective valuation of money (or your strength of preference for different amounts of money); it is a strictly increasing cardinal function $v : \mathcal{Z} \to \mathfrak{R}$, such that $v(z)$ is the utility you attach to outcome $z \in \mathcal{Z}$.

These functions are as yet unknown; they are personal; they would have to be assessed by you — the decider. Let us bypass the assessment process, and instead imagine possible shapes of g and v (Figure 15.1). The density function g may be of any proper form. The utility function v must be strictly increasing, but otherwise may assume any shape. Intuitively, when the asking price is fixed, the shapes of the right tails of g and v beyond the point a should matter most because any outcome $z = w$ is possible if $w > a$, whereas only one outcome $z = e$ is possible if $w \leq a$. The probability of the latter outcome is

$$P(Z = e|a) = P(W \leq a) = \int_0^a g(w)\,dw,$$

which equals the area under the shaded portion of the density function g.

15.1.3 Optimal Decision Procedure

Any feasible decision $a \in \mathcal{A}$ is evaluated in terms of its utility. The *utility of decision*, $U(a)$, is defined as the expected utility of outcome resulting from that decision:

$$\begin{aligned} U(a) &= E[v(Z)|a] \\ &= E[v(h(a, W))] \\ &= \int_0^\infty v(h(a, w))g(w)\,dw \\ &= \int_0^a v(e)g(w)\,dw + \int_a^\infty v(w)g(w)\,dw. \end{aligned} \tag{15.2}$$

The three steps of this derivation follow from equation (15.1), where for a fixed decision a, the outcome z is a two-piece function of the bid w; this necessitates partitioning the integration interval $(0, \infty)$ into two subintervals.

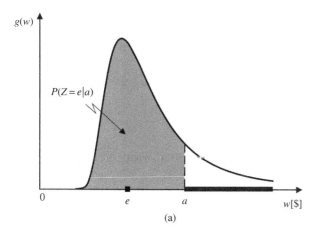

(a)

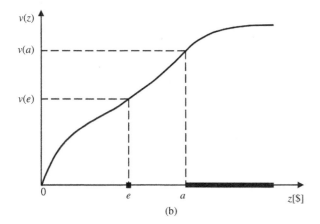

(b)

Figure 15.1 Elements of the asking price model: (a) the density function g of the bid, and (b) the utility function v of the outcome; when the asking price a is fixed, only outcome $z = e$ and outcomes $z = w > a$ matter.

Next, the *optimal decision* a^* is defined as one that maximizes the utility:

$$U(a^*) = \max_{a \in A} U(a). \tag{15.3}$$

To determine a^*, the first derivative of $U(a)$ with respect to a is found:

$$\frac{dU(a)}{da} = v(e)g(a) - v(a)g(a).$$

The necessary condition for a^* to be the optimal decision is that $dU(a^*)/da = 0$. Hence

$$v(e)g(a^*) - v(a^*)g(a^*) = 0,$$

wherefrom

$$v(a^*) = v(e).$$

Inasmuch as the utility function v is strictly increasing, it follows that

$$a^* = e. \tag{15.4}$$

15.1.4 Implications for the Seller

The optimal solution to the asking price problem with the uncertain bid has three practical implications.

1. The optimal asking price equals the value of the property.
2. This optimal decision rule is *absolute* — it is independent of the personal inputs from the decider: (i) the density function g of the bid, and (ii) the utility function v of the money.
3. Ergo, unlike in most decision problems, the decider need not quantify the uncertainty and the strength of preference, but should concentrate on assessing the **fair market value** of the property.

This simple theoretic solution also has a broader implication. It highlights the congruence between a *principle of Bayesian rationality* (maximize the expected utility of outcome) and a *principle of utilitarian ethics* (be honest about the value: do not overprice purposely — that is greed — but do not underprice either). Hence the following advice.

> **Advice** (Asking price). If you wish to maximize your utility, then ask for no more and no less than the value of the property you own.

Of course, this advice does not preclude you from accepting a bid higher than your optimal asking price, or from inducing the buyer to bid higher (via advertising the attributes of your property, for example). After all, the buyer is free to assign a value to your property that reflects his strength of preference for acquiring it.

15.1.5 Free Market Mechanism

Now switch the role of a seller for that of a buyer. Exercise 2 leads you (i) to formulate a bidding model with the asking price being uncertain, and (ii) to prove that the optimal bid b^* equals the value e' of the property to the buyer. Again, this optimal decision rule is *absolute* — it is independent of the density function and the utility function, which the decider might have assessed.

The implication of the two individually optimal decisions is this: The *transaction*, in which the property is exchanged for money, takes place if

$$a^* = e < e' = b^*. \tag{15.5}$$

This transaction rule highlights the congruence between a *principle of Bayesian rationality* (each individual maximizes the expected utility of outcome) and a *principle of free market* ("both sides to a **voluntary** and **informed** economic transaction can benefit from it"). To wit, the buyer gains the property he values (e'), the seller gains the money she wants (e), and the gain of each does not come at the expense of a loss to the other. Of course, this is feasible only when individuals can exercise their **freedom**: to decide and to transact without coercion (by other individuals, or groups, or the state — as by price control, or subsidy, or political dictum).

Bibliographical Notes

The asking price model is due to Jacob Marschak (1979), professor in the Western Management Science Institute, University of California at Los Angeles. The notion of congruence between the principles of Bayesian rationality and the principles of utilitarian ethics comes from John C. Harsanyi (1978), professor of business administration and of economics, University of California at Berkeley, who in 1994 received the Nobel Prize in economics for his contributions to game theory. The principle of the free market and the quotation come from Milton Friedman (1955), professor of economics, University of Chicago, and the 1976 laureate of the Nobel Prize in economics for his achievements in consumption analysis.

15.2 Yield Control Model: Airline Reservations

15.2.1 Yield Control Problem

Airlines tend to overbook flights (above the capacities of airplanes), presumably because such decisions increase the expected profits. Universities tend to overadmit applicants (above the capacities of classrooms and dormitories), presumably because such decisions lead to preferred outcomes in the long run.

If every traveler who had purchased a ticket showed up for boarding, and every applicant who had been admitted matriculated, then the decisions would be trivial: sell the number of tickets, and admit the number of applicants, equal to the capacity. But in reality, there exists uncertainty about the *yield ratio* — the fraction of people who purchased or reserved a service and then show up to receive it. It is the existence of this uncertainty that creates a decision problem, called the *yield control problem*: What is the optimal magnitude of over-booking or over-admitting?

A basic decision model, formulated for an airline, helps us to understand the nuances of the yield control problem, and to rationalize the over-booking tendency. This model is a sequel to the capacity planning model of Chapter 13.

15.2.2 Decision Paradigm

A regularly scheduled flight on a fixed route between two cities is serviced by an airplane with a fixed capacity (the number of passenger seats). The demand for tickets on any flight is unlimited; that is, any number of tickets offered will be sold. Independently of this number, only a fraction of travelers who purchased the tickets show up for boarding. This fraction varies from flight to flight, but its distribution is stationary (across the flights). When the number of boarding passengers is smaller than the airplane capacity, the seats left empty could have been filled in if more tickets were sold. When that number is larger than the airplane capacity, the excess passengers must be rebooked on another flight.

15.2.3 Mathematical Formulation

This yield control model comprises the following elements:

k — airplane capacity (the number of passenger seats).

a — target number of tickets to be sold; it is the *decision variable* assumed to be continuous and positive, $a > k$.

α — target yield fraction: airplane capacity as a fraction of tickets sold, $\alpha = k/a$, where $\alpha \in (\eta_L, \eta_U)$, with $0 \leq \eta_L < \eta_U \leq 1$; it is the *decision variable*, equivalent to a.

V — number of ticket holders arriving to board; it is a *random variable* assumed to be continuous and positive, with realization $v < a$.

Θ — actual yield fraction: fraction of ticket holders arriving to board, $\Theta = V/a$; it is a *random variable* assumed to be continuous on the sample space (η_L, η_U), and having the realization $\theta = v/a$, where $\theta \in (\eta_L, \eta_U)$, with $0 \leq \eta_L < \eta_U \leq 1$. Its distribution is assumed to be independent of a.

H — distribution function of the actual yield fraction Θ, such that $H(\theta) = P(\Theta \leq \theta)$ for any realization $\theta \in (\eta_L, \eta_U)$.

c_s — cost of one seat per flight (on a particular route).

c_r — cost of rebooking one excess passenger with ticket when the airplane is already full.

p_t — price of one ticket for a flight (on a particular route).

The economic parameters must satisfy two nontriviality conditions:

$p_t > c_s$ because otherwise the flight is not economical;
$c_r > p_t$ because otherwise an infinite over-booking would be optimal.

The decision problem is to find the optimal target yield fraction α^* that maximizes the expected profit from a flight. Then, for an airplane with capacity k, the optimal target number of tickets to be sold is $a^* = k/\alpha^*$.

15.2.4 Refinements

Variables

There are two equivalent formulations of the decision model: (i) in terms of the number of passengers a, v; and (ii) in terms of the yield fractions α, θ. The second formulation, to be presented here, utilizes the relationships

$$a = \frac{k}{\alpha}, \qquad v = \theta a;$$

$$k < v \iff \alpha < \theta. \tag{15.6}$$

However, the first formulation is necessary for collecting data: with a fixed, each flight n ($n = 1, \ldots, N$) supplies the count $v(n)$, from which the realization $\theta(n) = v(n)/a$ can be calculated; thus N flights supply a sample $\{\theta(n) : n = 1, \ldots, N\}$ for modeling and estimation of the distribution function H of Θ.

Economic parameters

To estimate the costs c_s and c_r, their definitions must be refined; here is one such a refinement:

c_s — cost of one seat per flight (on a particular route) is the sum of the *capital cost* and the *operating cost*:

$$c_s = c_p + c_h.$$

c_p — capital cost of providing one seat for a flight on a particular route, calculated from the airplane purchase price plus the lifetime maintenance cost, amortized over the airplane lifetime (the total number of flying hours), and scaled to the duration of one flight (in hours).

c_h — operating cost per seat of one flight on a particular route (e.g., the costs of fuel, crew, janitors, passenger services).

c_r — cost of rebooking one excess passenger with a ticket, when the airplane is already full, equals the cost of booking on another flight, plus the cost of reimbursing a stranded passenger for meals and lodging (if necessary), plus the cost of perks to soothe a displeased passenger (e.g., by offering a discount on a future airfare).

Example 15.1 The Canadian-made Bombardier CS100 jet, with two quiet engines, wide seats, and large overhead bins, which is flown by the Latvian Air Baltic airline on short-haul routes in Europe, has 100 passenger seats, the purchase price of about $80 million, and the operating cost of about $7000 per hour.

15.2.5 Criterion Function

Profit per flight

The *profit per flight*, $r(\alpha, \theta)$, depends upon the two yield fractions, target α and actual θ. When no overbooking is allowed, $\alpha = 1$, the number of tickets sold equals the airplane capacity, $a = k$. Consequently, $r(1, \theta) = k(p_t - c_s)$, regardless of the yield fraction θ, and, hence, regardless of the number of ticket holders arriving to board, which is $\theta a = \theta k < k$.

When over-booking is allowed, $0 < \alpha < 1$, two sets of outcomes are possible. If the number of ticket holders arriving to board does not exceed the airplane capacity, $\theta a \le k$, equivalently if $\theta \le \alpha$, the profit depends only on the number of tickets sold, $a = k/\alpha$. If $k < \theta a$, equivalently if $\alpha < \theta$, the profit from selling $a = k/\alpha$ tickets is reduced by the cost of rebooking the excess passengers, who number $\theta a - k = \theta k/\alpha - k$. Thus,

$$r(\alpha, \theta) = \begin{cases} p_t \dfrac{k}{\alpha} - c_s k & \text{if } \theta \le \alpha, \\[2mm] p_t \dfrac{k}{\alpha} - c_s k - c_r \left(\theta \dfrac{k}{\alpha} - k \right) & \text{if } \alpha < \theta, \end{cases}$$

$$= \begin{cases} k\, (p_t \dfrac{1}{\alpha} - c_s) & \text{if } \theta \le \alpha, \\[2mm] k \left[p_t \dfrac{1}{\alpha} - c_s - c_r \left(\dfrac{\theta}{\alpha} - 1 \right) \right] & \text{if } \alpha < \theta. \end{cases} \tag{15.7}$$

Nontriviality conditions

When $\theta \le \alpha$, the flight is profitable, for any $\alpha \le 1$, if $p_t > c_{s_.}$. When $\alpha < \theta$, the profit may be rewritten in the form

$$r(\alpha, \theta) = k \left[(p_t - c_r \theta) \frac{1}{\alpha} - c_s + c_r \right].$$

If $p_t > c_r$ were true, so that $(p_t - c_r \theta) > 0$ for every $\theta \in (0, 1]$, then the profit would be maximized by setting $\alpha = 0$ to obtain $r(0, \theta) = \infty$. This trivial (and unrealistic) solution vanishes if the reverse inequality is true: $p_t < c_r$. Combining the two nontriviality conditions yields the necessary order, $c_s < p_t < c_r$; in other words,

{cost of a seat} < {price of a ticket} < {cost of rebooking}.

Properties of profit function

First, the airplane capacity k plays the role of a scaling constant. Consequently, the optimal α will be independent of k. (Note, however, that c_s may depend on k due to the economy of scale: as k increases, c_s decreases.)

Second, the total cost of rebooking per flight increases with θ/α — the ratio of the two yield fractions, the actual to the target — because $\alpha < \theta$ implies $(\theta/\alpha - 1) > 0$.

Third, as Figure 15.2 illustrates, for a fixed target yield fraction α, the profit is a *two-piece, constant-linear* function of the actual yield fraction θ. This function has the vertex at $\theta = \alpha$, at which the slope changes from 0 to $-c_r/\alpha$, and the minimum at $\theta = 1$ (or at $\theta = \eta_U$ if $\eta_U < 1$); this minimum may be negative (as it is for $\alpha = 0.5$ in Figure 15.2), implying a loss.

Perfect forecast

The graphs of the profit function r in Figure 15.2 imply this: If the actual yield fraction θ were known (were revealed by a clairvoyant), then (i) the optimal target yield fraction would be $\alpha = \theta$, and (ii) the *maximum profit per flight* would be

$$\max_{\alpha} r(\alpha, \theta) = r(\theta, \theta)$$

$$= k \left(p_t \frac{1}{\theta} - c_s \right). \tag{15.8}$$

15.2.6 Optimal Decision Procedure

The optimal decision procedure is developed in three steps. First, any feasible decision α is evaluated in terms of the *expected profit* per flight resulting from that decision:

$$R(\alpha) = E[r(\alpha, \Theta)]. \tag{15.9}$$

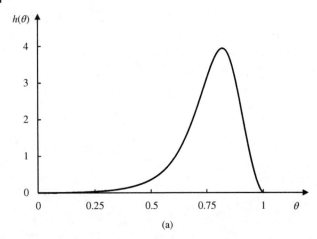

(a)

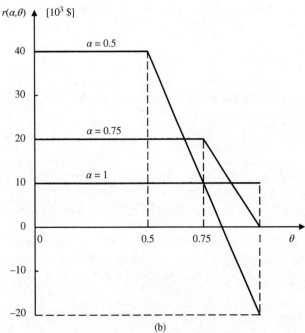

(b)

Figure 15.2 Elements of the yield control model: (a) the density function h of the actual yield fraction Θ, when it is log-ratio logistic LR1(0, 1)-LG(1.3, 0.4); and (b) the profit function $r(\alpha, \cdot)$ of θ for fixed target yield fraction α ($\alpha = 0.5, 0.75, 1$); when $k = 100$ seats, $c_s = \$200/\text{seat}$, $c_r = \$600/\text{passenger}$, $p_t = \$300/\text{ticket}$.

Next, the *optimal decision* α^* is found, which maximizes the expected profit per flight:

$$\max_{\alpha} R(\alpha). \tag{15.10}$$

Finally, the *maximum expected profit* per flight is determined,

$$R^* = R(\alpha^*)$$
$$= \max_{\alpha} R(\alpha). \tag{15.11}$$

The essence of this decision procedure may be visualized with the aid of Figure 15.2. Each of the profit functions $r(\alpha, \cdot)$ of θ offers a different trade-off: the decision $\alpha = 0.5$ may result in the highest profit (when $\theta \leq 0.5$), but also in the lowest profit — a loss (when $\theta > 0.83$); the decision $\alpha = 0.75$ results in a profit within a narrower range, and never in a loss; the decision $\alpha = 1$ results in a small but constant, and therefore certain, profit (regardless of θ). The question: When θ varies randomly from flight to flight, according to the density function h, which decision α maximizes the average profit across many flights? The analytic decision procedure outlined above, and detailed next, allows α to be any number from an open interval (η_L, η_U), evaluates an infinite family of profit functions $r(\alpha, \cdot)$, and finds the one that answers the question.

15.2.7 Expected Profit

Given the two-piece profit function r (expression (15.7)), and a density function h of the actual yield fraction Θ on a bounded open interval (η_L, η_U), the *expected profit* per flight resulting from any feasible decision $\alpha \in (\eta_L, \eta_U)$ is

$$R(\alpha) = E[r(\alpha, \Theta)]$$

$$= \int_{\eta_L}^{\eta_U} r(\alpha, \theta) h(\theta)\, d\theta$$

$$= k \int_{\eta_L}^{\alpha} \left(p_t \frac{1}{\alpha} - c_s \right) h(\theta)\, d\theta + k \int_{\alpha}^{\eta_U} \left[p_t \frac{1}{\alpha} - c_s - c_r \left(\frac{\theta}{\alpha} - 1 \right) \right] h(\theta)\, d\theta.$$

Exercise 4 directs you to prove that the above expression simplifies to

$$R(\alpha) = k \left[(p_t - c_r E(\Theta)) \frac{1}{\alpha} + c_r(1 - H(\alpha)) - c_s + c_r \frac{1}{\alpha} \int_{\eta_L}^{\alpha} \theta h(\theta)\, d\theta \right]. \tag{15.12}$$

The integral in the last summand is called the *incomplete expectation* of Θ. It appeared in Chapters 12 and 13. For some distributions, closed-form expressions can be derived (see Section 12.3); for other distributions, numerical integration may be necessary.

15.2.8 Optimal Decision

Optimality equation

In order to find the optimal decision, the first derivative of $R(\alpha)$ with respect to α must be obtained. Exercise 5 directs you to prove that

$$\frac{dR(\alpha)}{d\alpha} = \frac{k}{\alpha^2} \left[c_r E(\Theta) - p_t - c_r \int_{\eta_L}^{\alpha} \theta h(\theta)\, d\theta \right]. \tag{15.13}$$

The necessary condition for α^* to be the optimal decision is that $dR(\alpha^*)/d\alpha = 0$. Inasmuch as $k/\alpha^2 > 0$ for any $\alpha \in (\eta_L, \eta_U)$, it is the term in the brackets that necessarily must equal zero. It follows that α^* is the *optimal target yield fraction* if it satisfies the following *optimality equation*:

$$\int_{\eta_L}^{\alpha^*} \theta h(\theta)\, d\theta = E(\Theta) - \frac{p_t}{c_r}. \tag{15.14}$$

The method of finding α^* depends on the distribution of Θ. (i) For a distribution that admits a closed-form expression for the incomplete expectation (Section 12.3), equation (15.14) can be solved for α^* either analytically (Exercise 9, distribution UN), or via an iteration method (Exercise 11, distributions P1, P2). (ii) For other distributions, finding α^* requires the combination of an iteration method (or a search method) and a numerical integration method (Exercise 12, distributions of the LR type).

Existence condition

Does α^* exist for any p_t, c_r, and h? Inasmuch as the range of the incomplete expectation of Θ is the open interval $(0, E(\Theta))$, the optimal decision exists if

$$0 < E(\Theta) - \frac{p_t}{c_r} < E(\Theta).$$

The right inequality holds if $0 < p_t/c_r$, which is always true; the left inequality holds if

$$\frac{p_t}{c_r} < E(\Theta). \tag{15.15}$$

This is then a necessary condition for the existence of α^*: the ratio of the ticket price to the rebooking cost must be smaller than the expected yield fraction. Because $E(\Theta) < 1$, constraint (15.15) is tighter than constraint $p_t/c_r < 1$, which follows from the nontriviality condition $p_t < c_r$. Embedding constraint (15.15) in the nontriviality conditions yields the order necessary for the existence of a nontrivial optimal decision:

$$c_s < p_t < p_t/E(\Theta) < c_r. \tag{15.16}$$

This is an interesting feature of the yield control problem with probabilistic forecast: a statistical parameter imposes a constraint on the economic parameters. For instance, $p_t = \$350$ and $c_r = \$500$ give $p_t/c_r = 0.7$. Consequently, when the expected fraction of ticket holders arriving to board is $E(\Theta) = 0.8$, the optimal target fraction α^* exists; but when $E(\Theta) = 0.69$, the decision problem is unsolvable. Of course, the airline does not control the expectation $E(\Theta)$, but it does control the economics of a flight, so it could either decrease p_t or increase c_r to satisfy constraint (15.15).

Expected rebooking

When $\alpha^* < \theta$, rebooking of the excess passengers is needed. This entails not only a cost that reduces the profit, but also an administrative process. To organize this process efficiently, one wishes to know X — the number of excess passengers who must be rebooked; it is a variate, of course, because Θ is random. Exercise 6 directs you to prove that

$$E(X) = k \left[\frac{1}{\alpha^*} \frac{p_t}{c_r} - (1 - H(\alpha^*)) \right]. \tag{15.17}$$

15.2.9 Maximum Expected Profit

When equation (15.14) is utilized inside expression (15.12), a formula for the *maximum expected profit* per flight is obtained. Exercise 7 directs you to prove that

$$R^* = R(\alpha^*)$$
$$= k \left[c_r(1 - H(\alpha^*)) - c_s \right], \tag{15.18}$$

a seemingly simple expression, considering the route we had to take to derive it. Let us examine its properties and implications.

1. The airplane capacity plays the role of a scaling constant, as expected in Section 15.2.5. The maximum expected profit R^* is determined directly by the cost of a seat, c_s, and the cost of rebooking, c_r, but also indirectly by the price of a ticket, p_t, which affects the optimal target yield fraction α^*.
2. The uncertainty about the actual yield fraction Θ determines R^* in two ways: (i) directly through $1 - H(\alpha^*) = P(\Theta > \alpha^*)$, the probability of the actual yield fraction exceeding the optimal target yield fraction, and (ii) indirectly through α^*, which depends on the density function h of Θ, as equation (15.14) shows. The event $\Theta > \alpha^*$ is central to R^* because its occurrence causes the airline to incur the rebooking cost c_r.
3. The cost of rebooking, c_r, is an especially important determinant of R^* because it causes the range of R^* to be bounded.

Theorem (*Bounds on profit*) *Suppose the sample space of Θ is an interval (η_L, η_U); the parameters $E(\Theta)$, c_s, p_t are fixed and satisfy inequalities (15.16); and c_r is any number from the open interval $(p_t/E(\Theta), \infty)$. Then the maximum expected profit per seat is bounded:*

$$\frac{p_t}{\eta_U} - c_s < \frac{R^*}{k} < \frac{p_t}{E(\Theta)} - c_s.$$

Exercise 8 directs you to prove this fact, while we turn to its practical implications.

1. Each bound is greater than $p_t - c_s$, the profit from selling one ticket per seat, simply because $E(\Theta) < \eta_U < 1$.
2. The bounds arise as the limits of R^*/k; the upper bound when $c_r \to p_t/E(\Theta)$; the lower bound when $c_r \to \infty$. Clearly, a lower cost of rebooking produces a higher maximum expected profit.
3. But no matter how high c_r rises, R^*/k does not decline to zero (or turn into a loss). The optimal decision procedure adapts α^* to c_r, so that R^*/k remains positive.

15.2.10 Value of Perfect Forecaster

How much at most should the airline be willing to pay for a perfect forecast of the actual yield fraction? The procedure that answers this question may be rationalized similarly to the procedure of Section 13.2.6, formulated for the basic inventory model. Therefore, we proceed here directly to the expressions.

With the maximum profit per flight specified by expression (15.8), the *expected maximum profit* per flight is

$$R^{**} = E\left[\max_\alpha r(\alpha, \Theta)\right]$$
$$= E\left[k\left(p_t\frac{1}{\Theta} - c_s\right)\right]$$
$$= k\left[p_t E\left(\frac{1}{\Theta}\right) - c_s\right]. \tag{15.19}$$

The *value of the perfect forecaster* of the actual yield fraction Θ per flight is defined as the difference between the expected maximum profit per flight R^{**} and the maximum expected profit per flight R^*:

$$VPF = R^{**} - R^*$$
$$= k\left[p_t E\left(\frac{1}{\Theta}\right) - c_r(1 - H(\alpha^*))\right]. \tag{15.20}$$

The *VPF* admits two useful interpretations. First, it is the maximum price the airline should be willing to pay for one perfect forecast of the actual yield fraction. Second, it is an upper bound on the value of any imperfect forecast.

Note that *VPF* is unaffected by c_s because the cost kc_s is incurred in each flight, regardless of the number of tickets sold or the number of passengers rebooked.

The calculation of *VPF* entails the expectation $E(1/\Theta)$. Formulae for this expectation may be derived for some distributions (see Exercises 9 and 11); for other distributions, one must resort to numerical integration. (Note that $E(1/\Theta) \neq 1/E(\Theta)$.)

15.3 Yield Control Model: College Admissions

The basic decision model, formulated in Section 15.2 for yield control in the context of airline reservations, is now adapted to the context of college admissions.

15.3.1 Decision Paradigm

Every year, a college receives applications whose number vastly exceeds the freshman class capacity. But only a fraction of students admitted in January accept the admission offer in April, and then matriculate in August. Thus the yield control problem: What is the optimal number of applicants to be offered admission?

Three aspects distinguish this problem from the airline reservation problem. (i) The interpretation of model elements. (ii) The source and timing of the revenue: tuition plus fees are collected not from all admitted applicants in January, but only from matriculants in August. (iii) The nonprofit status of the college.

Capturing these three aspects mathematically is the objective of this section. Aspects of application are deferred to mini-projects (Exercises 14, 15, 16).

15.3.2 Mathematical Formulation

This yield control model comprises the following elements:

k — freshman class capacity.
a — target number of admission offers, $a > k$.
α — target yield fraction: $\alpha = k/a$, $\alpha \in (\eta_L, \eta_U)$, $0 \le \eta_L < \eta_U \le 1$.
V — number of admitted applicants who matriculate, $v < a$.
Θ — actual yield fraction: $\Theta = V/a$, $\theta = v/a$, $\theta \in (\eta_L, \eta_U)$.
H — distribution function of Θ.
c_s — annual cost of education and services per freshman.
c_r — annual cost of over-admitting one freshman, in excess of k.
p_t — annual revenue per freshman (tuition plus fees).

The economic parameters must satisfy the nontriviality conditions:

$$c_s < p_t < c_r.$$

Assumptions
Once the freshman class capacity k is fixed: (i) the annual cost $c_s k$ is committed; (ii) the annual cost of over-admitting, $c_r(v - k)$, is incurred only if $k < v$; (iii) the annual revenue is $p_t v$. (iv) Application fees and forfeited deposits (from applicants who accept the admission offer but do not matriculate) are ignored.

The decision problem is to find the optimal target yield fraction α^* that maximizes the expected profit per year from the freshman class. Then the optimal target number of admissions to be offered is $a^* = k/\alpha^*$.

15.3.3 Nonprofit Principle

Whereas for-profit colleges (established by entrepreneurs and investors) exist, most colleges claim a nonprofit status. In principle, such colleges should set tuition and fees to recoup the necessary costs — but nothing else. Assuming a college adheres to this principle when it sets the economic parameters (c_s, c_r, p_t), the maximization of the expected profit per freshman class, max $R(\alpha)$, may be viewed as the maximization of economic efficiency of college admissions. But our decision model goes further: for the specified costs (c_s, c_r), it allows us to find the amount p_t^o of tuition plus fees and the target yield fraction α^o that guarantee $R(\alpha^o) = \max R(\alpha) = 0$. In other words, we can find the joint solution (p_t^o, α^o) that is economically optimal under uncertainty about the actual yield fraction, and is simultaneously nonprofit.

15.3.4 Criterion Function

Profit per freshman class
The *profit per freshman class*, $r(\alpha, \theta)$, depends on α and θ. When no over-admitting is allowed, $\alpha = 1$, $a = k$, and $v = k\theta$. Consequently, $r(1, \theta) = p_t v - c_s k = k(p_t \theta - c_s)$. The expected profit is

$$E[r(1, \Theta)] = k(p_t E(\Theta) - c_s).$$

For it to be nonnegative, $c_s/E(\Theta) \leq p_t$ is necessary. This lower bound on p_t is higher than the one stemming from the nontriviality conditions because $E(\Theta) \in (0,1)$.

When over-admitting is allowed, $0 < \alpha < 1$, $a = k/\alpha$, and $v = k\theta/\alpha$. Consequently, two sets of outcomes are possible: those when the freshman class capacity is not exceeded, $v \leq k$, equivalently $\theta \leq \alpha$; and those when over-crowding occurs and must be dealt with at a cost of c_r per freshman in excess of capacity, $k < v$, equivalently $\alpha < \theta$. Thus,

$$
r(\alpha, \theta) = \begin{cases} p_t \theta \dfrac{k}{\alpha} - c_s k & \text{if } \theta \leq \alpha, \\[2ex] p_t \theta \dfrac{k}{\alpha} - c_s k - c_r \left(\theta \dfrac{k}{\alpha} - k \right) & \text{if } \alpha < \theta, \end{cases}
$$

$$
= \begin{cases} k\left(p_t \dfrac{\theta}{\alpha} - c_s\right) & \text{if } \theta \leq \alpha, \\[2ex] k\left[(p_t - c_r)\dfrac{\theta}{\alpha} + c_r - c_s\right] & \text{if } \alpha < \theta. \end{cases} \tag{15.21}
$$

The implied nontriviality condition is $p_t < c_r$, for if $p_t > c_r$ were true, then the profit would be maximized by setting $\alpha = 0$ to obtain $r(0, \theta) = \infty$ for every $\theta \in (0,1)$.

Perfect forecast

If the actual yield fraction θ were known (were revealed by a clairvoyant), then (i) the optimal target yield fraction would be $\alpha = \theta$, and (ii) the *maximum profit per freshman class* would be

$$
\max_{\alpha} r(\alpha, \theta) = r(\theta, \theta)
$$
$$
= k(p_t - c_s). \tag{15.22}
$$

For this maximum profit to be positive, $c_s < p_t$ is required, which is the second nontriviality condition. For a nonprofit college receiving a perfect forecast of Θ, setting $p_t = c_s$ would be required.

The reader should compare the functions (15.21)–(15.22) with the corresponding functions (15.7)–(15.8) for the airline reservation problem. The distinction is caused by the impact of the actual yield fraction θ (and the uncertainty associated with it): while θ affects only the over-booking cost ($c_r \theta k/\alpha$) for the airline, it affects both the over-admitting cost ($c_r \theta k/\alpha$) and the revenue ($p_t \theta k/\alpha$) for the college. This distinction propagates to the optimal decision.

15.3.5 Optimal Decision Procedure

This procedure follows the steps defined in Section 15.2.6, and parallels the implementation explained in Sections 15.2.7–15.2.10. Therefore, the explanations are not repeated, and the derivations are deferred to Exercise 13. Here are the results.

Expected profit per freshman class, for any $\alpha \in (\eta_L, \eta_U)$:

$$
R(\alpha) = E[r(\alpha, \Theta)]
$$
$$
= \int_{\eta_L}^{\eta_U} r(\alpha, \theta) h(\theta)\, d\theta
$$
$$
= k\left[(p_t - c_r)E(\Theta)\frac{1}{\alpha} + c_r(1 - H(\alpha)) - c_s + c_r \frac{1}{\alpha} \int_{\eta_L}^{\alpha} \theta h(\theta)\, d\theta \right]. \tag{15.23}
$$

Optimality equation for $\alpha^* \in (\eta_L, \eta_U)$:

$$\int_{\eta_L}^{\alpha^*} \theta h(\theta) \, d\theta = \left(1 - \frac{p_t}{c_r}\right) E(\Theta). \tag{15.24}$$

Expected number of excess matriculants who must be accommodated:

$$E(X) = k \left[\frac{1}{\alpha^*} \frac{p_t}{c_r} E(\Theta) - (1 - H(\alpha^*))\right]. \tag{15.25}$$

Maximum expected profit per freshman class:

$$R^* = R(\alpha^*)$$
$$= k \left[c_r(1 - H(\alpha^*)) - c_s\right]. \tag{15.26}$$

Value of the perfect forecaster of Θ per freshman class:

$$VPF = k \left[p_t - c_r(1 - H(\alpha^*))\right]. \tag{15.27}$$

Two properties of these results are noteworthy. (i) The *optimal target yield fraction* α^*, equation (15.24), depends on the uncertainty about Θ, which is quantified by the density function h, and on the ratio p_t/c_r. This may be interpreted as the ratio of the marginal opportunity losses: p_t — the unit revenue lost due to the *shortage* of matriculants (when $\theta < \alpha$); and c_r — the unit cost incurred due to the *excess* of matriculants (when $\alpha < \theta$). (ii) Once α^* is found, the only relevant measure of uncertainty, which affects $E(X)$, R^*, and VPF, is $P(\Theta > \alpha^*) = 1 - H(\alpha^*)$, the probability of excess matriculants.

15.3.6 Nonprofit Optimal Solution

The nonprofit principle, introduced in Section 15.3.3, prescribes $R(\alpha^*) = 0$. This equality may not hold when all four input elements, which determine $R(\alpha^*)$, are specified beforehand. But suppose the college provided h, and set only the costs (c_s, c_r), each being necessary to achieve its educational goals. Then we might ask: Is there a tuition plus fees amount p_t^o such that the optimal target fraction α^o yields the maximum expected profit $R(\alpha^o)$ equal to zero?

To find the answer, start with expression (15.26), substitute α^* by α^o, set $R(\alpha^o) = 0$, and solve the resultant equation for α^o. Next, take expression (15.24), replace α^* by the calculated α^o, substitute p_t by p_t^o, and solve the resultant equation for p_t^o. The details of this deduction may be arranged into an algorithm.

Step 0. Given are $c_s, c_r, h, H^{-1}, E(\Theta)$.

Step 1. Calculate the nonprofit optimal target yield fraction:

$$\alpha^o = H^{-1}\left(1 - \frac{c_s}{c_r}\right).$$

Step 2. Calculate the incomplete expectation of Θ:

$$I^o = \int_{\eta_L}^{\alpha^o} \theta h(\theta) \, d\theta.$$

Step 3. Calculate the nonprofit tuition plus fees:

$$p_t^o = c_r \left(1 - \frac{I^o}{E(\Theta)}\right).$$

Properties

First, α^o is the quantile of the actual yield fraction Θ corresponding to probability $1 - c_s/c_r$; the nontriviality condition, $c_s < c_r$, ensures that $0 < c_s/c_r < 1$. Second, p_t^o is smaller than the over-admission cost, as required by the nontriviality condition, $p_t^o < c_r$, because $0 < I^o < E(\Theta)$.

Interpretations

The pair (p_t^o, α^o) constitutes the simultaneous solution to two decision problems arising from the uncertainty that exists at the admission time about the actual freshman class size: p_t^o is the *nonprofit tuition plus fees* amount to be charged the matriculated freshman by the college bursar; α^o is the *nonprofit optimal target yield fraction* to be used in the college admissions.

Note on Principles

The principles of economic efficiency and nonprofit (Section 15.3.3) underlying this decision model are idealistic. Unlike businesses that always strive to keep the costs low, American colleges and universities do not. In 30 years (1975–2005), they increased the number of full-time professors by 51%, in sync with the growth of student enrollments (which held the student to faculty ratio constant), but they also increased the number of administrators and staffers by 85% and 240%, respectively (Ginsberg, 2011). This bloated administrative apparatus, and the indulgent services it creates, cost. In 25 years (1990–2015), the cost of attending a four-year college or university rose about four times faster than the rate of inflation (*The Wall Street Journal*, 20 May 2013). In 20 years (2002–2022), the median (in terms of spending increases) of the 50 best-known public universities (one per state) raised the average cost of attending it by 64% above inflation (*The Wall Street Journal*, 11 August 2023: "State Colleges 'Devour' Money, and Students Foot the Bill"). Do these statistics suggest an economically efficient operation of colleges and universities? A broader question: "Is College Worth It?" (Bennett and Wilezol, 2013).

Note on Bargaining Market

Increasingly, American colleges and universities emulate for-profit enterprises. They employ enrollment managers who set prices as high as most families would pay, then assess the "price sensitivity" of applicants, and offer discounts (on tuition, room and board) to the highly valued ones, while parents of applicants admitted to several colleges bargain for discounts, often with the help of paid consultants (*The Wall Street Journal*, 30 June 2020). This creates a market between a seller (who sets the asking price) and a buyer (who responds with a bid) — not unlike the decision paradigm analyzed in Section 15.1. Here is the empirical evidence of this market: In the 2019–20 academic year, the difference between the published undergraduate tuition and the actually paid tuition could be over $22 000 at a private college and over $6000 for an in-state student at a public college. In such a *bargaining market*, the optimal solution from our decision model, either (p_t, α^*) or (p_t^o, α^o), could offer an initial guidance to the college enrollment manager or to an admission consultant.

Exercises

1 **Asking price: value of perfect forecast**. Imagine you want to sell some property whose value you assessed to be $e = \$1300$. How much at most should you be willing to pay the clairvoyant for disclosing the bid b before you decide the asking price a?

2 **Bidding model**. Your objective is to formulate and to solve the bidding model, which is a *counterpart* to the asking price model (Section 15.1).
   2.1 Rewrite the decision paradigm (Section 15.1.1) from the viewpoint of a buyer.

2.2 Formulate the bidding model mathematically, in parallel to the asking price model (Section 15.1.2), under these assumptions. The buyer (i) assessed the value of the property to him as e', (ii) does not know the asking price W, (iii) can characterize the uncertainty about the asking price in terms of a density function g, and (iv) can characterize his subjective valuation of money in terms of a utility function v.

2.3 Formulate the optimal decision procedure that maximizes the expected utility of outcome to the buyer, in parallel to the procedure for the seller (Section 15.1.3), and derive the expression for the optimal bid b^*.

3 **Bid: value of perfect forecast**. Imagine you want to bid for some property whose value to you is $e' = \$800$. How much at most should you be willing to pay the clairvoyant for disclosing the asking price a before you decide the bid b?

4 **Expected profit**. Derive expression (15.12), for the expected profit per flight, from the expression that precedes it.

5 **Optimal decision**. Derive expression (15.13), for the derivative of the expected profit per flight, from expression (15.12).

6 **Expected rebooking**. Derive expression (15.17). *Hint*: Follow the rationalization of the profit function, and express x as a two-piece function of θ, in parallel to equation (15.7).

7 **Maximum expected profit**. Derive expression (15.18), for the maximum expected profit per flight, from expressions (15.12) and (15.14).

8 **Bounds on profit**. Prove the theorem of Section 15.2.9.
Hints. (i) Frame the proof as the task of finding first $\lim \alpha^*$, and next $\lim R(\alpha^*)$, as c_r approaches an endpoint of its domain. (ii) Use either expression (15.18) for $R(\alpha^*)$, or expression (15.12) for $R(\alpha)$. Think thusly: $\lim R(\alpha) = R(\alpha^*)$, as α approaches α^*.

9 **Uniform yield fraction**. A regular daily flight on a particular route is serviced by an airplane with 134 passenger seats. An airline economist estimates c_s [\$/seat] and c_r [\$/passenger], and then sets $p_t = c_r E(\Theta) - 100$ [\$/ticket]. The distribution of the actual yield fraction Θ is $UN(\eta_L, \eta_U)$.

Options:	(A)	$\eta_L = 0.75$,	$\eta_U = 0.95$;	$c_s = 200$,	$c_r = 500$.
	(B)	$\eta_L = 0.65$,	$\eta_U = 0.95$;	$c_s = 200$,	$c_r = 560$.
	(C)	$\eta_L = 0.7$,	$\eta_U = 0.9$;	$c_s = 250$,	$c_r = 600$.
	(D)	$\eta_L = 0.6$,	$\eta_U = 0.9$;	$c_s = 250$,	$c_r = 660$.
	(E)	$\eta_L = 0.65$,	$\eta_U = 0.85$;	$c_s = 300$,	$c_r = 700$.
	(F)	$\eta_L = 0.55$,	$\eta_U = 0.85$;	$c_s = 300$,	$c_r = 760$.

9.1 Write the expressions for the density function h of Θ, the distribution function H of Θ, the expectation of Θ, and the incomplete expectation of Θ.

9.2 Derive the expressions for (i) the optimal target yield fraction α^* (in terms of parameters and $E(\Theta)$), (ii) the exceedance probability $P(\Theta > \alpha^*)$, and (iii) the maximum expected profit per flight R^* (in terms of parameters and α^*). Calculate the three quantities; interpret them for an airline executive.

9.3 Calculate the optimal target number of tickets to be sold per flight, the expected number of ticket holders arriving to board, and the expected number of excess passengers who must be rebooked.

9.4 Calculate the bounds on the maximum expected profit per seat. Do they satisfy the theorem? What advice do these bounds imply for an airline executive?

9.5 Make a figure showing two graphs on the sample space (η_L, η_U): a graph of the density function h, and a graph of the profit function $r(\alpha^*, \cdot)$. Explain the figure to an airline executive.

9.6 Derive the expression for $E(1/\Theta)$. Calculate the value of perfect forecast of Θ per flight; interpret this value for an airline executive.

10 **Power distribution** — *preliminary* to Exercise 11. An expert hired by an airline analyzed a sparse record of data from flights on a particular route (the number of tickets sold per flight, the number of ticket holders arriving to board), as well as data from flights on other but similar routes. He then assessed the bounds (η_L, η_U) for the sample space of the actual yield fraction Θ, and the median $\theta_{0.5}$ of Θ for the particular route, and hypothesized that Θ has a power-type distribution.

Options for the median when $\Theta \sim P1(\beta, 0.43, 0.91)$:

(12)	$\theta_{0.5} = 0.769$,	(14)	$\theta_{0.5} = 0.834$,
(13)	$\theta_{0.5} = 0.811$,	(15)	$\theta_{0.5} = 0.848$.

Options for the median when $\Theta \sim P2(\beta, 0.51, 0.98)$:

(22)	$\theta_{0.5} = 0.648$,	(24)	$\theta_{0.5} = 0.585$,
(23)	$\theta_{0.5} = 0.607$,	(25)	$\theta_{0.5} = 0.571$.

10.1 Estimate the shape parameter β of the distribution specified in the option; round the estimate to the nearest integer.

10.2 Graph the density function h of Θ, and the distribution function H of Θ.

10.3 Calculate the standard deviation and the mean of Θ. Locate the mean $E(\Theta)$ and the median $\theta_{0.5}$ on the horizontal axis of the graph of h. Do the mean, the median, and the mode satisfy the common order? *Hint*: Recall Section 2.6.4.

Mini-Projects

11 **Power yield fraction**. A regular daily flight on a particular route is serviced by an airplane with 156 passenger seats. An airline economist estimates c_s [\$/seat] and c_r [\$/passenger], and then sets $p_t = c_r E(\Theta) - 50$ [\$/ticket]. The distribution of the actual yield fraction Θ and the parameter values are specified in the options. *Options for the distribution*:

(P1)	$P1(\beta, 0.43, 0.91)$.
(P2)	$P2(\beta, 0.51, 0.98)$.

Options for the shape parameter:

(B2)	$\beta = 2$,	(B4)	$\beta = 4$,
(B3)	$\beta = 3$,	(B5)	$\beta = 5$.

Options for the economic parameters:

(A)	$c_s = 210$,	$c_r = 510$.	(D)	$c_s = 260$,	$c_r = 670$.
(B)	$c_s = 210$,	$c_r = 570$.	(E)	$c_s = 310$,	$c_r = 710$.
(C)	$c_s = 260$,	$c_r = 610$.	(F)	$c_s = 310$,	$c_r = 760$.

11.1 Write the expressions for the density function h of Θ, the distribution function H of Θ, the expectation of Θ, and the incomplete expectation of Θ.

11.2 Write the optimality equation, and insert into it the expression for the incomplete expectation of Θ. Find the optimal target yield fraction α^* via an *iteration method* as follows. (i) Solve the optimality equation for α^* that appears in the term raised to power β; show every step toward the solution, which is: when $\Theta \sim \mathrm{P1}(\beta, \eta_L, \eta_U)$,

$$\alpha^* = \left(\frac{c(\beta + 1)m^\beta}{\beta\alpha^* + \eta_L} \right)^{\frac{1}{\beta}} + \eta_L;$$

when $\Theta \sim \mathrm{P2}(\beta, \eta_L, \eta_U)$,

$$\alpha^* = \eta_U - m \left(\frac{\eta_U + \beta\eta_L - (\beta + 1)c}{\eta_U + \beta\alpha^*} \right)^{\frac{1}{\beta}};$$

where $m = \eta_U - \eta_L$ and $c = E(\Theta) - p_t/c_r$. (ii) Each of the two equations is now in the form of a *contraction operator* ξ, such that $\alpha^* = \xi(\alpha^*)$. Choose an initial point α_0^* (say, $\alpha_0^* = E(\Theta)$), and calculate a sequence $\{\alpha_1^*, \alpha_2^*, \alpha_3^*, \ldots\}$ recursively as $\alpha_{i+1}^* = \xi(\alpha_i^*)$ for $i = 0, 1, 2, \ldots$ until $|\alpha_{i+1}^* - \alpha_i^*| < \varepsilon$ for some small $\varepsilon > 0$. Then take α_{i+1}^* as the optimal point α^*.

11.3 Write the expressions for the exceedance probability $P(\Theta > \alpha^*)$, and the maximum expected profit per flight R^*. Calculate the two quantities; interpret them for an airline executive.

11.4 Calculate the optimal target number of tickets to be sold per flight, the expected number of ticket holders arriving to board, and the expected number of excess passengers who must be rebooked.

11.5 Calculate the bounds on the maximum expected profit per seat. Do they satisfy the theorem? What advice do these bounds imply for an airline executive?

11.6 Make a figure showing two graphs on the sample space (η_L, η_U): a graph of the density function h, and a graph of the profit function $r(\alpha^*, \cdot)$. Explain the figure to an airline executive.

11.7 Write the integral that defines $E(1/\Theta)$, and solve it as follows. (i) Apply the substitution rule to arrive at an expression containing an integral whose indefinite form is

$$F_n(x) = \int \frac{x^n}{ax + b}\, dx, \quad n \in \{1, 2, 3, 4\}.$$

(ii) Employ the antiderivative for the given n (Bois, 1961):

$$F_1(x) = \frac{x}{a} - \frac{b}{a^2}\ln(ax + b),$$

$$F_2(x) = \frac{x^2}{2a} - \frac{bx}{a^2} + \frac{b^2}{a^3}\ln(ax + b),$$

$$F_3(x) = \frac{x^3}{3a} - \frac{bx^2}{2a^2} + \frac{b^2 x}{a^3} - \frac{b^3}{a^4}\ln(ax + b),$$

$$F_4(x) = \frac{x^4}{4a} - \frac{bx^3}{3a^2} + \frac{b^2 x^2}{2a^3} - \frac{b^3 x}{a^4} + \frac{b^4}{a^5}\ln(ax + b).$$

11.8 Calculate the value of perfect forecast of Θ per flight; interpret this value for an airline executive.

12 **Log-ratio yield fraction**. A regular daily flight on a particular route is serviced by an airplane with 178 passenger seats. An airline economist estimates the cost of a seat per flight c_s [\$/seat], sets the nominal profit from a seat per flight to $p_t - c_s = 127$ [\$/seat], and constrains the cost of rebooking an excess passenger to $c_r = p_t/E(\Theta) + 90$ [\$/passenger]. The actual yield fraction Θ has a log-ratio type distribution on the sample space $(\eta_L, \eta_U) = (0.31, 0.89)$; the type and the parameter values are specified in the options.

Options for the distribution type:

(LR-LG) $LR1(\eta_L, \eta_U)\text{-}LG(\beta, \alpha)$.

(LR-LP) $LR1(\eta_L, \eta_U)\text{-}LP(\beta, \alpha)$.

(LR-GB) $LR1(\eta_L, \eta_U)\text{-}GB(\beta, \alpha)$.

(LR-RG) $LR1(\eta_L, \eta_U)\text{-}RG(\beta, \alpha)$.

(LR-NM) $LR1(\eta_L, \eta_U)\text{-}NM(\beta, \alpha)$.

Options for the parameters:

(A) $\beta = 1.3$, $\alpha = 0.4$; $c_s = 170$.

(B) $\beta = 0.9$, $\alpha = 0.6$; $c_s = 170$.

(C) $\beta = 0.5$, $\alpha = 0.4$; $c_s = 190$.

(D) $\beta = -0.5$, $\alpha = 0.6$; $c_s = 190$.

(E) $\beta = -0.9$, $\alpha = 0.4$; $c_s = 210$.

(F) $\beta = -1.3$, $\alpha = 0.6$; $c_s = 210$.

12.1 Construct the expressions for the density function h of Θ, and the distribution function H of Θ. Graph each function.

12.2 Write the expressions for $E(\Theta)$ and $E(1/\Theta)$ in terms of the density function h. Evaluate each expectation. *Hint*: Employ a numerical integration method.

12.3 Write the optimality equation. Find the optimal target yield fraction α^*. *Hint*: Use an iteration method to solve the optimality equation; at each iteration, evaluate the incomplete expectation via a numerical integration method.

12.4 Write the expressions for the exceedance probability $P(\Theta > \alpha^*)$, and the maximum expected profit per flight R^*. Calculate the two quantities; interpret them for an airline executive.

12.5 Calculate the optimal target number of tickets to be sold per flight, the expected number of ticket holders arriving to board, and the expected number of excess passengers who must be rebooked.

12.6 Calculate the bounds on the maximum expected profit per seat. Do they satisfy the theorem? What advice do these bounds imply for an airline executive?

12.7 Make a figure showing two graphs on the sample space (η_L, η_U): a graph of the density function h, and a graph of the profit function $r(\alpha^*, \cdot)$. Explain the figure to an airline executive.

12.8 Calculate the value of perfect forecast of Θ per flight; interpret this value for an airline executive.

13 **College admissions: derivations**. The decision model is as described in Sections 15.3.4–15.3.6. The exercises below pertain to the derivations of expressions. Each exercise is independent of the other exercises.

13.1 Derive expression (15.23), for the expected profit per freshman class, from expression (15.21) for the profit function.

13.2 Find the derivative $dR(\alpha)/d\alpha$, and derive the optimality equation (15.24).

13.3 Derive expression (15.25) for the expected number of excess matriculants. *Hint*: Follow the rationalization of the profit function, and express x as a two-piece function of θ, in parallel to equation (15.21).

13.4 Derive expression (15.26), for the maximum expected profit per freshman class, from expressions (15.23) and (15.24).

13.5 Derive expression (15.27), for the value of perfect forecaster, from expressions (15.22) and (15.26).

13.6 Fix all input elements in the optimality equation (15.24), except the annual cost of over-admitting one freshman c_r. Consider the optimal target yield fraction α^* as a function of c_r on the interval (p_t, ∞). Prove that this function is strictly monotone, and that its range is the open interval (η_L, η_U). (A function is *strictly monotone* on an interval if it is either strictly increasing or strictly decreasing on that interval.)

13.7 Prove that the expression for α^o in the algorithm of Section 15.3.6 yields $R(\alpha^o) = 0$, where R is specified by expression (15.23).

14 **College admissions: economics.** The decision problem is as described in Sections 15.3.1–15.3.3. The exercises below pertain to the economic and system analysis of the problem. Each exercise is independent of the other exercises.

14.1 List all the assumptions behind the yield control model for college admissions. Examine each assumption: Is it reasonable? Or is it questionable, and if so, why?

14.2 Design a method of determining the freshman class capacity of a college. (What numbers to consider, and how to synthesize them into k.)

14.3 List the major consequences of under-admitting (enrollment below the capacity) and over-admitting (enrollment above the capacity).

14.4 Detail the summands of each economic parameter (c_s, c_r, p_t), so that it could be calculated from data collected in years past.

14.5 Find the values of (k, c_s, c_r, p_t) at your college. Graph the resultant profit function $r(\alpha, \cdot)$ on the domain $(0, 1)$ for $\alpha = 0.5, 0.75, 1$.

Hints. (i) If a value is unavailable, then assess it judgmentally (say, as your judgmental median) based on the available related information. (ii) If no sufficient information is available, then imagine being an entrepreneur: Design your own college for a community of your choice, and set the parameters to the values that you judge reasonable for the financial viability of your college and its attractiveness to prospective students.

15 **College admissions: forecasting.** This forecast problem arises within the decision paradigm described in Sections 15.3.1–15.3.2. The grand objective is to prepare a probabilistic forecast of the actual yield fraction Θ. The forecast should be available in January before the admission offers are decided; the realization of Θ becomes known in August; thus the required forecast lead time is 8 months. A statistical model developed from the Bayesian forecaster theory (Section 10.1) would quantify the uncertainty about Θ in terms of a posterior distribution, which revises a prior distribution (invariant from year to year) based on a predictand realization (known in January and varying from year to year). The objective of this exercise is narrower: to develop only a model for the prior distribution of Θ based on the admission data recorded in the years past and specified in the option. These data (Table 15.1) comprise joint realizations from $N = 19$ years of three variables: b — the number of applicants; a — the target number of admission offers (decided in January); v — the number of matriculants (observed in August).

Note. The variable b is redundant for the decision model, but its values are reported because they may explain, at least partially, the variability and trends of the other two variables.

Options for the data:

(U) University of Virginia.

(A) School of Architecture.

(E) School of Engineering.

15.1 *Exploratory data analysis.* For each year, calculate θ to obtain the joint sample $\{(b(n), a(n), v(n), \theta(n)) : n = 1, \dots, N\}$, where $n = 1$ is year 1990 and $n = 19$ is year 2008. The sample of each variable may be called a *time series*. To learn the behavior of these data, graph each of the four time series (e.g., $\theta(n)$ versus n for $n = 1, \dots, 19$). Characterize in words the overall pattern of each time series. Which time series are similar and which are not?

15.2 *Validation of fraction stationarity.* An assumption simplifying the modeling is that the prior distribution function H of the actual yield fraction Θ is identical for every year. This is true if the sample

Table 15.1 Record of fall admissions of the first-year undergraduate applicants to the University of Virginia and to its two schools: Architecture and Engineering; b − number of applicants; a − number of admission offers made; v − number of matriculants.

Year	University			Architecture			Engineering		
	b	a	v	b	a	v	b	a	v
1990	12 862	4943	2566	455	136	81	1984	950	411
1991	14 334	4861	2541	470	130	79	2181	937	398
1992	12 318	5282	2804	352	154	86	1911	961	439
1993	13 374	5384	2678	351	162	82	1949	1024	420
1994	12 845	5713	2764	329	166	86	1811	1102	444
1995	13 569	5712	2882	367	170	89	1901	1164	467
1996	16 898	5650	2834	433	167	85	2328	1225	483
1997	16 189	5755	2909	450	148	86	2297	1228	541
1998	15 955	5429	2908	434	155	96	2319	1079	501
1999	16 461	5588	2928	483	166	84	2558	1105	508
2000	14 145	5482	2930	380	162	97	2085	1085	510
2001	14 739	5534	2983	467	152	83	2138	1141	540
2002	14 320	5588	2999	424	160	87	1841	1052	494
2003	14 627	5775	3102	433	157	94	1836	1135	523
2004	14 822	5760	3097	448	159	87	1983	1144	545
2005	15 657	5898	3113	476	162	81	2066	1159	503
2006	16 086	6019	3091	493	145	72	2123	1192	544
2007	17 798	6273	3248	471	150	80	2442	1187	559
2008	18 363	6735	3257	503	163	72	2595	1412	582

Source: Office of Institutional Assessment and Studies, University of Virginia, 2014.

$\{\theta(n) : n = 1, \ldots, N\}$ exhibits properties of a *random sample* (for the definition, see Section 6.1.2). These properties can be validated approximately by examining the graph of the time series $\theta(n)$ versus n: judge visually whether or not the scatter of points about the horizontal line passing through the sample median of Θ is (i) random on each side, and (ii) independent of n. Report your observations and conclusion.

15.3 *Validation of fraction invariance.* An assumption embedded in the decision model (Section 15.3.2) is that the distribution function H of the actual yield fraction $\Theta = V/a$ is independent of the target number of admission offers a. To validate it, make a scatterplot of $\theta(n)$ versus $a(n)$ for all n ($n = 1, \ldots, N$). Next judge visually whether or not the scatter of points about the horizontal line passing through the sample median of Θ is (i) random on each side, and (ii) independent of a. Report your observations and conclusion.

15.4 *Model for prior distribution.* Under the hypothesis that the given data satisfy, at least approximately, the two assumptions tested above, the prior distribution function H of the actual yield fraction Θ can be modeled based on the given sample $\{\theta(n) : n = 1, \ldots, N\}$. Toward this end, apply the distribution modeling methodology of Chapter 3 to obtain a parametric model for H. Hypothesize any of the seven models on a bounded open interval. Estimate the parameters via the method of your choice. Document all steps of the methodology.

16 **College admissions: decision making**. The decision problem and its model are as described in Section 15.3. Given the inputs specified in the options, with c_s, c_r, p_t in [\$/academic year], perform a complete decision analysis for the admissions office of a college.

Options for the distribution:

(PD)	Prior distribution from Exercise 15.
(UN)	UN(0.53, 0.87).
(P1)	P1(3, 0.43, 0.89).
(P2)	P2(4, 0.51, 0.98).
(LR-LG)	LR1(0.39, 0.81)-LG(1.3, 0.4).
(LR-LP)	LR1(0.41, 0.83)-LP(0.9, 0.6).
(LR-GB)	LR1(0.43, 0.85)-GB(0.5, 0.4).
(LR-RG)	LR1(0.45, 0.87)-RG(−0.5, 0.6).
(LR-NM)	LR1(0.47, 0.89)-NM(−0.9, 0.4).

Options for the parameters:

(Y)	(k, c_s, c_r, p_t) are yours, from Exercise 14.5.
(A)	$k = 3360, c_s = 26\ 800, c_r = 33\ 700, p_t = 30\ 100.$
(B)	$k = 1490, c_s = 15\ 800, c_r = 22\ 600, p_t = 19\ 700.$

16.1 Write the expressions for the density function h of Θ, and the distribution function H of Θ. Graph each function.

16.2 Write the expression for $E(\Theta)$ and evaluate it.
Hints. (i) For the UN, P1, P2 distributions, there are closed-form expressions. (ii) For a LR-type distribution, numerical integration is needed.

16.3 Write the optimality equation. Find the optimal target yield fraction α^*.
Hints. (i) For the UN distribution, derive the closed-form expression for α^*. (ii) For the P1 or P2 distribution, formulate and use an iteration method parallel to that described in Exercise 11.2. (iii) For an LR-type distribution, use an iteration method to solve the optimality equation; at each iteration, evaluate the incomplete expectation via a numerical integration method.

16.4 Write the expressions for the exceedance probability $P(\Theta > \alpha^*)$, and the maximum expected profit per freshman class R^*. Calculate the two quantities; interpret them for a college admissions director.

16.5 Calculate the optimal target number of admission offers, the expected number of matriculants, and the expected number of excess matriculants who must be accommodated.

16.6 Make a figure showing two graphs on the sample space (η_L, η_U): a graph of the density function h, and a graph of the profit function $r(\alpha^*, \cdot)$. Explain the figure to a college admissions director.

16.7 Calculate VPF, the value of the perfect forecast of Θ per freshman class; interpret this value for a college admissions director.

16.8 Consider the nonprofit version of the college. (i) Find the simultaneous solution (p_t^o, α^o). (ii) Calculate $P(\Theta > \alpha^o)$, and verify that $R(\alpha^o) = 0$. (iii) Calculate the three numbers defined in Exercise 16.5. (iv) Calculate VPF.

16.9 Compare the seven quantities $\{p_t, \alpha^*, P(\Theta > \alpha^*), a^* = k/\alpha^*, E(V), E(X), VPF\}$ from the profit-maximizing version of the college with the corresponding seven quantities from the nonprofit version of the college. Explain the differences and draw conclusions.

A

Rationality Postulates

> One of the great intellectual achievements of the twentieth century is the Bayesian theory of rational behavior under risk and uncertainty. … [The] Bayesian rationality postulates are absolutely inescapable criteria of rationality for policy decisions. (Harsanyi, 1978).

A.1 Postulates

A.1.1 Preliminaries

The complete set of postulates (also called *axioms* — the self-evident truths), which underpin the Bayesian theory, actually comprises two sets.

1. Axioms regarding the *degree of certainty* (Section 4.1) from which the judgmental probability theory is derived.
2. Axioms regarding the *strength of preferences* (Section 14.3) from which the cardinal utility theory is derived, together with the maximization of expected utility as the decision criterion.

These two sets of axioms also underlie the mathematization of the two components that this book couples into the F–D system (Chapter 1): the probabilistic forecaster and the rational decider.

Only the latter axioms are stated herein, in two versions: (i) the degenerate version, sufficient for intuitive interpretation and for appreciation of the Bayesian concept of rationality; and (ii) the general version, necessary for proving the two theorems cited in Section 14.3.2.

With the symbols defined in Section 14.1, the generic decision problem is specified by $(\mathcal{Z}, \mathcal{G}, \succsim)$, with two amendments. The set of outcomes $\mathcal{Z}$ is a bounded interval. The set of distribution functions, or gambles, $\mathcal{G}$ is either finite or infinite.

In what follows, $p, q, \alpha, \beta, \gamma \in (0, 1)$ are probabilities, and $z, z_1, z_2, z_3 \in \mathcal{Z}$ are any outcomes.

A.1.2 Degenerate Axioms and Interpretations

Four axioms are stated in their degenerate version, each accompanied by an intuitive interpretation. Follow it with your own question and reflection: Would you accept the axiom as a principle for making decisions?

Probabilistic Forecasts and Optimal Decisions, First Edition. Roman Krzysztofowicz.
© 2025 John Wiley & Sons Ltd. Published 2025 by John Wiley & Sons Ltd.
Companion website: www.wiley.com/go/ProbabilisticForecastsandOptimalDecisions1e

Axiom 1 (***Weak order***).   The binary preference–indifference relation $\succsim$ on the set of outcomes $\mathcal{Z}$ is a *weak order*. Such order has two properties: it is *strongly complete* if $z_1 \succ z_2$, or $z_2 \succ z_1$, or $z_1 \sim z_2$; and it is *transitive* if $z_1 \succsim z_2$ and $z_2 \succsim z_3 \implies z_1 \succsim z_3$.

- Interpretation.   Given any two outcomes, the decider can state that either one is preferred over the other, or they are indifferent. And given any three outcomes, two comparisons are sufficient to infer the third one.

Axiom 2 (***Sure-thing principle***).   For a fixed outcome z and any probability p,

$$z_2 \succ z_1 \iff \begin{pmatrix} p, & z_2 \\ 1-p, z \end{pmatrix} \succ \begin{pmatrix} p, & z_1 \\ 1-p, z \end{pmatrix}.$$

Example A.1

$$\$1000 \succ \$200 \iff \begin{pmatrix} 0.4, \$1000 \\ 0.6, -\$400 \end{pmatrix} \succ \begin{pmatrix} 0.4, & \$200 \\ 0.6, -\$400 \end{pmatrix}.$$

Would this be your preference? If so, then change $p = 0.4$ to some other value, say $p = 0.1$. Does your preference between the gambles remain intact? Would it remain intact for any other value of p?

- *Interpretation.*   The **common outcome**, z, of the two gambles, to be obtained with a **common probability**, $1 - p$, is irrelevant to the preference.

Axiom 3 (***Continuity condition***).   Suppose $z_2 \succ z \succ z_1$. Then there exist probabilities p and q such that

$$\begin{pmatrix} p, & z_2 \\ 1-p, z_1 \end{pmatrix} \succ z \succ \begin{pmatrix} q, & z_2 \\ 1-q, z_1 \end{pmatrix}.$$

Example A.2

$$\begin{pmatrix} 0.8, & \$1000 \\ 0.2, & -\$300 \end{pmatrix} \succ \$500 \succ \begin{pmatrix} 0.3, & \$1000 \\ 0.7, & -\$300 \end{pmatrix}.$$

Would this be your preference order? If not, then change the probability values: increase p until you prefer the first gamble over the sure outcome z; decrease q until you prefer the sure outcome over the second gamble.

- *Interpretation.*   There always exists a **large probability** which, when assigned to the more desirable outcome, makes the gamble preferred.

    There always exists a **small probability** which, when assigned to the more desirable outcome, makes the sure outcome preferred.

Example A.3

$$\begin{pmatrix} 0.999, & \$8000 \\ 0.001, & -\$100\,000\,000 \end{pmatrix} \prec \$3000, \qquad \$3000 \prec \begin{pmatrix} 0.001, \$100\,000\,000 \\ 0.999, & -\$8000 \end{pmatrix}.$$

In the first choice, the individual may never prefer the gamble, no matter how large p is, because the loss would be ruinous. In the second choice, the individual may always prefer the gamble, no matter how small q is, because the gain is so alluring. In both cases, the axiom is violated. Hence the following complementarity.

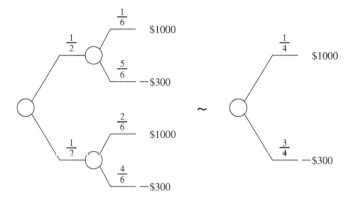

Figure A.1 Illustration of Axiom 4 (substitution principle) in the degenerate version: the two-stage gamble should be indifferent to the one-stage gamble. That is, the decider should take "a purely result-oriented attitude".

- *Interpretation.*   No outcome in $\mathcal{Z}$ can be so **undesirable**, relative to other outcomes, that assigning to it an arbitrarily small probability would not make the gamble preferred.

  No outcome in $\mathcal{Z}$ can be so **desirable**, relative to other outcomes, that assigning to it an arbitrarily small probability would not make the sure outcome preferred.

  To sum up: In an ordinary decision problem, there is no outcome that the decider would avoid or seek, regardless of the probability assigned to it. This delineates the *domain of applicability* of the decision theory: it excludes problems outside one's normal life decisions, and problems in which one's decisions would be guided by dogmas.

Axiom 4   (*Substitution principle*).   Let α, β, γ be any probabilities, and $p = \alpha\beta + (1 - \alpha)\gamma$. Then the *two-stage gamble* is indifferent to the *one-stage gamble*:

$$\left\langle \begin{matrix} \alpha, & \left\langle \begin{matrix} \beta, & z_2 \\ 1 - \beta, & z_1 \end{matrix} \right\rangle \\ 1 - \alpha, & \left\langle \begin{matrix} \gamma, & z_2 \\ 1 - \gamma, & z_1 \end{matrix} \right\rangle \end{matrix} \right\rangle \sim \left\langle \begin{matrix} p, & z_2 \\ 1 - p, & z_1 \end{matrix} \right\rangle .$$

Example A.4   It is depicted in Figure A.1 using event trees. The outcomes are a loss, $z_1 = -\$300$, or a gain, $z_2 = \$1000$.

(i) The outcome from the two-stage gamble is generated by the following mechanism. A coin is tossed. If heads appears ($\alpha = 1/2$), then a die is rolled; if a 1 appears ($\beta = 1/6$), you gain; otherwise ($1 - \beta = 5/6$) you lose. If tails appears ($1 - \alpha = 1/2$), then a die is rolled; if a 1 or 2 appears ($\gamma = 2/6$), you gain; otherwise ($1 - \gamma = 4/6$) you lose. The total probability of the gain is $p = 1/2 \times 1/6 + 1/2 \times 2/6 = 1/4$.

(ii) The outcome from the one-stage gamble is generated by tossing a coin twice. If two heads appear ($p = 1/4$), you gain; otherwise ($1 - p = 3/4$) you lose. Inasmuch as the probability of the gain is identical in both gambles, does the mechanism that generates the identical outcomes matter to you?

- *Interpretation.* The decider should be indifferent between any two gambles that specify the same probability function of the outcome variate, regardless of the *mechanism* (or *process*) that generates the outcome in each gamble. In particular:

  1. Any emotion, such as the pleasure or the distaste of gambling, or the allure or the fear of complexity (which the two-stage gamble may enhance) is precluded from effecting the preference.

  2. When comparing gambles, the decider should take "a purely *result-oriented* attitude".

A.1.3 General Axioms

Let $\mathcal{G}$ be a set of distribution functions on a bounded interval $\mathcal{Z}$, and let G, G_1, G_2, G_3 be distribution functions from that set.

The degenerate version of Axioms 1, 2, 3 can be transformed into the general version by replacing set $\mathcal{Z}$ with set $\mathcal{G}$, and outcomes z, z_1, z_2, z_3 with the distribution functions G, G_1, G_2, G_3, respectively.

Axiom 4 in the general version states this. For every $G \in \mathcal{G}$, there exists $p \in (0, 1)$ such that

$$G \sim \left\langle \begin{matrix} p, & z^s \\ 1 - p, & z^e \end{matrix} \right\rangle.$$

- *Interpretation.* The decider is always willing to substitute the distribution function G on $\mathcal{Z}$ by a suitable gamble between the most preferred outcome $z^s \in \mathcal{Z}$ and the least preferred outcome $z^e \in \mathcal{Z}$.

A.2 Implications

A.2.1 Utility Theorems

In common sense, being rational means being agreeable to reason. In Bayesian decision theory, being *rational* is also tantamount to accepting Axioms 1–4 as principles of making decisions under uncertainty. And if one accepts the axioms, then one should also accept their mathematical implications. There are two:

1. Axioms 1, 2, and 3 (in their general version) are sufficient to prove the theorem on the existence of the utility function (expressions (14.1)–(14.2)).
2. Axiom 4 (in its general version) and the existence of the utility function are sufficient to prove the theorem on the expected utility criterion (expressions (14.3)–(14.4)).

These theorems are *normative*: they prescribe how decisions under uncertainty should be made in complex situations if they are to be consistent with the rationality postulates that one has accepted for simple situations. Inasmuch as the utility function may be linear, the maximization of expected profit, or the minimization of expected opportunity loss, is also consistent with the rationality postulates.

A.2.2 Utilitarian Ethics

Whereas each axiom constitutes a normative postulate of rationality, Axiom 2 (sure-thing principle) and Axiom 4 (substitution principle) imply also two principles of *utilitarian ethics*.

1. ***Impartial predictive information***. The two axioms presuppose that the decider's strength of preferences for outcomes is independent of the events that generate the outcomes. This is a **decomposition principle**: the

quantification of the strength of preferences in terms of the utility function v should be done independently of the quantification of uncertainty in terms of the distribution function G. The concomitant **ethics principle**: a responsible financial advisor should impartially furnish to the investor all available predictive information for assessing G.

2. ***Result-oriented attitude***. The two axioms also presuppose that, when comparing gambles, the decider takes "a purely result-oriented attitude". The concomitant **ethics principle**, in the words of Harsanyi (1978): "responsible business executives using their shareholders' money, and responsible political leaders acting on behalf of their constituents, are expected to do their utmost to achieve the best possible results, rather than to gratify their own personal desire" for any sort of emotion.

A.3 Training for Rationality

A.3.1 Coherence of Multiple Decisions

Numerous experiments have demonstrated that individuals making decisions intuitively, in situations similar to those in the degenerate version of axioms, violate the axioms. Some authors conclude that the axioms are too restrictive — even as normative postulates. Others view these violations as the *prima facie* evidence of human cognitive limitations (or bounded rationality) — the very reason for designing decision support systems based in mathematical models that adhere to the axioms. We join this debate by considering just two choices.

Task A.1 (***Choice 1***). You are offered an opportunity to play one of two gambles, a or b. Which one do you choose? Circle your choice.

$$\underset{a}{\left\langle \begin{matrix} 1.0, & \$650 \\ 0.0, & \$0 \end{matrix} \right\rangle} \quad \underset{b}{\left\langle \begin{matrix} 0.3, & \$0 \\ 0.7, & \$1000 \end{matrix} \right\rangle}$$

Task A.2 (***Choice 2***). You are offered an opportunity to play one of two gambles, c or d. Which one do you choose? Circle your choice.

$$\underset{c}{\left\langle \begin{matrix} 0.6, & \$650 \\ 0.4, & \$0 \end{matrix} \right\rangle} \quad \underset{d}{\left\langle \begin{matrix} 0.58, & \$0 \\ 0.42, & \$1000 \end{matrix} \right\rangle}$$

Let us analyze the two choices with the aid of Figure A.2 for the decider with the preference $a \succ b$. Three inferences can be made, each invoking one axiom.

Inference 1 (Sure-thing principle). In gamble c, outcome \$650 can be exchanged for gamble a. The result is a seeming two-stage gamble e, but because \$650 obtains with probability 1.0, c and e are equivalent, $c \equiv e$. Next, gamble f is created from gamble e by exchanging gamble a for gamble b. Because $a \succ b$, by the sure-thing principle (outcome \$0 with probability 0.4 is common to both gambles) $e \succ f$.

Inference 2 (Substitution principle). The two-stage gamble f yields outcome \$0 with probability $0.6 \times 0.3 + 0.4 = 0.58$, or outcome \$1000 with probability $0.6 \times 0.7 = 0.42$, like the one-stage gamble d does. By the substitution principle, $f \equiv d$.

Inference 3 (Transitivity property). The preceding two inferences can now be concatenated via the transitivity property of the weak order:

$$a \succ b \text{ and } c \equiv e \implies e \succ f \text{ and } f \equiv d \implies c \succ d.$$

TASK A.1

TASK A.2

Figure A.2 A test of coherence of two simple decisions. Task A.1: the choice between gamble *a* and gamble *b*. Task A.2: the choice between gamble *c* (equivalent to gamble *e*) and gamble *d* (equivalent to gamble *f*).

Your task: repeat the three inferences for the decider with the preference $a \prec b$. You should arrive at the following:

$$a \prec b \text{ and } c \equiv e \implies e \prec f \text{ and } f \equiv d \implies c \prec d.$$

In summary, the decider's preferences expressed in the two choices obey Axioms 1, 2, and 4 only if they are coherent; that is, only if the preferred gambles are

    (*a* and *c*)   or   (*b* and *d*).

The incoherent choices are

    (*a* and *d*)   or   (*b* and *c*).

The distribution of choices observed in an experiment is reported in Section A.3.3. It may be read now, or at the conclusion of this appendix.

A.3.2 Demonstrable Irrationality

One may ask: Does the coherence of multiple decisions matter? For the sake of the argument, imagine being the decider who has expressed incoherent preferences ($a \succ b, c \prec d$). These preferences will now guide a sequence of computerized (automatic) trades on your behalf (Table A.1).

Trade 1. You own initially an investment (gamble) *c*. Because you prefer investment (gamble) *d*, the computerized trader exchanges *c* for *d*, and charges you a fee (say $2).

Trade 2. The trader finds that *d* is equivalent to *f*, which includes *b*. Because you prefer *a*, the trader exchanges *b* for *a*, and charges you $2. You end up owning *e*.

Trade 3. With *e* being equivalent to *c*, you own back your initial investment. The trader now repeats trades 1 and 2 — *ad infinitum*.

Table A.1 Demonstration that incoherence implies irrationality. The consequence of incoherent preferences ($a > b, c < d$) expressed in Tasks A.1 and A.2 for trading decisions: circular trades and needless cost.

Trade number	You own gamble	A trader exchanges	You pay the fee	Your accumulated cost is [$]
1	c	c for d	$2	2
2	$d \equiv f$	b for a	$2	4
3	$e \equiv c$	c for d	$2	6
4	$d \equiv f$	b for a	$2	8
5	$e \equiv c$	...	...	...
⋮	⋮	⋮	⋮	*ad infinitum*

In summary, an incoherent decider makes circular trades and ends up with a needless cost, or a needless bankruptcy.

Some studies (*The Wall Street Journal*, 22 September 1998) concluded that many investors trade stocks more frequently than justified, for instance, by the need to rebalance the portfolio (as discussed in Section 14.6.7). Two explanations were suggested: the overconfident forecasts (a bias discussed in Chapters 6 and 11), and the pleasure of gambling (the violation of Axiom 4 discussed in Section A.1.2). We may suggest a third explanation: the incoherence of intuitive decisions, which induces circular trades sans gains.

> **Advice** (Coherence principle). In investing, an incoherent decider becomes the dupe to a smart trader. To ensure coherence of multiple decisions, and thereby to prevent circular (and potentially wasteful) decisions, the decider's preferences should satisfy the rationality postulates (Axioms 1–4), as a matter of logical necessity. The assessment of a utility function and the maximization of expected utility are the analytic tools that ensure coherence.

Example A.5 *(Expected utility maximizer)*

Suppose a decider has exponential utility function (14.9) with parameter $\theta = \$555$. The utilities of the three outcomes are

$$v(0) = 0, \qquad v(650) = 0.690, \qquad v(1000) = 0.835;$$

and the utilities of the four gambles are

$$U(a) = 1v(650) + 0v(0) = 0.690,$$
$$U(b) = 0.3v(0) + 0.7v(1000) = 0.585,$$
$$U(c) = 0.6v(650) + 0.4v(0) = 0.414,$$
$$U(d) = 0.58v(0) + 0.42v(1000) = 0.351.$$

Hence, $a > b$ and $c > d$, which are coherent preferences. To illustrate the coherent reversed preferences, repeat the above analysis for a decider with $\theta = \$1100$, who is less risk averse than the first one (as interpreted in Section 14.3.8).

A.3.3 Experimental Results

Tasks A.1 and A.2 were performed by 248 students who took this course at the University of Virginia over 9 years. After an exploratory analysis of data, they were clustered into two groups having different response

Table A.2 Percentage of subjects in each group who chose the particular two gambles in Tasks A.1 and A.2. Group U: 189 undergraduate students and 19 master students of engineering. Group D: 40 doctoral students of engineering and sciences.

Group	Size	Coherent choices			Incoherent Choices		
		(a, c)	(b, d)	Sum	(a, d)	(b, c)	Sum
U	208	23	19	42	54	4	58
D	40	35	18	53	35	12	47

patterns: Group U, consisting of 189 undergraduate students and 19 master students, all 208 studying engineering; and Group D, consisting of 40 doctoral students of engineering and sciences. Table A.2 reports the percentage of subjects in each group who chose the particular two gambles. The percentage of incoherent choices is 58 in Group U and 47 in Group D. Although a positive effect of doctoral studies is noticeable, coherence is not the prevalent attribute of intuitive decisions under uncertainty, even in the simplest choice problems.

A cognitive explanation of such incoherences rests on the proposition that a two-outcome gamble presents a 4-dimensional stimulus. While evaluating it intuitively, the person may not integrate information but instead may focus intensely on just one or two dimensions (e.g., on one probability and one outcome), as a way of coping with information load. When comparing four gambles (four 4-dimensional stimuli), the dimensions focused upon may change from one gamble to the next. Such inconsistent mental information-processing may induce incoherent choices.

A.3.4 Training

"Training for rationality" is a succinct heading that we attach to the amalgam of approaches. They invariably involve an effort to gain the deep understanding of the rationality postulates and their implications. Cogent arguments help. When the financial peril of circular investment decisions is explained, it offers a reason for rethinking and reconciling one's initial choices. When the rationality postulates are internalized — become part of one's repertoire of ways to reason — one is likely to become a more reflective–logical intuitive decider. But foremost, one should become a good *meta-decider*: one who decides in which situations intuition should be supported, or replaced, by a decision model.

Bibliographical Notes

There exist several axiomatizations of decision rationality; some are reviewed by Fishburn (1982). The original axioms, which date to 1944, were published with proofs by von Neumann and Morgenstern (1947); later, Morgenstern (1979) recalled their genesis and their intended *domain of applicability*. The presentation herein condenses the development by DeGroot (1970, Chapter 7) and emulates some of his interpretations. The implications regarding the utilitarian ethics are due to Harsanyi (1978).

Likewise, there exist several axiomatic developments of subjective probability theory; DeGroot (1970, Chapter 6) presents one that is consistent with the perspective in this book.

The cognitive explanation (and a mathematical model) of information-processing by a person choosing the gamble were proposed by Slovic and Lichtenstein (1968). The review by Slovic (1982) of research on intuitive judgments and decisions in the face of uncertainty concluded that "we must look toward decision aids to help minimize errors" caused by our *cognitive limitations*. The pioneering study in this domain by Miller (1956) quantified the limits on people's capacity to process sensory signals sans errors. The theory of "bounded rationality" was proposed by Simon (1957). "Train rational decision makers!" was the response of Marschak (1979) to critiques of axioms as being too restrictive.

Seminal references to measurement theory and to measurement scales (utility being measured on a cardinal scale, and probability being measured on an absolute scale) are the books by Krantz et al. (1971) and Roberts (1979).

(Herbert A. Simon, professor of computer science and psychology at Carnegie Mellon University, received the Nobel Prize in economics in 1978 "for his pioneering research into the decision-making process within economic organizations".)

B

Parameter Estimation Methods

Three methods are detailed for estimating the parameters (α, β) of a hypothesized distribution function H on a specified sample space $\mathcal{X}$, when the shift parameters (if any) have already been assessed. Each method is applicable to every family of the 20 distributions in Appendix C, as well as to every family of the four Gaussian distributions (Section 2.6).

B.1 Method of Least Squares

The method of *least squares* (LS) is applicable whenever information about the variate X comes in the form of (i) a sample of X, as in Chapter 3, or (ii) quantiles of X assessed judgmentally by an expert, as in Chapter 9. Its drawback is that it does not minimize the MAD between the parametric distribution function H and the empirical distribution function $\check{H}$. Instead, it optimizes the fit between the corresponding linearized quantile functions; however, this formulation makes the method as easy as fitting the regression line.

B.1.1 Least Squares Algorithm

Step 0. Prepare a set of points $\{(x_{(n)}, p_n) : n = 1, \dots, N\}$, where $x_{(n)}$ is a value of X and p_n is the probability such that $p_n = P(X \leq x_{(n)})$; the set is ordered so that $x_{(1)} \leq \cdots \leq x_{(N)}$ and $p_1 < \cdots < p_N$. It is interpreted as either the *empirical distribution function* of X or the *assessed distribution function* of X. In the former case, $\{x_{(n)} : n = 1, \dots, N\}$ is the sample of X in the ascending order of realizations, and $\{p_n : n = 1, \dots, N\}$ are the plotting positions calculated according to a formula from Section 3.2. In the latter case, $\{x_{(n)} : n = 1, \dots, N\}$ are the quantiles of X corresponding to the probabilities $\{p_n : n = 1, \dots, N\}$. Thus, $(x_{(n)}, p_n)$ equals (y_p, p) and $N = I$ in the notation of Section 9.2.1.

Step 1. Transform the given set of points into the set of points $\{(v_n, u_n) : n = 1, \dots, N\}$, where v_n is obtained from $x_{(n)}$, and u_n is obtained from p_n. Use the transformations from x to v, and from p to u, which are specified for the hypothesized distribution.

Step 2. Calculate the least squares estimates $\hat{a}, \hat{b}$ of the coefficients a, b of the linearized quantile function. When this function takes the form $v = bu + a$,

$$\hat{b} = \frac{\sum_{n=1}^{N} v_n u_n - N \bar{v} \bar{u}}{\sum_{n=1}^{N} u_n^2 - N \bar{u}^2}, \qquad \hat{a} = \bar{v} - \hat{b} \bar{u}; \tag{B.1}$$

Probabilistic Forecasts and Optimal Decisions, First Edition. Roman Krzysztofowicz.
© 2025 John Wiley & Sons Ltd. Published 2025 by John Wiley & Sons Ltd.
Companion website: www.wiley.com/go/ProbabilisticForecastsandOptimalDecisions1e

when it takes the form $v = bu$,

$$\hat{b} = \frac{\sum_{n=1}^{N} v_n u_n}{\sum_{n=1}^{N} u_n^2};$$

(B.2)

when it takes the form $v = u + a$,

$$\hat{a} = \overline{v} - \overline{u}.$$

(B.3)

In equations (B.1) and (B.3),

$$\overline{v} = \frac{1}{N}\sum_{n=1}^{N} v_n, \quad \overline{u} = \frac{1}{N}\sum_{n=1}^{N} u_n.$$

(B.4)

Step 3. Transform the estimates $\hat{a}, \hat{b}$ of the coefficients a, b into the estimates $\hat{\alpha}, \hat{\beta}$ of the parameters α, β of the hypothesized distribution. Use the transformations between a, b and α, β, which are specified for the particular distribution.

Step 4. Graph and evaluate. (i) In the transformed space, graph the data points $\{(v_n, u_n)\}$ and the estimated linear function. Judge visually: Is the fit good? (ii) In the original space, graph the data points $\{(x_{(n)}, p_n)\}$ and the estimated distribution function H. Judge visually: Is the fit good?

Gaussian distributions

When the algorithm is used to estimate the location parameter (the mean) μ and the scale parameter (the standard deviation) σ of any of the four Gaussian distributions (Sections 2.6 and 9.3), the following substitutions should be made: $\alpha = \sigma, \beta = \mu$.

B.1.2 Goodness-of-Fit Evaluation

The least squares estimates $\hat{a}, \hat{b}$ are optimal in the sense that they minimize the sum of the squares of the differences between the transformed empirical quantiles v_n and their estimates $bu_n + a$. The criterion of optimality is

$$\min_{a,b} \sum_{n=1}^{N} [v_n - (bu_n + a)]^2.$$

(B.5)

The main reasons for employing this criterion are (i) the applicability of the least squares method to all distributions in this book, (ii) the ease of calculating the optimal estimates $\hat{a}, \hat{b}$, and (iii) the simplicity of the transformation between $\hat{a}, \hat{b}$ and the estimates $\hat{\alpha}, \hat{\beta}$ of the original parameters.

Once the estimates $\hat{\alpha}, \hat{\beta}$ of parameters α, β have been found, the goodness of fit should be judged visually and evaluated quantitatively in the original space, where it really matters. Toward this end, the hypothesized distribution function H, whose parameters are now set to $\hat{\alpha}, \hat{\beta}$, should be compared with the empirical distribution function $\{(x_{(n)}, p_n) : n = 1, \ldots, N\}$ according to the procedures of Section 3.6.

(It turns out that the least squares estimates yield sometimes a good fit and sometimes a poor fit in the original space. In general, their performance is not consistent. For this reason, in scientific works they are treated primarily as initial estimates that enter the uniform distance method.)

B.2 Method of Uniform Distance

The method of *uniform distance* (UD) is applicable whenever information about the variate X comes in the form of (i) a sample of X, as in Chapter 3, or (ii) quantiles of X assessed judgmentally by an expert, as in Chapter 9. The method optimizes the fit of the parametric distribution function H to the empirical distribution function $\check{H}$.

Because its execution requires various auxiliary procedures and specialized software, it is not required in this book. But when the software is accessible, it may be used. Here is the gist of the method.

Uniform estimates

Given is the empirical or the assessed distribution function $\{(x_{(n)}, p_n) : n = 1, \dots, N\}$ of the variate X, and the hypothesis that the distribution function H of X belongs to the family of parametric distribution functions of a particular form: $\{H(\cdot|\alpha, \beta) :$ all $\alpha, \beta\}$. The *uniform estimates* of the parameters α, β (also called the *minimax estimates* or the *Chebyshev estimates*), are obtained as the solution to the following optimization problem:

$$\min_{\alpha, \beta} \{ \max_{1 \leq n \leq N} |p_n - H(x_{(n)}|\alpha, \beta)| \}. \tag{B.6}$$

In other words, the uniform estimates are optimal in the sense that they minimize the MAD across all distribution functions in the family. Inasmuch as the optimization algorithm requires an initial solution, the LS estimates are found beforehand and then inserted into the algorithm. Consequently, the MAD from the UD method is never greater than the MAD from the LS method.

Distribution fitter

DFit (*Distribution Fitter*) is a web-based software — an aid to effective usage of the distribution modeling methodology of Chapter 3, together with the 20 special univariate distributions of Appendix C, and the 4 Gaussian distributions of Chapter 2. It guides the modeler through the six steps (Section 3.1), and it estimates the parameters of every hypothesized distribution via the UD method. It accepts data in two forms: either $\{x(n) : n = 1, \dots, N\}$, a sample of realizations of a continuous variate X; or $\{(x_{(n)}, p_n) : n = 1, \dots, N\}$, a set of quantiles of X and their corresponding probabilities. When the hypothesized distribution is one of those constructed from a transform of the given variate and a parametric distribution of another variate (as described in Sections C.4 and C.5), DFit performs all the necessary transformations. (DFit includes also procedures and distribution families not covered in this book — they should just be ignored.) Its web address is https://DFit.F-Dsystems.com.

B.3 Method of Moments

The method of *moments* (MS) is applicable whenever information about the variate X comes in the form of two estimates: m — an estimate of the mean $E(X)$, and s^2 — an estimate of the variance $Var(X)$. These estimates may be either (i) calculated from a sample of X, as sample mean and sample variance (Section 2.4), or (ii) specified based on various technical or engineering considerations for the problem at hand. The method is convenient for distributions under which the system of two equations for $E(X)$ and $Var(X)$, each of which is a function of the distribution parameters (α, β), can be solved analytically for α and β; otherwise, they must be solved numerically (e.g., via an iteration method — not discussed herein). Its drawback is that it does not guarantee the best fit of the parametric distribution function H to the empirical distribution function $\breve{H}$.

Step 0. Given are an estimate m of $E(X)$ and an estimate s^2 of $Var(X)$.

Step 1. Extract the expressions for $E(X)$ and $Var(X)$ under the hypothesized distribution. These expressions are in terms of the unknown distribution parameters (α, β); any shift parameter should be set to the value assessed beforehand.

Step 2. Equate the expression for $E(X)$ with m; equate the expression for $Var(X)$ with s^2:

$$E(X) = m, \quad Var(X) = s^2.$$

Step 3. Solve this system of two equations for the unknown α, β.

One-parameter distributions: $EX(\alpha, \eta)$, $P1(\beta, \eta_L, \eta_U)$, $P2(\beta, \eta_L, \eta_U)$. Under each of these distributions, a relationship is prescribed between $E(X)$ and $Var(X)$, which must also hold between m and s^2 if that distribution is a correct model. Solving each of the two equations yields two estimates of the unknown parameter. (i) If the difference is zero, or is small, then the hypothesized distribution may be a correct model; and the *geometric mean* of the two estimates may be taken as the final estimate. If the difference is large, then the hypothesized distribution is an incorrect model (one parameter being insufficient to fit both m and s^2). Of course, there is no dilemma when only m is specified.

Gaussian distributions: $LN(\mu, \sigma, \eta)$, $LR1(\eta_L, \eta_U)$-$NM(\mu, \sigma)$. Each of these is a distribution of variate Y, but the parameters (μ, σ^2) are the mean and the variance of variate X. Therefore, the transform between Y and X (equation (2.18) or (2.23)) should be used to map the given sample of Y into the sample of X, from which m and s^2 should be calculated. Then set $\mu = m$ and $\sigma^2 = s^2$ (per equation (2.20) or (2.25)).

Example B.1   *(Logistic distribution)*

Given are $m = 8$ and $s^2 = 25$. The hypothesized distribution of variate X is $LG(\beta, \alpha)$, with the location parameter β and the scale parameter α. From Section C.2.1,

$$E(X) = \beta, \qquad Var(X) = \frac{\pi^2 \alpha^2}{3}.$$

Setting and solving the two equations

$$\beta = m, \qquad \frac{\pi^2 \alpha^2}{3} = s^2,$$

yields the moment estimates of the parameters:

$$\hat{\beta} = m = 8,$$

$$\hat{\alpha} = \frac{\sqrt{3}}{\pi} s = \frac{\sqrt{3}}{\pi} 5 = 2.757.$$

B.4   Other Methods

The method of *maximum likelihood* (ML) is often favored because it has a firm theoretic foundation and good properties — the invariance of estimates being one of them (recall Section 2.4.3). It is not used herein for three reasons. (i) It is laborious — the equations for the estimates of parameters (α, β) must be derived for each distribution family. (ii) It is not applicable when the information about X comes in the form of quantiles. (iii) It does not guarantee the best fit of the parametric distribution function H to the empirical distribution function $\breve{H}$.

There are other specialized estimation methods, for instance, the method of probability weighted moments, and the method of L-moments. But they usually lack the good properties of the ML method, while sharing all three drawbacks.

C

Special Univariate Distributions

This is a catalogue of special univariate distributions of continuous variates. These distributions are special herein because they are easy to estimate and to use: each of these distributions is parametric, has closed-form expressions for the density function, the distribution function, and the quantile function (the inverse of the distribution function), and admits a linearization of the distribution function and the quantile function.

Section C.1 contains the usage guide. Sections C.2–C.4 catalogue the distributions, which are grouped according to the type of sample space: the unbounded interval $(-\infty, \infty)$, a bounded-below interval (η, ∞), and a bounded open interval (η_L, η_U), where bounds are real numbers, with $-\infty < \eta < \infty$ and $-\infty < \eta_L < \eta_U < \infty$. It is assumed that bounds are known, or specified beforehand. All distributions have two parameters to be estimated, with the exception of the exponential distribution and the power distributions (each of which has one parameter to be estimated). Section C.5 explains how to obtain distributions on a bounded-above interval $(-\infty, -\eta)$.

C.1 Usage Guide

C.1.1 Distribution Forms

The catalogue contains 20 distribution families (Table C.1). Each primary distribution family has a name and a mnemonic, which consists of two letters followed by the list of parameters in parentheses. A class of distributions on a bounded open interval (η_L, η_U) is obtained by (i) transforming the original variate into a new variate, and (ii) assuming that the new variate has a distribution of a particular primary form. Consequently, each distribution in this class has a two-part mnemonic: transformation and its parameters, hyphen, distribution form and its parameters.

The group of distributions on a bounded-above interval $(-\infty, -\eta)$ is obtained by reflecting (about the origin) each distribution defined on the bounded-below interval (η, ∞); hence the multiplier -1 is inserted at the end of the list of parameters.

C.1.2 Parameter Types

There are four types of distribution parameters: a *location parameter* β, a *scale parameter* α, a *shape parameter* β, and a *shift parameter*. A distribution has either a location parameter β (listed before α) or a shape parameter β (listed after α or without α); it has either no shift parameter, or one shift parameter η, or two shift parameters (η_L, η_U).

Probabilistic Forecasts and Optimal Decisions, First Edition. Roman Krzysztofowicz.
© 2025 John Wiley & Sons Ltd. Published 2025 by John Wiley & Sons Ltd.
Companion website: www.wiley.com/go/ProbabilisticForecastsandOptimalDecisions1e

Table C.1 Special parametric distributions of continuous variates.

Section	Name	Mnemonic	Sample space
C.2.1	Logistic	$LG(\beta, \alpha)$	$(-\infty, \infty)$
C.2.2	Laplace	$LP(\beta, \alpha)$	$(-\infty, \infty)$
C.2.3	Gumbel	$GB(\beta, \alpha)$	$(-\infty, \infty)$
C.2.4	Reflected Gumbel	$RG(\beta, \alpha)$	$(-\infty, \infty)$
C.3.1	Exponential	$EX(\alpha, \eta)$	(η, ∞)
C.3.2	Weibull	$WB(\alpha, \beta, \eta)$	(η, ∞)
C.3.3	Inverted Weibull	$IW(\alpha, \beta, \eta)$	(η, ∞)
C.3.4	Log-Weibull	$LW(\alpha, \beta, \eta)$	(η, ∞)
C.3.5	Log-logistic	$LL(\alpha, \beta, \eta)$	(η, ∞)
C.4.1	Power type I	$P1(\beta, \eta_L, \eta_U)$	(η_L, η_U)
C.4.2	Power type II	$P2(\beta, \eta_L, \eta_U)$	(η_L, η_U)
C.4.3	Log-ratio logistic	$LR1(\eta_L, \eta_U)\text{-}LG(\beta, \alpha)$	(η_L, η_U)
	Log-ratio Laplace	$LR1(\eta_L, \eta_U)\text{-}LP(\beta, \alpha)$	(η_L, η_U)
	Log-ratio Gumbel	$LR1(\eta_L, \eta_U)\text{-}GB(\beta, \alpha)$	(η_L, η_U)
	Log-ratio reflected Gumbel	$LR1(\eta_L, \eta_U)\text{-}RG(\beta, \alpha)$	(η_L, η_U)
C.5.1	Reflected exponential	$EX(\alpha, \eta, -1)$	$(-\infty, -\eta)$
	Reflected Weibull	$WB(\alpha, \beta, \eta, -1)$	$(-\infty, -\eta)$
	Reflected inverted Weibull	$IW(\alpha, \beta, \eta, -1)$	$(-\infty, -\eta)$
	Reflected log-Weibull	$LW(a, \beta, \eta, -1)$	$(-\infty, -\eta)$
	Reflected log-logistic	$LL(\alpha, \beta, \eta, -1)$	$(-\infty, -\eta)$

C.1.3 Density and Distribution Functions

For each distribution family, the expressions for the density function h and the distribution function H are given first. Wherever it is sensible, the expressions are written in such a way that the expression for the distribution function is a part of the expression for the density function. One can take advantage of this fact and reduce the number of calculations whenever both functions must be evaluated.

The graphs of functions h and H, for selected values of the parameters, are shown in the figures at the end of the appendix. Each figure bears the number of the section detailing the distribution. When a particular distribution is obtained from another distribution via a transformation of the variate, a graph of the transformation is shown as well.

C.1.4 Quantile Functions

For each distribution family, the quantile function H^{-1} is first given in the usual form, $x = H^{-1}(p)$, where H^{-1} is the inverse of the distribution function, p is the probability such that $p = P(X \leq x)$, and x is the corresponding quantile of variate X. Next, the *linearized quantile function* is given in the form

$$v = bu + a, \tag{C.1}$$

in which the variables u, v are obtained by transforming the variables p, x, and the coefficients a, b are obtained by transforming the distribution parameters α, β. For some distributions, equation (C.1) simplifies to one of its two special cases

$$v = bu, \tag{C.2}$$

$$v = u + a. \tag{C.3}$$

The advantage of the linearized quantile function is that it provides an equation through which the distribution parameters can be easily estimated. The task boils down to estimating the slope b and the intercept a of the linear equation by the least squares method. This method is described in Section B.1. Other methods of parameter estimation may be found in Appendix B and in the references under each distribution family.

C.1.5 Page Setup

The catalogue begins on the next page. Its organization is intended to facilitate the comparisons.

Each distribution family is characterized in the same format, and on a separate page. Figures showing graphs of the density and distribution functions follow, in the order of distribution families.

C.2   Distributions on the Unbounded Interval

Each distribution family catalogued in this section is defined on the set of real numbers, the unbounded interval $(-\infty, \infty)$.

C.2.1   The Logistic Distribution

Abbreviation: LG(β, α)
Parameters: location $-\infty < \beta < \infty$, scale $\alpha > 0$
Domains: $-\infty < x < \infty$, $0 < p < 1$
Density function

$$h(x) = \frac{1}{\alpha} \exp\left(-\frac{x-\beta}{\alpha}\right) \left[1 + \exp\left(-\frac{x-\beta}{\alpha}\right)\right]^{-2}$$

Distribution function $p = H(x)$

$$H(x) = \left[1 + \exp\left(-\frac{x-\beta}{\alpha}\right)\right]^{-1}$$

Quantile function $x = H^{-1}(p)$

$$H^{-1}(p) = \beta + \alpha \ln\left(\frac{p}{1-p}\right)$$

Linearized quantile function $v = bu + a$

$$v = x, \qquad u = \ln\left(\frac{p}{1-p}\right)$$

$$\alpha = b, \qquad \beta = a$$

Moments

$$E(X) = \beta, \qquad Var(X) = \frac{\pi^2 \alpha^2}{3}$$

Notes:

1. The density function is symmetric about β.
2. The mean of X, the median of X, and the mode of X are all equal to β.
3. When $\beta = 0$ and $\alpha = 1$, the distribution is called the *standard logistic distribution*.

Reference: Johnson and Kotz (1970, vol. 2)

C.2.2   The Laplace Distribution

Abbreviation: $LP(\beta, \alpha)$

Parameters: location $-\infty < \beta < \infty$, scale $\alpha > 0$

Domains: $-\infty < x < \infty$, $0 < p < 1$

Density function

$$h(x) = \frac{1}{2\alpha} \exp\left(-\frac{|x - \beta|}{\alpha}\right)$$

Distribution function $p = H(x)$

$$H(x) = \begin{cases} \dfrac{1}{2} \exp\left(\dfrac{x - \beta}{\alpha}\right) & \text{if } x \le \beta \\[2mm] 1 - \dfrac{1}{2} \exp\left(-\dfrac{x - \beta}{\alpha}\right) & \text{if } \beta \le x \end{cases}$$

Quantile function $x = H^{-1}(p)$

$$H^{-1}(p) = \begin{cases} \beta + \alpha \ln 2p & \text{if } p \le \frac{1}{2} \\[2mm] \beta - \alpha \ln 2(1 - p) & \text{if } \frac{1}{2} \le p \end{cases}$$

Linearized quantile function $v = bu + a$

$$v = x, \qquad u = \begin{cases} \ln 2p & \text{if } p \le \frac{1}{2} \\[2mm] -\ln 2(1 - p) & \text{if } \frac{1}{2} \le p \end{cases}$$

$$\alpha = b, \qquad \beta = a$$

Moments

$$E(X) = \beta, \qquad Var(X) = 2\alpha^2$$

Notes:

1. The density function is symmetric about β.
2. The mean of X, the median of X, and the mode of X are all equal to β.
3. In comparison with a normal density function (known as the second law of Laplace), the Laplace density function (also known as the first law of Laplace) has a sharper peak and heavier tails.
4. Other names for the Laplace distribution are the *two-tailed exponential* distribution, the *bilateral exponential* distribution, and the *double exponential* distribution.

Reference: Johnson and Kotz (1970, vol. 2)

C.2.3 The Gumbel Distribution

Abbreviation: $GB(\beta, \alpha)$

Parameters: location $-\infty < \beta < \infty$, scale $\alpha > 0$

Domains: $-\infty < x < \infty$, $0 < p < 1$

Density function

$$h(x) = \frac{1}{\alpha} \exp\left(-\frac{x-\beta}{\alpha}\right) \exp\left[-\exp\left(-\frac{x-\beta}{\alpha}\right)\right]$$

Distribution function $p = H(x)$

$$H(x) = \exp\left[-\exp\left(-\frac{x-\beta}{\alpha}\right)\right]$$

Quantile function $x = H^{-1}(p)$

$$H^{-1}(p) = \beta - \alpha \ln(-\ln p)$$

Linearized quantile function $v = bu + a$

$$v = x, \qquad u = -\ln(-\ln p)$$
$$\alpha = b, \qquad \beta = a$$

Moments

$$E(X) = \beta + \gamma\alpha = \beta + 0.5772157\alpha \qquad (\gamma \text{ is Euler's constant})$$

$$Var(X) = \frac{\pi^2\alpha^2}{6} = 1.6449341\alpha^2$$

Notes:

1. The density function is skew to the right (has positive skew).
2. The mode of X is equal to β.
3. A variate X has a Gumbel (β, α) distribution if the variate

$$Y = \exp\left(-\frac{x-\beta}{\alpha}\right)$$

has the exponential $(1, 0)$ distribution.
4. When $\beta = 0$ and $\alpha = 1$, the distribution is called the *standard Gumbel distribution*.
5. Other names for the Gumbel distribution are the *extreme value type I* distribution, and the *doubly exponential* distribution.

Reference: Johnson and Kotz (1970, vol. 1)

C.2.4 The Reflected Gumbel Distribution

Abbreviation: $RG(\beta, \alpha)$
Parameters: location $-\infty < \beta < \infty$, scale $\alpha > 0$
Domains: $-\infty < x < \infty$, $0 < p < 1$
Density function

$$h(x) = \frac{1}{\alpha} \exp\left(\frac{x - \beta}{\alpha}\right) \exp\left[-\exp\left(\frac{x - \beta}{\alpha}\right)\right]$$

Distribution function $p = H(x)$

$$H(x) = 1 - \exp\left[-\exp\left(\frac{x - \beta}{\alpha}\right)\right]$$

Quantile function $x = H^{-1}(p)$

$$H^{-1}(p) = \beta + \alpha \ln[-\ln(1 - p)]$$

Linearized quantile function $v = bu + a$

$$v = x, \qquad u = \ln[-\ln(1 - p)]$$
$$\alpha = b, \qquad \beta = a$$

Moments

$$E(X) = \beta - \gamma\alpha = \beta - 0.5772157\alpha \qquad (\gamma \text{ is Euler's constant})$$

$$Var(X) = \frac{\pi^2 \alpha^2}{6} = 1.6449341\alpha^2$$

Notes:

1. The density function is skew to the left (has negative skew); it is a horizontal reflection about the origin of the Gumbel density function.
2. The mode of X is equal to β.
3. A variate X has a reflected Gumbel (β, α) distribution if the variate $Y = -X$ has the Gumbel $(-\beta, \alpha)$ distribution.

C.3   Distributions on a Bounded-Below Interval

Each distribution family catalogued in this section is defined on an interval bounded below (η, ∞), where it is understood that the lower bound η, also called the *shift parameter*, is a real number, $-\infty < \eta < \infty$, and is specified. It is also understood that for any $x < \eta$, $h(x) = 0$ and $H(x) = 0$.

C.3.1   The Exponential Distribution

Abbreviation: $EX(\alpha, \eta)$
Parameters: scale $\alpha > 0$, shift $-\infty < \eta < \infty$
Domains: $\eta < x < \infty$, $0 < p < 1$
Density function

$$h(x) = \frac{1}{\alpha} \exp\left(-\frac{x - \eta}{\alpha}\right)$$

Distribution function $p = H(x)$

$$H(x) = 1 - \exp\left(-\frac{x - \eta}{\alpha}\right)$$

Quantile function $x = H^{-1}(p)$

$$H^{-1}(p) = -\alpha \ln(1 - p) + \eta$$

Linearized quantile function $v = u + a$

$$v = \ln(x - \eta), \qquad u = \ln\left[-\ln(1 - p)\right]$$
$$\alpha = \exp(a)$$

Moments

$$E(X) = \alpha + \eta, \qquad Var(X) = \alpha^2$$

Notes:

1. The exponential (α, η) distribution is a special case of the Weibull (α, β, η) distribution. The two distributions are identical when $\beta = 1$.
2. The mode of X is equal to η, the left endpoint of the sample space.
3. Unless some scientific law prescribes that a variate has an exponential distribution, a more general Weibull (α, β, η) distribution should be hypothesized. Should the estimate of the shape parameter β be close to one, the exponential (α, η) distribution may be considered as the parsimonious approximation.

Reference: Johnson and Kotz (1970, vol. 1)

C.3.2 The Weibull Distribution

Abbreviation: $WB(\alpha, \beta, \eta)$

Parameters: scale $\alpha > 0$, shape $\beta > 0$, shift $-\infty < \eta < \infty$

Domains: $\eta < x < \infty$, $0 < p < 1$

Density function

$$h(x) = \frac{\beta}{\alpha}\left(\frac{x-\eta}{\alpha}\right)^{\beta-1} \exp\left[-\left(\frac{x-\eta}{\alpha}\right)^{\beta}\right]$$

Distribution function $p - H(x)$

$$H(x) = 1 - \exp\left[-\left(\frac{x-\eta}{\alpha}\right)^{\beta}\right]$$

Quantile function $x = H^{-1}(p)$

$$H^{-1}(p) = \alpha\left[-\ln(1-p)\right]^{\frac{1}{\beta}} + \eta$$

Linearized quantile function $v = bu + a$

$$v = \ln(x-\eta), \qquad u = \ln\left[-\ln(1-p)\right]$$
$$\beta = \frac{1}{b}, \qquad\qquad \alpha = \exp(a)$$

Moments

$$E(X) = \alpha\Gamma\left(1+\frac{1}{\beta}\right) + \eta$$

$$Var(X) = \alpha^2\left[\Gamma\left(1+\frac{2}{\beta}\right) - \Gamma^2\left(1+\frac{1}{\beta}\right)\right]$$

Notes:

1. A variate X has a Weibull (α, β, η) distribution if the variate

$$Y = \left(\frac{X-\eta}{\alpha}\right)^{\beta}$$

 has the exponential $(1, 0)$ distribution.
2. When $\beta = 1$, the Weibull $(\alpha, 1, \eta)$ distribution is identical to the exponential (α, η) distribution.
3. When $\beta = 2$, the Weibull $(\alpha, 2, \eta)$ distribution is also called the *Rayleigh distribution* with parameters (α, η).
4. If a variate X has a Weibull $(\alpha, \beta, 0)$ distribution, then the variate $Y = -\ln X$ has the Gumbel $(-\ln\alpha, 1/\beta)$ distribution.
5. The density function has a reversed J-shape (and is unbounded at η) if $\beta < 1$; has an exponential shape if $\beta = 1$; has a mound shape if $1 < \beta$, with a positive skew if $1 < \beta < 3.6$, symmetry approximating a normal density function if $\beta = 3.6$, and a negative skew if $3.6 < \beta$.

References: Weibull (1951), Schütte et al. (1987), Rinne (2009)

C.3.3 The Inverted Weibull Distribution

Abbreviation: $\mathrm{IW}(\alpha, \beta, \eta)$

Parameters: scale $\alpha > 0$, shape $\beta > 0$, shift $-\infty < \eta < \infty$

Domains: $\eta < x < \infty, 0 < p < 1$

Density function

$$h(x) = \frac{\beta}{\alpha}\left(\frac{\alpha}{x-\eta}\right)^{\beta+1} \exp\left[-\left(\frac{\alpha}{x-\eta}\right)^{\beta}\right]$$

Distribution function $p = H(x)$

$$H(x) = \exp\left[-\left(\frac{\alpha}{x-\eta}\right)^{\beta}\right]$$

Quantile function $x = H^{-1}(p)$

$$H^{-1}(p) = \alpha(-\ln p)^{-\frac{1}{\beta}} + \eta$$

Linearized quantile function $v = bu + a$

$$v = \ln(x - \eta), \qquad u = -\ln(-\ln p)$$

$$\beta = \frac{1}{b}, \qquad\qquad \alpha = \exp(a)$$

Moments

$$E(X) = \alpha\Gamma\left(1 - \frac{1}{\beta}\right) + \eta \qquad\qquad \text{if } \beta > 1$$

$$Var(X) = \alpha^2\left[\Gamma\left(1 - \frac{2}{\beta}\right) - \Gamma^2\left(1 - \frac{1}{\beta}\right)\right] \quad \text{if } \beta > 2$$

Notes:

1. A variate X has an inverted Weibull (α, β, η) distribution if the variate

$$Y = \left(\frac{\alpha}{X-\eta}\right)^{\beta}$$

   has the exponential $(1, 0)$ distribution.

2. The mean exists if $\beta > 1$; the variance exists if $\beta > 2$.

C.3.4 The Log-Weibull Distribution

Abbreviation: $LW(\alpha, \beta, \eta)$
Parameters: scale $\alpha > 0$, shape $\beta > 0$, shift $-\infty < \eta < \infty$
Domains: $\eta < x < \infty$, $0 < p < 1$
Density function

$$h(x) = \frac{\beta}{\alpha(x - \eta + 1)} \left(\frac{\ln(x - \eta + 1)}{\alpha} \right)^{\beta - 1} \exp\left[-\left(\frac{\ln(x - \eta + 1)}{\alpha} \right)^{\beta} \right]$$

Distribution function $p = H(x)$

$$H(x) = 1 - \exp\left[-\left(\frac{\ln(x - \eta + 1)}{\alpha} \right)^{\beta} \right]$$

Quantile function $x = H^{-1}(p)$

$$H^{-1}(p) = \exp\left\{ \alpha\left[-\ln(1 - p) \right]^{\frac{1}{\beta}} \right\} + \eta - 1$$

Linearized quantile function $v = bu + a$

$$v = \ln[\ln(x - \eta + 1)], \qquad u = \ln\left[-\ln(1 - p) \right]$$
$$\beta = \frac{1}{b}, \qquad\qquad\qquad \alpha = \exp(a)$$

Moments

$$E(X) = \int_0^\infty \exp(\alpha t^{1/\beta} - t)\, dt + \eta - 1 \qquad\qquad \text{if } \beta > 1$$

$$Var(X) = \int_0^\infty \exp(2\alpha t^{1/\beta} - t)\, dt - [E(X) - \eta + 1]^2 \quad \text{if } \beta > 1$$

Notes:

1. A variate X has a log-Weibull (α, β, η) distribution if the variate

$$Y = \left(\frac{\ln(X - \eta + 1)}{\alpha} \right)^{\beta}$$

   has the exponential $(1, 0)$ distribution.
2. The mean and the variance exist if $\beta > 1$.
Reference: Krzysztofowicz and Kelly (2000)

C.3.5   The Log-Logistic Distribution

Abbreviation: $LL(\alpha, \beta, \eta)$
Parameters: scale $\alpha > 0$, shape $\beta > 0$, shift $-\infty < \eta < \infty$
Domains: $\eta < x < \infty$, $0 < p < 1$
Density function

$$h(x) = \frac{\beta}{\alpha}\left(\frac{x-\eta}{\alpha}\right)^{-\beta-1}\left[1+\left(\frac{x-\eta}{\alpha}\right)^{-\beta}\right]^{-2}$$

Distribution function $p = H(x)$

$$H(x) = \left[1+\left(\frac{x-\eta}{\alpha}\right)^{-\beta}\right]^{-1}$$

Quantile function $x = H^{-1}(p)$

$$H^{-1}(p) = \alpha\left(\frac{p}{1-p}\right)^{\frac{1}{\beta}}+\eta$$

Linearized quantile function $v = bu + a$

$$v = \ln(x-\eta), \qquad u = \ln\left(\frac{p}{1-p}\right)$$

$$\beta = \frac{1}{b}, \qquad\qquad \alpha = \exp(a)$$

Moments

$$E(X) = \alpha\Gamma\left(1+\frac{1}{\beta}\right)\Gamma\left(1-\frac{1}{\beta}\right)+\eta \qquad\qquad \text{if } \beta > 1$$

$$Var(X) = \alpha^2\left\{\Gamma\left(1+\frac{2}{\beta}\right)\Gamma\left(1-\frac{2}{\beta}\right)-\left[\Gamma\left(1+\frac{1}{\beta}\right)\Gamma\left(1-\frac{1}{\beta}\right)\right]^2\right\} \quad \text{if } \beta > 2$$

Notes:

1. A variate X has a log-logistic (α, β, η) distribution if the variate

$$Y = \ln\left[\left(\frac{X-\eta}{\alpha}\right)^{\beta}\right]$$

   has the logistic $(0, 1)$ distribution.
2. The mean exists if $\beta > 1$; the variance exists if $\beta > 2$.
Reference: Ahmad et al. (1988a, 1988b)

C.4 Distributions on a Bounded Interval

Each distribution family catalogued in this section is defined on a bounded open interval (η_L, η_U), where it is understood that the *lower bound* η_L and the *upper bound* η_U are real numbers, with $-\infty < \eta_L < \eta_U < \infty$, and are specified. It is also understood that for any $y < \eta_L$, $h(y) = 0$ and $H(y) = 0$; and for any $y > \eta_U$, $h(y) = 0$ and $H(y) = 1$.

The next two sections catalogue the two forms of the power distribution. Section C.4.3 defines a class of four log-ratio distributions.

Class of log-ratio distributions. Each particular distribution in this class is constructed according to the derived distribution theory (Section 3.8). Its two-part mnemonic (Table C.1) reflects the construction. For instance, $\text{LR1}(\eta_L, \eta_U)\text{-LG}(\beta, \alpha)$ means this: the log-ratio type I transformation of a variate on the bounded open interval (η_L, η_U) has a logistic distribution with parameters (β, α). In the construction, the original variate Y having the sample space $\mathcal{Y} = (\eta_L, \eta_U)$ is treated as the "output"; the transformed variate X having the sample space $\mathcal{X} = (-\infty, \infty)$ is treated as the "input"; the inverse transformation t^{-1}, such that $X = t^{-1}(Y)$, is the log-ratio type I transformation, as defined in Section C.4.3. To construct a particular distribution of Y, and to estimate its parameters, proceed as follows.

Step 1. Transform a given sample $\{y(n) : n = 1, \dots, N\}$ of variate Y into a sample $\{x(n) : n = 1, \dots, N\}$ of variate X.

Step 2. Use the transformed sample to estimate the parameters of a distribution of X. Any distribution form listed in Section C.2 may be hypothesized. Once the preferred distribution of X is chosen, the expressions for h_X, H_X, and H_X^{-1} are known.

Step 3. Use these expressions inside the expressions for h_Y, H_Y, and H_Y^{-1}, which are given in Section C.4.3.

Step 4. Evaluate the goodness of fit of the constructed distribution function H_Y to the empirical distribution function $\check{H}_Y$.

For an illustration of this procedure, see Section 3.8.6.

C.4.1 The Power Type I Distribution

Abbreviation: $P1(\beta, \eta_L, \eta_U)$
Parameters: shape $\beta > 0$, bounds $-\infty < \eta_L < \eta_U < \infty$
Domains: $\eta_L < y < \eta_U$, $0 < p < 1$
Density function

$$h(y) = \frac{\beta}{\eta_U - \eta_L} \left(\frac{y - \eta_L}{\eta_U - \eta_L} \right)^{\beta - 1}$$

Distribution function $p = H(y)$

$$H(y) = \left(\frac{y - \eta_L}{\eta_U - \eta_L} \right)^{\beta}$$

Quantile function $y = H^{-1}(p)$

$$H^{-1}(p) = (\eta_U - \eta_L)p^{\frac{1}{\beta}} + \eta_L$$

Linearized quantile function $v = bu$

$$v = \ln \frac{y - \eta_L}{\eta_U - \eta_L}, \qquad u = \ln p$$

$$\beta = \frac{1}{b}$$

Moments

$$E(Y) = \frac{\beta \eta_U + \eta_L}{\beta + 1}, \qquad Var(Y) = \frac{\beta (\eta_U - \eta_L)^2}{(\beta + 2)(\beta + 1)^2}$$

Notes:

1. The power type I distribution is a special case of the *beta* distribution, also known as a *Pearson type I* (or *II*) distribution.
2. It is especially suitable when the density function is negatively skew, or the distribution function is convex, in which case $\beta > 1$. The latter condition can be judged by graphing the empirical distribution function of Y. When $0 < \beta < 1$, the density function is unbounded at $y = \eta_L$, and the distribution function is concave.
3. When $\beta = 1$, the power type I $(1, \eta_L, \eta_U)$ distribution is identical to the uniform (η_L, η_U) distribution.

Reference: Johnson and Kotz (1970, vol. 2)

C.4.2   The Power Type II Distribution

Abbreviation: $P2(\beta, \eta_L, \eta_U)$
Parameters: shape $\beta > 0$, bounds $-\infty < \eta_L < \eta_U < \infty$
Domains: $\eta_L < y < \eta_U$, $0 < p < 1$
Density function

$$h(y) = \frac{\beta}{\eta_U - \eta_L} \left(\frac{\eta_U - y}{\eta_U - \eta_L} \right)^{\beta - 1}$$

Distribution function $p = H(y)$

$$H(y) = 1 - \left(\frac{\eta_U - y}{\eta_U - \eta_L} \right)^{\beta}$$

Quantile function $y = H^{-1}(p)$

$$H^{-1}(p) = \eta_U - (\eta_U - \eta_L)(1 - p)^{\frac{1}{\beta}}$$

Linearized quantile function $v = bu$

$$v = \ln \frac{\eta_U - y}{\eta_U - \eta_L}, \qquad u = \ln(1 - p)$$

$$\beta = \frac{1}{b}$$

Moments

$$E(Y) = \frac{\eta_U + \beta \eta_L}{\beta + 1}, \qquad Var(Y) = \frac{\beta (\eta_U - \eta_L)^2}{(\beta + 2)(\beta + 1)^2}$$

Notes:

1. The power type II distribution is a special case of the *beta* distribution, also known as a *Pearson type I* (or *II*) distribution.
2. It is especially suitable when the density function is positively skew, or the distribution function is concave, in which case $\beta > 1$. The latter condition can be judged by graphing the empirical distribution function of Y. When $0 < \beta < 1$, the density function is unbounded at $y = \eta_U$, and the distribution function is convex.
3. When $\beta = 1$, the power type II $(1, \eta_L, \eta_U)$ distribution is identical to the uniform (η_L, η_U) distribution.

Reference: Johnson and Kotz (1970, vol. 2)

C.4.3 The Log-Ratio Distributions

Abbreviation: LR1(η_L, η_U)-kl(β, α), kl $\in$ {LG, LP, GB, RG}

Domains: $\eta_L < y < \eta_U, -\infty < x < \infty, 0 < p < 1$

Log-ratio transformation

$$x = \ln \frac{y - \eta_L}{\eta_U - y}$$

Jacobian of the transformation

$$\frac{dx}{dy} = \frac{\eta_U - \eta_L}{(y - \eta_L)(\eta_U - y)}$$

Density function

$$h_Y(y) = \frac{\eta_U - \eta_L}{(y - \eta_L)(\eta_U - y)} \, h_X\left(\ln \frac{y - \eta_L}{\eta_U - y}\right)$$

Distribution function $p = H_Y(y)$

$$H_Y(y) = H_X\left(\ln \frac{y - \eta_L}{\eta_U - y}\right)$$

Quantile function $y = H_Y^{-1}(p)$

$$H_Y^{-1}(p) = \eta_U - \frac{\eta_U - \eta_L}{\exp[H_X^{-1}(p)] + 1}$$

Distribution of X: Any distribution listed in Section C.2: LG, LP, GB, RG.

Notes:

1. This is the log-ratio type I transformation. It is monotonically increasing; it is an odd function of the shifted variable $y - (\eta_U + \eta_L)/2$. It assigns value zero to the midpoint $(\eta_U + \eta_L)/2$ of the interval (η_L, η_U); it stretches the left half of the interval to $-\infty$, and the right half of the interval to $+\infty$.
2. When the distribution of X is either logistic or Laplace, the density function of Y is positively skew if $\beta < 0$, is symmetric if $\beta = 0$, and is negatively skew if $\beta > 0$.
3. When X has a normal distribution, the distribution of Y is sometimes called the *Johnson SB* distribution, where *SB* stands for the system of distributions on a bounded interval, which was studied by Johnson (1949).

References: Johnson (1949), Johnson and Kotz (1970, vol. 1)

C.5 Distributions on a Bounded-Above Interval

Each distribution family catalogued in this section is defined on an interval bounded above $(-\infty, -\eta)$, where it is understood that the upper bound $-\eta$, also called the *shift parameter*, is a real number, positive or negative, and is specified. It is also understood that for any $y > -\eta$, $h_Y(y) = 0$ and $H_Y(y) = 1$.

C.5.1 Reflection Transformation

Any original variate Y whose sample space is an interval bounded above, $\mathcal{Y} = (-\infty, -\eta)$, can be transformed into a variate X whose sample space is an interval bounded below, $\mathcal{X} = (\eta, \infty)$, where the bound η is a real number, $-\infty < \eta < \infty$. The transformation is simply a *reflection* about the origin

$$X = -Y.$$

Next, any of the distribution families listed in Section C.3 may be hypothesized for the variate X. Once the preferred distribution of X is chosen, the expressions for the density function h_X, the distribution function H_X, and the quantile function H_X^{-1} of variate X are known. Finally, the expressions for the density function h_Y, the distribution function H_Y, and the quantile function H_Y^{-1} of variate Y can be derived via the relationships:

$$h_Y(y) = h_X(-y),$$
$$H_Y(y) = 1 - H_X(-y),$$
$$H_Y^{-1}(p) = -H_X^{-1}(1 - p).$$

These relationships follow from the theory of derived distributions (Section 3.8) applied to the transformation $Y = t(X) = -X$. The relationships between the means and the variances of the two variates are

$$E(Y) = -E(X),$$
$$Var(Y) = Var(X).$$

C.5.2 Naming Convention

Two naming conventions are suggested for referring succinctly to a distribution family on an interval bounded above, $\mathcal{Y} = (-\infty, -\eta)$, which is obtained from a distribution family on an interval bounded below, $\mathcal{X} = (\eta, \infty)$. To illustrate them, suppose variate X has a Weibull distribution with parameters α, β, η; symbolically, $X \sim \mathrm{WB}(\alpha, \beta, \eta)$.

First convention: The *negative* of variate Y has a Weibull distribution with parameters α, β, η; symbolically, $-Y \sim \mathrm{WB}(\alpha, \beta, \eta)$.

Second convention: The variate Y has a *reflected* Weibull distribution with parameters α, β, η; symbolically, $Y \sim \mathrm{WB}(\alpha, \beta, \eta, -1)$. Here, -1 indicates that model $\mathrm{WB}(\alpha, \beta, \eta)$ must be used for h_X, H_X, and H_X^{-1} on the right side of the equations in Section C.5.1. Applying this naming convention to all distribution families from Section C.3, yields the last part of Table C.1.

C.5.3 Parameter Estimation

Step 0. Given is a sample of N realizations of the original variate Y:

$$\{y(n) : y(n) < -\eta, \ n = 1, \dots, N\}.$$

Step 1. Multiply the upper bound $-\eta$ and each realization $y(n)$ by -1. Denote $x(n) = -y(n)$ and form a sample of N realizations of the transformed variate X:

$$\{x(n) : x(n) > \eta, \, n = 1, \ldots, N\}.$$

Step 2. Use this sample to estimate the parameters of every hypothesized parametric model for H_X; any model listed in Section C.3 may be hypothesized. Choose the preferred model; at this moment, the expressions for h_X, H_X, and H_X^{-1} are known.

Step 3. Use these expressions in the equations for h_Y, H_Y, and H_Y^{-1}, which are given in Section C.5.1. If desired, carry out the derivations to obtain explicit formulae.

Step 4. Evaluate the goodness of fit of the constructed distribution function H_Y to the empirical distribution function $\check{H}_Y$.

For an illustration of this procedure, see Section 3.8.7.

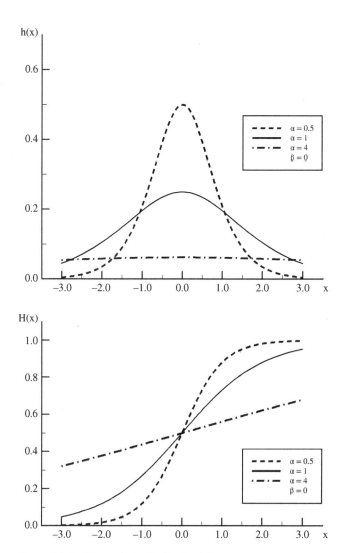

Figure C.2.1 The logistic distribution LG(β, α).

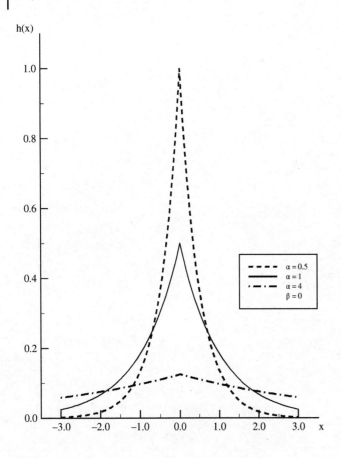

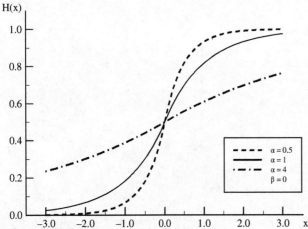

Figure C.2.2   The Laplace distribution LP(β, α).

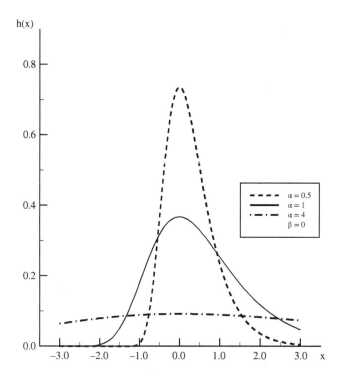

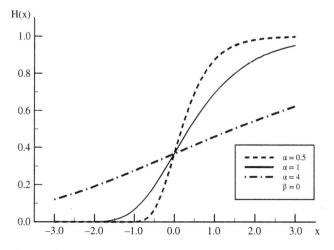

Figure C.2.3 The Gumbel distribution GB(β, α).

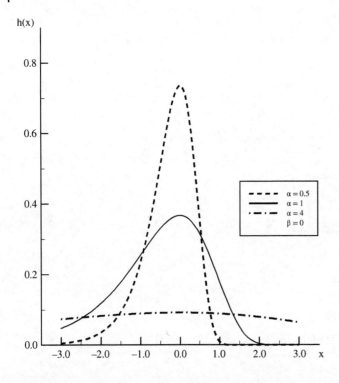

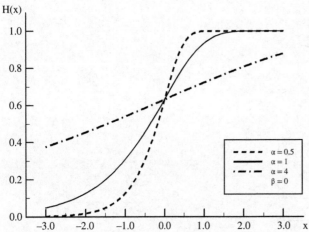

Figure C.2.4 The reflected Gumbel distribution RG(β, α).

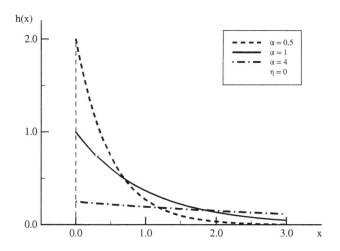

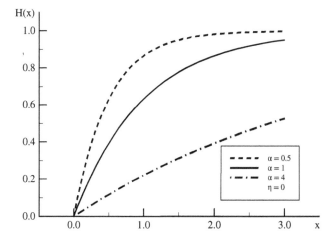

Figure C.3.1 The exponential distribution $EX(\alpha, \eta)$.

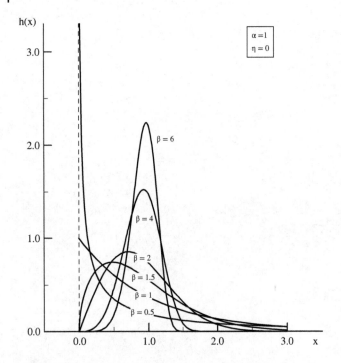

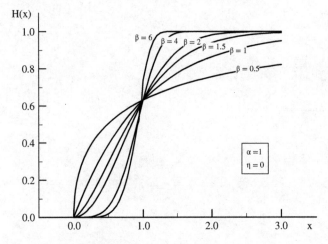

Figure C.3.2 The Weibull distribution $WB(\alpha, \beta, \eta)$.

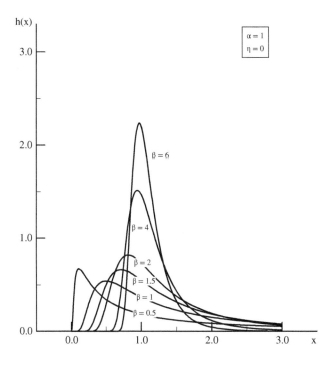

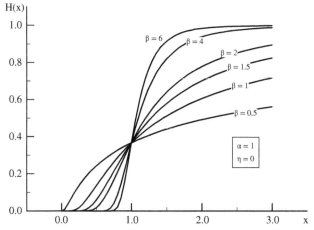

Figure C.3.3 The inverted Weibull distribution IW(α, β, η).

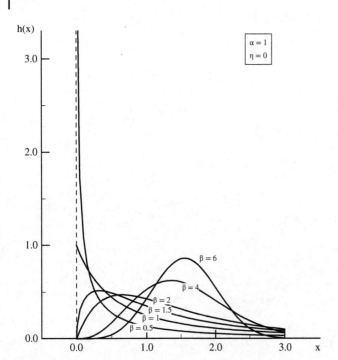

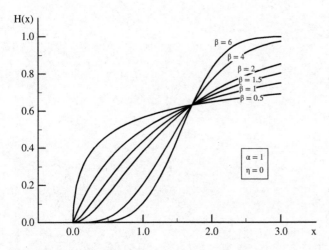

Figure C.3.4   The log-Weibull distribution $LW(\alpha, \beta, \eta)$.

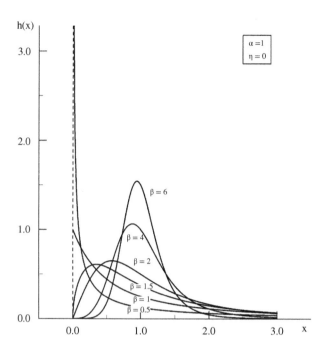

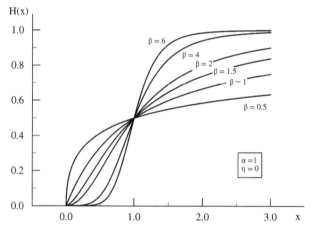

Figure C.3.5   The log-logistic distribution LL(α, β, η).

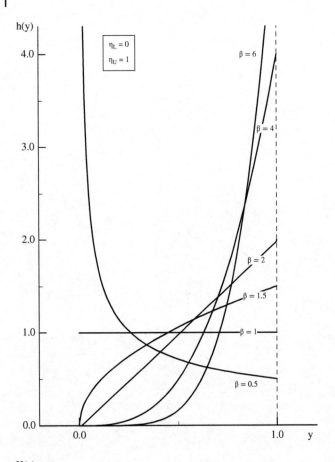

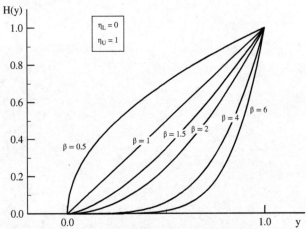

Figure C.4.1 The power type I distribution $P1(\beta, \eta_L, \eta_U)$.

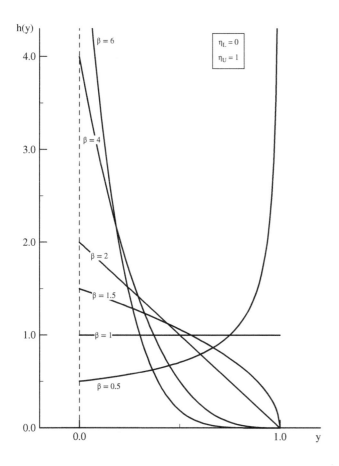

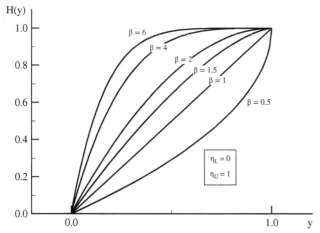

Figure C.4.2 The power type II distribution $P2(\beta, \eta_L, \eta_U)$.

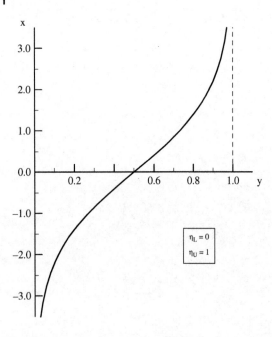

Figure C.4.3.1 The log-ratio transformation.

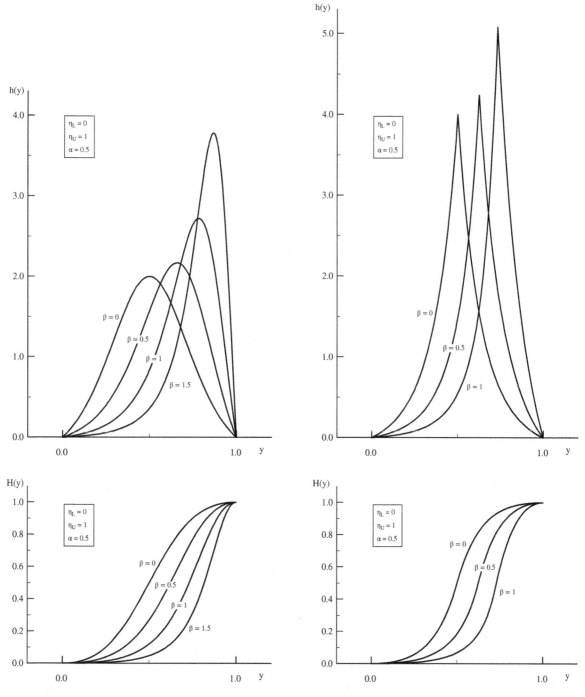

Figure C.4.3.2 Left panel: the log-ratio logistic distribution LR1(η_L, η_U)-LG(β, α); with $\alpha = 0.5$. Right panel: the log-ratio Laplace distribution LR1(η_L, η_U)-LP(β, α); with $\alpha = 0.5$. Each density function h with $\beta > 0$ has a negative skew; when this h is reflected about the vertical line $y = 1/2$, one obtains h with $-\beta < 0$ and a positive skew.

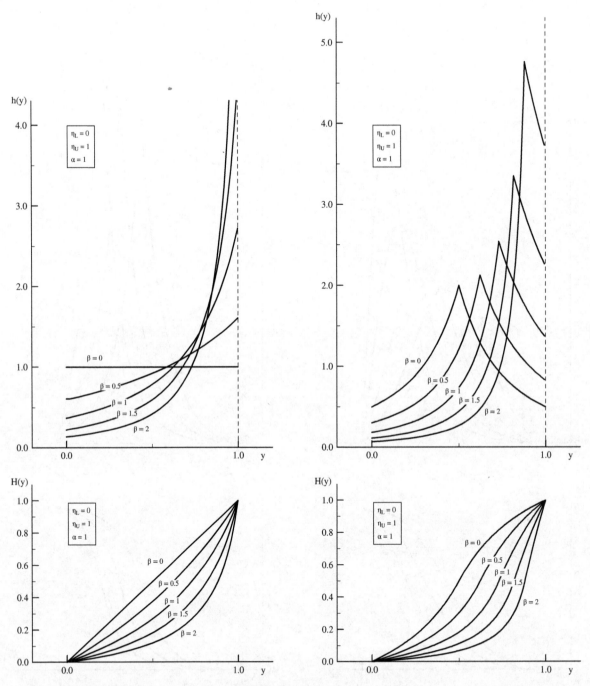

Figure C.4.3.3 Left panel: the log-ratio logistic distribution LR1(η_L, η_U)-LG(β, α); with $\alpha = 1$. Right panel: the log-ratio Laplace distribution LR1(η_L, η_U)-LP(β, α); with $\alpha = 1$. Each density function h with $\beta > 0$ has a negative skew; when this h is reflected about the vertical line $y = 1/2$, one obtains h with $-\beta < 0$ and a positive skew.

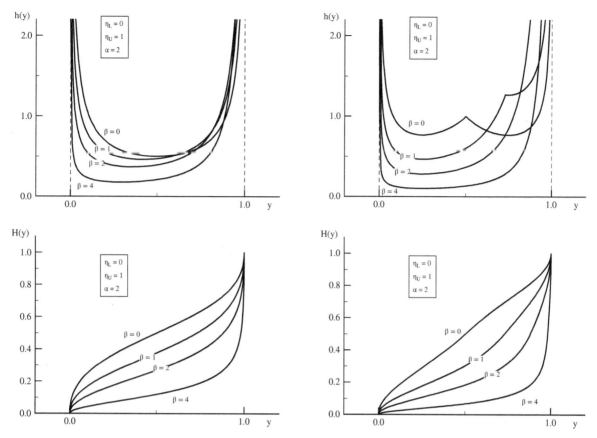

Figure C.4.3.4 Left panel: the log-ratio logistic distribution LR1(η_L, η_U)-LG(β, α); with $\alpha = 2$. Right panel: the log-ratio Laplace distribution LR1(η_L, η_U)-LP(β, α); with $\alpha = 2$. Each density function h with $\beta > 0$ has a negative skew; when this h is reflected about the vertical line $y = 1/2$, one obtains h with $-\beta < 0$ and a positive skew.

The Greek Alphabet

A	α	Alpha	N	ν	Nu
B	β	Beta	Ξ	ξ	Xi
Γ	γ	Gamma	O	o	Omicron
Δ	δ	Delta	Π	π	Pi
E	ε	Epsilon	P	ρ	Rho
Z	ζ	Zeta	Σ	σ	Sigma
H	η	Eta	T	τ	Tau
Θ	θ	Theta	Υ	υ	Upsilon
I	ι	Iota	Φ	ϕ	Phi
K	κ	Kappa	X	χ	Chi
Λ	λ	Lambda	Ψ	ψ	Psi
M	μ	Mu	Ω	ω	Omega

Probabilistic Forecasts and Optimal Decisions, First Edition. Roman Krzysztofowicz.
© 2025 John Wiley & Sons Ltd. Published 2025 by John Wiley & Sons Ltd.
Companion website: www.wiley.com/go/ProbabilisticForecastsandOptimalDecisions1e

References

Abramowitz, M., and Stegun, I.A. (1972). *Handbook of Mathematical Functions*, Dover Publications, New York.

Ahmad, M.I., Sinclair, C.D., and Spurr, B.D. (1988a). Assessment of flood frequency models using empirical distribution function statistics, *Water Resources Research*, **24**(8), 1323–1328.

Ahmad, M.I., Sinclair, C.D., and Werritty, A. (1988b). Log-logistic flood frequency analysis, *Journal of Hydrology*, **98**, 205–224.

Alpert, M., and Raiffa, H. (1982). A progress report on the training of probability assessors. In Kahneman, D., Slovic, P., and Tversky, A. (eds.), *Judgment Under Uncertainty: Heuristics and Biases*, Cambridge University Press, Cambridge, pp. 294–305.

Becker, S.W., and Siegel, S. (1958). Utility of grades: Level of aspiration in a decision theory context, *Journal of Experimental Psychology*, **55**(1), 81–85.

Bellman, R. (1957). *Dynamic Programming*, Princeton University Press, Princeton, NJ.

Benjamin, J.R., and Cornell, C.A. (1970). *Probability, Statistics, and Decision for Civil Engineers*, McGraw-Hill, New York.

Bennett, W.J., and Wilezol, D. (2013). *Is College Worth It?* Thomas Nelson, Nashville, TN.

Berger, J.O. (1985). *Statistical Decision Theory: Foundations, Concepts, and Methods*, 2nd edn, Springer-Verlag, New York.

Berling, R.L. (1978). Disaster response to flash flood, *Journal of the Water Resources Planning and Management Division*, **104**(1), 35–44.

Bernardo, J.M., and Smith, A.F.M. (1994). *Bayesian Theory*, Wiley, Chichester.

Bernoulli, D. (1738). Exposition of a new theory on the measurement of risk. *Papers of the Imperial Academy of Sciences in Petersburg*, **V**, 175–192. (Published in Latin. English translation published in *Econometrica*, **22**(1), 23–36, 1954.)

Blackwell, D. (1951). Comparison of experiments. In Neyman, J. (ed.), *Proceedings of the Second Berkeley Symposium on Mathematical Statistics and Probability*, University of California Press, Berkeley, CA, pp. 93–102.

Blackwell, D. (1953). Equivalent comparisons of experiments, *Annals of Mathematical Statistics*, **24**, 265–272.

Blackwell, D., and Girshick, M.A. (1954). *Theory of Games and Statistical Decisions*, Wiley, New York.

Bois, G.P. (1961). *Tables of Indefinite Integrals*, Dover Publications, New York.

Breznitz, S. (1984). *Cry Wolf: The Psychology of False Alarms*, Lawrence Erlbaum Associates, Hillsdale, NJ.

Curtis, R.B. (1972). Decision-rules and collective values in constitutional choice. In Niemi, R.G., and Weisberg, H.F. (eds.), *Probability Models of Collective Decision Making*, Charles E. Merrill Publishing Company, Columbus, OH, pp. 23–33.

Dalal, S.R., Fowlkes, E.B., and Hoadley, B. (1989). Risk analysis of the space shuttle: Pre-Challenger prediction of failure, *Journal of the American Statistical Association*, **84**(408), 945–957.

Davidson, L.B., and Cooper, D.O. (1980). Implementing effective risk analysis at Getty Oil Company, *Interfaces*, **10**(6), 62–75.

Dawid, A.P. (1982). The well-calibrated Bayesian, *Journal of the American Statistical Association*, **77**(379), 605–613.

Probabilistic Forecasts and Optimal Decisions, First Edition. Roman Krzysztofowicz.
© 2025 John Wiley & Sons Ltd. Published 2025 by John Wiley & Sons Ltd.
Companion website: www.wiley.com/go/ProbabilisticForecastsandOptimalDecisions1e

de Finetti, B. (1974). *Theory of Probability*, vol. **1**, Wiley, New York.

DeGroot, M.H. (1962). Uncertainty, information, and sequential experiments, *Annals of Mathematical Statistics*, **30**, 404–419.

DeGroot, M.H. (1970). *Optimal Statistical Decisions*, McGraw-Hill, New York.

DeGroot, M.H. (1986). *Probability and Statistics*, 2nd edn, Addison-Wesley, Reading, MA.

DeGroot, M.H., and Fienberg, S.E. (1982). Assessing probability assessors: Calibration and refinement. In Gupta, S.S., and Berger, J.O. (eds.), *Statistical Decision Theory and Related Topics III*, vol. **1**, Academic Press, New York, pp. 291–314.

DeGroot, M.H., and Fienberg, S.E. (1983). The comparison and evaluation of forecasters, *The Statistician*, **32**, 12–22.

DiGirolamo, V. (2019). *Crying the News: A History of America's Newsboys*, Oxford University Press, New York.

Dowswell, P. (2004). *True Stories of the First World War*, Scholastic Inc., New York, pp. 122–123.

Dweck, C.S. (2008). *Mindset: The New Psychology of Success*, Ballantine Books, New York.

Edwards, W., Phillips, L.D., Hays, W.L., and Goodman, B.C. (1968). Probabilistic information processing systems: Design and evaluation, *IEEE Transactions on Systems Science and Cybernetics*, **SSC-4**(3), 248–265.

Farquhar, P.H. (1984). Utility assessment methods, *Management Science*, **30**(11), 1283–1300.

Fishburn, P.C. (1977). Mean-risk analysis with risk associated with below-target returns, *American Economic Review*, **67**(2), 116–126.

Fishburn, P.C. (1982). *The Foundations of Expected Utility*, D. Reidel, Dordrecht, Holland.

Fishburn, P.C., and Kochenberger, G.A. (1979). Two-piece von Neumann-Morgenstern utility functions, *Decision Sciences*, **10**(4), 503–518.

Flavell, J.H. (1976). Metacognitive aspects of problem solving. In Resnick, L.B. (ed.), *The Nature of Intelligence*, Lawrence Erlbaum Associates, Hillsdale, NJ.

Friedman, M. (1955). Liberalism, old style. Reprinted under the heading Notable & Quotable: Liberalism and Freedom, *The Wall Street Journal*, 29 November 2019.

Ginsberg, B. (2011). *The Fall of the Faculty: The Rise of the All-Administrative University and Why It Matters*, Oxford University Press, New York.

Gneiting, T., Balabdaoui, F., and Raftery, A.E. (2007). Probabilistic forecasts, calibration and sharpness, *Journal of the Royal Statistical Society, Series B*, **69**(2), 243–268.

Good, I.J. (1950). *Probability and the Weighting of Evidence*, Charles Griffin & Company, London.

Good, I.J. (1961). Amount of deciding and decisionary effort, *Information and Control*, **4**, 271–281.

Good, I.J. (1983). *Good Thinking: The Foundations of Probability and Its Applications*, University of Minnesota Press, Minneapolis, MN.

Gruntfest, E.C. (1977). What people did during the Big Thompson flood, (Working Paper 32), Prepared for the Urban Drainage and Flood Control District, Denver; Institute of Behavioral Science, University of Colorado, Boulder, CO.

Gustafson, D.H., Shukla, R.K., Delbecq, A., and Walster, G.W. (1973). A comparative study of differences in subjective likelihood estimates made by individuals, interacting groups, Delphi groups, and nominal groups, *Organizational Behavior and Human Performance*, **9**, 280–291.

Harsanyi, J.C. (1978). Bayesian decision theory and utilitarian ethics, *American Economic Review*, **68**(2), 223–228.

Hosseini, J., and Ferrell, W.R. (1982). Detectability of correctness: A measure of knowing that one knows, *Instructional Science*, **11**(2), 113–127.

Howard, R.A. (1980). An assessment of decision analysis, *Operations Research*, **28**(1), 4–27.

Janis, I.L., and Mann, L. (1977). *Decision Making: A Psychological Analysis of Conflict, Choice, and Commitment*, Free Press, New York.

Johnson, N.L. (1949). Systems of frequency curves generated by methods of translation, *Biometrica*, **36**, 149–176.

Johnson, N.L., and Kotz, S. (1970). *Distributions in Statistics: Continuous Univariate Distributions*, vol. **1** and **2**, Wiley, New York.

Johnson, R.A. (2005). *Miller and Freund's Probability and Statistics for Engineers*, 7th edn, Pearson Prentice Hall, Upper Saddle River, NJ.

Kahneman, D., and Tversky, A. (1979). Prospect theory: An analysis of decision under risk, *Econometrica*, **47**(2), 263–291.

Kahneman, D., and Tversky, A. (1982). The simulation heuristic. In Kahneman, D., Slovic, P., and Tversky, A. (eds.), *Judgment Under Uncertainty: Heuristics and Biases*, Cambridge University Press, Cambridge, pp. 201–208.

Katz, R.W., Murphy, A.H., and Winkler, R.L. (1982). Assessing the value of frost forecasts to orchardists: A dynamic decision-making approach, *Journal of Applied Meteorology*, **21**(4), 518–531.

Keeney, R.L., and Raiffa, H. (1976). *Decisions with Multiple Objectives: Preferences and Value Tradeoffs*, Wiley, New York.

Kelly, K.S., and Krzysztofowicz, R. (1994). Probability distributions for flood warning systems, *Water Resources Research*, **30**(4), 1145–1152.

Kelly, K.S., and Krzysztofowicz, R. (1995). Bayesian revision of an arbitrary prior density, *Proceedings of the Section on Bayesian Statistical Science*, American Statistical Association, 50–53.

Kelly, K.S., and Krzysztofowicz, R. (1997). A bivariate meta-Gaussian density for use in hydrology, *Stochastic Hydrology and Hydraulics*, **11**(1), 17–31.

Kotz, S., and Seeger, J.P. (1991). A new approach to dependence in multivariate distributions. In Dall'Aglio, G., Kotz, S., and Salinetti, G. (eds.), *Advances in Probability Distributions with Given Marginals*, Kluwer, Dordrecht, The Netherlands, pp. 113–127.

Krantz, D.H., Luce, R.D., Suppes, P., and Tversky, A. (1971). *Foundations of Measurement*, vol. **1**, Academic Press, New York.

Kruskal, W.H. (1958). Ordinal measures of association, *Journal of the American Statistical Association*, **53**, 814–861.

Krzysztofowicz, R. (1985). Bayesian models of forecasted time series, *Water Resources Bulletin*, **21**(5), 805–814.

Krzysztofowicz, R. (1986a). Expected utility criterion for setting targets, *Large Scale Systems*, **10**(1), 21–37.

Krzysztofowicz, R. (1986b). Expected utility, benefit, and loss criteria for seasonal water supply planning, *Water Resources Research*, **22**(3), 303–312.

Krzysztofowicz, R. (1986c). Optimal water supply planning based on seasonal runoff forecasts, *Water Resources Research*, **22**(3), 313–321.

Krzysztofowicz, R. (1987). Markovian forecast processes, *Journal of the American Statistical Association*, **82**(397), 31–37.

Krzysztofowicz, R. (1990). Target-setting problem with exponential utility, *IEEE Transactions on Systems, Man, and Cybernetics*, **20**(3), 687–688.

Krzysztofowicz, R. (1992). Bayesian correlation score: A utilitarian measure of forecast skill, *Monthly Weather Review*, **120**(1), 208–219.

Krzysztofowicz, R. (1993). A theory of flood warning systems, *Water Resources Research*, **29**(12), 3981–3994.

Krzysztofowicz, R. (1994). Generic utility theory: Explanatory model, behavioral hypotheses, empirical evidence. In Allais, M., and Hagen, O. (eds.), *Cardinalism: A Fundamental Approach*, Kluwer Academic Publishers, Dordrecht, The Netherlands, pp. 249–288.

Krzysztofowicz, R., and Evans, W.B. (2008). Probabilistic forecasts from the National Digital Forecast Database, *Weather and Forecasting*, **23**(2), 270–289.

Krzysztofowicz, R., and Kelly, K.S. (2000). Hydrologic uncertainty processor for probabilistic river stage forecasting, *Water Resources Research*, **36**(11), 3265–3277.

Krzysztofowicz, R., and Long, D. (1990a). Fusion of detection probabilities and comparison of multisensor systems, *IEEE Transactions on Systems, Man, and Cybernetics*, **20**(3), 665–677.

Krzysztofowicz, R., and Long, D. (1990b). To protect or not to protect: Bayes decisions with forecasts, *European Journal of Operational Research*, **44**(3), 319–330.

Krzysztofowicz, R., and Long, D. (1991). Forecast sufficiency characteristic: Construction and application, *International Journal of Forecasting*, **7**, 39–45.

Krzysztofowicz, R., and Sigrest, A.A. (1999). Calibration of probabilistic quantitative precipitation forecasts, *Weather and Forecasting*, **14**(3), 427–442.

Krzysztofowicz, R., and Watada, L.M. (1986). Stochastic model of seasonal runoff forecasts, *Water Resources Research*, **22**(3), 296–302.

Krzysztofowicz, R., Drzal, W.J., Drake, T.R., Weyman, J.C., and Giordano, L.A. (1993). Probabilistic quantitative precipitation forecasts for river basins, *Weather and Forecasting*, **8**(4), 424–439.

Krzysztofowicz, R., Kelly, K.S., and Long, D. (1994). Reliability of flood warning systems, *Journal of Water Resources Planning and Management*, **120**(6), 906–926.

Kyle, C. (2012). *American Sniper*, Harper, New York, pp. 188–189.

Le Cam, L. (1964). Sufficiency and approximate sufficiency, *Annals of Mathematical Statistics*, **35**, 1419–1455.

Lichtenstein, S., and Fischhoff, B. (1977). Do those who know more also know more about how much they know?, *Organizational Behavior and Human Performance*, **20**, 159–183.

Lichtenstein, S., and Fischhoff, B. (1980). Training for calibration, *Organizational Behavior and Human Performance*, **26**, 149–171.

Lichtenstein, S., Fischhoff, B., and Phillips, L.D. (1982). Calibration of probabilities: The state of the art to 1980. In Kahneman, D., Slovic, P., and Tversky, A. (eds.), *Judgment Under Uncertainty: Heuristics and Biases*, Cambridge University Press, Cambridge, pp. 306–334.

Lindgren, B.W. (1993). *Statistical Theory*, 4th edn, Chapman & Hall, New York.

Lindley, D.V. (1982). The improvement of probability judgments, *Journal of the Royal Statistical Society, Series A*, **145**(1), 117–126.

Lindley, D.V. (1985). *Making Decisions*, 2nd edn, Wiley, London.

Maranzano, C.J., and Krzysztofowicz, R. (2008). Bayesian reanalysis of the Challenger O-ring data, *Risk Analysis*, **28**(4), 1053–1067.

Markowitz, H. (1952). The utility of wealth, *Journal of Political Economy*, **60**(2), 151–158.

Marschak, J. (1974). Optimal systems for information and decision, in *Economic Information, Decision, and Prediction*, vol. **2**, essay 32, D. Reidel, Dordrecht, Holland, pp. 342–355.

Marschak, J. (1979). Utilities, psychological values, and the training of decision makers. In Allais, M., and Hagen, O. (eds.), *Expected Utility Hypotheses and the Allais Paradox*, D. Reidel, Dordrecht, Holland, pp. 163–174.

Miller, G.A. (1956). The magical number seven, plus or minus two: Some limits on our capacity for processing information, *Psychological Review*, **63**, 81–97.

Misak, Ch. (2020). *Frank Ramsey: A Sheer Excess of Powers*, Oxford University Press, New York.

Morgenstern, O. (1979). Some reflections on utility. In Allais, M., and Hagen, O. (eds.), *Expected Utility Hypotheses and the Allais Paradox*, D. Reidel, Dordrecht, Holland, pp. 175–183.

Murphy, A.H. (1981). Subjective quantification of uncertainty in weather forecasts in the United States, *Meteorologische Rundschau*, **34**, 65–77.

Murphy, A.H. (1998). The early history of probability forecasts: Some extensions and clarifications, *Weather and Forecasting*, **13**, 5–15.

Murphy, A.H., and Daan, H. (1984). Impacts of feedback and experience on the quality of subjective probability forecasts: Comparison of results from the first and second years of the Zierikzee experiment, *Monthly Weather Review*, **112**, 413–423.

Murphy, A.H., and Winkler, R.L. (1974). Credible interval temperature forecasting: Some experimental results, *Monthly Weather Review*, **102**(11), 784–794.

Murphy, A.H., and Winkler, R.L. (1977). Reliability of subjective probability forecasts of precipitation and temperature, *Applied Statistics*, **26**(1), 41–47.

Murphy, A.H., and Winkler, R.L. (1982). Subjective probabilistic tornado forecasts: Some experimental results, *Monthly Weather Review*, **110**, 1288–1297.

Myers, D.G., and Lamm, H. (1975). The polarizing effect of group discussion, *American Scientist*, **63**, 297–303.

Olds, R. (2010). *Fighter Pilot*, St. Martin's Press, New York, pp. 281–282.

Olson, L. (2020). *Unsung Heroes of World War II: Europe*, The Great Courses, Chantilly, Virginia; Transcript Book, Lesson 2, pp. 25–47.

Payne, J.W., Laughhunn, D.J., and Crum, R. (1980). Translation of gambles and aspiration level effects in risky choice behavior, *Management Science*, **26**(10), 1039–1060.

Payne, J.W., Laughhunn, D.J., and Crum, R. (1981). Further tests of aspiration level effects in risky choice behavior, *Management Science*, **27**(8), 953–958.

Pierce, B. (1988). Introduction to "Formula for the field", *Response!*, **7**(3), 20–24.

Plott, C.R. (1972). Individual choice of a political-economic process. In Niemi, R.G., and Weisberg, H.F. (eds.), *Probability Models of Collective Decision Making*, Charles E. Merrill Publishing Company, Columbus, OH, pp. 83–97.

Pratt, J.W. (1964). Risk aversion in the small and in the large, *Econometrica*, **32**(1–2), 122–136.

Pratt, J.W., Raiffa, H., and Schlaifer, R. (1995). *Introduction to Statistical Decision Theory*, MIT Press, Cambridge, MA.

Presidential Commission (1986). *Report of the Presidential Commission on the Space Shuttle Challenger Accident*, vol. **1**, Washington, DC.

Rae, D. (1969). Decision-rules and individual values in constitutional choice, *American Political Science Review*, **63**, 40–56.

Raiffa, H., and Schlaifer, R. (1961). *Applied Statistical Decision Theory*, Graduate School of Business Administration, Harvard University, Boston, MA.

Ramsey, F.P. (1926). Truth and probability. In Braithwaite, R.B. (ed.), *The Foundations of Mathematics and other Logical Essays*, Humanities Press, New York, 1950; pp. 156–198.

Rinne, H. (2009). *The Weibull Distribution: A Handbook*, CRC Press, Boca Raton, FL.

Roberts, F.S. (1979). *Measurement Theory*, Addison-Wesley, Reading, MA.

Rubin, R.E. (2023). *The Yellow Pad: Making Better Decisions in an Uncertain World*, Penguin Press, New York.

Schofield, N.J. (1972). Is majority rule special? In Niemi, R.G., and Weisberg, H.F. (eds.), *Probability Models of Collective Decision Making*, Charles E. Merrill Publishing Company, Columbus, OH, pp. 60–82.

Schum, D.A. (1994). *The Evidential Foundations of Probabilistic Reasoning*, Wiley, New York.

Schütte, T., Salka, O., and Israelsson, S. (1987). The use of the Weibull distribution for thunderstorm parameters, *Journal of Climate and Applied Meteorology*, **26**, 457–463.

Selvidge, J.E. (1980). Assessing the extremes of probability distributions by the fractile method, *Decision Sciences*, **11**(3), 493–502.

Seneca, L.A. (2004). *Letters from a Stoic*, Penguin Books, London.

Shea, G. (1988). Formula for the field, *Response!*, **7**(3), 20–24.

Siegel, S. (1957). Level of aspiration and decision making, *Psychological Review*, **64**(4), 253–262.

Simon, H.A. (1957). *Models of Man: Social and Rational*, Wiley, New York.

Slovic, P. (1982). Toward understanding and improving decisions. In Howell, W.C., and Fleishman, E.A. (eds.), *Human Performance and Productivity*, vol. **2**: *Information Processing and Decision Making*, Lawrence Erlbaum Associates, Hillsdale, NJ.

Slovic, P., and Lichtenstein, S. (1968). Relative importance of probabilities and payoffs in risk taking, *Journal of Experimental Psychology Monograph*, **78**(3) Part 2, 1–18.

Solomon, I. (1982). Probability assessment by individual auditors and audit teams: An empirical investigation, *Journal of Accounting Research*, **20**(2), Pt. II, 689–710.

Spetzler, C.S., and Staël von Holstein, C.-A.S. (1975). Probability encoding in decision analysis, *Management Science*, **22**(3), 340–358.

Swalm, R.O. (1966). Utility theory — Insights into risk taking, *Harvard Business Review*, November–December, 123–136.

Szidarovszky, F., and Yakowitz, S. (1978). *Principles and Procedures of Numerical Analysis*, Plenum Press, New York.

Tversky, A., and Kahneman, D. (1982). Judgment under uncertainty: Heuristics and biases. In Kahneman, D., Slovic, P., and Tversky, A. (eds.), *Judgment Under Uncertainty: Heuristics and Biases*, Cambridge University Press, Cambridge, pp. 3–20.

von Neumann, J., and Morgenstern, O. (1947). *Theory of Games and Economic Behavior*, 2nd edn, Princeton University Press, Princeton, NJ.

Weibull, W. (1951). A statistical distribution function of wide applicability, *Journal of Applied Mechanics*, **73**, 293–297.

Winkler, R.L., and Murphy, A.H. (1979). The use of probabilities in forecasts of maximum and minimum temperatures, *Meteorological Magazine*, **108**, 317–329.

Wiper, M.P., French, S., and Cooke, R. (1994). Hypothesis-based calibration scores, *The Statistician*, **43**(2), 231–236.

Wymore, A.W. (1976). *Systems Engineering Methodology for Interdisciplinary Teams*, Wiley, New York.

Index

Probabilistic Forecasts and Optimal Decisions, First Edition. Roman Krzysztofowicz.
© 2025 John Wiley & Sons Ltd. Published 2025 by John Wiley & Sons Ltd.
Companion website: www.wiley.com/go/ProbabilisticForecastsandOptimalDecisions1e